AF595322

Innovative Target Tracking Techniques for Modern Radar and Sonar Systems

Innovative Target Tracking Techniques for Modern Radar and Sonar Systems

Editors

Alfonso Farina
Wei Yi

MDPI • Basel • Beijing • Wuhan • Barcelona • Belgrade • Manchester • Tokyo • Cluj • Tianjin

Editors

Alfonso Farina
Selex ES (retired), Consultant
Italy

Wei Yi
University of Electronic Science and Technology of China, School of Communication and Information Engineering
China

Editorial Office
MDPI
St. Alban-Anlage 66
4052 Basel, Switzerland

This is a reprint of articles from the Special Issue published online in the open access journal *Sensors* (ISSN 1424-8220) (available at: http://www.mdpi.com).

For citation purposes, cite each article independently as indicated on the article page online and as indicated below:

LastName, A.A.; LastName, B.B.; LastName, C.C. Article Title. *Journal Name* **Year**, *Volume Number*, Page Range.

ISBN 978-3-0365-3537-1 (Hbk)
ISBN 978-3-0365-3538-8 (PDF)

Contents

About the Editors

Alfonso Farina

Alfonso Farina received his Laurea degree in electronic engineering from the University of Rome, Rome, Italy, in 1973. In 1974, he joined SELENIA S.P.A., then Selex ES, where he became the Director of the Analysis of Integrated Systems Unit and, subsequently, the Director of Engineering of the Large Business Systems Division. In 2012, he was the Senior VP and the Chief Technology Officer (CTO) of the Company, reporting directly to the President. From 2013 to 2014, he was a Senior Advisor to the CTO. He retired in October 2014. Currently, he is President of the Radar & Sensors Academy of Leonardo S.p.A. Electronic Division, Italy. From 1979 to 1985, he was also a Professor of Radar Techniques with the University of Naples, Italy. He is currently a Visiting Professor with the Department of Electronic and Electrical Engineering at University College London and with the Centre of Electronic Warfare, Information and Cyber at Cranfield University, and a Distinguished Lecturer of the IEEE Aerospace and Electronic Systems Society. He is the author of about 1000 publications. He has won the Fred Nathanson Memorial Radar award (1987), the IEEE Dennis J. Picard Medal for Radar Technologies and Applications (2010), the IET Achievement Medal (2014), C. Hülsmeyer Award, German Institute Navigation (2019) and the IEEE AESS Pioneer Award (2020). Alfonso is LFIEEE, FIET, Fellow of The Royal Academy of Engineering (UK), F-EURASIP, European Academy Science, Académico Correspondiente Real Academia de Ingeniería de España. In February 2023, he was elected as an International Member of the United States National Academy of Engineering (NAE) "for contributions to the development and deployment of advanced radar systems and technology". He is ranked in the list of 2% top scientists in the World.

Wei Yi

Wei Yi (full professor) received his BE and PhD degrees in 2006 and 2012, respectively, both in electronic engineering from the University of Electronic Science and Technology of China. From 2010 to 2012, he was a visiting student at the Melbourne Systems Laboratory, University of Melbourne.Since 2012, he has been with the University of Electronic Science and Technology of China, where he is currently a full professor at the School of Information and Communication Engineering. His research interests include target detection and tracking, radar signal processing, multi-sensor information fusion, and resource management. Dr. Yi was the recipient of the "Best Student Paper Competition—First place winner" at the 2012 IEEE Radar Conference, Atlanta, and the "Best Student Paper Award" at the 15th FUSION Conference, Singapore, 2012, and was the co-recipient of the "Best Student Paper Award" at the 21st FUSION Conference, Cambridge, 2018. He is a member of the editorial board of the Journal of Radars. He served as a guest editor for MDPI Sensors and Frontiers of Information Technology & Electronic Engineering. He also served as the general co-chair of ICCAIS 2019, and the TPC member of international conferences such as IEEE Radar Conference and FUSION Conference.

Preface to "Innovative Target Tracking Techniques for Modern Radar and Sonar Systems"

Due to rapid advances in wireless communication and sensor technology, modern monitor systems, including modern radar and sonar systems, play important roles in both civil and defence applications. An important example is target tracking. Nowadays, radar and sonar systems constantly encounter a large number of targets of many different types, such as extended targets and low signal-to-noise ratio targets. These types of targets bring new challenges to target-tracking techniques. To cope with these challenges, conventional target tracking methods need to be significantly upgraded. There are several theoretical and practical problems, such as multi-sensor fusion architecture, track-before-detect methods, classification and identification methods, and other open problems yet to be addressed. The corresponding applications of machine learning in this field are also interesting and meaningful topics of investigation.

The aim of this Special Issue is to gather recent advances and development in target tracking techniques to determine how they can be adapted for modern radar and sonar systems. After peer review, 17 articles in related areas have been accepted for publishing in this Special Issue. The published articles cover a range of topics and applications central to target tracking, such as extended target tracking, resource management, and multi-target smoother. Although this Special Issue has been closed, the need for future research and development related to innovative target-tracking techniques for modern radar and sonar systems remains. We hope that this Special Issue and the published papers can inspire many more innovative theoretical and practical efforts in the field of target tracking.

Alfonso Farina, Wei Yi

Editors

Article

Analysis of Polynomial Nonlinearity Based on Measures of Nonlinearity Algorithms

Mahendra Mallick [1,*,†,‡] and Xiaoqing Tian [2]

1 Independent Consultant, Anacortes, WA 98221, USA
2 School of Automation Science and Engineering, Xi'an Jiaotong University, Xi'an 710049, China; tianxiaoqing2017@stu.xjtu.edu.cn
* Correspondence: mmallick.us@gmail.com
† Current address: Anacortes, WA 98221, USA.
‡ The corresponding author contributed most to this work. M. Mallick: Algorithm and Software Development, Analysis of Algorithms and Results, Manuscript Preparation. X. Tian: Software Development, Analysis of Algorithms and Results, Manuscript Preparation.

Received: 31 March 2020; Accepted: 10 June 2020; Published: 17 June 2020

Abstract: We consider measures of nonlinearity (MoNs) of a polynomial curve in two-dimensions (2D), as previously studied in our Fusion 2010 and 2019 ICCAIS papers. Our previous work calculated curvature measures of nonlinearity (MoNs) using (i) extrinsic curvature, (ii) Bates and Watts parameter-effects curvature, and (iii) direct parameter-effects curvature. In this paper, we have introduced the computation and analysis of a number of new MoNs, including Beale's MoN, Linssen's MoN, Li's MoN, and the MoN of Straka, Duník, and Šimandl. Our results show that all of the MoNs studied follow the same type of variation as a function of the independent variable and the power of the polynomial. Secondly, theoretical analysis and numerical results show that the logarithm of the mean square error (MSE) is an affine function of the logarithm of the MoN for each type of MoN. This implies that, when the MoN increases, the MSE increases. We have presented an up-to-date review of various MoNs in the context of non-linear parameter estimation and non-linear filtering. The MoNs studied here can be used to compute MoN in non-linear filtering problems.

Keywords: polynomial curve in 2D; measures of nonlinearity (MoNs); extrinsic curvature; Beale's MoN; Linssen's MoN; Bates and Watts parameter-effects curvature; direct parameter-effects curvature; Li's MoN; MoN of Straka, Duník, and Šimandl; maximum likelihood estimator (MLE); Cramér-Rao lower bound (CRLB)

1. Introduction

The Kalman filter (KF) [1–4] is an optimal estimator (in the minimum mean square error (MMSE) sense) for a filtering problem with linear dynamic and measurement models with additive Gaussian noise. However, many real-world filtering problems are non-linear due to nonlinearity in the dynamic and measurement models. Common real-world non-linear filtering (NLF) problems are bearing-only filtering [5–8], ground moving target indicator (GMTI) filtering [9], passive angle-only filtering in three-dimensions (3D) using an infrared search and track sensor [10–12], etc.

In the early stages of NLF, the extended Kalman filter (EKF) [1–4] was widely used. It was observed in some problems, e.g., falling of a body in earth's atmosphere with high velocity [13,14] and bearing-only filtering [5,7,8] that the EKF performs poorly due to linearization. The high degree of nonlinearity in these problems was the attributed cause for the poor performance of the problem without a quantitative measure of nonlinearity (MoN). To overcome the poor accuracy and convergence problems of the EKF, a number of improved approximate non-linear filters, such as the unscented

Kalman filter (UKF) [14,15], cubature KF (CKF) [16], and particle filter (PF) [8,17] have been proposed during the last two decades.

It is important to address the following questions for NLF problems:

1. Is it possible to find a quantitative MoN for a nonlinear filtering problem?
2. Can we establish a correspondence between the MoN of a NLF problem and the performance of a filtering algorithm?
3. Can we show that the UKF, CKF, or PF gives better results than the EKF, when the degree of nonlinearity (DoN) is high?

Remark 1. *In this paper we consider a parameter estimation problem with polynomial nonlinearity. We hope that insights and results from this analysis would encourage further study of MoN in NLF problems. Next, we describe some historical developments in the field of parameter estimation and NLF.*

Beale in his pioneering work [18] proposed four MoNs for the static non-random parameter estimation problem. Two MoNs were empirical and two were theoretical. Guttman and Meeter [19] and Linssen [20] observed that Beale's method gives lower MoN for highly non-linear problems and proposed a modified MoN. Using differential geometry based curvature measures, Bates and Watts [21,22] and Goldberg et al. [23] extended Beale's work and developed curvature measures of nonlinearity (CMoN) for the static non-random parameter estimation problem. Bates and Watts formulated two CMoN, the parameter-effects curvature and intrinsic curvature [21,24–26].

In [27], we first extended the method of Bates and Watts to the non-linear filtering problem with unattended ground sensor (UGS) to calculate CMoN. Next, we computed the parameter-effects curvature and intrinsic curvature for the bearing-only filtering (BOF) problem [28–31], GMTI filtering problem [30,32,33], video tracking problem [34], and polynomial nonlinearity [35].

In our previous work [35], we considered a polynomial curve in two-dimensions (2D) and calculated CMoN using differential geometry (e.g., extrinsic curvature) [36–38], Bates and Watts parameter-effects curvature [21,25,26], and direct parameter-effects curvature [29]. The computation of these curvatures requires the Jacobian and Hessian of the measurement function [2] evaluated at the true or estimated parameter. The extrinsic curvature uses the true parameter, whereas the other two CMoN use the estimated parameter.

In [35], we obtained the maximum likelihood (ML) estimate [2,39] of the parameter x while using a vector measurement by numerical minimization. In [40], we derived analytic expressions for the ML estimator (MLE) [2,39] and associated variance using a vector measurement. This approach is simple and efficient, since it does not require numerical minimization. We also showed through Monte Carlo simulations in [40] that the variance of the MLE and the Cramér-Rao lower bound (CRLB) [2,41] are nearly the same for different powers of x. We also found that the bias error was small and the mean square error (MSE) [2] was close to the CRLB and variance of the MLE. Our numerical results showed that the average normalized estimation error squared (ANEES) [42] was within the 99% confidence interval most of the time. Hence, the variance of the MLE was in agreement with the estimation error.

Li constructed a combined non-linear function while using the non-linear time evolution function and measurement function in a discrete-time nonlinear filtering problem, and he proposed a global MoN at each measurement time [43]. This MoN minimizes the mean square distance between the combined non-linear function and the set of all affine functions with the same dimension at each measurement time. An un-normalized MoN and a normalized MoN were proposed in [43]. These MoNs can also be unconditional or conditional. The normalized MoN lies in the interval $[0,1]$. A journal version of the paper with enhancements was published in [44].

The normalized MoN that was proposed in [43] was calculated for non-linear filtering problems, including one with the nearly constant turn motion and a non-linear measurement model [45], a video tracking problem using PF [46], and a hypersonic entry vehicle state estimation problem [47]. In these cases, the normalized MoN were rather low. In [33], we compared the normalized MoN for the BOF and GMTI filtering problems. Contrary to our expectation, we found that the GMTI filtering problem had a higher conditional normalized MoN than that of the BOF problem in the examples that we investigated.

Using the current mean (e.g., predicted mean) and associated covariance, Duník et al. [48] generate a number of sample points (e.g., sigma points using unscented transform [14]) and transform these points using a non-linear function (e.g., non-linear measurement function or time evolution function). Subsequently, they try to predict the transformed points using a linear transformation and estimate the parameter of the transformation using linear weighted least squares (WLS) [39]. They use the cost function of the WLS evaluated at the estimated parameter as a local MoN.

In [35], we showed analytically and through Monte Carlo simulations that affine mappings with positive slopes exist among the logarithm of the extrinsic curvature, Bates and Watts parameter-effects curvature, direct parameter-effects curvature, MSE, and CRLB. For completeness, we have included these key results from [35] in Section 4. New contributions in this paper include the computation and analysis of following MoNs:

- Beale's MoN [18],
- Least squares based Beale's MoN,
- Linssen's MoN [20],
- Least squares based Linssen's MoN,
- Li's MoN [43,44], and
- MoN of Straka, Duník, and Šimandl [48,49].

It is not possible to derive a mapping analytically between the logarithm of Beale's MoN, Linssen's MoN, Li's MoN, MoN of Straka, Duník, and Šimandl, and the logarithm of the MSE. The numerical results from Monte Carlo simulations also show that affine mappings with positive slopes exist among the logarithm of the MSE and the logarithm of two of these MoNs.

The paper is organized, as follows. Section 2.1 describes the measurement model for polynomial nonlinearity. The MLE for parameter estimation and CRLB using polynomial nonlinearity and a vector measurement is presented in Section 2. Section 3 presents different types of MoN, such as extrinsic curvature based on differential geometry, Beale's MoN, Linssen's MoN, Bates and Watts parameter-effects curvature, direct parameter-effects curvature, Li's MoN, and MoN of Straka, et al. Section 4 discusses mappings among logarithms of extrinsic curvature, parameter-effects curvature, CRLB, and MSE. Section 5 presents the numerical simulation and results. Finally, Section 6 summarizes our contribution and concludes with future work.

Notation Convention: For clarity, we use italics to denote scalar quantities and boldface for vectors and matrices. A lower or upper case Roman letter represents a name (e.g., "s" for "sensor", "RMS" for "root mean square", etc.). We use ":=" to define a quantity and $\mathbf{A}'$ denotes the transpose of the vector or matrix $\mathbf{A}$. The $n-$dimensional identity matrix, $m-$dimensional null matrix, and $m \times n$ null matrix are denoted by $\mathbf{I}_n$, $\mathbf{0}_m$, and $\mathbf{0}_{m\times n}$, respectively.

2. MLE Parameter Estimation and CRLB

2.1. Measurement Model

We studied CMoN of a polynomial smooth scalar function h of a non-random variable x in [35], where

$$h(x) = ax^n, \tag{1}$$

and a is a non-zero scalar. In scenarios considered, $x > 0$ and $n = 2, 3, 4, 5$.

Remark 2. *For MoN of other forms of nonlinearity, such as the bearing-only [27], GMTI [32], and video filtering [34] problems in radar communities, we shall discuss in detail in the end of Section 3.*

The measurement model for the polynomial function is given by

$$z_i = h(x) + v_i,\ i = 1, \ldots, N, \tag{2}$$

where v_i is a zero-mean white Gaussian measurement noise with variance σ^2,

$$v_i \sim \mathcal{N}(0, \sigma^2). \tag{3}$$

We assume that the measurement noises are independent.
The measurement model can be written in the vector form

$$\mathbf{z} = \mathbf{h}(x) + \mathbf{v}, \tag{4}$$

where

$$\mathbf{z} := \begin{bmatrix} z_1 & z_2 & \ldots & z_N \end{bmatrix}', \tag{5}$$

$$\mathbf{v} := \begin{bmatrix} v_1 & v_2 & \ldots & v_N \end{bmatrix}', \tag{6}$$

$$\mathbf{h}(x) := h(x)\mathbf{d}, \tag{7}$$

$$\mathbf{d} := \begin{bmatrix} 1 & 1 & \ldots & 1 \end{bmatrix}', \tag{8}$$

$$\mathbf{v} \sim \mathcal{N}(\mathbf{0}, \mathbf{R}), \quad \mathbf{R} = \mathbf{I}_N\sigma^2. \tag{9}$$

2.2. ML Estimate of Parameter

The likelihood function of x is [2,50,51]

$$\Lambda(x;\mathbf{z}) = p(\mathbf{z}|x) = [(2\pi)^N|\mathbf{R}|]^{-1/2}\exp\{-(1/2)[\mathbf{z}-\mathbf{h}(x)]'\mathbf{R}^{-1}[\mathbf{z}-\mathbf{h}(x)]\}. \tag{10}$$

The maximization of the likelihood in (10) is equivalent to the minimization of the cost function [2,51]

$$J(x) = [\mathbf{z}-\mathbf{h}(x)]'\mathbf{R}^{-1}[\mathbf{z}-\mathbf{h}(x)] = [\mathbf{z}-\mathbf{h}(x)]'[\mathbf{z}-\mathbf{h}(x)]/\sigma^2. \tag{11}$$

The maximum likelihood (ML) estimate $\hat{x}$ of x is obtained by setting the derivative of $J(x)$ to zero [2,51],

$$\frac{dJ(x)}{dx} = 0. \tag{12}$$

From (11) and (12), we obtain

$$[\mathbf{z}-\mathbf{h}(\hat{x})]'\frac{d\mathbf{h}(\hat{x})}{dx} = 0. \tag{13}$$

Because the derivative of $\mathbf{h}(x)$ with respect to x is not zero, we obtain

$$\mathbf{z}-\mathbf{h}(\hat{x}) = \mathbf{0}_{N\times 1}. \tag{14}$$

Hence, the ML estimate satisfies,

$$h(\hat{x})\mathbf{d} = \mathbf{z}. \tag{15}$$

Left-multiplying both sides of (15) by $\mathbf{d}'$, we obtain

$$h(\hat{x})\mathbf{d}'\mathbf{d} = \mathbf{d}'\mathbf{z} = \sum_{i=1}^{N} z_i. \tag{16}$$

We note that

$$\mathbf{d}'\mathbf{d} = N. \tag{17}$$

Using (1) and (17) in (16) we get

$$a\hat{x}^n = \bar{z}, \tag{18}$$

where $\bar{z}$ is the sample mean of z,

$$\bar{z} = \frac{1}{N}\sum_{i=1}^{N} z_i. \tag{19}$$

Thus, from (18), the ML estimate of x is given by

$$\hat{x} = (\bar{z}/a)^{1/n}, \quad n = 2, 3, \ldots. \tag{20}$$

Remark 3. *In general, the MLE for a nonlinear measurement model is biased [51]. We can calculate the variance of $\hat{x}$ under the small error assumption using the linearization approximation. To guarantee the validity of the variance, the bias in the MLE must be calculated. The bias can be numerically calculated using Monte Carlo simulation.*

The bias in the MLE is defined by [2,51]

$$b(x) := x - \hat{x}. \tag{21}$$

Remark 4. *The ML estimate of x in [35] was obtained by minimizing the cost function in (11) numerically. The estimator in (20) provides simple and efficient way of estimating x using a vector measurement $\mathbf{z}$ without numerical optimization.*

2.3. Variance of the MLE

The variance of $\hat{x}$ is given by [51]

$$\sigma_{\hat{x}}^2 = (\dot{\mathbf{H}}'\mathbf{R}^{-1}\dot{\mathbf{H}})^{-1}, \tag{22}$$

where

$$\dot{\mathbf{H}} = \frac{d\mathbf{h}(x)}{dx}\bigg|_{x=\hat{x}}. \tag{23}$$

Using the special form of $\mathbf{R}$ from (9) in (22), we get

$$\sigma_{\hat{x}}^2 = \sigma^2(\dot{\mathbf{H}}'\dot{\mathbf{H}})^{-1}. \tag{24}$$

Using (7) in (23), we get

$$\dot{\mathbf{H}} = \frac{dh(x)}{dx}\bigg|_{x=\hat{x}}\mathbf{d}. \tag{25}$$

Differentiating (1) with respect to x, we obtain

$$\frac{dh(x)}{dx} = anx^{n-1}. \tag{26}$$

Using (26) in (25), we get

$$\dot{\mathbf{H}} = an\hat{x}^{n-1}\mathbf{d}. \tag{27}$$

From (27), we obtain

$$\dot{\mathbf{H}}'\dot{\mathbf{H}} = (an\hat{x}^{n-1})^2\,\mathbf{d}'\mathbf{d}, \tag{28}$$

Using (28) and (17) in (24), we obtain

$$\sigma_{\hat{x}}^2 = \sigma^2(\dot{\mathbf{H}}'\dot{\mathbf{H}})^{-1} = \frac{\sigma^2}{N(an\hat{x}^{n-1})^2}, \tag{29}$$

$$\sigma_{\hat{x}} = \frac{\sigma}{\sqrt{N}an\hat{x}^{n-1}}. \tag{30}$$

2.4. Cramér-Rao Lower Bound

The CRLB [2,41] for the MSE in the current problem is given by

$$\text{CRLB}_x = \left[\frac{d\mathbf{h}'(x)}{dx}\mathbf{R}^{-1}\frac{d\mathbf{h}(x)}{dx}\right]^{-1}. \tag{31}$$

Remark 5. *Calculation of the variance $\sigma_{\hat{x}}^2$ and* CRLB$_x$ *are similar. For $\sigma_{\hat{x}}^2$, we use the estimate $\hat{x}$ while calculating the Jacobian of the measurement function, whereas, for* CRLB$_x$*, we use the true x while calculating the Jacobian of the measurement function.*

Using similar procedure, we obtain

$$\text{CRLB}_x = \frac{\sigma^2}{N(anx^{n-1})^2}, \tag{32}$$

$$\sqrt{\text{CRLB}_x} = \frac{\sigma}{\sqrt{N}anx^{n-1}}. \tag{33}$$

From (30) and (33), we find that, for a given x, the standard deviation (SD) and square root of CRLB are inversely proportional to the power n. Secondly (33) shows that, for a given power, the square root of CRLB decreases as x increases.

3. Measures of Nonlinearity

To explain the key concepts of nonlinearity, consider the scalar function $h(x) = 5\sin(4x)/x$ shown in Figure 1. We observe in Figure 1 that the function is nearly linear at A and E. If we draw a tangent to the curve at A and E, then the curve is close to the tangent in the neighborhood of A and E. However, tangents to the curve at points B, C, and D differ by large amounts from the curve in the neighborhood of these points. The tangent represents an affine approximation to the curve at a point. We observe that, among points B, C, and D, the curve bends the most at B and the least at point D. If we draw a circle (called the osculating circle) at these points, then the radius of the circle can be used to judge nonlinearity. The rate of bending is high when the radius of the circle is small. In differential geometry [37,38], the curvature κ is inverse of the radius of the osculating circle and, hence, curvature can be viewed as a measure of nonlinearity. The radii of the osculating circles at A and E are nearly infinity and, hence, the curvatures are nearly zero. From Figure 1, we observe that, in general, the nonlinearity of a function can vary with x. Hence, the nonlinearity is a local measure. If the second derivative of a function is non-zero, then the function is non-linear.

In [35,40], we analyzed the CMoN of a polynomial scalar function h of a non-random variable x, as described in Section 2.1. The CMoN were based on the extrinsic curvature using differential geometry, Bates and Watts parameter-effects curvature, and direct parameter-effects curvature. In this paper, we study the following MoNs:

- extrinsic curvature using differential geometry [36–38],
- Beale's MoN [18],
- least squares based Beale's MoN,
- Linssen's MoN [20],

- Least squares based Linssen's MoN,
- parameter-effects curvatures [21,25,29],
- Li's MoN [43,44], and
- MoN of Straka, Duník, and Šimandl [48,49].

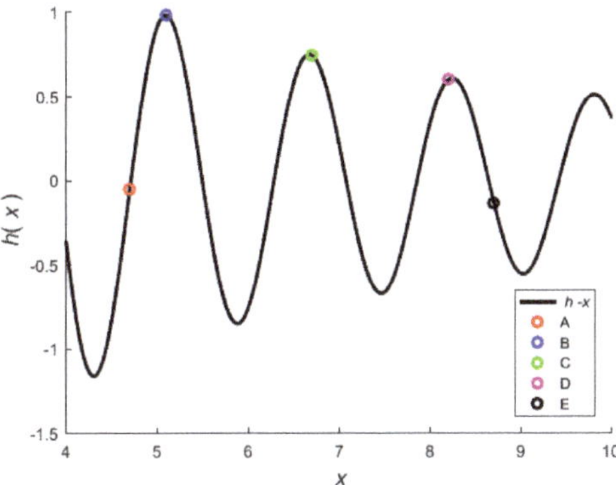

Figure 1. $y = 5\sin(4x)/x$ versus x.

If a MoN has a high value, then the nonlinearity is high and if it has a low value, then the Therefore, it is impossible to compare them based on numerical values. We can only study their variations.

Consider the m-dimensional vector non-linear function $\mathbf{h}$ of the non-random $n-$dimensional parameter $\mathbf{x}$. Let $\hat{\mathbf{x}}$ be a known estimate of $\mathbf{x}$. Using the Taylor series expansion of $\mathbf{h}(\mathbf{x})$ about $\hat{\mathbf{x}}$ and keeping the first order term gives

$$\mathbf{h}(\mathbf{x}) \approx \mathbf{T}(\mathbf{x}) = \mathbf{h}(\hat{\mathbf{x}}) + \dot{\mathbf{H}}(\mathbf{x} - \hat{\mathbf{x}}), \tag{34}$$

where $\mathbf{T}(\mathbf{x})$ represents the tangent plane approximation (an affine mapping) to $\mathbf{h}(\mathbf{x})$ and

$$\dot{\mathbf{H}} = \frac{d\mathbf{h}(\mathbf{x})}{d\mathbf{x}}\bigg|_{\mathbf{x}=\hat{\mathbf{x}}}. \tag{35}$$

If $m > n$, then $\mathbf{h}$ is an $n-$dimensional manifold embedded in an $m-$dimensional space [37,38]. The tangent plane is tangent to the surface $\mathbf{h}$ at $\hat{\mathbf{x}}$. The concept of tangent plane is used in Beale's MoN, Linssen's MoN, Bates and Watts parameter-effects curvatures [21,25], and direct parameter-effects curvature [44].

For polynomial nonlinearity, the CMoN using differential geometry is calculated at the true value x and, hence, it is non-random. The Bates and Watts parameter-effects curvature, direct parameter-effects curvature, Beale's MoN, Li's MoN, and the MoN of Straka et al. are calculated while using an estimate $\hat{x}$ of x. The estimate $\hat{x}$ is obtained from a measurement model involving the measurement function h. Since x is a scalar, we need one or more scalar measurements to estimate x. Table 1 summarizes features of various MoNs.

Table 1. Features of Various MoNs

MoN	Parameters Used	Local/Global	Need?			Basic Idea	Random?
			Measurements	Jacobian/Hessian	Covariances		
Extrinsic Curvature	True	Local	No	Jacobian and Hessian at true value	No	Differential Geometry	Non-random
Beale's MoN	True & estimated	Local	Yes	Jacobian	No	Scaled sum square distance	Random
Linssen's MoN	True & estimated	Local	Yes	Jacobian	No	Root scaled sum square distance	Random
Parameter-effects Curvature	True & estimated	Local	Yes	Jacobian and Hessian at estimated value	No	Differential Geometry	Random
Li's MoN	True	Global	Yes	No	Yes	min. mean square distance	Random
MoN by Straka et al.	True & estimated	Local	Yes	No	No	WLS cost function	Random

The CMoN using differential geometry [36–38] is calculated at the true value x, whereas the Bates and Watts parameter-effects curvature [21,25,26], direct parameter-effects curvature [29], Beale's MoN, Li's MoN, and the MoN of Straka et al. are calculated while using an estimate $\hat{x}$ of x. The estimate $\hat{x}$ is obtained from a measurement model involving the measurement function h. Since x is a scalar, we need one or more scalar measurements to estimate x. Next, we describe various MoN.

3.1. Extrinsic Curvature Using Differential Geometry

The curvature of a circle at every point on the circumference is equal to the inverse of the radius of the circle. Thus, the curvature of a circle is a constant. A circle with a smaller radius bends more sharply and, therefore, has a higher curvature.

We assume that the first and second derivatives of the nonlinear smooth scalar function h exist. The curvature of the curve $y = h(x)$ at a point x is equal to the curvature of the osculating circle at that point. The extrinsic curvature at the point x is defined by [36–38],

$$\kappa(x) := \frac{\left|\frac{d^2h(x)}{dx^2}\right|}{[1+(\frac{dh(x)}{dx})^2]^{3/2}} = \frac{|\ddot{h}(x)|}{[1+\dot{h}(x)^2]^{3/2}}. \tag{36}$$

The first derivative of h at a point x is given in (26). The second derivative of h with respect to x is given by

$$\ddot{h}(x) = \frac{d^2h(x)}{dx^2} = an(n-1)x^{n-2}, \quad n = 2, 3, \ldots. \tag{37}$$

Thus, using $\dot{h}(x)$ and $\ddot{h}(x)$ in (36), we can calculate the extrinsic curvature $\kappa(x)$ at any point x by

$$\kappa(x) = \frac{an(n-1)x^{n-2}}{[1+(an)^2x^{2(n-1)}]^{3/2}}. \tag{38}$$

3.2. Beale's MoN

Consider the nonlinear measurement model for the non-random n-dimensional parameter $\mathbf{x}$

$$\mathbf{z} = \mathbf{h}(\mathbf{x}) + \mathbf{v}, \tag{39}$$

where $\mathbf{z}$, $\mathbf{h}$, and $\mathbf{v}$ are the measurement, non-linear measurement function, and measurement noise, respectively. Let $\hat{\mathbf{x}}$ be an estimate of $\mathbf{x}$. Subsequently, a Taylor series expansion of $\mathbf{h}(\mathbf{x})$ about

$\hat{\mathbf{x}}$ and keeping the first order term is as (34). Suppose we choose m vectors $\mathbf{x}_i, i = 1, 2, \ldots, m$ in the neighborhood of $\mathbf{x}$. Then Beale's first empirical MoN [18] is given by

$$\hat{N}_{\mathbf{x}} = \rho^2 \frac{\sum_{i=1}^m \|\mathbf{h}(\mathbf{x}_i) - \mathbf{T}(\mathbf{x}_i)\|^2}{\sum_{i=1}^m \|\mathbf{h}(\mathbf{x}_i) - \mathbf{h}(\hat{\mathbf{x}})\|^4}, \tag{40}$$

where ρ is the standard radius and it is defined by

$$\rho^2 := \|\mathbf{z} - \mathbf{h}(\hat{\mathbf{x}})\|^2 / (n(N-n)). \tag{41}$$

Guttman and Meeter [19] observed that the empirical MoN underestimates severe nonlinearity. When m approaches infinity, the empirical MoN $\hat{N}_{\mathbf{x}}$ approaches the theoretical MoN $N_{\mathbf{x}}$.

3.3. Least Squares Based Beale's MoN

Consider the scalar function h for polynomial nonlinearity, as described in (1). As described in Beale's MoN, we choose m points $x_i, i = 1, 2, \ldots, m$ in th neighborhood of x. Let

$$y_i = a x_i^n, \quad i = 1, 2, \ldots, m. \tag{42}$$

An affine mapping as approximation to y_i is given by

$$L(x_i) = A + B x_i, \quad i = 1, 2, \ldots, m. \tag{43}$$

We compute A and B by minimizing the cost function

$$J(A, B) := \sum_{i=1}^m (y_i - A - B x_i)^2 \tag{44}$$

by the method of least squares (LS) [2,39]. The LS minimization of the cost function yields [52]

$$\hat{B} = (C_{xy} - \bar{x}\,\bar{y}) / (C_{xx} - \bar{x}^2), \tag{45}$$

$$\hat{A} = \bar{y} - \hat{B}\bar{x}, \tag{46}$$

where

$$\bar{x} = \frac{1}{m}\sum_{i=1}^m x_i, \quad \bar{y} = \frac{1}{m}\sum_{i=1}^m y_i, \tag{47}$$

$$C_{xx} = \frac{1}{m}\sum_{i=1}^m x_i^2, \quad C_{xy} = \frac{1}{m}\sum_{i=1}^m x_i y_i. \tag{48}$$

Then we can use the affine mapping with $\hat{A}$ and $\hat{B}$ in Beale's MoN.

3.4. Linssen's MoN

In order the correct the deficiency in Beale's MoN, Linssen proposed a modification to obtain the following MoN [20]

$$M^* = \sqrt{\rho^2 \frac{\sum_{i=1}^m \|\mathbf{h}(\mathbf{x}_i) - \mathbf{T}(\mathbf{x}_i)\|^2}{\sum_{i=1}^m \|\mathbf{h}(\hat{\mathbf{x}}) - \mathbf{T}(\mathbf{x}_i)\|^4}}. \tag{49}$$

3.5. Least Squares Based Linssen's MoN

Using the same procedure as in Section 3.3, we can use an affine mapping with $\hat{A}$ and $\hat{B}$ as an approximation to y_i in computing Linssen's MoN.

3.6. Parameter-Effects Curvatures

The parameter-effects curvature and intrinsic curvature defined by Bates and Watts [21,25,26] are associated with a non-linear parameter estimation problem and are defined at the estimated parameter. We note that in (1), $h : \mathbb{R} \to \mathbb{R}$. Since h is a scalar function, the intrinsic curvature of Bates and Watts $K^{\mathrm{N}}(\hat{x})$ [21] or the direct intrinsic curvature $\beta_{\delta}^{\mathrm{N}}(\hat{x})$ [29] is zero. Thus, only the parameter-effects curvature of Bates and Watts $K^{\mathrm{T}}(\hat{x})$ and the direct parameter-effects curvature $\beta_{\delta}^{\mathrm{T}}(\hat{x})$ are non-zero. Since the intrinsic curvature is zero, for simplicity in notation, we drop the superscript "T" from the parameter-effects curvature and they are given by

$$K(\hat{x}) := \frac{||\ddot{\mathbf{H}}\delta^2||}{||\dot{\mathbf{H}}\delta||^2} = \frac{||\ddot{\mathbf{H}}||}{||\dot{\mathbf{H}}||^2}, \tag{50}$$

$$\beta_{\delta}(\hat{x}) := \frac{||\ddot{\mathbf{H}}\delta^2||}{||\dot{\mathbf{H}}\delta||} = \frac{|\ddot{\mathbf{H}}||\delta|}{||\dot{\mathbf{H}}||}, \tag{51}$$

where

$$\ddot{\mathbf{H}} = \frac{d^2\mathbf{h}(x)}{dx^2}\Big|_{x=\hat{x}}, \tag{52}$$

$$\delta := x - \hat{x}. \tag{53}$$

From (26), we get

$$\ddot{\mathbf{H}} = an(n-1)\hat{x}^{n-2}\mathbf{d}. \tag{54}$$

Hence, from (27) and (52), we obtain

$$||\dot{\mathbf{H}}|| = an\hat{x}^{n-1}\sqrt{N}. \tag{55}$$

$$||\ddot{\mathbf{H}}|| = an(n-1)\hat{x}^{n-2}\sqrt{N}. \tag{56}$$

Substitution of results from (55) and (56) in (50) and (51) gives

$$K(\hat{x}) = \frac{n-1}{na\sqrt{N}}\frac{1}{\hat{x}^n}, \tag{57}$$

$$\beta_{\delta}(\hat{x}) = \frac{(n-1)|\delta|}{\hat{x}}. \tag{58}$$

We note that the extrinsic curvature in (36) is evaluated at the true x, while the parameter-effects curvatures $K(\hat{x})$ in (50) and $\beta_{\delta}(\hat{x})$ in (51) are evaluated at the estimate $\hat{x}$. Because $\hat{x}$ is a random variable, $K(\hat{x})$ and $\beta_{\delta}(\hat{x})$ are random variables. When we perform Monte Carlo simulations and estimate x from measurements, $\hat{x}$ varies among Monte Carlo runs. Therefore, $K(\hat{x})$ and the set of all linear $\beta_{\delta}(\hat{x})$ vary with Monte Carlo runs.

3.7. Li's MoN

For a scalar random variable x, the un-normalized MoN proposed by Li [43,44] represents the square root of the minimum mean square distance between the nonlinear measurement function h and the set of all affine functions L,

$$J = \min(E\{(L(x) - h(x))^2\})^{1/2}, \tag{59}$$

where $L(x) = Ax + B$. The scalar parameters A and B are determined in the minimization process. For the current problem, where x is non-random, the un-normalized MoN J and normalized MoN ν ar given, respectively, by

$$J = \sigma_h \sqrt{1 - \frac{c_{hx}^2}{\sigma_h^2 \sigma_x^2}}. \tag{60}$$

$$\nu = J/\sigma_h = \sqrt{1 - \frac{c_{hx}^2}{\sigma_h^2 \sigma_x^2}}. \tag{61}$$

Given $\hat{x}$ and σ_x (30), the unscented transformation (UT) [14,15], cubature transformation (CT) [16], or Monte Carlo method [8] can be used to compute σ_h^2 and c_{hx}. We find that the UT gives good results in calculating the two MoNs. Next we dscribe computing J and ν using the UT. We use $\kappa_{\text{UT}} = 2$ [14]. The three weights and sigma points are given, respectively, by

$$w_0 = 2/3, \quad w_1 = 1/6, \quad w_2 = 1/6, \tag{62}$$

$$\chi_0 = \hat{x}, \quad \chi_1 = \hat{x} + \sqrt{3}\sigma_x, \quad \chi_2 = \hat{x} - \sqrt{3}\sigma_x. \tag{63}$$

The measurement transformed points are

$$h_i = a\, \chi_i^n, \quad i = 0, 1, 2. \tag{64}$$

Then the mean and variance of h a given by

$$\bar{h} = \sum_{i=0}^{2} w_i h_i, \tag{65}$$

$$\sigma_h^2 = \sum_{i=0}^{2} w_i (h_i - \bar{h})^2, \tag{66}$$

The cross-covariance c_{hx} is computed by

$$c_{hx} = \sum_{i=0}^{2} w_i (h_i - \bar{h})(\chi_i - \hat{x}). \tag{67}$$

3.8. MoN of Straka, Duník, and Šimandl

Straka, Duník, and Šimandl presented two local MoNs in [48,49]. Given the estimate $\hat{x}$ and variance σ_x^2, these MoNs use a number of points $\chi_i, i = 1, 2, \ldots, m$ in the neighborhood of $\hat{x}$. We analyze the first MoN proposed by the authors. The transformed points by the non-linear function h are given by

$$z_i = h(\chi_i),\ i = 1, 2, \ldots, m. \tag{68}$$

Define

$$\mathbf{Z} := \begin{bmatrix} z_1 & z_2 & \ldots & z_m \end{bmatrix}', \tag{69}$$

$$\mathbf{X} := \begin{bmatrix} \chi_1 & \chi_2 & \ldots & \chi_m \end{bmatrix}'. \tag{70}$$

A linear approximation to $\mathbf{Z}$ is $\mathbf{X}\theta$, where θ is a scalar parameter to be estimated. The cost function that is proposed in [48,49] to determine θ is given by

$$J_1(\theta) := (\mathbf{Z} - \mathbf{X}\theta)'\mathbf{W}(\mathbf{Z} - \mathbf{X}\theta), \tag{71}$$

where the weight-matrix $\mathbf{W}$ is given by

$$\mathbf{W} = \text{diag}(d_1, d_2, \ldots, d_m), \tag{72}$$

$$d_i = (\chi_i - \hat{x})^2, \; i = 1, 2, \ldots, m. \tag{73}$$

The LS estimate [39] that minimizes the cost function is given by

$$\hat{\theta}_{\text{LS}} = (\mathbf{X}'\mathbf{W}\mathbf{X})^{-1}\mathbf{X}'\mathbf{W}\mathbf{Z}. \tag{74}$$

For this problem, the LS estimate in (74) reduces to

$$\hat{\theta}_{\text{LS}} = \left(\sum_{i=1}^{m} d_i \chi_i^2\right)^{-1} \sum_{i=1}^{m} d_i \chi_i z_i. \tag{75}$$

The cost function J_1 evaluated at $\hat{\theta}_{\text{LS}}$ is treated as a local MoN η, given by

$$\eta = J_1(\hat{\theta}_{\text{LS}}). \tag{76}$$

Remark 6. *We have calculated the average MoN for the bearing-only filtering [27], GMTI [32], and video filtering [34] problems. The MoN is presented in the table below (Table 2). From this table we find that the degree of nonlinearity of the bearing-only filtering problem is about two orders of magnitude higher than that of the GMTI or video filtering problem. This implies that a simple filter, such as the EKF or UKF, is sufficient for the GMTI or video filtering problem, but an advanced filter, such as the PF, is needed for the BOF [17] problem.*

Table 2. MoNs for the bearing-only, GMTI, and video filtering problems.

Curvature Type	Bearing-Only	GMTI	Video
Parameter-effects	(300–1200) $\times 10^{-4}$	(0.8–1.2) $\times 10^{-4}$	0.245 $\times 10^{-4}$
Intrinsic	(69–149) $\times 10^{-4}$	0.2 $\times 10^{-4}$	0.066 $\times 10^{-4}$
Total	(369–1349) $\times 10^{-4}$	(1.0–1.4) $\times 10^{-4}$	0.312 $\times 10^{-4}$

4. Mapping between CMoN and MSE in Polynomial NonLinearity

The nonlinearity of the problem imposes challenges in parameter estimation. We analyze the CMoN and MSE of the non-linear estimation problem to discover relationships among them. For the current problem, CMoN are measured by the parameter-effects curvature in (57) and the direct parameter-effects curvature in (58). In general, CMoN depend on the first and second derivatives of the non-linear function calculated at the parameter estimate and on the norm of the estimation error for $\beta_\delta(\hat{x})$. Therefore, the CMoN will depend the type of estimator (e.g., ML) used to obtain parameter estimate. The extrinsic curvature (38) depends on the first and second derivatives of the non-linear function evaluated at the true x.

4.1. MSE and Sample MSE

We estimate the x coordinate using noisy measurements at a discrete set $\{x_k\}_{k=1}^{N_x}$ of values. Let $\hat{x}_{k,m}$ denote the estimate of x_k in the mth Monte Carlo run. Subsequently, the error $\tilde{x}_{k,m}$ in $\hat{x}_{k,m}$ is defined by

$$\tilde{x}_{k,m} := x_k - \hat{x}_{k,m}, \; k = 1, 2, \ldots, N_x, \; m = 1, 2, \ldots, M, \tag{77}$$

where M is the number of Monte Carlo runs. The MSE at x_k is given by

$$\text{MSE}_k = E[(\tilde{x}_{k,m})^2], \quad k = 1, 2, \ldots, N_x. \tag{78}$$

The sample MSE (SMSE) at x_k is defined by

$$\mathrm{SMSE}_k := \frac{1}{M}\sum_{m=1}^{M}(\tilde{x}_{k,m})^2, \quad k = 1, 2, \ldots, N_x. \tag{79}$$

Let $L_{\mathrm{CRLB}}(x)$ denote the $\log_{10}$ of the CRLB,

$$L_{\mathrm{CRLB}}(x) := \log_{10}\mathrm{CRLB}_x. \tag{80}$$

Taking the log of CRLB_x in (32) we get

$$L_{\mathrm{CRLB}}(x) = \log_{10}\left(\frac{\sigma^2}{Nn^2a^2}\right) - 2(n-1)\log_{10}x. \tag{81}$$

4.2. MSE and Parameter-Effects Curvature

Let $L_K(x)$ denote the log of the expected value of $K(\hat{x})$ in (57). Then

$$L_K(x) := \log_{10}\{E[K(\hat{x})]\}. \tag{82}$$

In order to compute $L_K(x)$, we first approximate the expectation in (82) by assuming $\sigma_{\hat{x}} \ll x$, which holds for the case investigated in our paper,

$$E\{K(\hat{x})\} = \frac{(n-1)}{na}E\left(\frac{1}{\hat{x}^n}\right) \approx \frac{(n-1)}{na}\frac{1}{[E(\hat{x})]^n} \approx \frac{(n-1)}{na}\frac{1}{x^n}. \tag{83}$$

The last step of the above equation follows from an assumption that the estimator is nearly unbiased. Now, taking the logarithm, we have

$$L_K(x) = \log_{10}\left(\frac{n-1}{na}\right) - n\log_{10}x. \tag{84}$$

Now, from Equations (84) and (81), we can see that there is an affine mapping between $L_{\mathrm{CRLB}}(x)$ and $L_K(x)$. That is,

$$L_{\mathrm{CRLB}}(x) = \alpha_1^K L_K(x) + \alpha_0^K, \tag{85}$$

where

$$\begin{aligned}\alpha_1^K &= \frac{2(n-1)}{n},\\ \alpha_0^K &= \log_{10}\left(\frac{\sigma^2}{Nn^2a^2}\right) - \frac{2(n-1)}{n}\log_{10}\left(\frac{n-1}{na}\right).\end{aligned} \tag{86}$$

We observe that α_1^K is positive and, hence, $L_K(x)$ and $L_{\mathrm{CRLB}}(x)$ have the same sign of the non-zero slopes. As a result, $K(\hat{x})$ and CRLB have the same sign of the non-zero slopes.

4.3. MSE and Direct Parameter-Effects Curvature

The expression for the direct parameter-effects curvature $\beta_\delta(\hat{x})$ [29,30] is given by (58). Similar to the previous section, we define

$$L_\beta(x) := \log_{10}\left(E\{\beta_\delta(\hat{x})\}\right). \tag{87}$$

Now, taking the expected value of β, we have

$$E\{\beta(\hat{x})\} \approx \frac{(n-1)}{x}E\{|\delta|\} = \frac{(n-1)}{x}E\{|\hat{x}-x|\}. \tag{88}$$

The RHS of (88) can be simplified by assuming that $\hat{x}$ is unbiased and that it achieves the CRLB. Additionally, we approximate this error to be Gaussian and the variance of $\hat{x}$ is given in (29). Then,

$$E\{|\hat{x} - x|\} = \sqrt{\mathrm{CRLB}_x}\sqrt{\frac{2}{\pi}}. \tag{89}$$

Substituting (89) into (88) and using (32) for CRLB_x we have

$$\mathbb{E}[\beta(\hat{x})] \approx \frac{(n-1)}{na}\sigma\sqrt{\frac{2}{N\pi}}x^{-n}. \tag{90}$$

Thus,

$$L_\beta(x) \approx \log_{10}\left[\frac{(n-1)\sigma\sqrt{\frac{2}{N\pi}}}{na}\right] - n\log_{10} x. \tag{91}$$

From (91) and (81) we can write the affine mapping

$$L_{\mathrm{CRLB}}(x) = \alpha_1^\beta L_\beta(x) + \alpha_0^\beta, \tag{92}$$

where

$$\begin{aligned} \alpha_1^\beta &= \frac{2(n-1)}{n}, \\ \alpha_0^\beta &= \log_{10}\left(\frac{\sigma^2}{Nn^2a^2}\right) - \frac{2n-2}{n}\log_{10}\left[\frac{(n-1)\sigma\sqrt{2}}{na\sqrt{N\pi}}\right]. \end{aligned} \tag{93}$$

We also observe that α_1^β is positive and, hence, $L_\beta(x)$ and $L_{\mathrm{CRLB}}(x)$ have the same sign of the non-zero slopes. As a result, $\beta_\delta(\hat{x})$ and CRLB have the same sign of the non-zero slopes.

4.4. Extrinsic Curvature

The expression for extrinsic curvature for our problem is given in (38). Similar to previous sections, we define

$$L_\kappa(x) := \log_{10}(\kappa(x)). \tag{94}$$

Taking the log of (94), we have

$$\begin{aligned} L_\kappa(x) &= \log_{10}(\kappa(x)) \\ &= \log_{10}\left[an(n-1)x^{n-2}\right] - \frac{3}{2}\log_{10}\left[1 + (anx^{n-1})^2\right] \\ &\approx \log_{10}\left[an(n-1)x^{n-2}\right] - \frac{3}{2}\log_{10}\left[(anx^{n-1})^2\right] \\ &= \log_{10}\left[\frac{n-1}{(an)^2}\right] - (2n-1)\log_{10} x. \end{aligned} \tag{95}$$

Note that the second last expression is a valid approximation for $x > 2$. From (95) and (84) it is easy to establish the affine mapping

$$L_K(x) = \gamma_1^K L_\kappa(x) + \gamma_0^K, \tag{96}$$

where

$$\begin{aligned}\gamma_1^K &= \frac{n}{2n-1}, \\ \gamma_0^K &= \log_{10}\left(\frac{n-1}{na}\right) - \frac{n}{2n-1}\log_{10}\left[\frac{n-1}{(an)^2}\right]\end{aligned} \tag{97}$$

Similarly, from (95) and (91) we can establish the affine relationship

$$L_\beta(x) = \gamma_1^\beta L_\kappa(x) + \gamma_0^\beta, \tag{98}$$

where

$$\begin{aligned}\gamma_1^\beta &= \frac{n}{2n-1}, \\ \gamma_0^\beta &= \log_{10}\left[\frac{(n-1)\sigma\sqrt{2}}{na\sqrt{N\pi}}\right] - \frac{n}{2n-1}\log_{10}\left[\frac{n-1}{(an)^2}\right].\end{aligned} \tag{99}$$

Using similar arguments used in previous sections, we infer that the extrinsic curvature and parameter-effects curvature have the same sign of the non-zero slopes. Similarly, the extrinsic curvature and direct parameter-effects curvature have the same non-zero slopes.

4.5. Estimation of CMoN and SMSE by Monte-Carlo Simulations

Let $\overline{K}(\hat{x}_k)$ and $\overline{\beta}_\delta(\hat{x}_k)$ denote the sample means of the Bates and Watts and direct parameter-effects curvatures calculated from M Monte Carlo runs. Subsequently,

$$\overline{K}(\hat{x}_k) := \frac{1}{M}\sum_{m=1}^{M} K(\hat{x}_{k,m}), \quad k = 1, 2, \ldots, N_x, \tag{100}$$

$$\overline{\beta}_\delta(\hat{x}_k) := \frac{1}{M}\sum_{m=1}^{M} \beta_\delta(\hat{x}_{k,m}), \quad k = 1, 2, \ldots, N_x. \tag{101}$$

Correspondingly, we define

$$b_k := \log_{10}\mathrm{SMSE}_k, \quad k = 1, 2, \ldots, N_x, \tag{102}$$

$$c_k := \log_{10}\overline{K}(\hat{x}_k), \quad k = 1, 2, \ldots, N_x, \tag{103}$$

$$d_k := \log_{10}\beta_\delta(\hat{x}_k), \quad k = 1, 2, \ldots, N_x. \tag{104}$$

Define

$$\mathbf{b} := \begin{bmatrix} b_1 & b_2 & \ldots & b_{N_x} \end{bmatrix}', \tag{105}$$

$$\mathbf{c} := \begin{bmatrix} c_1 & c_2 & \ldots & c_{N_x} \end{bmatrix}', \tag{106}$$

$$\mathbf{d} := \begin{bmatrix} d_1 & d_2 & \ldots & d_{N_x} \end{bmatrix}'. \tag{107}$$

Suppose that an affine mapping exists between $\mathbf{b}$ and $\mathbf{c}$. Subsequently,

$$b_k = \hat{\alpha}_1^K c_k + \hat{\alpha}_0^K + e_k, \quad k = 1, 2, \ldots, N_x, \tag{108}$$

where e_k is a random noise. Afterwards, we can write (108) in the matrix-vector form by

$$\mathbf{b} = \mathbf{H}_c\boldsymbol{\alpha} + \mathbf{e}, \tag{109}$$

where

$$\boldsymbol{\alpha} := \begin{bmatrix} \hat{\alpha}_1^K & \hat{\alpha}_0^K \end{bmatrix}', \tag{110}$$

$$\mathbf{e} := \begin{bmatrix} e_1 & e_2 & \dots & e_{N_x} \end{bmatrix}', \tag{111}$$

$$\mathbf{H}_c := \begin{bmatrix} c_1 & 1 \\ c_2 & 1 \\ \dots & \dots \\ c_{N_x} & 1 \end{bmatrix}. \tag{112}$$

Given $\mathbf{b}$ and $\mathbf{H}_c$, we can estimate $\boldsymbol{\alpha}$ using the linear least squares (LLS).

We can similarly define the affine mapping between other variable pairs. Altogether, we consider the following four:

1. between $\mathbf{b}$ ($\log_{10}(\text{SMSE}_k)$) and $\mathbf{c}$ ($\log_{10}(\overline{K}(\hat{x}_k))$ for each power of the polynomial function, as in (85),
2. between $\mathbf{c}$ ($\log_{10}(\overline{K}(\hat{x}_k))$ and $\log_{10}(\kappa(x_k))$ (94) for each power of the polynomial function, as in (96), and
3. between $\mathbf{d}$ ($\log_{10}(\overline{\beta}_\delta(\hat{x}_k))$ and $\log_{10}(\kappa(x_k))$ (94) for each power of the polynomial function, as in (98).

5. Numerical Simulation and Results

We follow the same simulation scenario as used in our previous work [35]. We use $a = 0.6$ and $n = 2, 3, 4, 5$ and a number of uniformly spaced x coordinates with the spacing of 0.1 in the interval $[2, 7]$. The measurement noise standard deviation (σ) is 0.5. The dimension of the measurement vector is 10 or 20. The results are based on 1000 Monte Carlo runs. Figure 2 shows $\log_{10}(h(x))$ versus x.

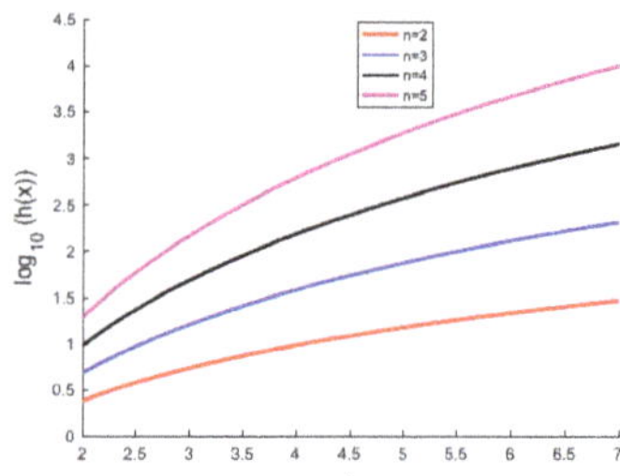

Figure 2. $\log_{10}(h(x))$ versus x.

To assess the accuracy of the MLE, we compute the sample bias, sample MSE, ANEES [42], and CRLB [2,41,51]. Let $x_{k,i} = x_k$, $\hat{x}_{k,i}$, and $\sigma_{k,i}^2$ denote the true parameter, ML estimate, and associated variance, respectively, at the kth point in the ith Monte Carlo run. The sample bias in the estimate at the kth point is defined by [9]

$$\hat{b}_k := \frac{1}{M} \sum_{i=1}^{M} (x_{k,i} - \hat{x}_{k,i}), \tag{113}$$

where M is the number of Monte Carlo runs. The sample root MSE (RMSE) [9] and ANEES [2,9,42] at the kth point are defined, respectively, by

$$\text{RMSE}_k := \left[\frac{1}{M} \sum_{i=1}^{M} (x_{k,i} - \hat{x}_{k,i})^2) \right]^{1/2}, \tag{114}$$

$$\mathrm{ANEES}_k := \frac{1}{M}\sum_{i=1}^{M}(x_{k,i} - \hat{x}_{k,i})^2/\sigma_{k,i}^2. \tag{115}$$

Figure 3 presents the sample bias for different powers of x. We observe from Figure 3 that the bias is small when compared with the true value of x and the bias decreases with increase in the power of x. In Figure 4, we have plotted the $\sqrt{\mathrm{CRLB}}$ and the average of σ_x over Monte Carlo runs. Figure 4 shows that, for each power of x, the $\sqrt{\mathrm{CRLB}}$ and the average of σ_x are on top of each other and it is not possible to distinguish them in the figure.

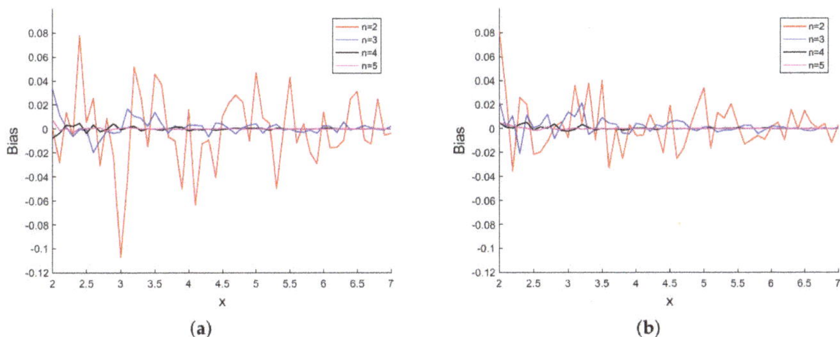

Figure 3. (**a**) Sample bias vs. x using 10 scalar measurements and (**b**) sample bias vs. x using 20 scalar measurements.

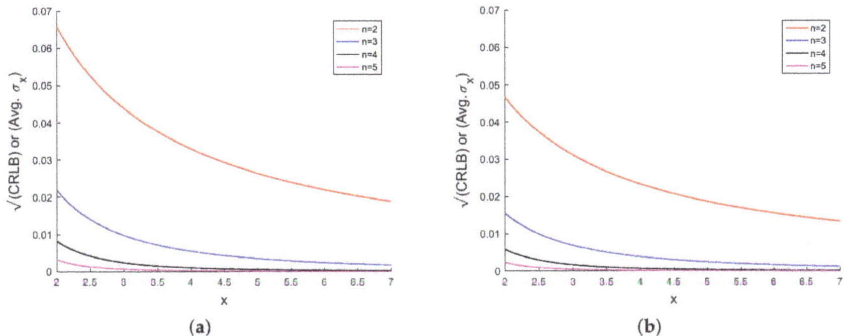

Figure 4. (**a**) $\sqrt{\mathrm{CRLB}}$ or (Avg. σ_x) vs. x using 10 scalar measurements and (**b**) $\sqrt{\mathrm{CRLB}}$ or (Avg. σ_x) vs. x using 20 scalar measurements.

Figure 5 presents $\sqrt{\mathrm{CRLB}}$ and RMSE for each power of x. Solid and dashed lines in Figure 5 represent the $\sqrt{\mathrm{CRLB}}$ and RMSE, respectively, for each power of x. We see from Figure 5 that corresponding values of $\sqrt{\mathrm{CRLB}}$ and RMSE are close to each other for each power of x. In Figures 3–5, the bias, $\sqrt{\mathrm{CRLB}}$, σ_x, and RMSE for 20 measurements are smaller than corresponding values for 10 measurements.

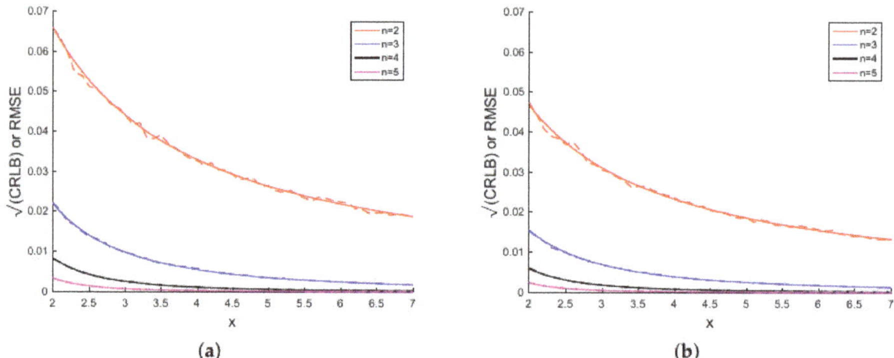

Figure 5. (**a**) $\sqrt{\text{CRLB}}$ or RMSE vs. x using 10 scalar measurements and (**b**) $\sqrt{\text{CRLB}}$ or RMSE vs. x using 20 scalar measurements.

We present the ANEES [42] in Figure 6 for different powers of x with 99% confidence bounds. We see from Figure 6 that the ANEES lies within the 99% confidence bounds. This shows that the variance σ_x^2 calculated using the MLE is consistent with the estimation error.

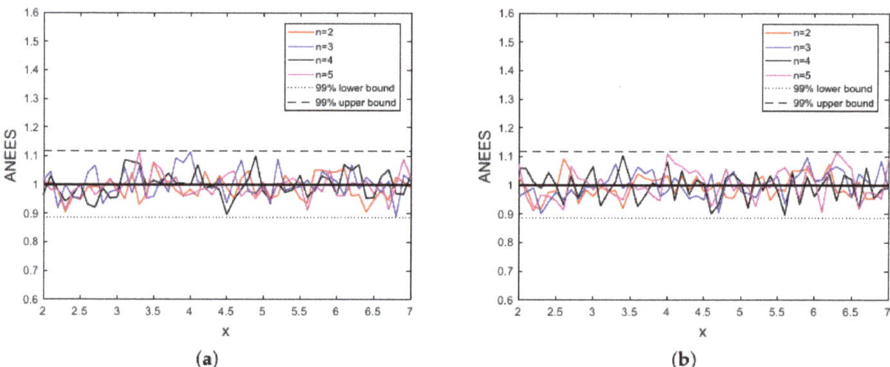

Figure 6. (**a**) ANEES vs. x using 10 scalar measurements and (**b**) ANEES vs. x using 20 scalar measurements.

Figure 7 presents the logarithm of the extrinsic curvature $\log_{10}(\kappa(x))$ versus x. The extrinsic curvature is completely determined by the first and second derivatives of the non-linear function h and it is evaluated while using the true x.

In Figures 8–18, we present results using 10 scalar measurements. We have also generated results using 20 scalar measurements. In order to limit the number of figures, we have not presented figures with 20 scalar measurements. The CRLB, variance of estimation error, all MoNs, and MSE follow the same trend. However, the corresponding values compared with 20 measurements are reduced due to improved estimation accuracy.

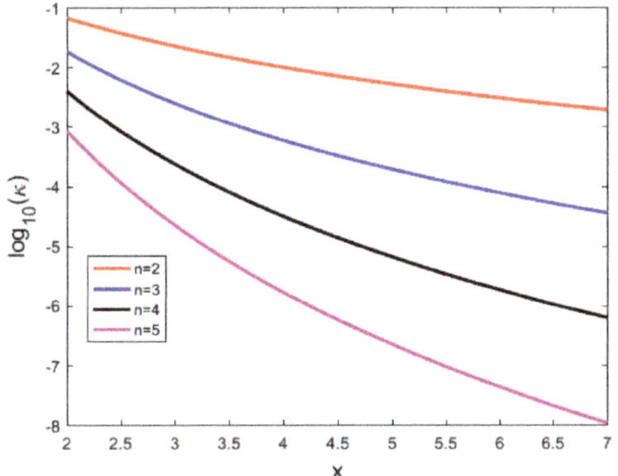

Figure 7. Logarithm of the extrinsic curvature $\log_{10}(\kappa(x))$ versus x.

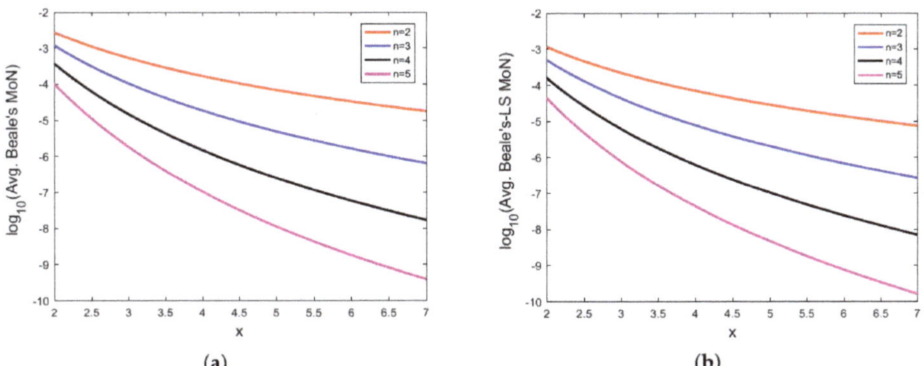

Figure 8. (**a**) Logarithm of Beale's MoN ($\log_{10}$(Avg. Beale's MoN)) vs. x and (**b**) logarithm of Beale's MoN using LS ($\log_{10}$(Avg. Beale's-LS MoN)) vs. x with 10 scalar measurements.

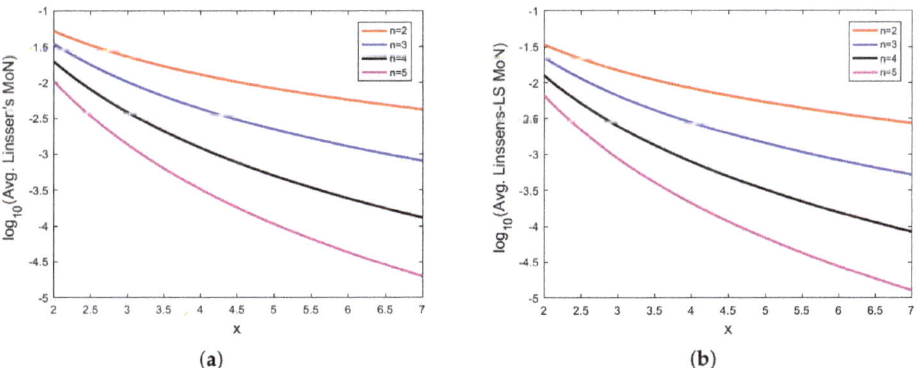

Figure 9. (**a**) Logarithm of Linssen's MoN ($\log_{10}$(Avg. Linssen's MoN)) vs. x and (**b**) logarithm of Linssen's MoN using LS ($\log_{10}$(Avg. Linssen's-LS MoN) vs. x with 10 scalar measurements.

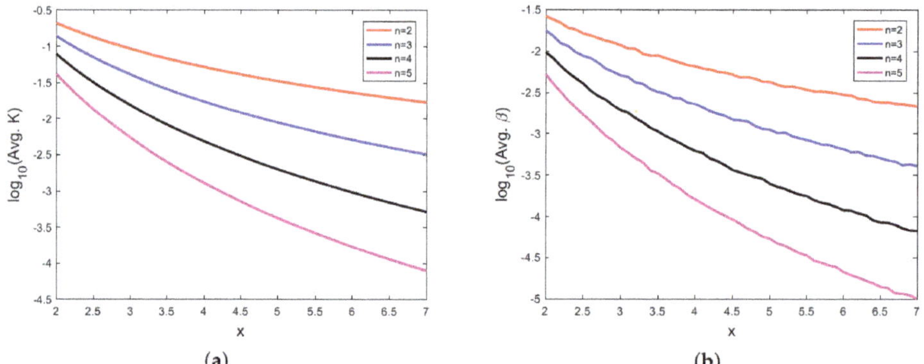

Figure 10. (**a**) Logarithm of Bates and Watts parameter-effects curvature ($\log_{10}$(Avg. K)) vs. x and (**b**) logarithm of direct parameter-effect curvature ($\log_{10}$(Avg. β)) vs. x using 10 scalar measurements.

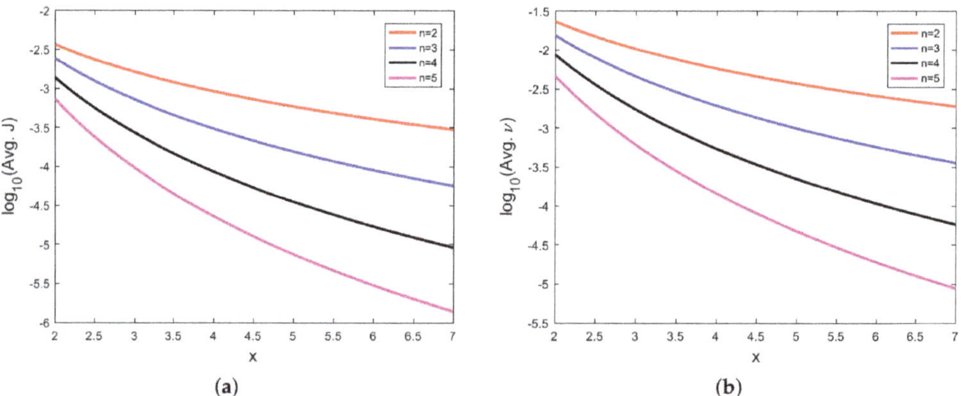

Figure 11. (**a**) Logarithm of Li's un-normalized MoN ($\log_{10}$(Avg. J)) vs. x and (**b**) logarithm of Li's normalized MoN ($\log_{10}$(Avg. ν)) vs. x using 10 scalar measurements.

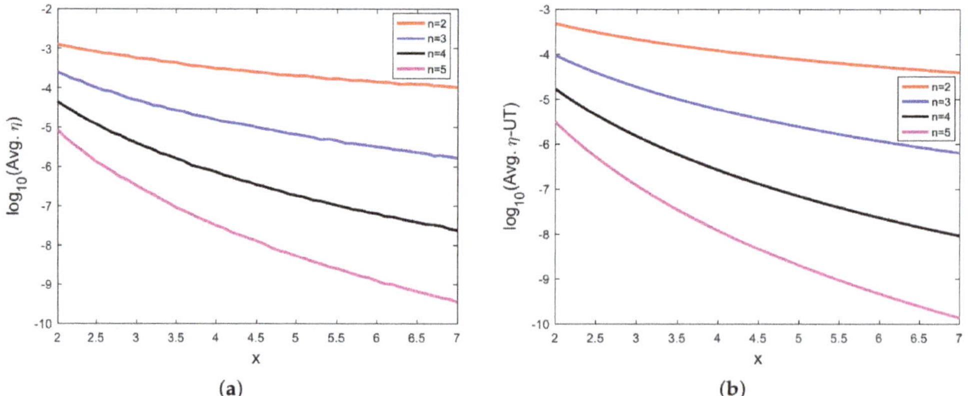

Figure 12. (**a**) Logarithm of MoN of Straka et al. ($\log_{10}$(Avg. η)) vs. x and (**b**) logarithm of MoN of Straka et al. with UT ($\log_{10}$(Avg. η-UT)) vs. x using 10 scalar measurements.

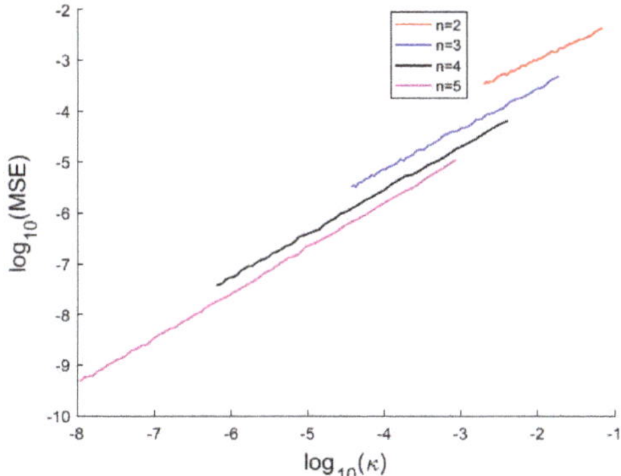

Figure 13. $\log_{10}$(MSE) vs. logarithm of extrinsic curvature ($\log_{10}(\kappa)$) using 10 scalar measurements.

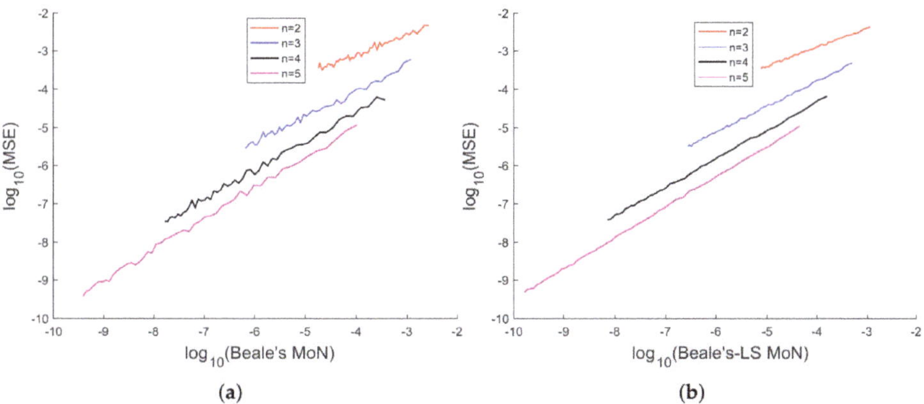

Figure 14. (**a**) $\log_{10}$(MSE) vs. $\log_{10}$(Avg. Beale's MoN) and (**b**) $\log_{10}$(MSE) vs. $\log_{10}$ (Avg. Beale's MoN using LS) using 10 scalar measurements.

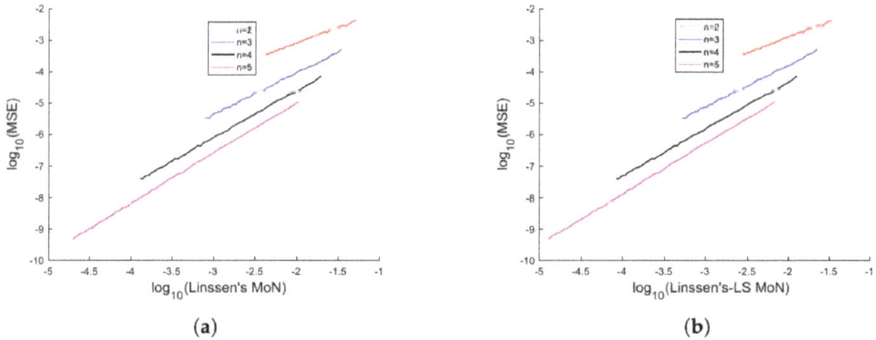

Figure 15. (**a**) $\log_{10}$(MSE) vs. $\log_{10}$(Linssen's MoN) and (**b**) $\log_{10}$(MSE) vs. $\log_{10}$ (Linssen's-LS) using 10 scalar measurements.

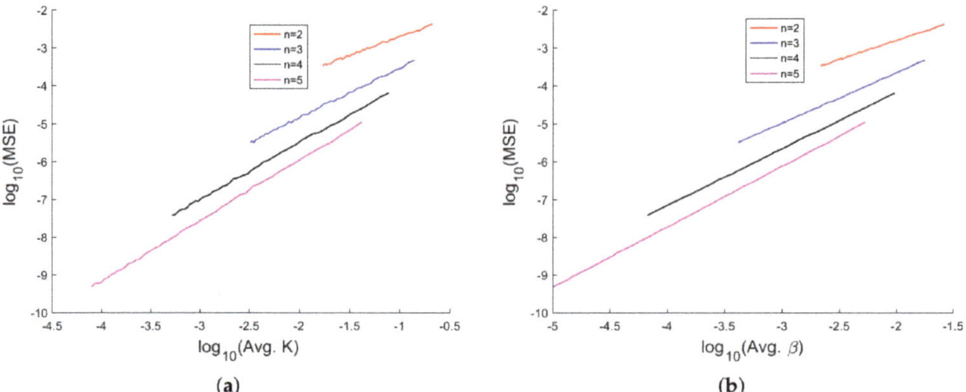

Figure 16. $\log_{10}$(MSE) vs. logarithm of parameter-effects curvatures. (**a**) $\log_{10}$(MSE) vs. $\log_{10}$(Avg. K) and (**b**) $\log_{10}$(MSE) vs. $\log_{10}$ (Avg. β) using 10 scalar measurements.

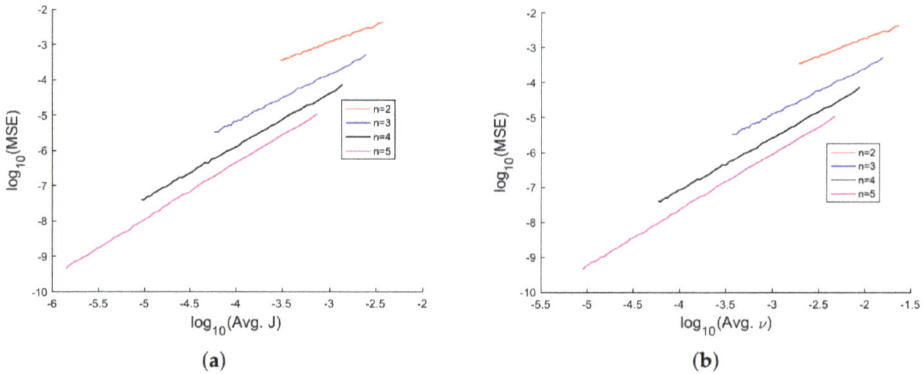

Figure 17. $\log_{10}$(MSE) vs. logarithm of Li's MoN. (**a**) $\log_{10}$(MSE) vs. $\log_{10}$(Avg. J) and (**b**) $\log_{10}$(MSE) vs. $\log_{10}$ (Avg. ν) using 10 scalar measurements.

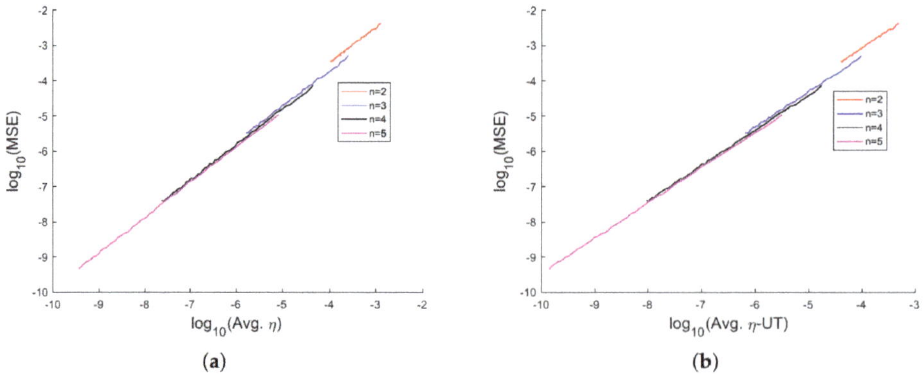

Figure 18. $\log_{10}$(MSE) vs. logarithm of MoN of Straka et al. (**a**) $\log_{10}$(MSE) vs. $\log_{10}$(Avg. η) and (**b**) $\log_{10}$(MSE) vs. $\log_{10}$ (Avg. η-UT) using 10 scalar measurements.

In [35], we had shown analytically, and through Monte Carlo simulation, that affine mappings exist among $\log_{10}$(MSE), $\log_{10}$ (κ), $\log_{10}$(Avg. K), and $\log_{10}$ (Avg. β). In Figures 13–18, we have plotted the $\log_{10}$(MSE) versus $\log_{10}$ of various MoNs using 10 scalar measurements. These figures show that the $\log_{10}$(MSE) varies with $\log_{10}$ (MoN) according to an affine mapping with a positive slope. This implies that the MSE increases as an MoN increases. We obtain similar results for the case of 20 scalar measurements.

The above results demonstrate that, for the polynomial nonlinearity problem analyzed, any of the seven MoNs analyzed is suitable metrics to quantify the MSE, which represents the complexity of a parameter estimation problem. Further research is needed to study the applicability of these MoNs in real-world non-linear filtering problems.

6. Conclusions

We considered a polynomial curve in 2D and derived analytic expressions for the ML estimate and associated variance of the independent variable x using a vector measurement. The ML estimate is used to evaluate the Jacobian and Hessian of the measurement function appearing in the computation of Bates and Watts and direct parameter-effects curvatures, Beale's MoN, and Linssen's MoN. Our numerical results show that the variance of the estimated parameter and the Cramér-Rao lower bound (CRLB) are nearly the same for different powers of x. The average normalized estimation error squared (ANEES) lies within the 99% confidence interval, which indicates that the ML based variance is consistent with the estimation error.

We used seven MoNs, including the extrinsic curvature using differential geometry, Beale's MoN (and its least squares variant), Linssen's MoN (and its least squares variant), Bates and Watts parameter-effects curvature, direct parameter-effects curvature, Li's MoN, and the MoN of Straka, Duník, and Šimandl. If a MoN has a high value, then the nonlinearity is high. All of the MoNs show the same type of variation with x and the power of of the polynomial. Secondly, as the logarithm of a MoN increases, the logarithm of the MSE also increases linearly for each MoN. This implies that, as a MoN increases, and then the MSE increases. These results are quite surprising, given the fact that these MoNs are derived based on completely different theoretical considerations. The second feature of our analysis is useful in establishing that a MoN in our study can be considered as a candidate metric for quantifying the MSE that represents the complexity of a parameter estimation problem. Our future work will study other practical parameter estimation and non-linear filtering problems.

Author Contributions: Formal analysis, M.M. and X.T.; methodology, M.M.; software, M.M. and X.T.; writing, M.M. and X.T. All authors have read and agreed to the published version of the manuscript.

Funding: This research received no external funding.

Acknowledgments: The author thanks Sanjeev Arulampalam of Defence Science and Technology Organisation, Edinburgh SA, Australia for useful discussions and insightful comments in improving the manuscript.

Conflicts of Interest: The authors declare no conflict of interest.

References

1. Anderson, B.D.O.; Moore, J.B. *Optimal Filtering*; Prentice-Hall: Englewood Cliffs, NJ, USA, 1979.
2. Bar-Shalom, Y.; Li, X.; Kirubarajan, T. *Estimation with Applications to Tracking and Navigation*; Wiley: New York, NY, USA, 2001.
3. Gelb, A. (Ed.) *Applied Optimal Estimation*; MIT Press: Cambridge, MA, USA, 1974.
4. Jazwinski, A.H. *Stochastic Processes and Filtering Theory*; Academic Press: New York, NY, USA, 1970.
5. Aidala, V.J. Kalman Filter Behaviour in Bearings-only Tracking Applications. *IEEE Trans. Autom. Control* **1979**, *AES-15*, 29–39.
6. Aidala, V.J.; Hammel, S.E. Utilization of modified polar coordinates for bearings-only tracking. *IEEE Trans. Autom. Control* **1983**, *28*, 283–294. [CrossRef]

7. Arulampalam, S.; Ristic, B. Comparison of the particle filter with range-parameterised and modified polar EKFs for angle-only tracking. In Proceedings of the SPIE Conference on Signal and Data Processing of Small Targets, Orlando, FL, USA, 24–28 April 2000; Volume 4048, pp. 288–299.
8. Ristic, B.; Arulampalam, S.; Gordon, N. *Beyond the Kalman Filter*; Artech House: Boston, MA, USA, 2004.
9. Mallick, M.; Arulampalam, S. Comparison of Nonlinear Filtering Algorithms in Ground Moving Target Indicator (GMTI) Target Tracking. In Proceedings of the SPIE Conference on Signal and Data Processing of Small Targets, San Diego, CA, USA, 5 January 2004; Volume 5204, pp. 630–647.
10. Blackman, S.S.; White, T.; Blyth, B.; Durand, C. Integration of Passive Ranging with Multiple Hypothesis Tracking (MHT) for Application with Angle-Only Measurements. In Proceedings of the SPIE Conference on Signal and Data Processing of Small Targets, Orlando, FL, USA, 15 April 2010; Volume 7698, pp. 769815-1–769815-11.
11. Mallick, M.; Morelande, M.; Mihaylova, L.; Arulampalam, S.; Yan, Y. Comparison of Angle-only Filtering Algorithms in 3D using Cartesian and Modified Spherical Coordinates. In Proceedings of the 15th International Conference on Information Fusion, Singapore, 9–12 July 2012.
12. Mallick, M.; Morelande, M.; Mihaylova, L.; Arulampalam, S.; Yan, Y. Angle-only Filtering in Three Dimensions, Chapter 1. In *Integrated Tracking, Classification, and Sensor Management: Theory and Applications*; Mallick, M., Krishnamurthy, V., Vo, B.-N., Eds.; Wiley/IEEE: Hoboken, NJ, USA, 2013.
13. Athans, M.; Wishner, R.P.; Bertolini, A. Suboptimal state estimation for continuous-time nonlinear systems from discrete noisy measurements. *IEEE Trans. Autom. Control* **1968**, *13*, 504–518. [CrossRef]
14. Julier, S.; Uhlmann, J.; Durrant-Whyte, H.F. A new method for the nonlinear transformation of means and covariances in filters and estimators. *IEEE Trans. Autom. Control* **2000**, *AC-45*, 477–482. [CrossRef]
15. Julier, S.J.; Uhlmann, J.K. Unscented filtering and nonlinear estimation. *Proc. IEEE* **2004**, *92*, 401–422. [CrossRef]
16. Arasaratnam, I.; Haykin, S. Cubature Kalman filters. *IEEE Trans. Autom. Control* **2009**, *54*, 1254–1269. [CrossRef]
17. Arulampalam, S.; Maskell, S.; Gordon, N.; Clapp, T. A tutorial on particle filters for on-line non-linear/non-Gaussian Bayesian tracking. *IEEE Trans. Signal Process.* **2002**, *50*, 174–188. [CrossRef]
18. Beale, E.M.L. Confidence regions in non-linear estimation. *J. R. Stat. Soc. Ser. B* **1960**, *22*, 41–76. [CrossRef]
19. Guttman, I.; Meeter, D.A. On Beale's measures of non-linearity. *Technometrics* **1965**, *7*, 623–637. [CrossRef]
20. Linssen, H.N. Nonlinearity measures: A case study. *Stat. Neerl.* **1975**, *29*, 93–99. [CrossRef]
21. Bates, D.M.; Watts, D.G. Relative curvature measures of nonlinearity. *J. R. Stat. Society. Ser. B (Methodol.)* **1980**, *42*, 1–25. [CrossRef]
22. Bates, D.M.; Watts, D.G. Parameter transformations for improved approximate confidence regions in nonlinear least squares. *Ann. Stat.* **1981**, *9*, 1152–1167. [CrossRef]
23. Goldberg, M.L.; Bates, D.M.; Watts, D.G. Simplified Methods of Assessing Nonlinearity. *Am. Stat. Assoc. Proc. Bus. Eco. Statistics Section* **1983**, 67–74.
24. Bates, D.M.; Hamilton, D.C.; Watts, D.G. Calculation of intrinsic and parameter-effects curvatures for nonlinear regression models. *Commun. Stat. Simul. Comput.* **1983**, *12*, 469–477. [CrossRef]
25. Bates, D.M.; Watts, D.G. *Nonlinear Regression Analysis and its Applications*; John Wiley: New York, NY, USA, 1988.
26. Seber, G.A.; Wild, C.J. *Nonlinear Regression*; John Wiley: New York, NY, USA, 1989.
27. Mallick, M. Differential geometry measures of nonlinearity with application to ground target tracking. In Proceedings of the 7th International Conference on Information Fusion, Stockholm, Sweden, 28 June–1 July 2004.
28. Scala, B.F.; Mallick, M.; Arulampalam, S. Differential Geometry Measures of Nonlinearity for Filtering with Nonlinear Dynamic and Linear Measurement Models. In Proceedings of the SPIE Conference on Signal and Data Processing of Small Targets, San Diego, CA, USA, 28–30 August 2007.
29. Mallick, M.; Scala, B.F.L.; Arulampalam, M.S. Differential geometry measures of nonlinearity for the bearing-only tracking problem. In Proceedings of the SPIE Conference on Signal Processing, Sensor Fusion, and Target Recognition XIV, Orlando, FL, USA, 25 May 2005; Volume 5809, pp. 288–300.
30. Mallick, M.; Scala, B.F.L.; Arulampalam, S. Differential geometry measures of nonlinearity with applications to target tracking. In Proceedings of the Fred Daum Tribute Conference, Monterey, CA, USA, 24 May 2007.

31. Duník, J.; Straka, O.; Mallick, M.; Blasch, E. Survey of nonlinearity and non-Gaussianity measures for state estimation. In Proceedings of the 19th International Conference on Information Fusion, Heidlberg, Germany, 5 July 2016; pp. 1845–1852.
32. Mallick, M.; Scala, B.F.L. Differential geometry measures of nonlinearity for ground moving target indicator (GMTI) filtering. In Proceedings of the 2005 8th International Conference on Information Fusion, Philadelphia, PA, USA, 25–28 July 2005; pp. 219–226.
33. Mallick, M.; Ristic, B. Comparison of Measures of Nonlinearity for Bearing-only and GMTI Filtering. In Proceedings of the 20th International Conference on Information Fusion, Xi'an, China, 10–13 July 2017.
34. Mallick, M.; Scala, B.F.L. Differential geometry measures of nonlinearity for the video tracking problem. In Proceedings of the SPIE Conference on Signal Processing, Sensor Fusion, and Target Recognition XV, Orlando (Kissimmee), FL, USA, 17 May 2006; Volume 6235, pp. 246–258.
35. Mallick, M.; Yan, Y.; Arulampalam, S.; Mallick, A. Connection between Differential Geometry and Estimation Theory for Polynomial Nonlinearity in 2D. In Proceedings of the 13th International Conference on Information Fusion, Edinburgh, UK, 26–29 July 2010.
36. Available online: http://en.wikipedia.org/wiki/curvature (accessed on 5 March 2020).
37. Carmo, M.D. *Differential Geometry of Curves and Surfaces*; Prentice-Hall: Englewood Cliffs, NJ, USA, 1976.
38. Struik, D.J. *Lectures on Classical Differential Geometry*, 2nd ed.; Dover Publications: Mineola, NY, USA, 1988.
39. Mendel, J.M. *Lessons in Estimation Theory for Signal Processing Communications, and Control*; Prentice-Hall: Englewood Cliffs, NJ, USA, 1995.
40. Mallick, M. Improvements to Measures of Nonlinearity Computation for Polynomial Nonlinearity. In Proceedings of the 8th International Conference on Control, Automation and Information Sciences (ICCAIS 2019), Chengdu, China, 23–26 October 2019.
41. Trees, H.L.V.; Bell, K.L.; Tian, Z. *Detection, Estimation, and Modulation Theory, Part I: Detection, Estimation, and Filtering Theory*, 2nd ed.; Wiley: New York, NY, USA, 2013.
42. Li, X.R.; Zhao, Z.; Jilkov, V.P. Estimator's credibility and its measures. In Proceedings of the IFAC 15th World Congress, Barcelona, Spain, 21–26 July 2002.
43. Li, X.R. Measure of Nonlinearity for Stochastic Systems. In Proceedings of the 15th International Conference on Information Fusion, Singapore, 9–12 July 2012; pp. 1073–1080.
44. Liu, Y.; Li, X.R. Measure of Nonlinearity for Estimation. *IEEE Trans. Signal Process.* **2015**, *63*, 2377–2388. [CrossRef]
45. Sun, T.; Xin, M.; Jia, B. Nonlinearity-based adaptive sparse-grid quadrature filter. In Proceedings of the American Control Conference (ACC), Chicago, IL, USA, 1–3 July 2015; pp. 2499–2504.
46. Wang, P.; Blasch, E.; Li, X.R.; Jones, E.; Hanak, R.; Yin, W.; Beach, A.; Brewer, P. Degree of nonlinearity (DoN) measure for target tracking in videos. In Proceedings of the 9th International Conference on Information Fusion, Heidlberg, Germany, 5 July 2016; pp. 1390–1397.
47. Sun, T.; Xin, M. Hypersonic entry vehicle state estimation using nonlinearity-based adaptive cubature Kalman filters. *Acta Astronaut.* **2017**, *134*, 221–230. [CrossRef]
48. Duník, J.; Straka, O.; Šimandl, M. Nonlinearity and non-Gaussianity measures for stochastic dynamic systems. In Proceedings of the 16th International Conference on Information Fusion, Istanbul, Turkey, 9 July 2013; pp. 204–211.
49. Straka, O.; Duník, J.; Šimandl, M. Unscented Kalman filter with controlled adaptation. In Proceedings of the 16th IFAC Symposium on System Identification, Brussels, Belgium, 11–13 July 2012; pp. 906–911.
50. Lu, Q.; Bar-Shalom, Y.; Willett, P.; Zhou, S. Nonlinear observation models with additive Gaussian noises and efficient MLEs. *IEEE Signal Process. Lett.* **2017**, *24*, 545–549. [CrossRef]
51. Schweppe, F.C. *Uncertain Dynamic Systems*; Prentice Hall: Englewood Cliffs, NJ, USA, 1973.
52. Walpole, R.E.; Myers, R.H.; Myers, S.L.; Ye, K. *Probability and Statistics for Engineers and Scientists*, 9th ed.; Prentice Hall: Englewood Cliffs, NJ, USA, 2011.

Article

Innovative Multi-Target Estimating with Clutter-Suppression Technique for Pulsed Radar Systems

Jo-Yen Nieh [1,*] and Yuan-Pin Cheng [2]

1 Department of Electrical and Electronic Engineering, Chung-Cheng Institute of Technology, National Defense University, Taoyuan 335, Taiwan

2 Electronic Systems Research Division, National Chung-Shang Institute of Technology, Taoyuan 335, Taiwan; cyoungbeen@gmail.com

* Correspondence: ccitb03011@ndu.edu.tw

Received: 15 February 2020; Accepted: 24 April 2020; Published: 25 April 2020

Abstract: Linear frequency modulation (LFM) waveforms have high Doppler-shift endurance because of the relative wide modulation bandwidth to the Doppler variation. The Doppler shift of the moving objects, nevertheless, constantly introduces obscure detection range offsets despite the exceptional Doppler tolerance in detection energy loss from LFM. An up-down-chirped LFM waveform is an efficient scheme to resolve the true target location and velocity by averaging the detection offset of two detection pairs from each single chirp LFM in opposite slopes. However, in multiple velocity-vary-target scenarios, without an efficient grouping scheme to find the detection pair of each moving target, the ambiguous detection results confine the applicability of precise target estimation by using these Doppler-tolerated waveforms. A succinct, three-multi-Doppler-shift-compensation (MDSC) scheme is applied to resolve the range and velocity of two moving objects by sorting the correct LFM detection pair of each target, even though the unresolvable scenarios of two close-by targets imply a fatal disability of detecting objects under a cluttered background. An innovative clutter-suppressed multi-Doppler-shift compensation (CS-MDSC) scheme is introduced in this research to compensate for the critical insufficient of resolving two overlapping objects with different velocities by solely MDSC. The CS-MDSC has been shown to successfully overcome this ambiguous scenario by integrating Doppler-selective moving target indication (MTI) filters to mitigate the distorting of near-zero-Doppler objects.

Keywords: linear frequency modulation; pulse compression; matched filter; Doppler shift compensation; Pulse-Doppler radar; moving target indication; comb filter; clutter suppression

1. Introduction

During the targets searching stage, velocity and location information of the potential targets are unknown to radar systems. The unknown velocity can significantly degrade the target acquisition capability of the systems due to the Doppler shift distortion introduced by the object's movement. Thus, a Doppler shift tolerated scheme is appreciative to deal with a variety of velocity object scenarios. In [1], a novel approach is proposed to integrated a linear frequency modulation pulse compression radar system with a time compression overlap-add technique to increase the signal-to-noise ratio. The transmitter divides a discrete linear frequency modulation chirp signal into overlapping segments and provides a significant processing gain. In [2], a new design of fast measurement to a linear frequency modulation is presented based on a linear amplitude comparison function that can ensure the accuracy of the measurement of multiple parameters. A study [3] verified the design of moving objects utilizing pulse compression technique and matched filter algorithm in linear frequency modulation in tracking the launch vehicle to follow the predetermined path or not. In [4], an 8 mm-range Gunn-diode oscillator was used in the experiment when the autodyne signal period duration was much longer than the delay time. The results of an autodyne short-range radar system with LFM in detecting moving reflecting objects were investigated. In [5], a new eigen-waveform design scheme was proposed to combine with the Range-Doppler map to identify moving targets where the detection performance was significantly improved over the wideband waveform and rectangular waveform. [6] announced a novel method to boost the detection probability of a radar system integrating eigen-waveform and pulse compression scheme. The hardware limitations were discussed under the scenarios of various waveforms. In [7], a new Doppler estimation method using space-variant synthetic aperture radar (SAR) imaging to enhance the performance of ship images was purposed and verified with GF-3 satellite SAR data. In [8], a new estimation method exploited moving target's two-dimensional velocity parameters from SAR imaging for velocity compensation. The 2D motion parameters can be effectively computed by the matched compression. In [9], sea surface velocity estimation with the SAR technique is presented based on environmental satellite and an interferometric airborne SAR data-set.

In modern Pulse-Doppler radar systems, the coherent pulse train is commonly applied for power accumulation under the limitation of the maximum instantaneous power in the transmitters' end. The ambiguity function $|\chi(\tau, \nu)|$ is widely used to exhibit the waveform characteristics in terms of the object time delay τ (location) vs. the Doppler frequency (f_d) created by the velocity ν in the following section. In this paper, the ambiguity function of the LFM waveform is investigated and it shows a robust velocity tolerance of the LFM after the matched filtering (MF) in Section 2. The sidelobe level of LFM waveform after MF is high due to the rectangular modulation waveform. Nonlinear frequency modulation (NLFM) applies waveform smoothing techniques such as cosine spectrum shape [10] or Tayler windows [11] to mitigate the discontinuous transition region of the LFM. MLFM has advantages on lower side lobe level over LFM, but sacrifices wider main beam width and shaping power loss. However, this study focuses on the relationship between the target velocity and the detection range offset after MF of LFM and it is presented in Section 3. A novel MDSC method of sorting LFM detection pair out of multiple targets for overcoming the unknown range offset and its false estimation scenarios are shown in Sections 4 and 5, respectively. Section 6 discusses an efficient Doppler frequency selective scheme, moving target indication (MTI), for pulse radar systems in order to overcome the MDSC's disability in heavy clutter background scenarios. Section 7 presents case studies to an innovative CS-MDSC scheme and show how the MDSC can be improved and be functional under heavy clutter background scenarios. The comprehensive discussion is summarized in the conclusion section.

2. Ambiguity Function of LFM Waveform

Linear frequency modulation (LFM) waveforms have wide modulation bandwidth compared with the relative narrow Doppler frequency (f_d) variation of moving targets. It withstands Doppler interference by only sacrificing minimal energy loss in MF operation and it is also called pulse compression (PC) in the study.

Figure 1 shows the ambiguity plot at the zero-time delay of an LFM waveform with bandwidth = 1 MHz, pulse width = 1 s. In this chart, the frequency domain is normalized by the signal bandwidth for studying the relationship between the Doppler shift and the modulation bandwidth. The energy loss by the Doppler shift of the LFM wave is computed as follows

$$\text{Loss} = |\nu|/\text{BW}. \quad (1)$$

where ν is the Doppler frequency shift from the moving target. The sign is positive when the target is moving toward the observation point and the sign is negative otherwise.

The PC amplitude vs. f_d /BW is defined as follows [12]

$$\text{Amp} = 1 - |\nu|/\text{BW}. \quad (2)$$

The Doppler frequency f_d caused by a moving target from a stationary observation point is calculated by

$$f_d = 2 \times fc \times Vel/\text{speed of light}. \quad (3)$$

where the carrier frequency is *fc* and the object velocity is in the shorthand of *Vel*.

For covering object velocity up to 20 Mach, from Equations (1) and (2), the Doppler frequency is calculated as f_d = 136 kHz with the carrier frequency *fc*= 3GHz and the f_d/BW = 0.136. The energy under this velocity is

$$\text{Amp} = 10 \times \log_{10}(0.868) = -0.6148(\text{dB}). \quad (4)$$

Unlike single tone pulse train waveform, the amplitude coverage of LFM has a linear decay without a vicious variation.

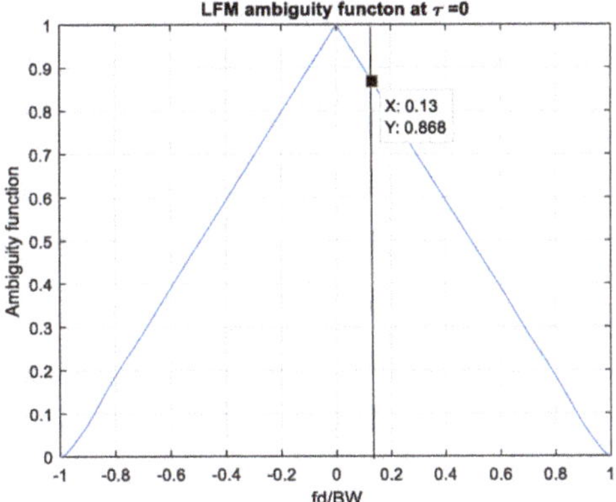

Figure 1. Normalized zero-time-delay ambiguity plot of linear frequency modulation (LFM) pulse. The vertical line marks the Doppler to bandwidth ratio of a moving target at f_d/BW = 0.13.

Despite the robustly Doppler-shift endurance of the LFM waveform, however, the time delay response also contains a linear offset along with the Doppler-frequency shift. The time delay offset can be observed in Figure 2, the contour plot of the LFM ambiguity function. The layout of the time delay diagonally corresponds to the Doppler shift. This offset results in the range error to Pulse-Doppler radars processing non-stationary moving target detection by matched filter detectors.

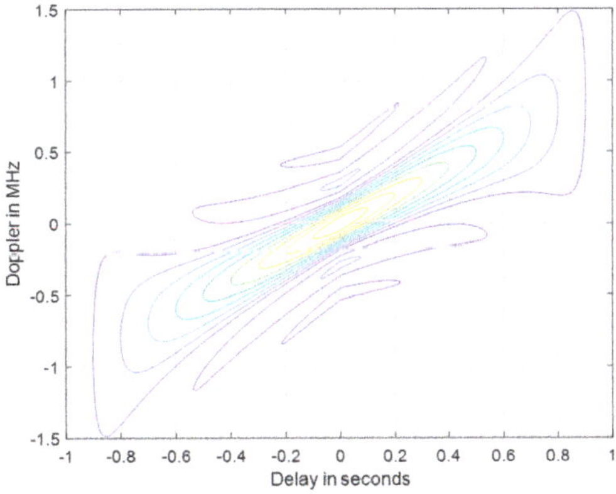

Figure 2. The contour plot of the LFM ambiguity function.

3. Range, Velocity Estimation by Offset of LFM Waveform

A single-chirped LFM waveform is defined as [12]

$$x_1(t) = \text{Rect}(\frac{t}{\tau_0})e^{j2\pi(f_0+\frac{\mu}{2}t^2)} \tag{5}$$

where the pulse duration is τ_0, the center frequency is f_0, and the frequency chirping slope is μ. The sign of the slope μ indicates descending or ascending of the frequency increment. The steeper the slope, the more frequency difference in a fixed time period.

Figure 3 illustrates the phenomena of the range offset caused by the convolution of a reference simulated LFM waveform with a Doppler-shifted returning signal. The range offset is proportional to the amount of the Doppler shift (Δf_d).

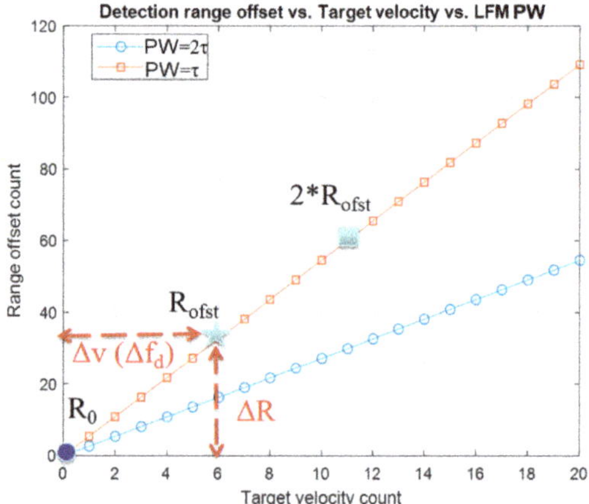

Figure 3. The linearity between target velocity and convolution range offset of a single chirped LFM waveform. The orange-squared line is the steeper chirping slope signal with pulse width (PW) = 2τ while the shows the flatter chirping slope signal with PW = τ in blue-circle line.

The range offset of convolution is linearly proportional to the moving object Δf_d or Δv, which can be converted by Equation (3). The range offset changing rate is higher with the steeper chirping slope (pulse width (PW) = τ, orange-squared line) than the flatter chirping slope, PW = 2τ, LFM waveform (blue circle) under the same Δf_d. Due to the linearity of a single chirp, the target velocity and detection range offset can be resolved from one another by the leaner ratio between f_d/BW versus range offset R_{oft}/μ. The equation is set up as follow [13]

$$\frac{(\text{PW} \times \text{C}/2)}{\text{BW}} = \frac{\text{R}_{\text{ofst}}}{\text{fd}} \tag{6}$$

where the pulse width is PW, speed of light is C, bandwidth is BW, the detection range offset is R_{ofst} due to the Doppler shift f_d.

Derived from Equation (6)

$$\text{R}_{\text{ofst}} = \frac{\text{fd} \times (\text{PW} \times \text{C}/2)}{\text{BW}} \tag{7}$$

R_{ofst} is proportional to f_d in a single chirped LFM waveform. With a stationary target, the detecting range offset is zero, which is shown as the blue dot in the origin in Figure 3.

However, in the target searching stage, without further target information, it is difficult to resolve the true moving target location by a single chirped LFM waveform. Therefore, a two chirped LFM scheme is introduced for resolving the true location of a non-stationary moving target.

Extended from Equation (5), a two chirped LFM waveform with two opposite slopes is derived as follows [14]

$$
\begin{aligned}
&x_{2chirp}(t) \\
&= \mathrm{Rect}\left(\frac{t-0.25\tau_0}{0.5\tau_0}\right)e^{j2\pi\left[f_0(t-0.25\tau_0)+\frac{\mu}{2}(t-0.25\tau_0)^2\right]} \\
&+\mathrm{Rect}\left(\frac{t+0.25\tau_0}{0.5\tau_0}\right)e^{j2\pi\left[f_0(t+0.25\tau_0)+\frac{\mu}{2}(t+0.25\tau_0)^2\right]}
\end{aligned} \quad (8)
$$

where the waveform is composed of half of positive slope μ and a half of negative slope $-\mu$ LFM within a PW = τ_0.

In Figure 4, the example illustrates the phenomena of a detection pair of two equal range offset along with the true target position in the opposite direction after the matched filter detectors of a moving target with Δf_d. When an up-down chirp referenced LFM signal (blue line) is shifted up by Δf_d (red line), the matched filter detector has two correlation points at $\pm\Delta R$ locations offset. The up-chirp signal matches the reference signal at the yellow dotted line position on the left while the down-chirp one has a matched point at the purple dotted line on the right. With the detection pair of two chirps, the true location of the target, which is zero, can be resolved unambiguously by the mean of two locations of the detection pair [14].

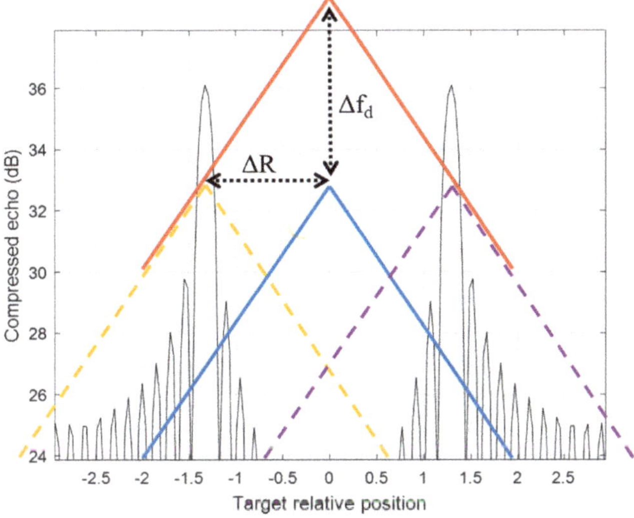

Figure 4. An up-down LFM waveform resolves two detection peaks from a non-stationary target by matched filter detectors.

Nevertheless, applying a two-chirped LFM waveform in multiple non-stationary targets scenarios, how to find the right pairs of the targets can be obscure without correct pairing information. There are three cases of ambiguous detection pair scenarios of two non-stationary moving targets.

Case 1 in Figure 5 demonstrates the detection pairs of two targets with the same velocity are right next to each other and the detection pairs are without crossover. Target one is at position zero, while target two is at position 3.3.

Case 2 in Figure 6 shows the detections of each target have one crossover with each other. Target one is at position zero, while target two is at position 0.86.

In Case 3 in Figure 7, the target one pair is enclosed by the target two detection pairs. The two targets are overlapped at position zero.

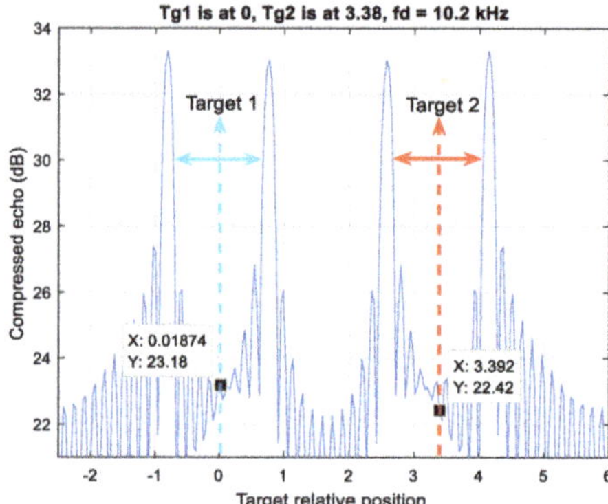

Figure 5. Case 1, an up-down chirp LFM waveform resolves four detection peaks by two moving targets. The detection pair of each target has no crossover.

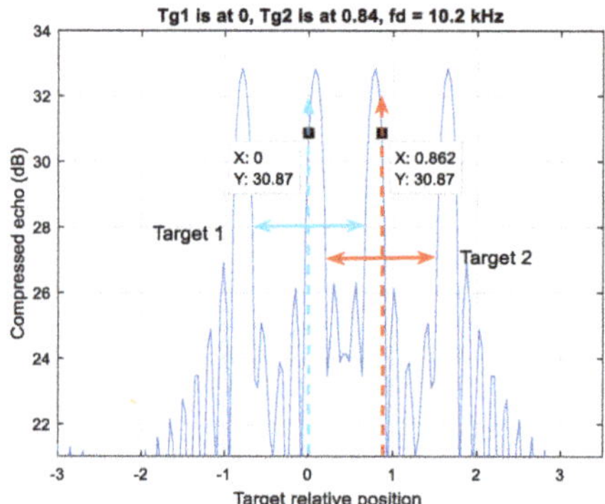

Figure 6. Case 2, an up-down chirp LFM waveform resolves four detection peaks by two moving targets. The detection pair of each target has one side crossover.

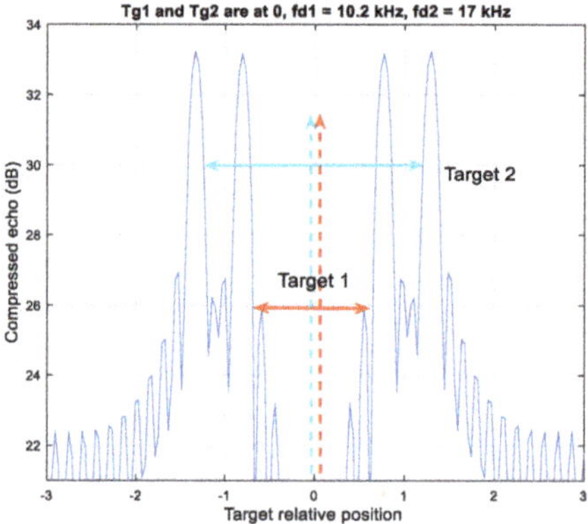

Figure 7. Case 3, an up-down chirp LFM waveform resolves four detection peaks by two moving targets. The detection pair of target 1 is enclosed by the pair of target 2.

Observing from the examples in Figures 5–7, how to determine the right detection pair resolve the true location of each target is vague without the illustrative target-location marks. Thus, Doppler shift compensation (DSC) is applied to distinguish moving targets pair.

The DSC operation is computed as follows

$$S_{offset}(t) = S_0(t)^* \exp(j2\pi \times fd_{offset} \times t) \quad (9)$$

where the complex signal before DSC is $S_0(t)$, the DSC frequency step is f_{d_offset} I, and the signal after DSC is $S_{offse}t(t)$.

Figures 8 and 9 demonstrate the behavior of a detection pair after a series of DSC operations. The detection of up-chirp LFM moves toward a positive direction by ΔR in each DSC, while the up-chirp LFM moves toward left each time with the same amount of ΔR. Even if the pair position has crossover in the Figure 9 scenario, the ΔR movement rule for each detection in this DSC operation is still valid.

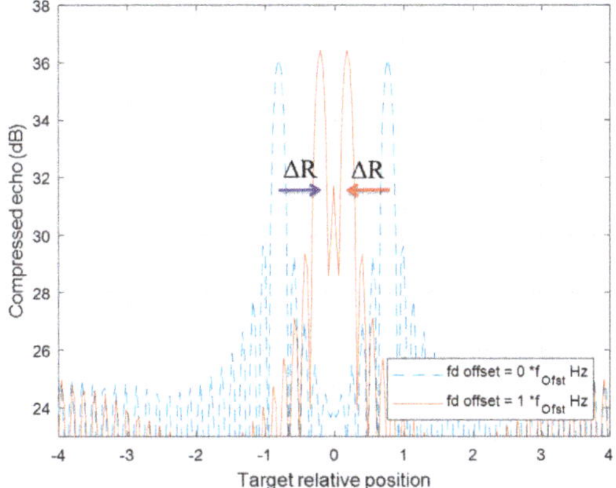

Figure 8. The range offset of a two-chirp detection pair of a moving target is processed by one fd_offset DSC. Two detections both have ΔR offset, but in the opposite direction.

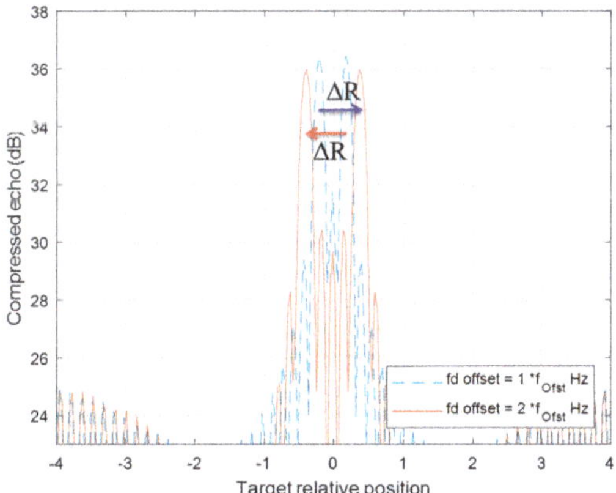

Figure 9. The range offset of a two-chirp detection pair of a moving target is processed by two f_d_offset DSC. Two detection lines crossover each other from one f_d_offset DSC (blue-dotted line) to two f_d_offset DSC (red line) with the same ΔR offset in opposite direction.

4. Multiple-Doppler-Shift-Compensation (MDSC) Scheme for Sorting LFM Detection Pair of a Target

Following the movement regulation of up-down chirp pair in DSC operations, a multiple-Doppler-shift-compensation (MDSC) scheme is applied to sort the detection pair out of multiple objects. In a two-moving-target scenario, one DSC operation flowchart is presented in Figure 10. There are five procession steps of one DSC operation:

Step 1.: An LFM signal containing two non-stationary targets after the correlation process yields four detection peaks located at Det1_fi to Det4_fi at the initial stage (left blue box) and an illustrative example is shown in Figure 11 as the blue-dash line.

Step 2.: An i-time f_d_offset DSC operation is applied to the two-target LFM signal, indicated as the yellow bold arrow in Figure 10.

Step 3.: Four detection peaks locate at Det1_fi_{+1} to Det4_fi_{+1} (right blue box) after the DSC operation, and a consequential example is shown in Figure 11 red line.

Step 4.: The red boxed arrow is the moving trend finding procedures, which associate Det1_fi to Det4_fi with Det1_fi_{+1} to Det4_fi_{+1} by finding the range offset matches the ±ΔR. Since the DSC operation only introduces a fix ΔR to each detection peak in the direction along with the LFM chirping slope. The ΔR directions after a DSC operation can be observed clearly in Figure 11. Those detection peaks with a positive ΔR offset are from up-chirped LFM, while the down-chirped ones introduce negative direction ΔR offsets.

Step 5.: Detection peak grouping (the green-dotted box) by finding: (1) these two peaks have opposite ΔR offset due to the up-down chirped LFM waveform being used; (2) the minimum distance between these presumed peaks. Figure 11 shows two possible detection pairs screened by rule (1), but the ΔRmin1 is accounted as the result in this DSC process because of the shortest distance selection assumption in the grouping procedure.

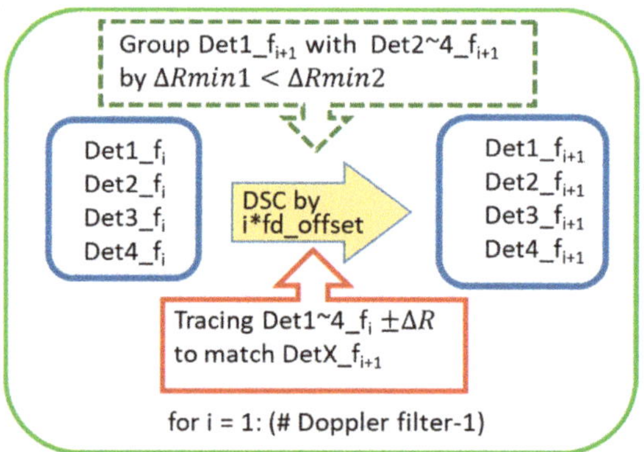

Figure 10. The processing flowchart of one Doppler filter in the multi-Doppler-shift-compensation (MDSC) scheme.

The overall flowchart of three-MDSC is presented in Figure 12. There are three identical DSC operations with one to three times f_d_offset applied respectively to estimate the pairing indexes of two objects, as shown in the green dash-line box scheme. During each DSC operation, the location of DetX_f_{i+1} can only be found by a specific DetX_f_i ΔR, each DetX always has one possible match location at the next DSC (red box), and the grouping index is chosen by the condition of ΔRmin1 < ΔRmin2 (green-dotted box).

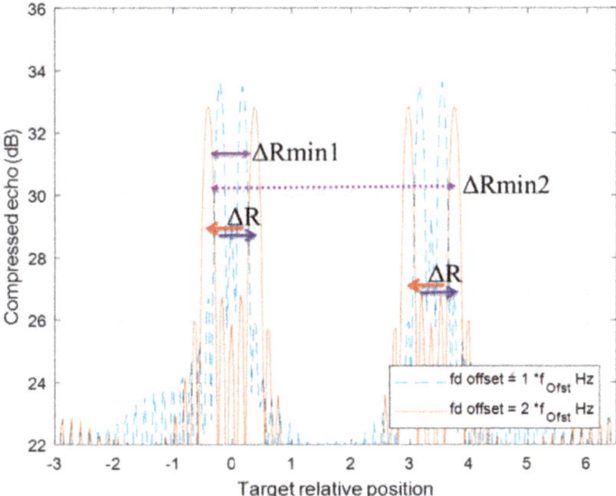

Figure 11. Two moving target MDSC of Figure 5 Case 1 scenario. The signal is traced from one f_d_offset Hz DSC (blue-dotted line) to two f_d_offset Hz DSC (red line).

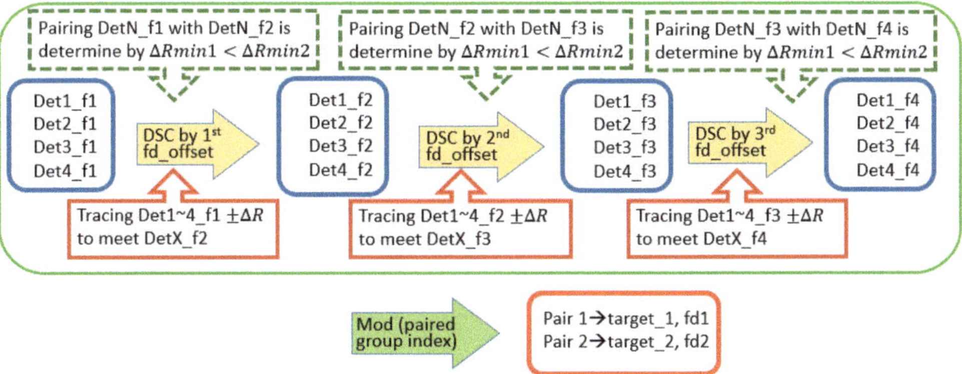

Figure 12. The decision flowchart of the multi-Doppler-shift-compensation (MDSC) scheme. A succinct three-Doppler-filter scenario is applied in this research.

The three-MDSC operation yields three presumed detection pairs of these two targets from each DSC operation. The final target grouping pairs are determined by the majority grouping presumed results out of three DSC trials (green bold arrow). The true target location now can be resolved firmly by the mean of the detection pair, which is processed in the red box in the three-MDSC scheme.

The illustrative scenario of Case 1 in Figure 5 has three out of three trials correct detection peak estimation in three MDSC operation.

The Case 2 scenario in Figure 5, on the other hand, has one false pairing estimating result out of three DSC operations in three-MDSC when the detection pairs have crossover in a DSC shown in Figure 13. Nevertheless, the correct pairing index count is two out of three processes. The final decision is eventually correct (green arrow in Figure 10).

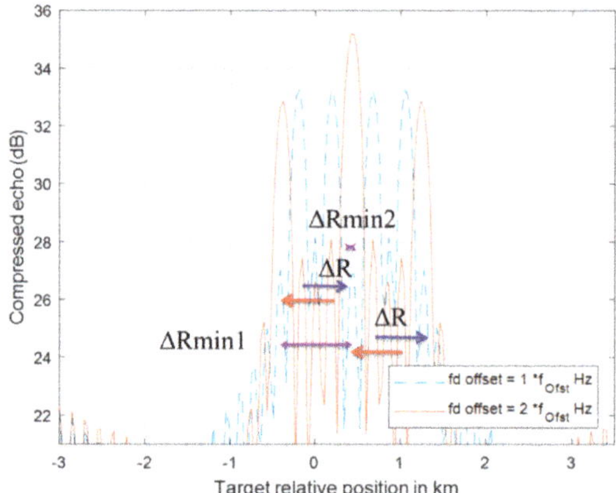

Figure 13. Two moving target MDSC of Figure 6 Case 2 scenario. The signal is traced from one f_d_offset Hz DSC (blue-dotted line) to two f_d_offset Hz DSC (red line). It is the special case, which the grouping rule finds a wrong pair out of three determining processes.

In prior research, five Doppler filters were used in the MDSC to resolve two target locations, respectively [14]. Since the pairing decision making is based on the majority grouping results of MDSC, a three-Doppler-filter MDSC yields three grouping pairs while the five-MDSC provides five grouping results. Both cases provide an odd number of pairing results, which means there is no ambiguity to make a majority decision out of the grouping results. Figure 14 shows the estimation accuracy comparison between three-MDSC and five-MDSC. Both MDSC schemes have accurate estimation results, as the SNR is above −15 dB while the PC ratio is 30 dB. As SNR < −15 dB, the detections are just too random for the MDSC scheme to have a correct matching pair to process, so the estimation cannot resolve targets' location without corresponding information. The overall results prove that the succinct three-Doppler-filter MDSC has an equally-likely estimation capability as the original five-Doppler-filter MDSC scheme with less computational complexity by reduction in the DSC by two in each calculation cycle.

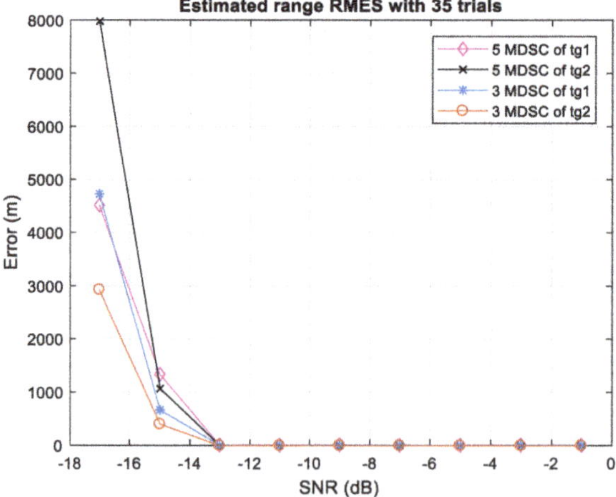

Figure 14. The 3-MDSC and 5-MDSC scheme evaluation error vs. SNR. The estimation error of target-one is marked as blue asterisk line and target-two is the red-circled line. The signal is with 30 dB matched filter pulse compression gain.

5. False Estimation Scenarios in MDSC

The MDSC scheme has a deficiency of correctly estimating the locations when the distance between two targets is smaller than the LFM range offset difference of two moving targets, as shown in Figure 15. In this case, the grouping results are never able to generate the correction pairing information.

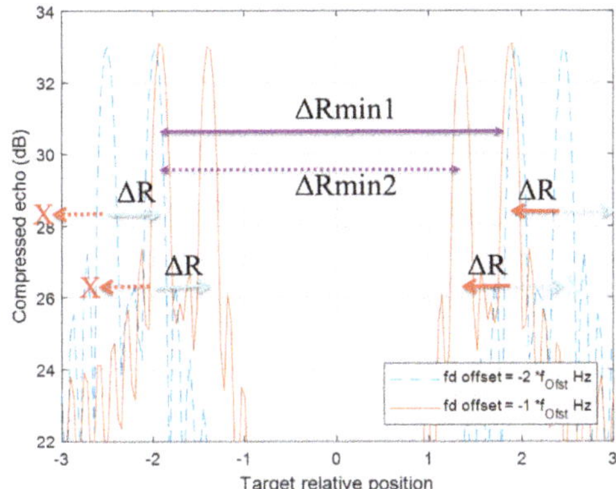

Figure 15. Two moving target MDSC of Figure 7 Case 3. The signal is traced by minus two f_{d}_offset Hz DSC (blue-dotted line) and minus one f_{d}_offset Hz DSC (red line). Since the target 1 pair is always enclosed by target 2 in such scenarios, the grouping rule always pairs the target 2 left detection with target 1 right detection at each DSC.

The ambiguous estimation scenarios of MDSC are grandly evaluated in Figure 16 with the targets velocity sweep from zero up to Mach 15 and distance is between zero and 80 km for a comprehensive analysis of unambiguous range estimation and possible false scenarios. The MDSC demonstrates a robust and accurate range estimation with zero root-mean-square error (RMSE) in unambiguous scenarios in Figures 5 and 6. These two cases have a grand-coverage of most of the two-moving-target scenarios in terms of relative distances and velocities variation. However, the ambiguous Case 3 scenario in Figure 7 introduces the high RMSE in the range estimation due to the constantly miss-matching pairing results of two targets in MDSC, which is shown in the left of Figure 16. The Case 3 scenario indicates that the space between targets is insufficient to prevent one target detection pair to enclosure the other target detection pair, such as the distance (x label) < 22.74 km and the velocity difference = Mach 14.06 (y label) or distance = 14.2 km and the velocity difference > Mach 0.93. The higher the velocity difference, the longer the distance required to avoid these misleading scenarios [14].

These misleading results are introduced by the false pairing outcomes in MDSC processes in the Figure 7 Case 3 condition. These ambiguous conditions can be classified as two real-life scenarios:

(1) Two targets with a high-velocity difference are in a relatively close range. It resembles two high-velocity aircraft coming across each other at that specific detection timing. The estimation offset only results in a short-term discontinuity within a long-term detection trace which can be mitigated by simple smoothing and predicting schemes.

(2) A moving target is close by a near-stationary target, such as a target traveling through a cluster of low-velocity clutter. The slow-moving clutter creates a vast nearly-zero Doppler clutter background and gives MDSC a constant false estimating result under such conditions. These clutter distortion scenarios cause long duration and non-negligible misleading results for MDSC and need to be eliminated for a more practical application.

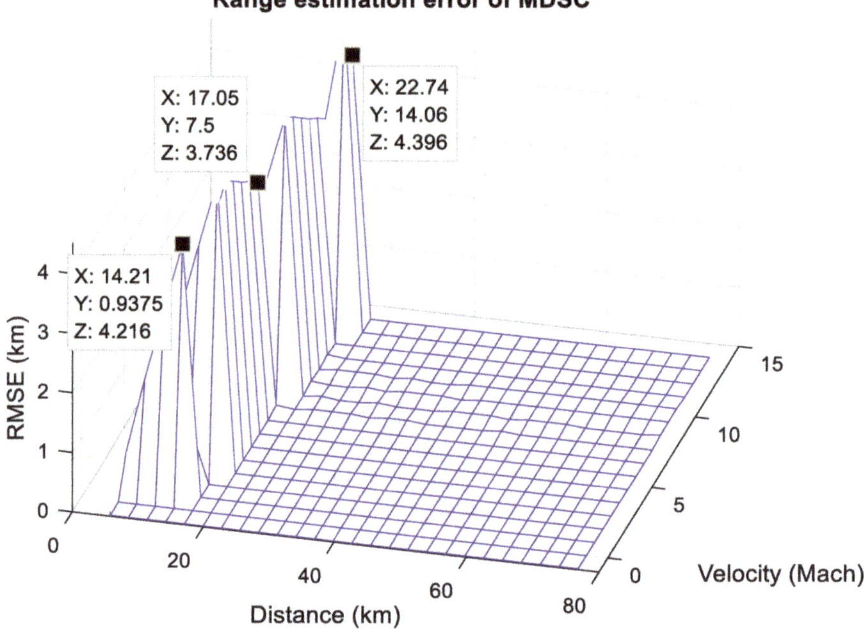

Figure 16. The range estimation error chart. Applying multi-Doppler-shift-compensation (MDSC) scheme estimates two non-stationary targets. The target distance is swept from 0 to 80 km while the velocity difference is swept from 0 to 15 Mach.

6. Low Velocity Target Suppression: Moving Target Indication (MTI)

To adapt the scheme to be practicable in the case (2) scenarios above, which resemble a crucially tactical scenario in which targets are blended into a heavy clutter background, integrating efficient clutter suppression schemes into MDSC is a substantial improvement to resolve this ambiguous scene. The techniques for implementing clutter filtering are the basis of the moving-target indication (MTI) scheme, which removes near-zero Doppler clutter spectrum and depth, and the width of the cut-off frequency of the filter is the factor of the number of pulses integrated and weighting coefficients of delay line applied on the pulse train.

Since this study is dealing with pulses transmitted at a pulse repetition frequency (PRF) f_r, the received signal, from a given range, consists of one PRI = $1/f_r$ apart. The spectrum of such a signal is folded around $f_r/2$ and centered around zero Doppler repeating every $\pm n^*f_r$, n = 0, 1, 2 ... With this zero-Doppler canceling response, the nearly stationary clutters are the subject to be removed out of the non-stationary targets the periodicity in the filter response. The periodic frequency response of the filter resembles a comb; hence it is a so-called comb filter [15].

The two-pulse MIT filter is also called a single delay line canceler which can be implemented as shown in Figure 17. It requires two distinct input pulse to yield out output. These sequential pulse trains are merged into a single pulse by feedback loops of the delay lines with specific coefficients weighing applying on each echo pulse. The output $y(t)$ is defined as follows [16]

$$y(t) = C_1 x(t) + C_2 x(t - T) \tag{10}$$

where the input $x(t)$ is an n-pulse pulse train, delay T = PRI = $(1/f_r)$, and C_1, C_2 the weighting coefficients of each delay-line pulse summation operation.

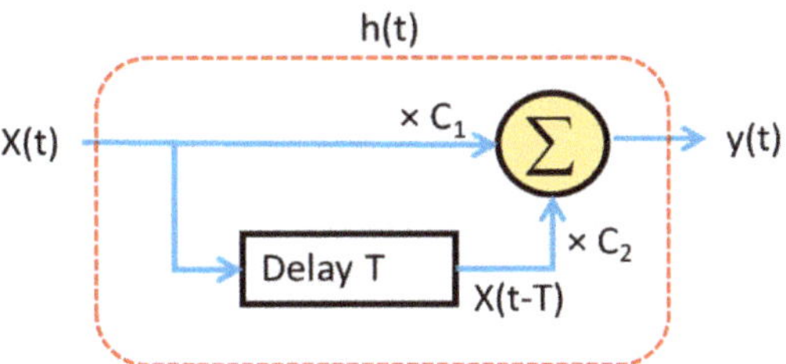

Figure 17. Single delay line canceler for two-pulse moving target indication (MTI).

The impulse response of the canceler is given by

$$h(t) = C_1\delta(t) + C_2\delta(t-T) \tag{11}$$

The double-delay-line cancelers are shown in Figure 18 and it is also called the three-pulse MTI filter. There are three consecutive pulse train the impulse response is given by

$$h(t) = C_1\delta(t) + C_2\delta(t-T) + C_3\delta(t-2T) \tag{12}$$

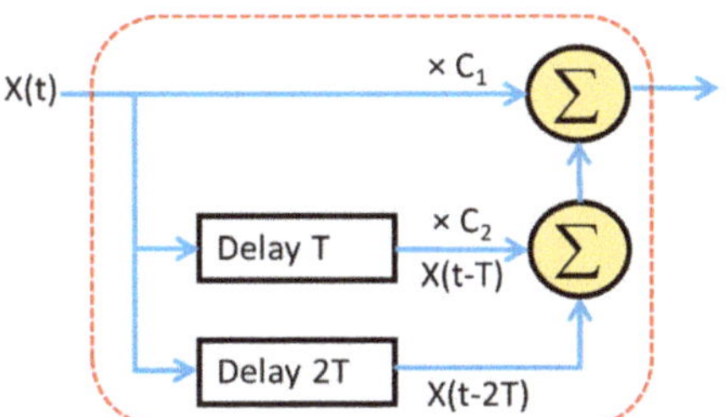

Figure 18. Double delay line canceler for three-pulse moving target indication (MTI).

The output signal of a three-pulse input signal processed by a double-delay-line canceler is calculated as follows

$$y(t) = C_1x(t) + C_2x(t-T) + C_3x(t-2T) \tag{13}$$

An MTI filter could be implemented using as little as two pulses and the filter high-pass response is determined by the number of pulses applied and weighting coefficients of the impulse response. Binomial coefficients are applied in the MIT filter frequency responses and a three-pulse delay line canceller is integrated with MDSC schemes to suppress the stationary object distortion to compensate for the ambiguous estimation in this study.

The binomial coefficients of multi-pulse MIT filters are given by

$$C_n = \begin{pmatrix} n \\ k \end{pmatrix} \times (-1)^k = \frac{n!}{k!(n-k)!} \times (-1)^k, 0 \le k \le n, \tag{14}$$

where the number of pulse n and the sign of the coefficient is toggled by every k^{th} index.

The formula is also called Pascal's rule or Pascal's triangle [17]. The frequency response of the MIT filter with two to four pulses is shown in Figure 19. The zero-Doppler has a deep null point response repeated in multiples of f_r = PRF, by which the ambiguous low-velocity target factors are eliminated in the original MDSC. The cut-off bandwidth of near-zero-Doppler in MIT filter frequency response is proportional to the number of canceling line tabs, i.e., the more identical pulse trains integrated, the wider and deeper the cut-off bandwidth. Figure 19 shows the cut-off bandwidth of 2-tab MTI (blue-dash line) < 3-tab MTI (red line) < 4-tab MTI (yellow-dotted-dash line).

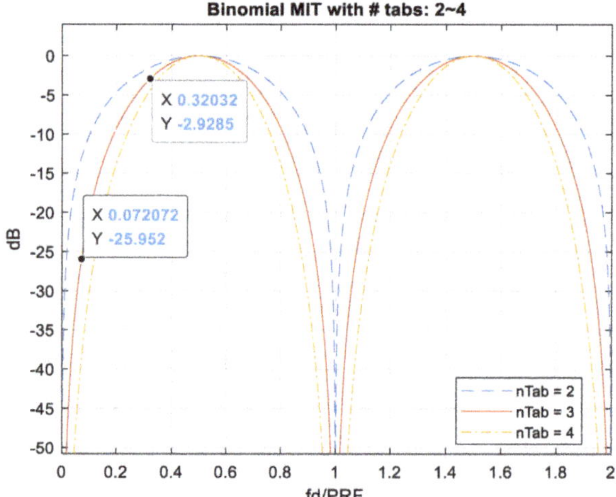

Figure 19. Frequency response of MTI comb filters applied binomial coefficients with the number of integrating pulses from two to four. The cut-off bandwidths are proportional to the number of delay lines applied.

7. Two-Target, Clutter-Suppressed Multi-Doppler-Shift-Compensation (CS-MDSC)

The ambiguous scenario with one near-stationary target has been discussed in Section 5 case (2). Figure 20 shows a typical unresolvable scenario to MDSC, which contains two targets at the same range cell with extremely velocity difference the target-one is stationary with zero-Doppler offset at the center, whereas the target-two velocity difference is at 15 Mach, which introduces a 102 kHz Doppler shift at S-band. The pulse compressed echo of up-down chirp LFM waveform has two peak values due to the Doppler shift at range cell ±6.25.

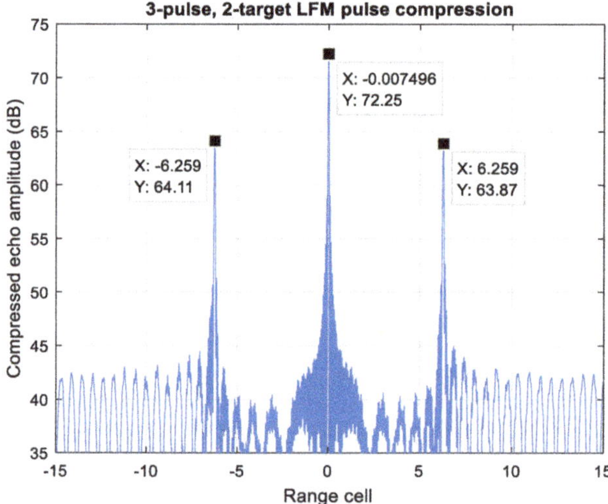

Figure 20. Pulse compressed echo of two overlapped targets. Target-one is zero Doppler, while target-two velocity is at Mach 15.

For an MIT comb filter application, Figure 21 demonstrates a three-pulse LFM echo with two targets with such extreme location and velocity conditions described above with f_d/PRF = 0 and 0.32, respectively. The frequency response of three-pulse MTI filter is shown in Figure 19; the target-1 echo with f_d1/PRF = 0 gets suppressed significantly while that of target-2, f_d2/PRF = 0.32, is in the passband of the filter.

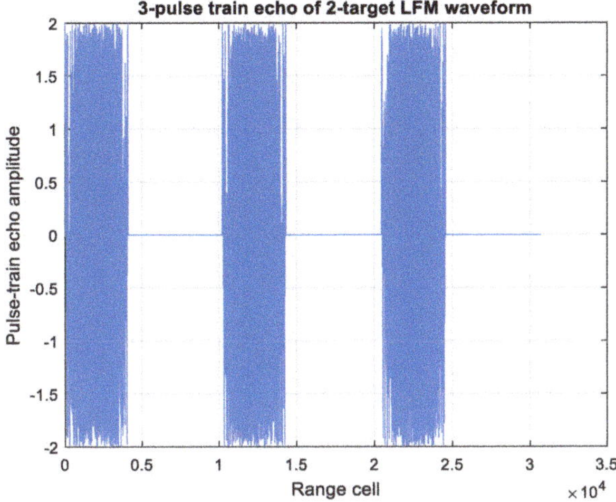

Figure 21. A 3-pulse LFM echo of two overlapped targets. Target-one is zero Doppler, while target-two velocity is at Mach 15.

Figure 22 displays the MTI processed echo of Figure 20, the zero-Doppler clutter is eliminated while the Mach 15 moving target echo remains, which leaves no ambiguity of the correctly pairing estimation.

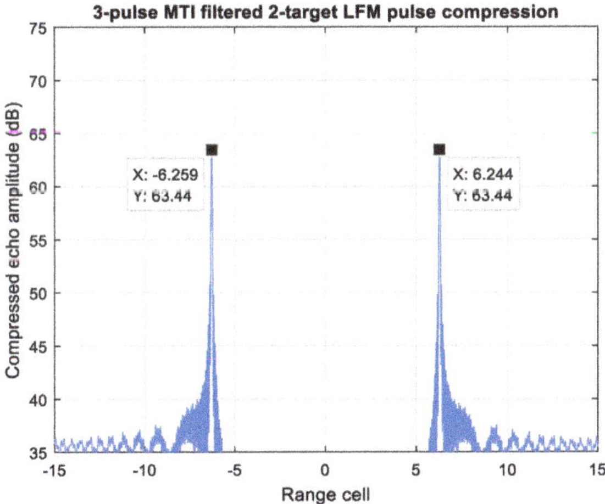

Figure 22. A 3-pulse MTI filtered echo of two overlapped targets. Target one is zero Doppler, while target two is at f_d2/PRIL = 0.32.

Another case study of a slow-moving target overlapping with a high-velocity target, which is also an ambiguous scenario for MDSC, is shown in Figure 23. Both moving targets introduce a pair of detection peaks with range offset after PC of the LFM waveform. The slow-moving target resembles a near-zero-Doppler clutter, such as cloud or sea with the f_d1/PRF = 0.072, while the second target resembles a high-velocity target, $f_d2/PRF = 0.32$, travelling across this strong clutter background. Figure 19 shows that the near-zero-Doppler clutter also suffers significant attenuation at $f_d1/PRF = 0.072$ for about 26 dB, whereas the second target with high velocity retains a strong response level at $f_d2/PRF = 0.32$.

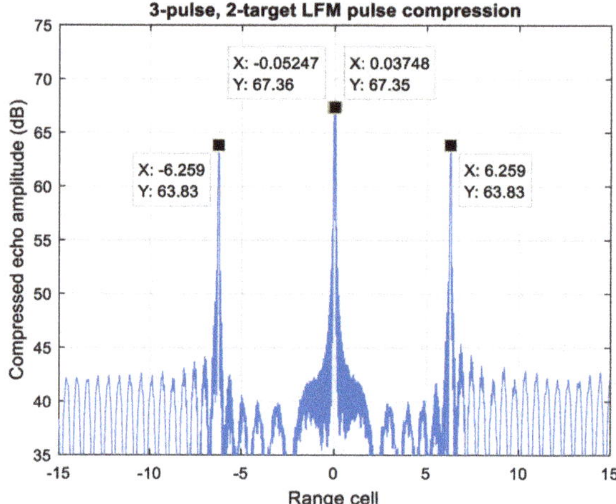

Figure 23. Pulse compressed echo of two overlapped targets. Target-one is Near-zero-Doppler, while target-two velocity is at 15 Mach.

Figure 24 shows the three-pulse MTI filtered echo of Figure 23. The −26 dB frequency response substantially drops down the target-one energy at $f_d1/PRF = 0.072$. Since the outcome energy of target-one, which is implied as a low velocity clutter at range cell = ±0.037, has been significantly deteriorated and leaves 16 dB power difference between the detection pairs of two targets, this artifact signal can be removed completely by setting up a reasonable detection threshold, such as a constant false alarm rate threshold. Therefore, with the prior state of a properly designed clutter suppression scheme, the remaining detection pair at range cell ±6.2 can resolve the target-two range and velocity information unambiguously.

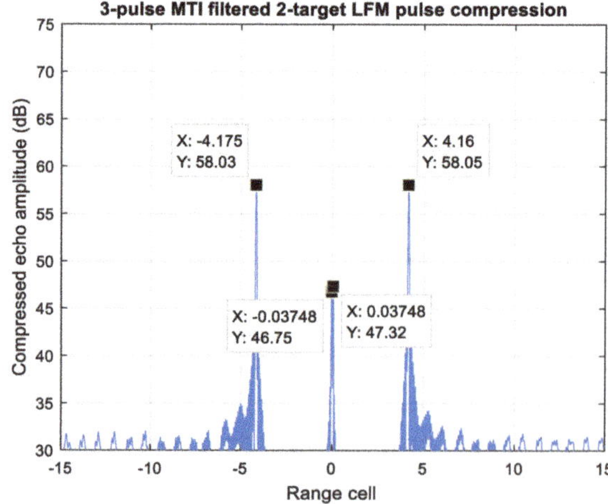

Figure 24. A 3-pulse MTI filtered echo of two overlapped targets. Target one is near-zero-Doppler, while the target-two velocity is at Mach 15.

To compensate the inapplicable scenarios of MDSC, a two-target, clutter suppression multi-Doppler-shift-compensation (CS-MDSC) workflow is illustrated in Figure 25. The process procedures begin with:

(1) An n-pulse LFM echo under a proper cut-off near-zero Doppler response MTI filter.
(2) Pulse compression and detection process.
(3) Number of detection count determination.

   i. If the detection count is greater than two, then process MDSC to find the correct detection pair of LFM waveform for resolving these two targets' locations and velocity.
   ii. If the number of detection is less than and equal to two due to the clutter suppression process, then there is no ambiguity on finding the right pair of the target. The location and velocity information of the moving target can be resolved by the remaining two or one detected signals.

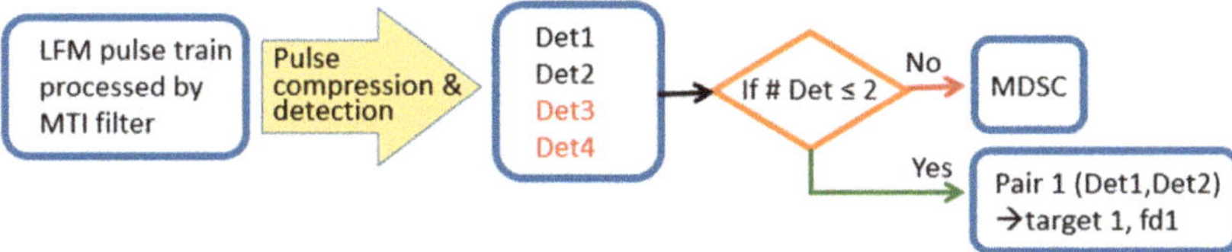

Figure 25. Clutter-suppression, multi-Doppler-shift-compensation (CS-MDSC) workflow.

8. Conclusions

The computational complexity and possible ambiguous estimation scenarios of two targets close by are the two noticeable drawbacks of the prior MDSC, which may create unwilling latency in the real-time system and false range/velocity estimation in inevitable heavy clutter background detection scenarios. There are two significant achievements in this study:

(1) Three-Doppler-offset MDSC operation has been proven as already providing sufficient range pairing information to have equally likely reliability as five-Doppler-offset MDSC, which was presented in the prior study. This succinct three-Doppler-offset MDSC workflow reduces the computational complexity by 40% as compared to five-Doppler-offset MDSC.
(2) The MTI comb filter clutter suppression scheme has been successfully integrated to prior MDSC by its pulsed Doppler periodic characteristics to eliminate the misleading pairing peaks from unwanted clutter-like signals. An innovative clutter-suppression multi-Doppler-shift-compensation (CS-MDSC) scheme has been introduced in this study and demonstrates the capability not only of maintaining the precise range and velocity estimation in most two moving targets scenarios as the original MDSC, but extract the moving target out of a clutter background, which is a magnificent improvement to adapt the MDSC scheme to broader and more realistic application scenarios.

Despite these advantages of CS-MDSC, further study of cognitive threshold selection algorithms and different coefficients applied on MIT comb filter should improve the scheme to adapt to more complex scenarios with more efficient processing algorithms. Also, the high sidelobe introduced by LFM rectangle waveform after MF leads out unresolvable detection pair of targets as the peaks are close. The shortage is more obvious, especially in dissimilar amplitude target scenarios. Thus, the future work is to look deep into the study of algorithm performance in presence of targets with dissimilar amplitudes and explore means to mitigate small target capture from larger target response.

Author Contributions: Conceptualization, J.-Y.N. and Y.-P.C.; writing—original draft preparation, J.-Y.N.; writing—review and editing, J.-Y.N. and Y.-P.C. All authors have read and agreed to the published version of the manuscript.

Funding: This research received no external funding.

Conflicts of Interest: The authors declare no conflict of interest.

References

1. Youssep, A.; Driessen, P.F.; Gebali, F.; Moa, B. On Time Compression Overlap-Add Technique in Linear Frequency Modulation Pulse Compression Radar Systems: Design and Performance Evaluation. *IEEE Access* **2017**, *5*, 27525–27537. [CrossRef]
2. Zhang, B.; Wang, X.; Pan, S. Photonics-Based Instantaneous Multi-Parameter Measurement of a Linear Frequency Modulation Microwave Signal. *J. Lightwave Technol.* **2018**, *36*, 921–926. [CrossRef]
3. Arya, V.J.; Subha, V. Pulse Compression using Linear Frequency Modulation Technique. In Proceedings of the International Conference on Intelligent Computing, Instrumentation and Control Technologies (ICICICT), Kannur, India, 6–7 July 2017.
4. Kryzhanovskyi, V.S.; Ermak, G.P.; Vasiliev, A.S. Signals from a Moving Object of Autodyne Radars with Linear Frequency Modulation. In Proceedings of the IEEE Microwaves, Radar and Remote Sensing Symposium (MRRS), Kiev, Ukraine, 29–31 August 2017.
5. Nieh, J.; Romero, R. Comparison of Ambiguity Function of Eigenwaveform to Wideband and Pulsed Radar Waveforms: A Comprehensive tutorial. *IET J. Eng.* **2018**. [CrossRef]
6. Wu, Y.; Nieh, J. Performance Analysis of Radar Eigenwaveform Design Integrated with Pulse Compression Scheme. In Proceedings of the International Conference on Radar, Antenna, Microwave, Electronics, and Telecommunications (ICRAMET), Tangerang, Indonesia, 23–24 October 2019.
7. Zhu, Y.; He, F.; Dong, Z. Moving Ships Refocusing for Spaceborne SAR Based on Doppler Parameters Setimation. *J. Eng.* **2019**, *2019*, 5905–5908.
8. Gu, D.; Yue, H.; Lu, B.; Xu, X. Velocity Estimation for Moving Target Refocusing in the Long-Time Coherent Integration SAR Imaging. In Proceedings of the IEEE China International SAR Symposium (CISS), Shanghai, China, 10–12 October 2018.
9. Jackson, G.; Fornaro, G.; Berardino, P.; Esposito, C.; Lanari, R.; Pauciullo, A.; Reale, D.; Zamparelli, V.; Perna, S. Experiments of Sea Surface Currents Estimation with Space and Airborne SAR Systems. In Proceedings of the IEEE International Geoscience and Remote Sensing Symposium (IGARSS), Milan, Italy, 26–31 July 2015.
10. Boukeffa, S.; Jiang, Y.; Jiang, T. Sidelobe Reduction with Nonlinear Frequency Modulated Waveforms. In Proceedings of the IEEE 7th International Colloquium on Signal Processing and its Applications, Penang, Malaysia, 4–6 March 2011.
11. Jin, G.; Deng, Y.; Wang, R.; Wang, W.; Wang, P.; Long, Y.; Zhang, Z.; Zhang, Y. An Advanced Nonlinear Frequency Modulation Waveform for Radar Imaging with Low Sidelobe. *IEEE Trans. Geosci. Remote Sens.* **2019**, *57*, 6155–6168. [CrossRef]
12. Mahafza, B. *Radar Systems Analysis and Design Using Matlab*, 3rd ed.; CRC Press: Boca Raton, FL, USA, 2012; pp. 175–181.
13. Cheng, Y.B.; Pace, P.E.; Brutzman, A.H.; Buss, A.H. Hybrid High-Idelity Modeling of Radar Scenarios using Stemporal Discrete Event Ad Time-step Simulation. In Proceedings of the World Congress on Engineering and Computer Science, San Francisco, CA, USA, 19–21 October 2016; Available online: https://calhoun.nps.edu/handle/10945/51670 (accessed on 25 April 2020).
14. Cheng, Y.P.; Nieh, J. Precise Range and Doppler Estimation of Multi-Nonstationary Targets by LFM Pulse-Doppler Radars. In Proceedings of the International Conference on Radar, Antenna, Microwave, Electronics, and Telecommunications (ICRAMET), Tangerang, Indonisia, 1–2 November 2018.
15. Levanon, N. *Radar Principles*; Wiley-Interscience: New York, NY, USA, 1988; p. 224.
16. Mahafza, B. *Radar Systems Analysis and Design Using Matlab*, 3rd ed.; CRC Press: Boca Raton, FL, USA, 2012; Chapter 10; pp. 363–381.
17. Binomial Coefficient. Available online: https://en.wikipedia.org/wiki/Binomial_coefficient (accessed on 15 February 2020).

Article

Joint Tracking and Classification of Multiple Targets with Scattering Center Model and CBMeMBer Filter †

Ronghui Zhan *, Liping Wang and Jun Zhang

Science and Technology on Automatic Target Recognition Laboratory, National University of Defense Technology, Changsha 410073, Hunan, China; wangliping17@nudt.edu.cn (L.W.); zhangjun@nudt.edu.cn (J.Z.)

* Correspondence: zhanrh@nudt.edu.cn

† This paper is an extended version of R. Zhan, L. Wang, and J. Zhang, Joint Target Tracking and Classification with Scattering Center Model for Radar Sensor, In Proceedings of ICCAIS 2019 Conference, Chengdu, China, 24–27 October 2019.

Received: 15 February 2020; Accepted: 10 March 2020; Published: 17 March 2020

Abstract: This paper deals with joint tracking and classification (JTC) of multiple targets based on scattering center model (SCM) and wideband radar observations. We first introduce an SCM-based JTC method, where the SCM is used to generate the predicted high range resolution profile (HRRP) with the information of the target aspect angle, and target classification is implemented through the data correlation of observed HRRP with predicted HRRPs. To solve the problem of multi-target JTC in the presence of clutter and detection uncertainty, we then integrate the SCM-based JTC method into the CBMeMBer filter framework, and derive a novel SCM-JTC-CBMeMBer filter with Bayesian theory. To further tackle the complex integrals' calculation involved in targets state and class estimation, we finally provide the sequential Monte Carlo (SMC) implementation of the proposed SCM-JTC-CBMeMBer filter. The effectiveness of the presented multi-target JTC method is validated by simulation results under the application scenario of maritime ship surveillance.

Keywords: joint tracking and classification; scattering center model; high range resolution profile; CBMeMBer filter; sequential Monte Carlo

---

1. Introduction

Traditionally, target tracking and target classification are treated as two independent problems, and they are usually solved separately. However, these two problems are closely related. For example, tracking affects classification by providing flight envelope information for different air target classes, while classification affects tracking via selecting appropriate class-dependent kinematic models. Therefore, a good classification may benefit tracking and vice versa. For this reason, the joint tracking and classification (JTC) method is receiving more and more attention.

By now, many JTC methods have been proposed [1–10], and these methods can roughly be divided into three categories. The first category is the most popular one and is dedicated to point targets. In this case, the resolution of the tracking sensor is very limited, and realization of target classification has to exploit attribute/identity sensor (such as electronic support measure) information or target dynamics (such as class-dependent maneuverability) [1–5].

The second category treats the target as an extended target, and the measurement of the target is modeled as the extent in down-range direction (length of the target) or both in down-range and lateral-range directions (i.e., size of the observed target contour), based on the assumption that the target has an ellipsoidal shape [6–8]. Targets are classified with the feature information of different length or size.

The third category treats the target as a rigid body, and the measurements are the target geometric shapes, which correspond to the projection of the target computer-aided design (CAD) models on the charge-coupled device (CCD) sensor [9,10]. Target classification is realized by image features.

For the wideband radar, the high range resolution profile (HRRP) serves as an important signature (feature) for target classification. However, as is known to all, the HRRP is very sensitive to target pose and the length (down-range extent) is not a stable feature in the dynamic environment, especially when the relative state between target and sensor changes rapidly. Moreover, the constraint of ideal ellipsoidal shape assumption imposed on the target limits the real application of the extent-based JTC methods. In view of the fact that the 3D scattering center model (3D-SCM) [11–13] is very convenient to create a classification feature according to the pose and sensor parameters, we proposed a novel SCM-based single-target JTC method in the conference paper [14]. The presented method exploited the 3D-SCM to predict the pose-dependent HRRP classification feature, together with the observation data of target's bearing, range and HRRP, to jointly infer the target state and class.

This paper is an extension of the SCM-based JTC method in [14] to the multi-target scenario, where targets with different classes may appear or disappear in the surveillance area, and false alarm/missed detection may exist. For the treatment of multi-target tracking, there are generally two main methods, i.e., the data-association-based method [15–17] and the random finite set (RFS)-based method [18,19]. The data-association-based method involves explicit associations. For example, the joint probabilistic data association (JPDA) algorithm [15] weights all the observations by association probabilities, and the multiple hypotheses tracking (MHT) algorithm [17] propagates association hypothesis. However, with the increase of the considered target number, the data-association-based method will suffer from a large computational burden. The RFS-based method models the multi-target state and the observations as RFSs and can avoid the explicit data association. Compared with the data-association-based method, the RFS-based method propagates the posterior density of the multi-target state recursively by means of the multi-target Bayes filter, and can be implemented through approximation approaches with a lower computational load. Therefore, the RFS-based method can serve as a good alternative to implement multi-target JTC.

Due to the intractability of Bayesian multi-target filter, two main approximation approaches, i.e., moment approximation (such as the probability hypothesis density (PHD) filter [20] and the cardinalized PHD (CPHD) filter [21]) and posterior density approximation (such as the multi-target multi-Bernoulli (MeMBer) filter, the cardinality balanced CBMeMBer filter [22] and the labeled multi-Bernoulli filter [23], etc.) are proposed, and have been widely used to various fields such as image processing [24,25] and multi-sensor fusion [26–28]. These approximation filters involve multiple integrals, and the implementation of the filters mainly depends on the characteristic of the system model. Generally, the Gaussian mixture (GM) implementation is suitable for the Gaussian linear system. However, for the non-Gaussian nonlinear system, sequential Monte Carlo (SMC) implementation has to be considered [29–32].

To deal with the problem of multi-target JTC, the SCM-based JTC method is integrated into the CBMeMBer filter framework in this paper, and the resulting filter is called SCM-JTC-CBMeMBer. Additionally, consider the high nonlinearity of the kinematic (range and bearing) and feature (HRRP) observation models, the proposed filter is implemented via the SMC technique.

The rest of this paper is organized as follows. The system model and a brief review of the CBMeMBer filter are provided in Section 2. The SCM-based JTC method and the proposed SCM-JTC-CBMeMBer filter are described in Section 3. The details on SMC implementation of the SCM-JTC-CBMeMBer filter are given in Section 4, followed by the simulation results in Section 5. A conclusion closes the paper.

2. Background

In this section, the system model (including state model and sensor observation model) will first be introduced. Then, a brief review of the CBMeMBer filter is given.

2.1. System Model

2.1.1. Target Motion Model

Without loss of generality, the considered target moves on the 2D plane with a nearly constant velocity, and the evolvement process of the target state is given as

$$\boldsymbol{x}_k = \boldsymbol{F}\boldsymbol{x}_{k-1} + \boldsymbol{w}_k \tag{1}$$

where $\boldsymbol{x}_k = [x_k\ y_k\ \dot{x}_k\ \dot{y}_k]^T$ represents target state including position component $\boldsymbol{pos}_k = [x_k\ y_k]^{\mathrm{T}}$ and velocity component $\boldsymbol{vel}_k = [\dot{x}_k\ \dot{y}_k]^{\mathrm{T}}$. The subscript k denotes sampling time and T is the sign for vector transpose. $\boldsymbol{F}$ is the state transition matrix. $\boldsymbol{w}_k \sim \mathcal{N}(\boldsymbol{w}; \boldsymbol{0}, \boldsymbol{Q})$ represents the multi-dimensional Gaussian process noise vector, where $\mathcal{N}(\zeta; \boldsymbol{\mu}, \boldsymbol{\Sigma})$ denotes Gaussian function with variable ζ, mean $\boldsymbol{\mu}$ and covariance matrix $\boldsymbol{\Sigma}$. $\boldsymbol{F}$ and $\boldsymbol{Q}$ can be further written as

$$\boldsymbol{F} = \begin{bmatrix} 1 & 0 & t & 0 \\ 0 & 1 & 0 & t \\ 0 & 0 & 1 & 0 \\ 0 & 0 & 0 & 1 \end{bmatrix}, \boldsymbol{Q} = q^2 \begin{bmatrix} t^4/4 & 0 & t^3/2 & 0 \\ 0 & t^4/4 & 0 & t^3/2 \\ t^3/2 & 0 & t^2 & 0 \\ 0 & t^3/2 & 0 & t^2 \end{bmatrix} \tag{2}$$

where t is the sampling interval and q is the acceleration variance.

2.1.2. Sensor Observation Model

The radar can provide range and bearing measurements of the target's centroid, and the observation model of kinematic (position) measurement can be described by

$$z_k = \begin{bmatrix} r_k \\ \beta_k \end{bmatrix} = \begin{bmatrix} \sqrt{x_k^2 + y_k^2} + v_{1,k} \\ \tan^{-1}(y_k/x_k) + v_{2,k} \end{bmatrix} = h(\boldsymbol{x}_k) + \boldsymbol{v}_k \tag{3}$$

where $h(\cdot)$ is the kinematic observation function and r_k and β_k represent noisy measurements of target range and bearing at time k, respectively. $\boldsymbol{v}_k = \begin{bmatrix} v_{1,k}\ v_{2,k} \end{bmatrix}^{\mathrm{T}}$ denotes the corresponding zero-mean observation noise with covariance matrix $\boldsymbol{R} = E[\boldsymbol{v}_k\boldsymbol{v}_k^{\mathrm{T}}] = diag[\sigma_r^2, \sigma_\beta^2]$.

As shown in Figure 1, for the low-speed maritime target, the heading is almost aligned with the axial direction of the target body because of its limited maneuverability. Under this condition, the aspect angle ϕ_k of the target can be obtained as

$$\phi_k = \theta_k - \beta_k \tag{4}$$

where $\theta_k = \tan^{-1}(\dot{y}_k/\dot{x}_k)$ is the heading angle.

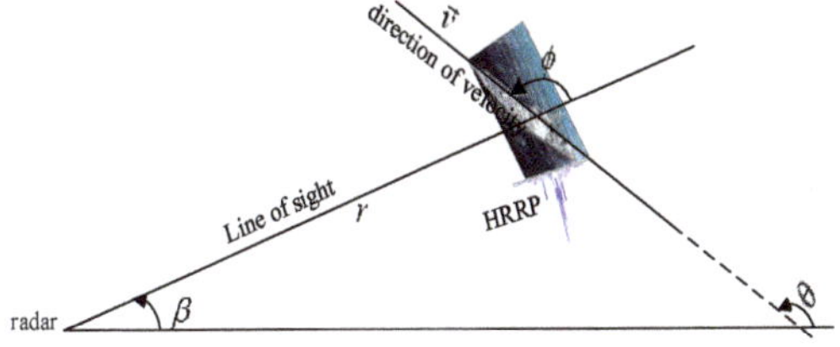

Figure 1. Illustration of target aspect angle.

An equivalent expression of aspect angle is in the form of

$$\phi_k = \cos^{-1}\left(\frac{\langle \boldsymbol{pos}_k, \boldsymbol{vel}_k \rangle}{||\boldsymbol{pos}_k|| \cdot ||\boldsymbol{vel}_k||}\right) = \cos^{-1}\left(\frac{x_k \dot{x}_k + y_k \dot{y}_k}{\sqrt{x_k^2 + y_k^2}\sqrt{\dot{x}_k^2 + \dot{y}_k^2}}\right) \tag{5}$$

The 3D-SCM is an equivalent of the target in geometry space to the radar response in the electromagnetic field. It provides a concise and physically relevant description of the target's scattering through a set of representative scattering parts, and thus a more effective way to characterize the target's electromagnetic scattering behavior. The 3D-SCM consists of a set of scattering center with a specific position, amplitude and type parameters, and it can be represented by

$$\mathbf{S} = \{a_n, \alpha_n, x_n, y_n, z_n\}_{n=1}^{N} \tag{6}$$

where a_n is the amplitude of nth scattering center, α_n is a frequency-dependent factor, (x_n, y_n, z_n) is the corresponding 3D spatial position in the target body coordinates and N is the number of scatters involved in the model. For a specific target class c, the associated 3D-SCM can be denoted as $\mathbf{S}_c$.

The whole target's backscattering with respect to radar instantaneous frequency f, viewing angles (i.e., azimuth angle ϕ and elevation angle γ) can be expressed as [13]

$$E(f, \phi, \gamma, \mathbf{S}) = \sum_{n=1}^{N} (\mathrm{j}f/f_c)^{\alpha_n} a_n(\phi, \gamma) \cdot \exp(-\mathrm{j}4\pi(x_n \cos\gamma\cos\phi + y_n \cos\gamma\sin\phi + z_n \sin\gamma)/\lambda) \tag{7}$$

where λ is the wavelength, f_c is the central frequency of signal, $\mathrm{j} = \sqrt{-1}$ is the imaginary unit and $a_n(\phi, \gamma)$ represents the amplitude of nth scattering center which may change with ϕ and γ.

At a specific viewing angle of (ϕ, γ), the corresponding projection position of the nth scattering center at the down-range direction is

$$r_n(\phi, \gamma) = x_n \cos\gamma\cos\phi + y_n \cos\gamma\sin\phi + z_n \sin\gamma \tag{8}$$

Assuming that the bandwidth of the radar signal is B and the ith discrete frequency point is

$$f_i = f_c - B/2 + i \cdot \Delta F, \quad i = 0, 1, \cdots, I \tag{9}$$

where ΔF is the frequency interval and $I + 1$ is the total number of frequency points.

Then, the frequency response of the ith frequency point can be written as $E_i = E(f_i, \phi, \gamma, \mathbf{S})$. After a direct operation of inverse discrete Fourier transform (IDFT) on frequency response sequence $\mathbf{E} = [E_0, E_1, \cdots, E_I]$, the desired HRRP can be immediately obtained.

When the target's motion is restricted on the 2D plane, the observation model of HRRP is represented as

$$\begin{aligned} \boldsymbol{d} &= g(\phi, \mathbf{S}) + \boldsymbol{n} \\ &= IDFT[E_i = E(f_i, \phi, \gamma, \mathbf{S}), i = 0, 1, \cdots, I] + \boldsymbol{n} \\ &= IDFT(\mathbf{E}) + \boldsymbol{n} \end{aligned} \tag{10}$$

where $\boldsymbol{d}$ denotes the (I+1)-dimensional measurement of the HRRP and each component of $\boldsymbol{d}$ corresponds to one range resolution cell, $g(\phi, \mathbf{S}) \triangleq IDFT(\mathbf{E})$ denotes the compact form of observation function, f_i and $\mathbf{S}$ are known parameters, $\gamma \approx 0$, ϕ can be obtained through Equations (4) or (5) and $\boldsymbol{n}$ is the observation noise vector.

2.2. CBMeMBer Filter

The CBMeMBer filter is outlined as follows and the details can be found in [18,21]. It approximates the posterior multi-target density by a multi-Bernoulli RFS. The multi-Bernoulli RFS consists of M independent Bernoulli RFSs $\boldsymbol{X}^{(i)}$, that is, $\boldsymbol{X} = \cup_{i=1}^{M} \boldsymbol{X}^{(i)}$. The probability density of Bernoulli RFS $\boldsymbol{X}^{(i)}$ is

$$\pi(\boldsymbol{X}^{(i)}) = \begin{cases} 1 - r^{(i)}, \boldsymbol{X}^{(i)} = \varnothing \\ r^{(i)} p^{(i)}(\boldsymbol{x}), \boldsymbol{X}^{(i)} = \{\boldsymbol{x}\} \end{cases} \tag{11}$$

where $r^{(i)} \in [0,1]$ is the target existence probability and $p^{(i)}(\cdot)$ is a spatial distribution. Therefore, the probability density of multi-Bernoulli RFS $\boldsymbol{X}$ is given by

$$\pi(\boldsymbol{X}) = \prod_{i=1}^{M} (1 - r^{(i)}) \sum_{1 \le i_1 \ne \cdots \ne i_n \le M} \prod_{j=1}^{n} \frac{r^{(i_j)} p^{(i_j)}(\boldsymbol{x}_j)}{1 - r^{(i_j)}} \tag{12}$$

where n is the number of targets.

The multi-Bernoulli RFS $\boldsymbol{X}$ is completely described by the multi-Bernoulli parameter set $\left\{(r^{(i)}, p^{(i)}(\boldsymbol{x}))\right\}_{i=1}^{M}$, and the probability density of the multi-Bernoulli RFSs $\boldsymbol{X}$ can be abbreviated by $\left\{(r^{(i)}, p^{(i)}(\boldsymbol{x}))\right\}_{i=1}^{M}$. The CBMeMBer filter consists of a prediction step and an update step.

2.2.1. Prediction Step

If the posterior probability density at time $k-1$ is $\pi_{k-1}(\boldsymbol{X}) = \{(r_{k-1}^{(i)}, p_{k-1}^{(i)}(\boldsymbol{x}_{k-1}))\}_{i=1}^{M_{k-1}}$, then the predicted multi-target density is also a multi-Bernoulli formed by the union of the multi-Bernoulli for the surviving targets and target births

$$\pi_{k|k-1} = \left\{(r_{P,k|k-1}^{(i)}, p_{P,k|k-1}^{(i)}(\boldsymbol{x}_k))\right\}_{i=1}^{M_{k-1}} \cup \left\{(r_{\Gamma,k|k-1}^{(i)}, p_{\Gamma,k|k-1}^{(i)}(\boldsymbol{x}_k))\right\}_{i=1}^{M_{\Gamma,k}} \tag{13}$$

where $\left\{(r_{\Gamma,k|k-1}^{(i)}, p_{\Gamma,k|k-1}^{(i)}(\boldsymbol{x}_k))\right\}_{i=1}^{M_{\Gamma,k}}$ is the predicted multi-Bernoulli for the target births and it is usually assumed to be known. $\left\{(r_{P,k|k-1}^{(i)}, p_{P,k|k-1}^{(i)}(\boldsymbol{x}_k))\right\}_{i=1}^{M_{k-1}}$ is the predicted multi-Bernoulli for the surviving targets and it is given by

$$r_{P,k|k-1}^{(i)} = r_{k-1}^{(i)} \left\langle p_{k-1}^{(i)}(\boldsymbol{x}_{k-1}), p_{S,k}(\boldsymbol{x}_{k-1}) \right\rangle \tag{14}$$

$$p_{P,k|k-1}^{(i)}(\boldsymbol{x}_k) = \frac{\left\langle f_{k|k-1}(\boldsymbol{x}_k|\boldsymbol{x}_{k-1}), p_{k-1}^{(i)}(\boldsymbol{x}_{k-1}) p_{S,k}(\boldsymbol{x}_{k-1}) \right\rangle}{\left\langle p_{k-1}^{(i)}(\boldsymbol{x}_{k-1}), p_{S,k}(\boldsymbol{x}_{k-1}) \right\rangle} \tag{15}$$

where $\langle f, g \rangle = \int f(x) g(x) dx$ denotes the inner product operation, $p_{S,k}(\boldsymbol{x}_{k-1})$ is the survival probability of the surviving targets, $f_{k|k-1}(\boldsymbol{x}_k|\boldsymbol{x}_{k-1})$ is the single-target transition density. There are $M_{k|k-1} = M_{k-1} + M_{\Gamma,k-1}$ predicted hypothesized tracks.

2.2.2. Update Step

Assuming that $n_{k,z}$ measurements are collected as $\boldsymbol{Z}_k = \left\{z_{k,1}, \cdots, z_{k,n_{k,z}}\right\}$ and the predicted probability density is $\pi_{k|k-1}(\boldsymbol{X}) = \{(r_{k|k-1}^{(i)}, p_{k|k-1}^{(i)}(\boldsymbol{x}_k))\}_{i=1}^{M_{k|k-1}}$, then the posterior multi-target density at time k can be approximated by a multi-Bernoulli as

$$\pi_k \approx \left\{(r_{L,k}^{(i)}, p_{L,k}^{(i)}(\boldsymbol{x}_k))\right\}_{i=1}^{M_{k|k-1}} \cup \left\{(r_{U,k}^{*}(z), p_{U,k}^{*}(\boldsymbol{x}_k; z))\right\}_{z \in \boldsymbol{Z}_k} \tag{16}$$

The first term in Equation (16) corresponds to the multi-Bernoulli density for the legacy tracks and it can be given by

$$r_{L,k}^{(i)} = r_{k|k-1}^{(i)} \frac{1 - \left\langle p_{k|k-1}^{(i)}(\boldsymbol{x}_k), p_{D,k}(\boldsymbol{x}_k) \right\rangle}{1 - r_{k|k-1}^{(i)} \left\langle p_{k|k-1}^{(i)}(\boldsymbol{x}_k), p_{D,k}(\boldsymbol{x}_k) \right\rangle} \tag{17}$$

$$p_{L,k}^{(i)}(\boldsymbol{x}_k) = p_{k|k-1}^{(i)}(\boldsymbol{x}_k) \frac{1 - p_{D,k}(\boldsymbol{x}_k)}{1 - \left\langle p_{k|k-1}^{(i)}(\boldsymbol{x}_k), p_{D,k}(\boldsymbol{x}_k) \right\rangle} \tag{18}$$

where $p_{D,k}(\boldsymbol{x}_k)$ is the detection probability.

The second term in Equation (16) corresponds to the multi-Bernoulli density for measurement-corrected tracks and it can be given as

$$r_{U,k}^{*}(\boldsymbol{z}) = \frac{\sum_{i=1}^{M_{k|k-1}} \frac{r_{k|k-1}^{(i)}(1-r_{k|k-1}^{(i)})\left\langle p_{k|k-1}^{(i)}(\boldsymbol{x}_k), g_k(\boldsymbol{z}|\boldsymbol{x}_k) p_{D,k}(\boldsymbol{x}_k)\right\rangle}{\left(1-r_{k|k-1}^{(i)}\left\langle p_{k|k-1}^{(i)}(\boldsymbol{x}_k), p_{D,k}(\boldsymbol{x}_k)\right\rangle\right)^2}}{\kappa(\boldsymbol{z}) + \sum_{i=1}^{M_{k|k-1}} \frac{r_{k|k-1}^{(i)}\left\langle p_{k|k-1}^{(i)}(\boldsymbol{x}_k), g_k(\boldsymbol{z}|\boldsymbol{x}_k) p_{D,k}(\boldsymbol{x}_k)\right\rangle}{1-r_{k|k-1}^{(i)}\left\langle p_{k|k-1}^{(i)}(\boldsymbol{x}_k), p_{D,k}(\boldsymbol{x}_k)\right\rangle}} \tag{19}$$

$$p_{U,k}^{*}(\boldsymbol{x}_k; \boldsymbol{z}) = \frac{\sum\limits_{i=1}^{M_{k|k-1}} \frac{r_{k|k-1}^{(i)} p_{k|k-1}^{(i)}(\boldsymbol{x}_k) p_{D,k} g_k(\boldsymbol{z}|\boldsymbol{x}_k)}{1-r_{k|k-1}^{(i)}}}{\sum\limits_{i=1}^{M_{k|k-1}} \frac{r_{k|k-1}^{(i)}}{1-r_{k|k-1}^{(i)}} \left\langle p_{k|k-1}^{(i)}(\boldsymbol{x}_k), p_{D,k}(\boldsymbol{x}_k) g_k(\boldsymbol{z}|\boldsymbol{x}_k) \right\rangle} \tag{20}$$

where $g_k(\boldsymbol{z}|\boldsymbol{x}_k)$ is the likelihood function, $\kappa(\boldsymbol{z})$ is the clutter intensity function. There are $M_k = M_{k|k-1} + n_{k,z}$ updated hypothesized tracks.

3. JTC Method Based on SCM and CBMeMBer Filter

In this section, the SCM-based JTC method is first presented by using the HRRP as the feature for target classification. Then, the SCM-JTC-CBMeMBer filter is derived for multi-target JTC.

3.1. SCM-Based JTC Method: Single-Target Case

The joint target state can be modeled as $\xi_{k-1} \triangleq (\boldsymbol{x}_{k-1}, c)$, where $\boldsymbol{x}_{k-1}$ is the kinematic state and c is the class label that can be taken from the set of the target classes $\boldsymbol{C} = \{c^1, c^2, \cdots, c^{n_c}\}$. n_c and c^m represent the total number of the target class and the mth target class, respectively. In the SCM-based JTC method, the available measurement at time k consists of kinematic (position) measurement $z_k^{\mathrm{P}} = [r, \theta]^{\mathrm{T}}$ and signature (HRRP) measurement $z_k^{\mathrm{c}} = \boldsymbol{d}$, and the joint measurement is denoted as $\widetilde{z}_k \triangleq (z_k^{\mathrm{P}}, z_k^{\mathrm{c}})$. The measurement set up to time k is represented by $\widetilde{\boldsymbol{Z}}^k = \{\widetilde{z}_\tau\}_{\tau=0}^k$.

The purpose of Bayesian JTC is to estimate the target state and class simultaneously at time k, under the condition that the distribution $p(\boldsymbol{x}_{k-1}, c|\widetilde{\boldsymbol{Z}}^{k-1})$ at time $k-1$ and the measurement $\widetilde{z}_k$ at time k are available. That is, to obtain the posterior probability-mass distribution

$$p(\boldsymbol{x}_k, c|\widetilde{\boldsymbol{Z}}^k) = p(\boldsymbol{x}_k|c, \widetilde{\boldsymbol{Z}}^k) p(c|\widetilde{\boldsymbol{Z}}^k) \tag{21}$$

For target tracking, the class-dependent probability density function (PDF) for a specific target class c^m can be represented as

$$p(\boldsymbol{x}_k|c^m, \widetilde{\boldsymbol{Z}}^k) = \frac{p(\widetilde{\boldsymbol{Z}}_k|\boldsymbol{x}_k, c^m) p(\boldsymbol{x}_k|c^m, \widetilde{\boldsymbol{Z}}^{k-1})}{p(\widetilde{z}_k|c^m, \widetilde{\boldsymbol{Z}}^{k-1})} \tag{22}$$

where $p(\widetilde{z}_k|c^m, \widetilde{Z}^{k-1}) = \int p(\widetilde{z}_k|x_k, c^m)p(x_k|c^m, \widetilde{Z}^{k-1})dx_k$ is the normalized factor.

Accordingly, for target classification, the probability function can be obtained by

$$\mu_k^m \triangleq p(c^m|\widetilde{Z}^k) = \frac{p(\widetilde{z}_k|c^m, \widetilde{Z}^{k-1})p(c^m|\widetilde{Z}^{k-1})}{p(\widetilde{z}_k|\widetilde{Z}^{k-1})} \tag{23}$$

where $p(\widetilde{z}_k|\widetilde{Z}^{k-1}) = \sum_{m=1}^{n_c} p(\widetilde{z}_k|c^m, \widetilde{Z}^{k-1})p(c^m|\widetilde{Z}^{k-1})$ is the normalized factor.

To obtain the recursive equations of the SCM-based JTC method, two assumptions should be followed.

Assumption 1. *All the targets have the same motion model, i.e., the single state transition function* $f_{k|k-1}(x_k, c^j|x_{k-1}, c^m)$ *is*

$$f_{k|k-1}(x_k, c^j|x_{k-1}, c^m) = f_{k|k-1}^{\mathrm{k}}(x_k|x_{k-1})f_{k|k-1}^{\mathrm{c}}(c^j|c^m) \tag{24}$$

where $f_{k|k-1}^{\mathrm{k}}(x_k|x_{k-1})$ *is the kinematic state transition function and is decided by the system model and* $f_{k|k-1}^{\mathrm{c}}(c^j|c^m)$ *is the class state transition function and can be represented by the Dirac function* $\delta(\cdot)$ *as*

$$f_{k|k-1}^{\mathrm{c}}(c^j|c^m) = \delta_m(j) = \begin{cases} 1, \textbf{if } j = m \\ 0, \textbf{if } j \neq m \end{cases} \tag{25}$$

Assumption 2. *The kinematic measurement and HRRP measurement are independent of each other, and the kinematic measurement is independent of the target class, so the measurement likelihood can be written as*

$$p(\widetilde{z}_k|x_k, c^m, \widetilde{Z}^{k-1}) = p(z_k^{\mathrm{P}}|x_k)p(z_k^{\mathrm{c}}|x_k, c^m) \tag{26}$$

where $p(z_k^{\mathrm{P}}|x_k) \triangleq g_k^{\mathrm{k}}(x_k) = N(z_k; h(x_k), R)$ *is the likelihood function of kinematic measurement.* $p(z_k^{\mathrm{c}}|x_k, c^m) \triangleq g_k^{\mathrm{c}}(x_k, c^m) = \langle d_k, g(\phi_k, S_{c^m})\rangle / (\|d_k\| \cdot \|g(\phi_k, S_{c^m})\|)$ *is the likelihood function of HRRP measurement, and is defined as normalized correlation coefficient of observed HRRP with model-predicted HRRP.* S_{c^m} *is the SCM corresponding to target class* c^m.

Therefore, the SCM-based JTC method can be constructed through the following two steps.

The prediction steps of the target state and class are given by

$$p(x_k|c^m, \widetilde{Z}^{k-1}) = \int f_{k|k-1}^{\mathrm{k}}(x_k|x_{k-1})p(x_{k-1}|c^m, \widetilde{Z}^{k-1})dx_{k-1} \tag{27}$$

$$\mu_{k|k-1}^m = \mu_{k-1}^m \tag{28}$$

Similarly, the update steps of target state and class are

$$p(x_k|c^m, \widetilde{Z}^k) = \frac{g_k^{\mathrm{k}}(x_k)g_k^{\mathrm{c}}(x_k, c^m)p(x_k|c^m, \widetilde{Z}^{k-1})}{p(\widetilde{z}_k|c^m, \widetilde{Z}^{k-1})} \tag{29}$$

$$\mu_k^m = p(c^m|\widetilde{Z}^k) = \frac{p(\widetilde{z}_k|c^m, \widetilde{Z}^{k-1})p(c^m|\widetilde{Z}^{k-1})}{p(\widetilde{z}_k|\widetilde{Z}^{k-1})} \tag{30}$$

with

$$p(\widetilde{z}_k|c^m, Z^{k-1}) = \int g_k^{\mathrm{k}}(x_k)g_k^{\mathrm{c}}(x_k, c^m)p(x_k|c^m, Z^{k-1})dx_k \tag{31}$$

$$p(\widetilde{z}_k|\widetilde{Z}^{k-1}) = \sum_{m=1}^{n_c} p(\widetilde{z}_k|c^m, \widetilde{Z}^{k-1})p(c^m|\widetilde{Z}^{k-1}) \tag{32}$$

Because of the complexity and high nonlinearity of the observation model, there is no analytic form to obtain the recursive estimation of target state and class, so we have to resort to the SMC technique (also known as particle filter, PF). Given the particle set $\left\{w_{k-1}^{j}, \boldsymbol{x}_{k-1}^{j}, l^{j}\right\}_{j=1}^{n_{k-1,p}}$, where the superscript j denotes the index of the particles, $l^j \in C$ represents the class label corresponding to the jth particle and $n_{k-1,p}$ is the number of particles at time $k-1$. The posterior target state and associated class probability can be represented by

$$p(\boldsymbol{x}_{k-1}|c^m, \boldsymbol{\mathcal{Z}}^{k-1}) = \sum_{j=1}^{n_{k-1,p}} w_{k-1}^{m,j} \delta_{\boldsymbol{x}_{k-1}^{m,j}}(\boldsymbol{x}_{k-1}) \tag{33}$$

$$p(c^m|\boldsymbol{\mathcal{Z}}^{k-1}) = \sum_{j=1}^{n_{k-1,p}} w_{k-1}^{j} \delta_{l^j}(c^m) / \sum_{j=1}^{n_{k-1,p}} w_{k-1}^{j} \tag{34}$$

with

$$w_{k-1}^{m,j} = w_{k-1}^{j} \delta_{l^j}(c^m) / \sum_{j=1}^{n_{k-1,p}} w_{k-1}^{j} \delta_{l^j}(c^m) \tag{35}$$

$$\boldsymbol{x}_{k-1}^{m,j} = \boldsymbol{x}_{k-1}^{j} \delta_{l^j}(c^m) \tag{36}$$

Then, a complete recursive procedure from time $k-1$ to k can be summarized as Algorithm 1.

Algorithm 1 Single-time step recursion of the scattering center model (SCM)-based joint tracking and classification (JTC) method

---

Step 1. Model prediction

1) Target state prediction: $\boldsymbol{x}_k^j = \boldsymbol{F}\boldsymbol{x}_{k-1}^j + \boldsymbol{w}_{k-1}^j$
2) Kinematic observation prediction: $\hat{\boldsymbol{z}}_k^j = h(\boldsymbol{x}_k^j)$
3) Aspect angle prediction: $\phi_k^j = \cos^{-1}\left(\frac{x_k^j \dot{x}_k^j + y_k^j \dot{y}_k^j}{\sqrt{(x_k^j)^2 + (y_k^j)^2}\sqrt{(\dot{x}_k^j)^2 + (\dot{y}_k^j)^2}}\right)$
4) HRRP prediction: $\hat{\boldsymbol{d}}_k^j = g(\phi_k^j, \boldsymbol{S}_{l^j})$ where $\boldsymbol{S}_{l^j}$ represents the 3D-SCM corresponding to target class l^j.

Step 2. Likelihood evaluation

1) Kinematic observation likelihood: $g_k^{\mathrm{k}}(\boldsymbol{x}_k^j) = \mathcal{N}(\boldsymbol{z}_k; h(\boldsymbol{x}_k^j), \boldsymbol{R})$
2) HRRP correlation coefficient: $g_k^{\mathrm{c}}(\boldsymbol{x}_k^j, l^j) = \left\langle \boldsymbol{d}_k, \hat{\boldsymbol{d}}_k^j \right\rangle / \left(\|\boldsymbol{d}_k\| \cdot \|\hat{\boldsymbol{d}}_k^j\| \right)$

Step 3. Particle weight evaluation

1) Joint weight calculation: $w_k^j = w_{k-1}^j \cdot g_k^{\mathrm{k}}(\boldsymbol{x}_k^j) \cdot g_k^{\mathrm{c}}(\boldsymbol{x}_k^j, l^j)$
2) Normalization of weights: $\widetilde{w}_k^j = w_k^j / \sum_{j=1}^{n_{k-1,p}} w_k^j$

The posterior target state estimation $\hat{\boldsymbol{x}}_k$ and class probabilities $p(c^m|\widetilde{\boldsymbol{Z}}^k)$ at time k can be obtained by

$$\hat{\boldsymbol{x}}_k = \sum_{m=1}^{n_c} p(c^m|\widetilde{\boldsymbol{Z}}^k) \hat{\boldsymbol{x}}_k^m \tag{37}$$

$$p(c^m|\widetilde{\boldsymbol{Z}}^k) = \sum_{j=1}^{n_{k-1,p}} \widetilde{w}_k^j \delta_{l^j}(c^m) / \sum_{j=1}^{n_{k-1,p}} \widetilde{w}_k^j \tag{38}$$

with

$$\hat{x}_k^m = \sum_{j=1}^{n_{k-1,p}} \widetilde{w}_k^{m,j} \delta_{x_k^{m,j}}(x_k) \tag{39}$$

$$\widetilde{w}_k^{m,j} = \widetilde{w}_k^j \delta_{l^j}(c^m) / \sum_{j=1}^{n_{k-1,p}} \widetilde{w}_k^j \delta_{l^j}(c^m) \tag{40}$$

$$x_k^{m,j} = x_k^j \delta_{l^j}(c^m) \tag{41}$$

To reduce the effect of particle degeneracy, the resampling operation [33] should be considered in method implementation. Specifically, the class-dependent resampling strategy is adopted to avoid particle degeneracy caused by an incorrect target classification. In this strategy, the maximum number of particles for the SCM-based JTC method is set as $L_{\max}$, while the minimum number of particles for each target class is set as $L_{\min}$. In this paper, the standard resampling operation is used to resample each class, that is, for the particle with $l^j = c^m$, $\left\{x_k^{j'}, 1/n_{k,p,m}, l^{j'}\right\}_{j'=1}^{n_{k,p,m}} = \text{resample}\left\{x_k^j, \widetilde{w}_k^j, l^j\right\}_{l^j=c^m}$, $n_{k,p,m} \triangleq \max(L_{\min}, n_{k-1,p} \cdot p(c^m|\widetilde{Z}^k))$.

3.2. SCM-JTC-CBMeMBer Filter: Multi-Target Case

For the JTC of multi-target, the available measurement set at time k is denoted as $\mathcal{Z}_k = \left\{\widetilde{z}_{k,l}\right\}_{l=0}^{n_{k,z}}$ and the measurement set up to time k is $\mathcal{Z}^k = \{\mathcal{Z}_l\}_{l=1}^k$. The posterior multi-target density at time $k-1$ is modeled as a multi-Bernoulli

$$\begin{aligned} \pi_{k-1} &= \left\{\left(r_{k-1}^{(i)}, p_{k-1}^{(i)}(x_{k-1}|\mathcal{Z}^{k-1})\right)\right\}_{i=1}^{M_{k-1}} \\ &= \left\{\left(r_{k-1}^{(i)}, \sum_{m=1}^{n_c} p_{k-1}^{(i)}(x_{k-1}|c^m, \mathcal{Z}^{k-1}) p_{k-1}^{(i)}(c^m|\mathcal{Z}^{k-1})\right)\right\}_{i=1}^{M_{k-1}} \end{aligned} \tag{42}$$

In addition to Assumption 1 and Assumption 2, the following assumptions should also be followed to obtain the SCM-JTC-CBMeMBer filter.

Assumption 3. *Each target evolves motion and generates measurements independently.*

Assumption 4. *The clutter is modeled as Poisson RFS with Poisson average rate λ_c, and it is independent of target-originated measurements. The spatial distribution of the clutter is a uniform distribution, denoted by $C(\widetilde{z})$. The clutter intensity function is $\kappa(\widetilde{z}) = \lambda_c C(\widetilde{z})$.*

Assumption 5. *The survival and detection probabilities are state-independent, i.e., $p_{S,k}(x,c) = p_{S,k}$, $p_{D,k}(x,c) = p_{D,k}$.*

Assumption 6. *The PDF of birth targets at time $k-1$ is also a multi-Bernoulli, namely*

$$\begin{aligned} \pi_{B,k-1} &= \left\{\left(r_{B,k-1}^{(i)}, p_{B,k-1}^{(i)}(x_{k-1}|\mathcal{Z}^{k-1})\right)\right\}_{i=1}^{M_{B,k-1}} \\ &= \left\{\left(r_{B,k-1}^{(i)}, \sum_{m=1}^{n_c} p_{B,k-1}^{(i)}(x_{k-1}|c^m, \mathcal{Z}^{k-1}) p_{B,k-1}^{(i)}(c^m|\mathcal{Z}^{k-1})\right)\right\}_{i=1}^{M_{B,k-1}} \end{aligned} \tag{43}$$

Proposition 1. *If the posterior multi-target density at time $k-1$ is a multi-Bernoulli, as shown in Equation (42), then the predicted multi-target density is also a multi-Bernoulli and is given by*

$$\begin{aligned} \pi_{k|k-1} &= \left\{\left(r_{P,k|k-1}^{(i)}, p_{P,k|k-1}^{(i)}(x_k|\mathcal{Z}^{k-1})\right)\right\}_{i=1}^{M_{k-1}} \cup \left\{\left(r_{\Gamma,k|k-1}^{(i)}, p_{\Gamma,k|k-1}^{(i)}(x_k|\mathcal{Z}^{k-1})\right)\right\}_{i=1}^{M_{\Gamma,k}} \\ &= \left\{\left(r_{P,k|k-1}^{(i)}, \sum_{m=1}^{n_c} p_{P,k|k-1}^{(i)}(x_k|c^m, \mathcal{Z}^{k-1}) p_{P,k|k-1}^{(i)}(c^m|\mathcal{Z}^{k-1})\right)\right\}_{i=1}^{M_{k-1}} \cup \left\{\left(r_{\Gamma,k|k-1}^{(i)}, p_{\Gamma,k|k-1}^{(i)}(x_k|\mathcal{Z}^{k-1})\right)\right\}_{i=1}^{M_{\Gamma,k}} \end{aligned} \tag{44}$$

with

$$r^{(i)}_{P,k|k-1} = r^{(i)}_{k-1} \sum_{m=1}^{n_c} p^{(i)}_{k-1}(c^m|\mathcal{Z}^{k-1}) \left\langle p^{(i)}_{k-1}(\boldsymbol{x}_{k-1}|c^m, \mathcal{Z}^{k-1}), p_{S,k} \right\rangle \tag{45}$$

$$p^{(i)}_{P,k|k-1}(c^m|\mathcal{Z}^{k-1}) = p^{(i)}_{k-1}(c^m|\mathcal{Z}^{k-1}) \tag{46}$$

$$p^{(i)}_{P,k|k-1}(\boldsymbol{x}_k|c^m, \mathcal{Z}^{k-1}) = \frac{\left\langle f^{\mathrm{k}}_{k|k-1}(\boldsymbol{x}_k|\boldsymbol{x}_{k-1}), p^{(i)}_{k-1}(\boldsymbol{x}_{k-1}|c^m, \mathcal{Z}^{k-1}) p_{S,k} \right\rangle}{\sum\limits_{m=1}^{n_c} p^{(i)}_{k-1}(c^m|\mathcal{Z}^{k-1}) \left\langle p^{(i)}_{k-1}(\boldsymbol{x}_{k-1}|c^m, \mathcal{Z}^{k-1}), p_{S,k} \right\rangle} \tag{47}$$

$$r^{(i)}_{\Gamma,k|k-1} = r^{(i)}_{B,k-1} \tag{48}$$

$$p^{(i)}_{\Gamma,k|k-1}(\boldsymbol{x}_k|\mathcal{Z}^{k-1}) = \sum_{m=1}^{n_c} p^{(i)}_{B,k-1}(\boldsymbol{x}_{k-1}|c^m, \mathcal{Z}^{k-1}) p^{(i)}_{B,k-1}(c^m|\mathcal{Z}^{k-1}) \tag{49}$$

The proof of Proposition 1 is given in Appendix A.

Proposition 2. *If the predicted multi-target density at time k is a multi-Bernoulli*

$$\begin{aligned} \pi_{k|k-1} &= \left\{ \left(r^{(i)}_{k|k-1}, p^{(i)}_{k|k-1}(\boldsymbol{x}_k|\mathcal{Z}^{k-1}) \right) \right\}_{i=1}^{M_{k|k-1}} \\ &= \left\{ \left(r^{(i)}_{k|k-1}, \sum_{m=1}^{n_c} p^{(i)}_{k|k-1}(\boldsymbol{x}_k|c^m, \mathcal{Z}^{k-1}) p^{(i)}_{k|k-1}(c^m|\mathcal{Z}^{k-1}) \right) \right\}_{i=1}^{M_{k|k-1}} \end{aligned} \tag{50}$$

Then, the posterior multi-target density can be approximated by a multi-Bernoulli as

$$\begin{aligned} \pi_k &\approx \left\{ \left(r^{(i)}_{L,k}, p^{(i)}_{L,k}(\boldsymbol{x}_k|\mathcal{Z}^k) \right) \right\}_{i=1}^{M_{k|k-1}} \cup \left\{ \left(r^*_{U,k}(\widetilde{z}), p^*_{U,k}(\widetilde{z}) \right) \right\}_{\widetilde{z}\in\mathcal{Z}_k} \\ &= \left\{ \left(r^{(i)}_{L,k}, \sum_{m=1}^{n_c} p^{(i)}_{L,k}(\boldsymbol{x}_k|c^m, \mathcal{Z}^k) p^{(i)}_{L,k}(c^m|\mathcal{Z}^k) \right) \right\}_{i=1}^{M_{k|k-1}} \cup \left\{ \left(r^*_{U,k}(\widetilde{z}), \sum_{m=1}^{n_c} p^{(i)}_{U,k}(\boldsymbol{x}_k|c^m, \mathcal{Z}^k) p^{(i)}_{U,k}(c^m|\mathcal{Z}^k) \right) \right\}_{\widetilde{z}\in\mathcal{Z}_k} \end{aligned} \tag{51}$$

with

$$r^{(i)}_{L,k} = r^{(i)}_{k|k-1} \frac{1 - \sum\limits_{m=1}^{n_c} p^{(i)}_{k|k-1}(c^m|\mathcal{Z}^{k-1}) \left\langle p^{(i)}_{k|k-1}(\boldsymbol{x}_k|c^m, \mathcal{Z}^{k-1}), p_{D,k} \right\rangle}{1 - r^{(i)}_{k|k-1} \sum\limits_{m=1}^{n_c} p^{(i)}_{k|k-1}(c^m|\mathcal{Z}^{k-1}) \left\langle p^{(i)}_{k|k-1}(\boldsymbol{x}_k|c^m, \mathcal{Z}^{k-1}), p_{D,k} \right\rangle} \tag{52}$$

$$p^{(i)}_{L,k}(\boldsymbol{x}_k|c^m, \mathcal{Z}^k) = \frac{(1 - p_{D,k}) p^{(i)}_{k|k-1}(\boldsymbol{x}_k|c^m, \mathcal{Z}^{k-1})}{1 - \sum\limits_{m=1}^{n_c} p^{(i)}_{k|k-1}(c^m|\mathcal{Z}^{k-1}) \left\langle p^{(i)}_{k|k-1}(\boldsymbol{x}_k|c^m, \mathcal{Z}^{k-1}), p_{D,k} \right\rangle} \tag{53}$$

$$p^{(i)}_{L,k}(c^m|\mathcal{Z}^k) = p^{(i)}_{k|k-1}(c^m|\mathcal{Z}^{k-1}) \tag{54}$$

$$r^*_{U,k}(\widetilde{z}) = \frac{\sum_{i=1}^{M_{k|k-1}} \frac{r^{(i)}_{k|k-1}(1 - r^{(i)}_{k|k-1}) \sum\limits_{m=1}^{n_c} p^{(i)}_{k|k-1}(c^m|\mathcal{Z}^{k-1}) \left\langle p^{(i)}_{k|k-1}(\boldsymbol{x}_k|c^m, \mathcal{Z}^{k-1}), g^{\mathrm{k}}_k(\boldsymbol{x}_k) g^{\mathrm{c}}_k(\boldsymbol{x}_k, c^m) p_{D,k} \right\rangle}{\left(1 - r^{(i)}_{k|k-1} \sum\limits_{m=1}^{n_c} p^{(i)}_{k|k-1}(c^m|\mathcal{Z}^{k-1}) \left\langle p^{(i)}_{k|k-1}(\boldsymbol{x}_k|c^m, \mathcal{Z}^{k-1}), p_{D,k} \right\rangle\right)^2}}{\kappa(\widetilde{z}) + \sum_{i=1}^{M_{k|k-1}} \frac{r^{(i)}_{k|k-1} \sum\limits_{m=1}^{n_c} p^{(i)}_{k|k-1}(c^m|\mathcal{Z}^{k-1}) \left\langle p^{(i)}_{k|k-1}(\boldsymbol{x}_k|c^m, \mathcal{Z}^{k-1}), g^{\mathrm{k}}_k(\boldsymbol{x}_k) g^{\mathrm{c}}_k(\boldsymbol{x}_k, c^m) p_{D,k} \right\rangle}{1 - r^{(i)}_{k|k-1} \sum\limits_{m=1}^{n_c} p^{(i)}_{k|k-1}(c^m|\mathcal{Z}^{k-1}) \left\langle p^{(i)}_{k|k-1}(\boldsymbol{x}_k|c^m, \mathcal{Z}^{k-1}), p_{D,k} \right\rangle}} \tag{55}$$

$$p_{U,k}^{(i)}(\mathbf{x}_k|c^m,\mathcal{Z}^k)=\frac{\sum_{i=1}^{M_{k|k-1}}\frac{r_{k|k-1}^{(i)}p_{k|k-1}^{(i)}(\mathbf{x}_k|c^m,\mathcal{Z}^{k-1})p_{k|k-1}^{(i)}(c^m|\mathcal{Z}^{k-1})p_{D,k}g_k^{\mathrm{k}}(\mathbf{x}_k)g_k^{\mathrm{c}}(\mathbf{x}_k,c^m)}{1-r_{k|k-1}^{(i)}}}{\sum_{i=1}^{M_{k|k-1}}\frac{r_{k|k-1}^{(i)}p_{k|k-1}^{(i)}(c^m|\mathcal{Z}^{k-1})\left\langle p_{k|k-1}^{(i)}(\mathbf{x}_k|c^m,\mathcal{Z}^{k-1}),p_{D,k}g_k^{\mathrm{k}}(\mathbf{x}_k)g_k^{\mathrm{c}}(\mathbf{x}_k,c^m)\right\rangle}{1-r_{k|k-1}^{(i)}}} \tag{56}$$

$$p_{U,k}^{(i)}(c^m|\mathcal{Z}^k)=\frac{\sum_{i=1}^{M_{k|k-1}}\frac{r_{k|k-1}^{(i)}p_{k|k-1}^{(i)}(c^m|\mathcal{Z}^{k-1})\left\langle p_{k|k-1}^{(i)}(\mathbf{x}_k|c^m,\mathcal{Z}^{k-1}),p_{D,k}g_k^{\mathrm{k}}(\mathbf{x}_k)g_k^{\mathrm{c}}(\mathbf{x}_k,c^m)\right\rangle}{1-r_{k|k-1}^{(i)}}}{\sum_{i=1}^{M_{k|k-1}}\frac{r_{k|k-1}^{(i)}}{1-r_{k|k-1}^{(i)}}\sum_{m=1}^{n_c}p_{k|k-1}^{(i)}(c^m|\mathcal{Z}^{k-1})\left\langle p_{k|k-1}^{(i)}(\mathbf{x}_k|c^m,\mathcal{Z}^{k-1}),p_{D,k}g_k^{\mathrm{k}}(\mathbf{x}_k)g_k^{\mathrm{c}}(\mathbf{x}_k,c^m)\right\rangle} \tag{57}$$

The proof of Proposition 2 is shown in Appendix B.

The state extraction step is similar to the CBMeMBer filter, and the details can be found in [22].

4. SMC Implementation of the SCM-JTC-CBMeMBer Filter

In what follows, SMC implementation of the SCM-JTC-CBMeMBer filter recursion will be presented.

Supposing that the posterior multi-target density $\pi_{k-1}=\left\{\left(r_{k-1}^{(i)},p_{k-1}^{(i)}(\mathbf{x}_{k-1}|\mathcal{Z}^{k-1})\right)\right\}_{i=1}^{M_{k-1}}$ is given, and each component $p_{k-1}^{(i)}(\mathbf{x}_{k-1}|\mathcal{Z}^{k-1})$ is comprised of n_{k-1}^{i} weighted particles $\left\{w_{k-1}^{i,j},\mathbf{x}_{k-1}^{i,j},l^{i,j}\right\}_{j=1}^{n_{k-1}^{i}}$, that is

$$p_{k-1}^{(i)}(\mathbf{x}_{k-1}|\mathcal{Z}^{k-1})=\sum_{j=1}^{n_{k-1}^{i}}w_{k-1}^{i,j}\delta_{\mathbf{x}_{k-1}^{i,j}}(\mathbf{x}_{k-1}) \tag{58}$$

Then the class probability $p_{k-1}^{(i)}(c^m|\mathcal{Z}^{k-1})$ and the kinematic state distribution conditioned on classification $p_{k-1}^{(i)}(\mathbf{x}_{k-1}|c^m,\mathcal{Z}^{k-1})$ can be obtained by

$$p_{k-1}^{(i)}(c^m|\mathcal{Z}^{k-1})=\sum_{j=1}^{n_{k-1}^{i}}w_{k-1}^{i,j}\delta_{l^{i,j}}(c^m)\Big/\sum_{j=1}^{n_{k-1}^{i}}w_{k-1}^{i,j} \tag{59}$$

$$p_{k-1}^{(i)}(\mathbf{x}_{k-1}|c^m,\mathcal{Z}^{k-1})=\sum_{j=1}^{n_{k-1}^{i}}w_{k\ 1}^{i,m,j}\delta_{\mathbf{x}_{k-1}^{i,m,j}}(\mathbf{x}_{k-1}) \tag{60}$$

with

$$w_{k-1}^{i,m,j}=w_{k-1}^{i,j}\delta_{l^{i,j}}(c^m)\Big/\sum_{j=1}^{n_{k-1}^{i}}w_{k-1}^{i,j}\delta_{l^{i,j}}(c^m) \tag{61}$$

$$\mathbf{x}_{k-1}^{i,m,j}=\mathbf{x}_{k-1}^{i,j}\delta_{l^{i,j}}(c^m) \tag{62}$$

Proposition 3. *Given the importance density $q_k^{(i)}(\cdot|\mathbf{x}_{k-1},\mathcal{Z}_k)$ of the posterior distribution and importance densities $b_k^{(i)}(\cdot|\mathbf{x}_{k-1},\mathcal{Z}_k)$ of the birth targets, according to Proposition 1, if the prior distribution is multi-Bernoulli $\pi_{k-1}=\left\{\left(r_{k-1}^{(i)},\sum_{m=1}^{n_c}p_{k-1}^{(i)}(\mathbf{x}_{k-1}|c^m,\mathcal{Z}^{k-1})p_{k-1}^{(i)}(c^m|\mathcal{Z}^{k-1})\right)\right\}_{i=1}^{M_{k-1}}$ and each*

$p_{k-1}^{(i)} = \sum_{m=1}^{n_c} p_{k-1}^{(i)}(\mathbf{x}_{k-1}|c^m, \mathcal{Z}^{k-1}) p_{k-1}^{(i)}(c^m|\mathcal{Z}^{k-1})$ *is comprised of* n_{k-1}^i *weighted particles* $\left\{w_{k-1}^{i,j}, \mathbf{x}_{k-1}^{i,j}, l^{i,j}\right\}_{j=1}^{n_{k-1}^i}$, *then the predicted multi-target density is also multi-Bernoulli and the SMC implementation is calculated as*

$$r_{P,k|k-1}^{(i)} = r_{k-1}^{(i)} \sum_{j=1}^{n_{k-1}^i} w_{k-1}^{i,j} p_{S,k} \tag{63}$$

$$p_{P,k|k-1}^{(i)}(c^m|\mathcal{Z}^{k-1}) = \sum_{j=1}^{n_{k-1}^i} w_{P,k|k-1}^{i,j} \delta_{l^{i,j}}(c^m) / \sum_{j=1}^{n_{k-1}^i} w_{P,k|k-1}^{i,j} \tag{64}$$

$$p_{P,k|k-1}^{(i)}(\mathbf{x}_k|c^m, \mathcal{Z}^{k-1}) = \sum_{j=1}^{n_{k-1}^i} w_{P,k-1}^{i,m,j} \delta_{\mathbf{x}_{P,k|k-1}^{i,m,j}}(\mathbf{x}_k) \tag{65}$$

$$r_{\Gamma,k|k-1}^{(i)} = r_{B,k-1}^{(i)} \tag{66}$$

$$p_{\Gamma,k|k-1}^{(i)}(\mathbf{x}_k|c^m, \mathcal{Z}^{k-1}) = \sum_{j=1}^{n_{k-1}^i} w_{B,k|k-1}^{i,m,j} \delta_{\mathbf{x}_{B,k|k-1}^{i,m,j}}(\mathbf{x}_k) \tag{67}$$

$$p_{\Gamma,k|k-1}^{(i)}(c^m|\mathcal{Z}^{k-1}) = \sum_{j=1}^{n_{k-1}^i} w_{B,k|k-1}^{i,j} \delta_{l^{i,j}}(c^m) / \sum_{j=1}^{n_{k-1}^i} w_{B,k|k-1}^{i,j} \tag{68}$$

with

$$\mathbf{x}_{P,k|k-1}^{i,j} \sim q_k^{(i)}(\cdot|\mathbf{x}_{k-1}, \mathcal{Z}_k) \tag{69}$$

$$\mathbf{x}_{B,k|k-1}^{i,j} \sim b_k^{(i)}(\cdot|\mathbf{x}_{k-1}, \mathcal{Z}_k) \tag{70}$$

$$\widetilde{w}_{P,k|k-1}^{i,j} = w_{k-1}^{i,j} p_{S,k} \tag{71}$$

$$w_{P,k-1}^{i,j} = \widetilde{w}_{P,k|k-1}^{i,j} / \sum_{j=1}^{n_{k-1}^i} \widetilde{w}_{P,k|k-1}^{i,j} \tag{72}$$

$$w_{P,k|k-1}^{i,m,j} = w_{P,k|k-1}^{i,j} \delta_{l^{i,j}}(c^m) / \sum_{j=1}^{n_{k-1}^i} w_{P,k|k-1}^{i,j} \delta_{l^{i,j}}(c^m) \tag{73}$$

$$\mathbf{x}_{P,k|k-1}^{i,m,j} = \mathbf{x}_{P,k|k-1}^{i,j} \delta_{l^{i,j}}(c^m) \tag{74}$$

$$w_{B,k|k-1}^{i,j} = w_{B,k-1}^{i,j} \tag{75}$$

$$w_{B,k|k-1}^{i,m,j} = w_{B,k|k-1}^{i,j} \delta_{l^{i,j}}(c^m) / \sum_{j=1}^{n_{k-1}^i} w_{B,k|k-1}^{i,j} \delta_{l^{i,j}}(c^m) \tag{76}$$

$$\mathbf{x}_{B,k|k-1}^{i,m,j} = \mathbf{x}_{B,k|k-1}^{i,j} \delta_{l^{i,j}}(c^m) \tag{77}$$

The proof of Proposition 3 is given in Appendix C.

Proposition 4. *If the predicted multi-target density is the multi-Bernoulli* $\pi_{k|k-1} = \left\{\left(r_{k|k-1}^{(i)}, \sum_{m=1}^{n_c} p_{k|k-1}^{(i)}(\mathbf{x}_k|c^m, \mathcal{Z}^{k-1}) p_{k|k-1}^{(i)}(c^m|\mathcal{Z}^{k-1})\right)\right\}_{i=1}^{M_{k|k-1}}$*, then the updated multi-target density is also a multi-Bernoulli, and the SMC implementation can be computed by*

$$r_{L,k}^{(i)} = r_{k|k-1}^{(i)} \frac{1-\rho_{L,k}^i}{1-r_{k|k-1}^{(i)}\rho_{L,k}^i} \tag{78}$$

$$p_{L,k}^{(i)}(c^m|\mathcal{Z}^k) = \sum_{j=1}^{n_{k-1}^i} w_{k|k-1}^{i,j}\delta_{l^{i,j}}(c^m) / \sum_{j=1}^{n_{k-1}^i} w_{k|k-1}^{i,j} \tag{79}$$

$$p_{L,k}^{(i)}(\mathbf{x}_k|c^m, \mathcal{Z}^k) = \sum_{j=1}^{n_{k-1}^i} w_{L,k}^{i,m,j}\delta_{\mathbf{x}_{k|k-1}^{i,m,j}}(\mathbf{x}_k) \tag{80}$$

$$r_{U,k}^{*}(\widetilde{\mathbf{z}}) = \frac{\sum_{i=1}^{M_{k|k-1}} \frac{r_{k|k-1}^{(i)}(1-r_{k|k-1}^{(i)})\rho_{U,k}^i}{\left(1-r_{k|k-1}^{(i)}\rho_{L,k}^i\right)^2}}{\kappa(\widetilde{\mathbf{z}}) + \sum_{i=1}^{M_{k|k-1}} \frac{r_{k|k-1}^{(i)}\rho_{U,k}^i}{1-r_{k|k-1}^{(i)}\rho_{L,k}^i}} \tag{81}$$

$$p_{U,k}^{(i)}(\mathbf{x}_k|c^m, \mathcal{Z}^k) = \sum_{i=1}^{M_{k|k-1}} \sum_{j=1}^{n_{k-1}^i} w_{U,k}^{i,m,j}\delta_{\mathbf{x}_{k|k-1}^{i,m,j}}(\mathbf{x}_k) \tag{82}$$

$$p_{U,k}^{(i)}(c^m|\mathcal{Z}^k) = \sum_{i=1}^{M_{k|k-1}} \sum_{j=1}^{n_{k-1}^i} w_{U,k}^{i,j}\delta_{l^{i,j}}(c^m) / \sum_{i=1}^{M_{k|k-1}} \sum_{j=1}^{n_{k-1}^i} w_{U,k}^{i,j} \tag{83}$$

with

$$\rho_{L,k}^i = \sum_{j=1}^{n_{k-1}^i} w_{k|k-1}^{i,j} p_{D,k} \tag{84}$$

$$\rho_{U,k}^i = \sum_{j=1}^{n_{k-1}^i} w_{k|k-1}^{i,j} g_k^{\mathrm{k}}(\mathbf{x}_k) g_k^{\mathrm{c}}(\mathbf{x}_k, c^m) p_{D,k} \tag{85}$$

$$\widetilde{w}_{L,k}^{i,j} = (1-p_{D,k}) w_{k|k-1}^{i,j} \tag{86}$$

$$w_{L,k}^{i,j} = \widetilde{w}_{L,k}^{i,j} / \sum_{j=1}^{n_{k-1}^i} \widetilde{w}_{L,k}^{i,j} \tag{87}$$

$$w_{L,k}^{i,m,j} = w_{L,k}^{i,j}\delta_{l^{i,j}}(c^m) / \sum_{j=1}^{n_{k-1}^i} w_{L,k}^{i,j}\delta_{l^{i,j}}(c^m) \tag{88}$$

$$\mathbf{x}_{L,k}^{i,m,j} = \mathbf{x}_{L,k}^{i,j}\delta_{l^{i,j}}(c^m) \tag{89}$$

$$\widetilde{w}_{U,k}^{i,j} = \frac{r_{k|k-1}^{(i)}}{1-r_{k|k-1}^{(i)}} p_{D,k} w_{k|k-1}^{i,j} g_k^{\mathrm{k}}(\mathbf{x}_k) g_k^{\mathrm{c}}(\mathbf{x}_k, c^m) \tag{90}$$

$$w_{U,k}^{i,j} = \widetilde{w}_{U,k}^{i,j} / \sum_{j=1}^{n_{k-1}^{i}} \widetilde{w}_{U,k}^{i,j} \tag{91}$$

$$w_{U,k}^{i,m,j} = \frac{w_{U,k}^{i,j} \delta_{l^{i,j}}(c^m)}{\sum_{i=1}^{M_{k|k-1}} \sum_{j=1}^{n_{k-1}^{i}} w_{U,k}^{i,j} \delta_{l^{i,j}}(c^m)} \tag{92}$$

The proof of Proposition 4 is detailed in Appendix D.

Particle resampling is also needed for SMC implementation of the SCM-JTC-CBMeMBer filter, and the same resampling strategy as introduced in Section 3.1 is used. Similar to the CBMeMBer filter, in the proposed SCM-JTC-CBMeMBer filter, the pruning and merge strategy (refer to [22] for the details) should be adopted to reduce the computational burden.

5. Simulation Results

The effectiveness of the proposed SCM-based JTC method and SMC-JTC-CBMeMBer filter is evaluated by simulations. The observation precisions of range and bearing are $\sigma_r = 1$ m and $\sigma_\beta = 0.3°$, respectively. The radar is located at the origin of the coordinates with center frequency $f_c = 35$ GHz, bandwidth $B = 150$ MHz, frequency interval $\Delta F = 1$ MHz. In this paper, three ($n_c = 3$) different ship target classes are considered. The maximum and minimum number of particles used in all the simulations are $L_{\max} = 2000$ and $L_{\min} = 300$, respectively. The CAD models and the corresponding 3D-SCMs (denoted by a red asterisk "*") are shown in Figure 2.

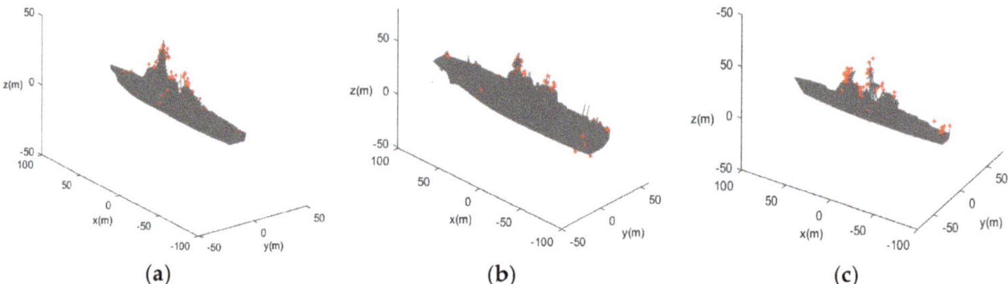

Figure 2. Target CAD models and the corresponding 3D scattering center models (3D-SCMs): (**a**) Target Class A; (**b**) Target Class B; (**c**) Target Class C.

5.1. SCM-Based JTC Method

In this simulation, the SCM-based JTC method will be evaluated under the scenario of single-target measurement without clutter. The process noise of the target motion state is characterized by $q = 0.5$ m/s^2. The three ship targets have the same motion model, and the initial state is $x_0 = [1.2 \text{ km } 1.5 \text{ km } -7.5 \text{ m/s } 5.0 \text{ m/s}]^{\mathrm{T}}$. The sampling interval is $t = 1$s and the total duration is 40 s. At a certain time, only one ship target is present, and the JTC performance of different target is analyzed separately.

The trajectory tracking (for a single run) and target classification (averaged by 20 Monte Carlo runs) results are shown in Figures 3–5. As can be seen from the figures, the proposed method can not only accurately estimate the state of the target but also correctly classify the target simultaneously. It is also seen that the classification probability curve, which matches with the true target class, increases rapidly, and can approach one (100%) within 30 s under all the tested conditions. Conversely, the classification probability curves mismatching the true target class decrease gradually and reach 0 after about 10–30 estimate cycles, meaning that a high confidence classification result is obtained.

However, it does not mean that we have to wait for a period of 10–30 estimate cycles to obtain a reliable decision on target class, since the classification probability matching the true target class is obviously higher than others within only a few cycles.

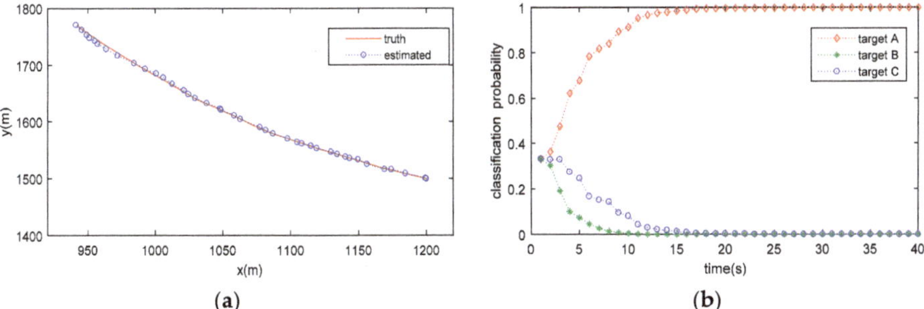

(a) **(b)**

Figure 3. Result of JTC when Target Class A is present: (**a**) estimated target trajectory; (**b**) estimated target class probability.

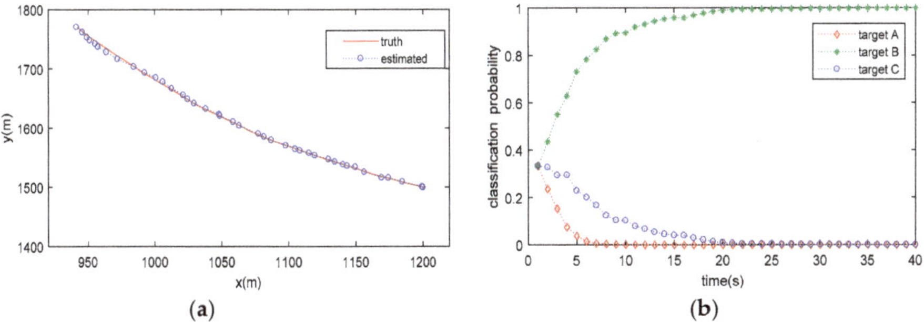

(a) **(b)**

Figure 4. Result of JTC when Target Class B is present: (**a**) estimated target trajectory; (**b**) estimated target class probability.

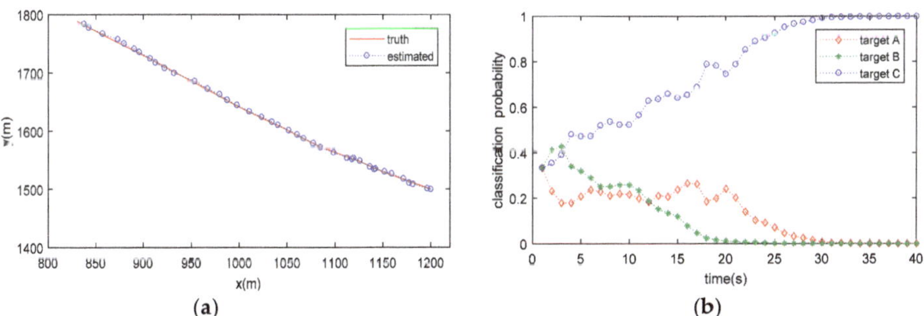

(a) **(b)**

Figure 5. Result of JTC when Target Class C is present: (**a**) estimated target trajectory; (**b**) estimated target class probability.

As a comparison, the simulation also considers the target classification result directly obtained from the HRRP correlation method, where the HRRP templates (training data) are generated from the CAD model with electromagnetic simulation tool, and the test data are predicted from the 3D-SCM. In the simulation, for each target, 360 HRRP training samples (which cover 0–360° in azimuth angle space) are generated with 1° interval. Accordingly, 1440 test samples are generated from each 3D-SCM at the same viewing angle with azimuth angle interval 0.2°. The classification results are shown in

Table 1. The corresponding probabilities of correct classification (PCC) for Target Classes A, B and C are 0.8986, 0.8993 and 0.8417, respectively. The over-all PCC (OA-PCC) for all the test samples is 0.8799. The metric index PCC and OA-PCC are defined as

$$PCC(m) = N_{correct}(m)/N_{total}(m), m = 1, 2, 3. \tag{93}$$

$$OA - PCC = \sum_{m=1}^{3} N_{correct}(m) / \sum_{m=1}^{3} N_{total}(m) \tag{94}$$

where $N_{total}(m)$ denotes the total test samples for the mth target class, $N_{correct}(m)$ represents the correctly classified test samples for the mth target class.

Table 1. Classification results of the ship targets.

True Target Class	Classified Results			PCC
	Ship A	Ship B	Ship C	
Ship A	**1294**	58	99	**0.8986**
Ship B	77	**1295**	129	**0.8993**
Ship C	69	87	**1212**	**0.8417**
OA-PCC		**0.8799**		

Compared the classification results in Table 1 (no tracking process involved) with those shown in Figures 3–5 (with JTC processing), it is seen that the SCM-based JTC method achieves a performance improvement of more than 0.1 (10%) in PCC after the tracking filter is stable, indicating the advantage of the proposed method in classification accuracy.

5.2. SCM-JTC-CBMeMBer Filter

In this simulation, all three classes of ship targets will appear in the surveillance area. Target A appears at time $k = 5$ and disappears at time $k = 55$ with initial state $x_0^{(1)} = [1000\text{m}\ 1000\text{m}\ -9.82\text{m/s}\ -9.82\text{m/s}]^{\mathrm{T}}$. Target B appears at time $k = 15$ and disappears at time $k = 65$ with initial state $x_0^{(2)} = [1000\text{m}\ -1000\text{m}\ -9.82\text{m/s}\ 9.82\text{m/s}]^{\mathrm{T}}$. Target C appears at time $k = 25$ and disappears at time $k = 75$ with initial state $x_0^{(3)} = [-1000\text{m}\ -1000\text{m}\ 9.82\text{m/s}\ 9.82\text{m/s}]^{\mathrm{T}}$. The Poisson average clutter rate is $\lambda_c = 3$. The surveillance area is $[-2000\text{m}, 2000\text{m}] \times [-2000\text{m}, 2000\text{m}]$. Target surviving probability and detection probability are $p_{S,k} = p_{D,k} = 0.99$. The sampling interval t is 1 s and the total simulation time is 100 s. The target births are modeled as multi-Bernoulli RFS with $\pi_B = \left\{\left(r_B^{(i)}, p_B^{(i)}(x|\mathcal{Z})\right)\right\}_{i=1}^{4}$, where

$$\begin{aligned}
& r_B^{(1)} = r_B^{(2)} = r_B^{(3)} = r_B^{(4)} = 0.02 \\
& p_B^{(i)}(c^1|\mathcal{Z}) = p_B^{(i)}(c^2|\mathcal{Z}) = p_B^{(i)}(c^3|\mathcal{Z}) = 1/3 \\
& p_B^{(i)}(x|\mathcal{Z}) = \mathcal{N}(x; m^{(i)}, P^{(i)}) \\
& P^{(1)} = P^{(2)} = P^{(3)} = P^{(4)} = diag([100\ \text{m}^2\ 100\ \text{m}^2\ 10\ \text{m}^2/\text{s}^2\ 10\ \text{m}^2/\text{s}^2]) \\
& m^{(1)} = [1000\ \text{m}\ 1000\ \text{m}\ -10\ \text{m/s}\ -10\ \text{m/s}]^{\mathrm{T}} \\
& m^{(2)} = [1000\ \text{m}\ -1000\ \text{m}\ -10\ \text{m/s}\ 10\ \text{m/s}]^{\mathrm{T}} \\
& m^{(3)} = [-1000\ \text{m}\ -1000\ \text{m}\ 10\ \text{m/s}\ 10\ \text{m/s}]^{\mathrm{T}} \\
& m^{(4)} = [-1000\ \text{m}\ 1000\ \text{m}\ 10\ \text{m/s}\ -10\ \text{m/s}]^{\mathrm{T}}
\end{aligned} \tag{95}$$

The trajectory tracking results for a single run are shown in Figure 6. As can be clearly seen from Figure 6, under the clutter environments, the proposed SCM-JTC-CBMeMBer filter can estimate target number and state correctly, and it can also obtain correct target classification, which will be further validated by the repeated Monte Carlo trials.

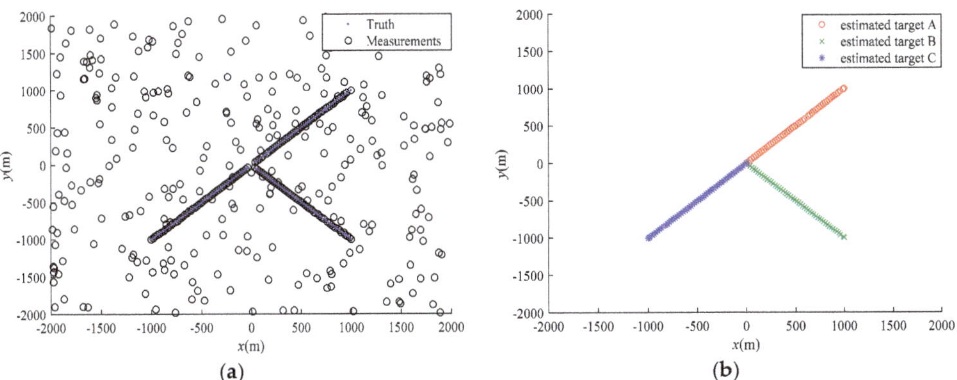

Figure 6. Multi-target tracking results for a single run: (**a**) the true target trajectories and received measurements; (**b**) estimated target trajectories.

To further test the performance of the proposed SCM-JTC-CBMeBer filter, 50 Monte Carlo runs are carried out under the same scenario as above. Specifically, the metric of Optimal Subpattern Assignment (OSPA) distance [34] is used to evaluate the multi-target tracking results, and the CBMeMBer filter is also considered as a comparison.

The OSPA distance and cardinality estimation are shown in Figure 7, and the target classification results are plotted in Figure 8.

From Figure 7, we can see that the SCM-JTC-CBMeMBer filter can effectively estimate the target state and target number. At the instant when the target appears and disappears, a slight degradation in estimation performance is observed, which is the normal phenomenon confronted in multi-target tracking. Compared with the conventional CBMeMBer filter (which can only be used for multi-target tracking purposes rather than targets classification), the SCM-JTC-CBMeMBer filter has almost the same performance in target tracking.

As can be seen from Figure 8, the SCM-JTC-CBMeMBer filter can also correctly classify multiple targets, and the classification probability of each target is very high (almost reaches one).

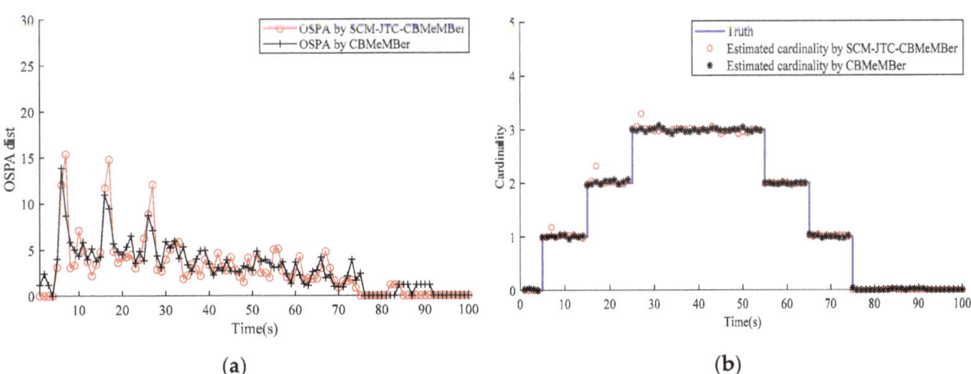

Figure 7. Estimated target state: (**a**) Optimal Subpattern Assignment (OSPA) distance; (**b**) the estimated cardinality.

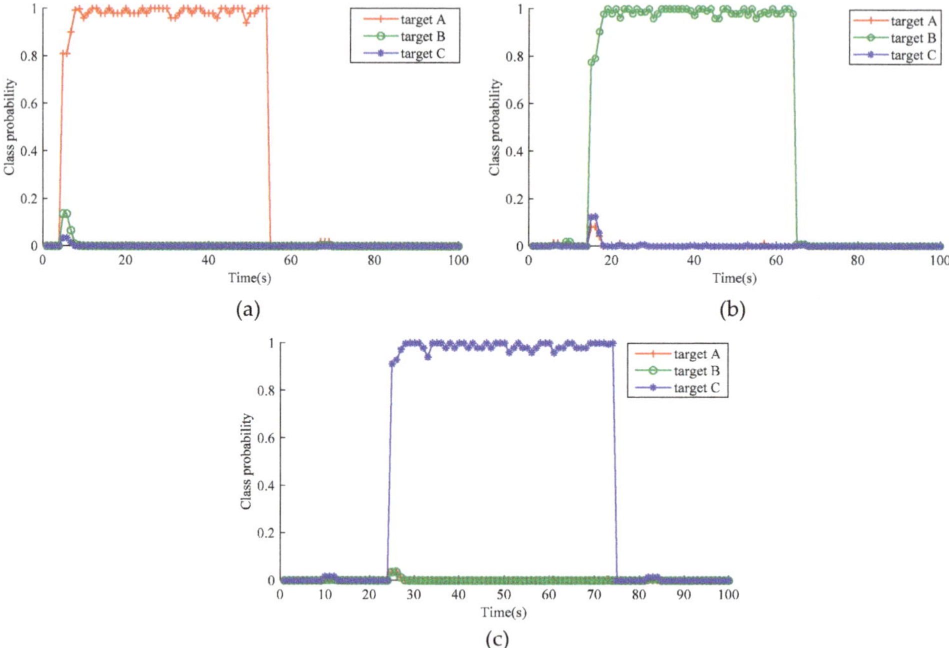

Figure 8. Results of targets classification: (**a**) Ship Target A; (**b**) Ship Target B; (**c**) Ship Target C.

6. Conclusions

In this paper, an SCM-based JTC method is first introduced. The presented method can implement target tracking and classification simultaneously by using a model based on HRRP prediction, target kinematic and HRRP measurements, and thus to alleviate the dependence of target classification on the requirements of target maneuvers or other support information (such as target attribute/identity) in the conventional methods. The SCM-based JTC is then integrated into the framework of the CBMeMBer filter, and the resulting SCM-JTC-CBMeMBer filter for multi-target JTC is derived under the condition with detection uncertainty. Finally, the SMC technique is adopted to implement the proposed filter in view of the complex calculation in multi-target state recursion. Simulations are carried out under the typical scenario with three different ship targets, and the results show that the developed method can not only effectively estimate the target state, but also obtain reliable target classification decision. Additionally, the proposed joint processing method can achieve better performance than separate HRRP classification without involving target tracking.

Author Contributions: For conceptualization, R.Z.; methodology, R.Z.; software, L.W.; validation, R.Z. and L.W.; writing-original draft preparation, R.Z. and L.W.; writing-review and editing, J.Z. All authors have read and agreed to the published version of the manuscript.

Funding: This work is supported by the National Natural Science Foundation of China (61471370).

Conflicts of Interest: The authors declare no conflicts of interest.

Appendix A

According to Equations (14), (15) and (42), we can obtain

$$\begin{aligned} r_{P,k|k-1}^{(i)} &= r_{k-1}^{(i)}\left\langle p_{k-1}^{(i)}(\boldsymbol{x}_{k-1}), p_{S,k}\right\rangle \\ &= r_{k-1}^{(i)}\left\langle \sum_{m=1}^{n_c} p_{k-1}^{(i)}(\boldsymbol{x}_{k-1}|c^m, \mathcal{Z}^{k-1})p_{k-1}^{(i)}(c^m|\mathcal{Z}^{k-1}), p_{S,k}\right\rangle \\ &= r_{k-1}^{(i)}\sum_{m=1}^{n_c} p_{k-1}^{(i)}(c^m|\mathcal{Z}^{k-1})\left\langle p_{k-1}^{(i)}(\boldsymbol{x}_{k-1}|c^m, \mathcal{Z}^{k-1}), p_{S,k}\right\rangle \end{aligned} \tag{A1}$$

$$\begin{aligned} p_{P,k|k-1}^{(i)}(\boldsymbol{x}_k|\mathcal{Z}^{k-1}) &= \frac{\left\langle f_{k|k-1}(\boldsymbol{x}_k|\boldsymbol{x}_{k-1}), p_{k-1}^{(i)}(\boldsymbol{x}_{k-1})p_{S,k}\right\rangle}{\left\langle p_{k-1}^{(i)}(\boldsymbol{x}_{k-1}), p_{S,k}\right\rangle} \\ &= \frac{\left\langle f_{k|k-1}^{k}(\boldsymbol{x}_k|\boldsymbol{x}_{k-1}), \sum_{m=1}^{n_c} p_{k-1}^{(i)}(\boldsymbol{x}_{k-1}|c^m, \mathcal{Z}^{k-1})p_{k-1}^{(i)}(c^m|\mathcal{Z}^{k-1})p_{S,k}\right\rangle}{\left\langle \sum_{m=1}^{n_c} p_{k-1}^{(i)}(\boldsymbol{x}_{k-1}|c^m, \mathcal{Z}^{k-1})p_{k-1}^{(i)}(c^m|\mathcal{Z}^{k-1}), p_{S,k}\right\rangle} \\ &= \frac{\sum_{m=1}^{n_c} p_{k-1}^{(i)}(c^m|\mathcal{Z}^{k-1})\left\langle f_{k|k-1}^{k}(\boldsymbol{x}_k|\boldsymbol{x}_{k-1}), p_{k-1}^{(i)}(\boldsymbol{x}_{k-1}|c^m, \mathcal{Z}^{k-1})p_{S,k}\right\rangle}{\sum_{m=1}^{n_c} p_{k-1}^{(i)}(c^m|\mathcal{Z}^{k-1})\left\langle p_{k-1}^{(i)}(\boldsymbol{x}_{k-1}|c^m, \mathcal{Z}^{k-1}), p_{S,k}\right\rangle} \end{aligned} \tag{A2}$$

Appendix B

According to Equations (17)–(20) and Equation (50), we have

$$\begin{aligned} r_{L,k}^{(i)} &= r_{k|k-1}^{(i)}\frac{1-\left\langle p_{k|k-1}^{(i)}(\boldsymbol{x}_k), p_{D,k}\right\rangle}{1-r_{k|k-1}^{(i)}\left\langle p_{k|k-1}^{(i)}(\boldsymbol{x}_k), p_{D,k}\right\rangle} \\ &= r_{k|k-1}^{(i)}\frac{1-\left\langle \sum_{m=1}^{n_c} p_{k|k-1}^{(i)}(\boldsymbol{x}_k|c^m, \mathcal{Z}^{k-1})p_{k|k-1}^{(i)}(c^m|\mathcal{Z}^{k-1}), p_{D,k}\right\rangle}{1-r_{k|k-1}^{(i)}\left\langle \sum_{m=1}^{n_c} p_{k|k-1}^{(i)}(\boldsymbol{x}_k|c^m, \mathcal{Z}^{k-1})p_{k|k-1}^{(i)}(c^m|\mathcal{Z}^{k-1}), p_{D,k}\right\rangle} \\ &= r_{k|k-1}^{(i)}\frac{1-\sum_{m=1}^{n_c} p_{k|k-1}^{(i)}(c^m|\mathcal{Z}^{k-1})\left\langle p_{k|k-1}^{(i)}(\boldsymbol{x}_k|c^m, \mathcal{Z}^{k-1}), p_{D,k}\right\rangle}{1-r_{k|k-1}^{(i)}\sum_{m=1}^{n_c} p_{k|k-1}^{(i)}(c^m|\mathcal{Z}^{k-1})\left\langle p_{k|k-1}^{(i)}(\boldsymbol{x}_k|c^m, \mathcal{Z}^{k-1}), p_{D,k}\right\rangle} \end{aligned} \tag{A3}$$

$$\begin{aligned} p_{L,k}^{(i)}(\boldsymbol{x}_k|\mathcal{Z}^{k}) &= p_{k|k-1}^{(i)}(\boldsymbol{x}_k|\mathcal{Z}^{k-1})\frac{1-p_{D,k}}{1-\left\langle p_{k|k-1}^{(i)}(\boldsymbol{x}_k|\mathcal{Z}^{k-1}), p_{D,k}\right\rangle} \\ &= \sum_{m=1}^{n_c} p_{k|k-1}^{(i)}(\boldsymbol{x}_k|c^m, \mathcal{Z}^{k-1})p_{k|k-1}^{(i)}(c^m|\mathcal{Z}^{k-1})\frac{1-p_{D,k}}{1-\left\langle \sum_{m=1}^{n_c} p_{k|k-1}^{(i)}(\boldsymbol{x}_k|c^m, \mathcal{Z}^{k-1})p_{k|k-1}^{(i)}(c^m|\mathcal{Z}^{k-1}), p_{D,k}\right\rangle} \\ &= \sum_{m=1}^{n_c} p_{k|k-1}^{(i)}(c^m|\mathcal{Z}^{k-1})\frac{1-p_{D,k}}{1-\sum_{m=1}^{n_c} p_{k|k-1}^{(i)}(c^m|\mathcal{Z}^{k-1})\left\langle p_{k|k-1}^{(i)}(\boldsymbol{x}_k|c^m, \mathcal{Z}^{k-1}), p_{D,k}\right\rangle}p_{k|k-1}^{(i)}(\boldsymbol{x}_k|c^m, \mathcal{Z}^{k-1}) \end{aligned} \tag{A4}$$

$$
\begin{aligned}
r^{*}_{U,k}(\widetilde{z}) &= \frac{\sum_{i=1}^{M_{k|k-1}} \frac{r^{(i)}_{k|k-1}(1-r^{(i)}_{k|k-1})\left\langle p^{(i)}_{k|k-1}(x_k|\mathcal{Z}^{k-1}), g_k(\widetilde{z}|x_k)p_{D,k}\right\rangle}{\left(1-r^{(i)}_{k|k-1}\left\langle p^{(i)}_{k|k-1}(x_k|\mathcal{Z}^{k-1}), p_{D,k}\right\rangle\right)^2}}{\kappa(\widetilde{z})+\sum_{i=1}^{M_{k|k-1}} \frac{r^{(i)}_{k|k-1}\left\langle p^{(i)}_{k|k-1}(x_k|\mathcal{Z}^{k-1}), g_k(\widetilde{z}|x_k)p_{D,k}\right\rangle}{1-r^{(i)}_{k|k-1}\left\langle p^{(i)}_{k|k-1}(x_k|\mathcal{Z}^{k-1}), p_{D,k}\right\rangle}} \\
&= \frac{\sum_{i=1}^{M_{k|k-1}} \frac{r^{(i)}_{k|k-1}(1-r^{(i)}_{k|k-1})\left\langle \sum_{m=1}^{n_c} p^{(i)}_{k|k-1}(x_k|c^m,\mathcal{Z}^{k-1})p^{(i)}_{k|k-1}(c^m|\mathcal{Z}^{k-1}), g^{\mathrm{k}}_k(x_k)g^{\mathrm{c}}_k(x_k,c^m)p_{D,k}\right\rangle}{\left(1-r^{(i)}_{k|k-1}\left\langle \sum_{m=1}^{n_c} p^{(i)}_{k|k-1}(x_k|c^m,\mathcal{Z}^{k-1})p^{(i)}_{k|k-1}(c^m|\mathcal{Z}^{k-1}), p_{D,k}\right\rangle\right)^2}}{\kappa(\widetilde{z})+\sum_{i=1}^{M_{k|k-1}} \frac{r^{(i)}_{k|k-1}\left\langle \sum_{m=1}^{n_c} p^{(i)}_{k|k-1}(x_k|c^m,\mathcal{Z}^{k-1})p^{(i)}_{k|k-1}(c^m|\mathcal{Z}^{k-1}), g^{\mathrm{k}}_k(x_k)g^{\mathrm{c}}_k(x_k,c^m)p_{D,k}\right\rangle}{1-r^{(i)}_{k|k-1}\left\langle \sum_{m=1}^{n_c} p^{(i)}_{k|k-1}(x_k|c^m,\mathcal{Z}^{k-1})p^{(i)}_{k|k-1}(c^m|\mathcal{Z}^{k-1}), p_{D,k}\right\rangle}} \\
&= \frac{\sum_{i=1}^{M_{k|k-1}} \frac{r^{(i)}_{k|k-1}(1-r^{(i)}_{k|k-1})\sum_{m=1}^{n_c} p^{(i)}_{k|k-1}(c^m|\mathcal{Z}^{k-1})\left\langle p^{\mathrm{k},(i)}_{k|k-1}(x_k|c^m,\mathcal{Z}^{k-1}), g^{\mathrm{k}}_k(x_k)g^{\mathrm{c}}_k(x_k,c^m)p_{D,k}\right\rangle}{\left(1-r^{(i)}_{k|k-1}\sum_{m=1}^{n_c} p^{(i)}_{k|k-1}(c^m|\mathcal{Z}^{k-1})\left\langle p^{(i)}_{k|k-1}(x_k|c^m,\mathcal{Z}^{k-1}), p_{D,k}\right\rangle\right)^2}}{\kappa(\widetilde{z})+\sum_{i=1}^{M_{k|k-1}} \frac{r^{(i)}_{k|k-1}\sum_{m=1}^{n_c} p^{(i)}_{k|k-1}(c^m|\mathcal{Z}^{k-1})\left\langle p^{(i)}_{k|k-1}(x_k|c^m,\mathcal{Z}^{k-1}), g^{\mathrm{k}}_k(x_k)g^{\mathrm{c}}_k(x_k,c^m)p_{D,k}\right\rangle}{1-r^{(i)}_{k|k-1}\sum_{m=1}^{n_c} p^{(i)}_{k|k-1}(c^m|\mathcal{Z}^{k-1})\left\langle p^{\mathrm{k},(i)}_{k|k-1}(x_k|c^m,\mathcal{Z}^{k-1}), p_{D,k}\right\rangle}}
\end{aligned} \tag{A5}
$$

$$
\begin{aligned}
p^{*}_{U,k}(x_k;\widetilde{z}) &= \frac{\sum_{i=1}^{M_{k|k-1}} \frac{r^{(i)}_{k|k-1}p^{(i)}_{k|k-1}(x_k|\mathcal{Z}^{k-1})p_{D,k}g^{\mathrm{k}}_k(x_k)g^{\mathrm{c}}_k(x_k,c^m)}{1-r^{(i)}_{k|k-1}}}{\sum_{i=1}^{M_{k|k-1}} \frac{r^{(i)}_{k|k-1}}{1-r^{(i)}_{k|k-1}}\left\langle p^{(i)}_{k|k-1}(x_k|\mathcal{Z}^{k-1}), p_{D,k}g^{\mathrm{k}}_k(x_k)g^{\mathrm{c}}_k(x_k,c^m)\right\rangle} \\
&= \frac{\sum_{i=1}^{M_{k|k-1}} \frac{r^{(i)}_{k|k-1}\sum_{m=1}^{n_c} p^{(i)}_{k|k-1}(x_k|c^m,\mathcal{Z}^{k-1})p^{(i)}_{k|k-1}(c^m|\mathcal{Z}^{k-1})p_{D,k}g^{\mathrm{k}}_k(x_k)g^{\mathrm{c}}_k(x_k,c^m)}{1-r^{(i)}_{k|k-1}}}{\sum_{i=1}^{M_{k|k-1}} \frac{r^{(i)}_{k|k-1}}{1-r^{(i)}_{k|k-1}}\left\langle \sum_{m=1}^{n_c} p^{(i)}_{k|k-1}(x_k|c^m,\mathcal{Z}^{k-1})p^{(i)}_{k|k-1}(c^m|\mathcal{Z}^{k-1}), p_{D,k}g^{\mathrm{k}}_k(x_k)g^{\mathrm{c}}_k(x_k,c^m)\right\rangle} \\
&= \frac{\sum_{m=1}^{n_c}\sum_{i=1}^{M_{k|k-1}} \frac{r^{(i)}_{k|k-1}p^{(i)}_{k|k-1}(x_k|c^m,\mathcal{Z}^{k-1})p^{(i)}_{k|k-1}(c^m|\mathcal{Z}^{k-1})p_{D,k}g^{\mathrm{k}}_k(x_k)g^{\mathrm{c}}_k(x_k,c^m)}{1-r^{(i)}_{k|k-1}}}{\sum_{i=1}^{M_{k|k-1}} \frac{r^{(i)}_{k|k-1}}{1-r^{(i)}_{k|k-1}}\left\langle \sum_{m=1}^{n_c} p^{(i)}_{k|k-1}(x_k|c^m,\mathcal{Z}^{k-1})p^{(i)}_{k|k-1}(c^m|\mathcal{Z}^{k-1}), p_{D,k}g^{\mathrm{k}}_k(x_k)g^{\mathrm{c}}_k(x_k,c^m)\right\rangle}
\end{aligned} \tag{A6}
$$

Appendix C

According to Equations (44)–(49) and Equation (58), we can obtain

$$
\begin{aligned}
r^{(i)}_{P,k|k-1} &= r^{(i)}_{k-1}\sum_{m=1}^{n_c} p^{(i)}_{k-1}(c^m|\mathcal{Z}^{k-1})\left\langle p^{(i)}_{k-1}(x_{k-1}|c^m,\mathcal{Z}^{k-1}), p_{S,k}\right\rangle \\
&= r^{(i)}_{k-1}\sum_{m=1}^{n_c} \frac{\sum_{j=1}^{n^i_{k-1}} w^{i,j}_{k-1}\delta_{l^{i,j}}(c^m)}{\sum_{j=1}^{n^i_{k-1}} w^{i,j}_{k-1}}\left\langle \sum_{j=1}^{n^i_{k-1}} w^{i,m,j}_{k-1}\delta_{x^{i,m,j}_{k-1}}(x_{k-1}), p_{S,k}\right\rangle \\
&= r^{(i)}_{k-1}\sum_{m=1}^{n_c} \frac{\sum_{j=1}^{n^i_{k-1}} w^{i,j}_{k-1}\delta_{c^{i,j}}(m)}{\sum_{j=1}^{n^i_{k-1}} w^{i,j}_{k-1}}\left\langle \sum_{j=1}^{n^i_{k-1}} \frac{w^{i,j}_{k-1}\delta_{c^{i,j}}(m)}{\sum_{j=1}^{n^i_{k-1}} w^{i,j}_{k-1}\delta_{c^{i,j}}(m)}\delta_{x^{i,m,j}_{k-1}}(x), p_{S,k}\right\rangle \\
&= r^{(i)}_{k-1}\sum_{m=1}^{n_c}\sum_{j=1}^{n^i_{k-1}} \frac{\sum_{j=1}^{n^i_{k-1}} w^{i,j}_{k-1}\delta_{l^{i,j}}(c^m)}{\sum_{j=1}^{n^i_{k-1}} w^{i,j}_{k-1}\delta_{l^{i,j}}(c^m)} w^{i,j}_{k-1}\delta_{l^{i,j}}(c^m)p_{S,k}(x^{i,m,j}_{k-1}) \\
&= r^{(i)}_{k-1}\sum_{j=1}^{n^i_{k-1}} w^{i,j}_{k-1}p_{S,k}
\end{aligned} \tag{A7}
$$

$$p_{P,k|k-1}^{(i)}(c^m|\mathbf{Z}) = p_{k-1}^{(i)}(c^m|\mathcal{Z}^{k-1}) = \frac{\sum_{j=1}^{n_{k-1}^i} w_{P,k|k-1}^{i,j}\delta_{l^{i,j}}(c^m)}{\sum_{j=1}^{n_{k-1}^i} w_{P,k|k-1}^{i,j}} \tag{A8}$$

$$\begin{aligned}
&p_{P,k|k-1}^{(i)}(\boldsymbol{x}_k \quad |c^m,\mathcal{Z}^{k-1})\\
&= \frac{\left\langle f_{k|k-1}^{\mathrm{k}}(\boldsymbol{x}_k|\boldsymbol{x}_{k-1}), p_{k-1}^{(i)}(\boldsymbol{x}_{k-1}|c^m,\mathcal{Z}^{k-1})p_{S,k}\right\rangle}{\sum_{m=1}^{n_C} p_{k-1}^{(i)}(c^m|\mathcal{Z}^{k-1})\left\langle p_{k-1}^{(i)}(\boldsymbol{x}_{k-1}|c^m,\mathcal{Z}^{k-1}),p_{S,k}\right\rangle}\\
&= \frac{\left\langle f_{k|k-1}^{\mathrm{k}}(\boldsymbol{x}_k|\boldsymbol{x}_{k-1}), \sum_{j=1}^{n_{k-1}^i} w_{k-1}^{i,m,j}\delta_{x_{k-1}^{i,m,j}}(\boldsymbol{x}_{k-1})p_{S,k}\right\rangle}{\sum_{m=1}^{n_C} \frac{\sum_{j=1}^{n_{k-1}^i} w_{k-1}^{i,j}\delta_{l^{i,j}}(c^m)}{\sum_{j=1}^{n_{k-1}^i} w_{k-1}^{i,j}}\left\langle \sum_{j=1}^{n_{k-1}^i} w_{k-1}^{i,m,j}\delta_{x_{k-1}^{i,m,j}}(\boldsymbol{x}),p_{S,k}\right\rangle}\\
&= \frac{\sum_{j=1}^{n_{k-1}^i} w_{k-1}^{i,m,j}p_{S,k}(\boldsymbol{x}_{k-1}^{i,m,j})\delta_{x_{P,k|k-1}^{i,m,j}}(\boldsymbol{x}_k)}{\sum_{m=1}^{n_C} \frac{\sum_{j=1}^{n_{k-1}^i} w_{k-1}^{i,j}\delta_{l^{i,j}}(c^m)}{\sum_{j=1}^{n_{k-1}^i} w_{k-1}^{i,j}}\sum_{j=1}^{n_{k-1}^i} w_{k-1}^{i,m,j}p_{S,k}(\boldsymbol{x}_{k-1}^{i,m,j})} = \frac{\sum_{j=1}^{n_{k-1}^i} w_{k-1}^{i,m,j}p_{S,k}\delta_{x_{P,k|k-1}^{i,m,j}}(\boldsymbol{x})}{\sum_{j=1}^{n_{k-1}^i} w_{k-1}^{i,j}p_{S,k}}
\end{aligned} \tag{A9}$$

Appendix D

According to Equations (52)–(57), we have

$$\begin{aligned}
r_{L,k}^{(i)} &= r_{k|k-1}^{(i)}\frac{1-\sum_{m=1}^{n_C} p_{k|k-1}^{(i)}(c^m|\mathcal{Z}^{k-1})\left\langle p_{k|k-1}^{(i)}(\boldsymbol{x}_k|c^m,\mathcal{Z}^{k-1}),p_{D,k}\right\rangle}{1-r_{k|k-1}^{(i)}\sum_{m=1}^{n_C} p_{k|k-1}^{(i)}(c^m|\mathcal{Z}^{k-1})\left\langle p_{k|k-1}^{(i)}(\boldsymbol{x}_k|c^m,\mathcal{Z}^{k-1}),p_{D,k}\right\rangle}\\
&= r_{k|k-1}^{(i)}\frac{1-\sum_{m=1}^{n_C}\frac{\sum_{j=1}^{n_{k-1}^i} w_{k|k-1}^{i,j}\delta_{l^{i,j}}(c^m)}{\sum_{j=1}^{n_{k-1}^i} w_{k|k-1}^{i,j}}\left\langle \sum_{j=1}^{n_{k-1}^i} w_{k|k-1}^{i,m,j}\delta_{x_{k|k-1}^{i,m,j}}(\boldsymbol{x}_k),p_{D,k}\right\rangle}{1-r_{k|k-1}^{(i)}\sum_{m=1}^{n_C}\frac{\sum_{j=1}^{n_{k-1}^i} w_{k|k-1}^{i,j}\delta_{l^{i,j}}(c^m)}{\sum_{j=1}^{n_{k-1}^i} w_{k|k-1}^{i,j}}\left\langle \sum_{j=1}^{n_{k-1}^i} w_{k|k-1}^{i,m,j}\delta_{x_{k|k-1}^{i,m,j}}(\boldsymbol{x}_k),p_{D,k}\right\rangle}\\
&= r_{k|k-1}^{(i)}\frac{1-\sum_{j=1}^{n_{k-1}^i} w_{k|k-1}^{i,j}p_{D,k}}{1-r_{k|k-1}^{(i)}\sum_{j=1}^{n_{k-1}^i} w_{k|k-1}^{i,j}p_{D,k}}
\end{aligned} \tag{A10}$$

$$
\begin{aligned}
p_{L,k}^{(i)}(\boldsymbol{x}_k|c^m,\mathcal{Z}^k) &= \frac{1-p_{D,k}}{1-\sum_{m=1}^{n_C} p_{k|k-1}^{(i)}(c^m|\mathcal{Z}^{k-1})\left\langle p_{k|k-1}^{(i)}(\boldsymbol{x}_k|c^m,\mathcal{Z}^{k-1}),p_{D,k}\right\rangle} p_{k|k-1}^{(i)}(\boldsymbol{x}_k|c^m,\mathcal{Z}^{k-1}) \\
&= \frac{1-p_{D,k}}{1-\sum_{m=1}^{n_C} \frac{\sum_{j=1}^{n_{k-1}^i} w_{k|k-1}^{i,j}\delta_{l^{i,j}}(c^m)}{\sum_{j=1}^{n_{k-1}^i} w_{k|k-1}^{i,j}} \left\langle \sum_{j=1}^{n_{k-1}^i} w_{k|k-1}^{i,m,j}\delta_{\boldsymbol{x}_{k|k-1}^{i,m,j}}(\boldsymbol{x}_k),p_{D,k}\right\rangle} \sum_{j=1}^{n_{k-1}^i} w_{k|k-1}^{i,m,j}\delta_{\boldsymbol{x}_{k|k-1}^{i,m,j}}(\boldsymbol{x}_k) \\
&= \frac{1-p_{D,k}}{1-\sum_{m=1}^{n_C}\sum_{j=1}^{n_{k-1}^i} \frac{\sum_{j=1}^{n_{k-1}^i} w_{k|k-1}^{i,j}\delta_{l^{i,j}}(c^m)}{\sum_{j=1}^{n_{k-1}^i} w_{k|k-1}^{i,j}} \frac{w_{k|k-1}^{i,j}\delta_{l^{i,j}}(c^m)}{\sum_{j=1}^{n_{k-1}^i} w_{k|k-1}^{i,j}\delta_{l^{i,j}}(c^m)} p_{D,k}(\boldsymbol{x}_{k|k-1}^{i,m,j})} \sum_{j=1}^{n_{k-1}^i} w_{k|k-1}^{i,m,j}\delta_{\boldsymbol{x}_{k|k-1}^{i,m,j}}(\boldsymbol{x}_k) \\
&= \frac{1-p_{D,k}}{1-\sum_{j=1}^{n_{k-1}^i} w_{k|k-1}^{i,j}p_{D,k}} \sum_{j=1}^{n_{k-1}^i} w_{k|k-1}^{i,m,j}\delta_{\boldsymbol{x}_{k|k-1}^{i,m,j}}(\boldsymbol{x}_k)
\end{aligned} \tag{A11}
$$

$$
\begin{aligned}
r_{U,k}^*(\widetilde{\boldsymbol{z}}) &= \frac{\sum_{i=1}^{M_{k|k-1}} \frac{r_{k|k-1}^{(i)}(1-r_{k|k-1}^{(i)})\sum_{m=1}^{n_C} p_{k|k-1}^{(i)}(c^m|\mathcal{Z}^{k-1})\left\langle p_{k|k-1}^{(i)}(\boldsymbol{x}_k|c^m,\mathcal{Z}^{k-1}),g_k^{\mathbf{k}}(\boldsymbol{x}_k)g_k^{\mathrm{c}}(\boldsymbol{x}_k,c^m)p_{D,k}\right\rangle}{\left(1-r_{k|k-1}^{(i)}\sum_{m=1}^{n_C} p_{k|k-1}^{(i)}(c^m|\mathcal{Z}^{k-1})\left\langle p_{k|k-1}^{(i)}(\boldsymbol{x}_k|c^m,\mathcal{Z}^{k-1}),p_{D,k}\right\rangle\right)^2}}{\kappa_k(\widetilde{\boldsymbol{z}})+\sum_{i=1}^{M_{k|k-1}} \frac{r_{k|k-1}^{(i)}\sum_{m=1}^{n_C} p_{k|k-1}^{(i)}(c^m|\mathcal{Z}^{k-1})\left\langle p_{k|k-1}^{(i)}(\boldsymbol{x}_k|c^m,\mathcal{Z}^{k-1}),g_k^{\mathbf{k}}(\boldsymbol{x}_k)g_k^{\mathrm{c}}(\boldsymbol{x}_k,c^m)p_{D,k}\right\rangle}{1-r_{k|k-1}^{(i)}\sum_{m=1}^{n_C} p_{k|k-1}^{(i)}(c^m|\mathcal{Z}^{k-1})\left\langle p_{k|k-1}^{(i)}(\boldsymbol{x}_k|c^m,\mathcal{Z}^{k-1}),p_{D,k}\right\rangle}} \\
&= \frac{\sum_{i=1}^{M_{k|k-1}} \frac{r_{k|k-1}^{(i)}(1-r_{k|k-1}^{(i)})\sum_{m=1}^{n_C} \frac{\sum_{j=1}^{n_{k-1}^i} w_{k|k-1}^{i,j}\delta_{l^{i,j}}(c^m)}{\sum_{j=1}^{n_{k-1}^i} w_{k|k-1}^{i,j}}\left\langle \sum_{j=1}^{n_{k-1}^i} w_{k|k-1}^{i,m,j}\delta_{\boldsymbol{x}_{k|k-1}^{i,m,j}}(\boldsymbol{x}_k),g_k^{\mathbf{k}}(\boldsymbol{x}_k)g_k^{\mathrm{c}}(\boldsymbol{x}_k,c^m)p_{D,k}\right\rangle}{\left(1-r_{k|k-1}^{(i)}\sum_{m=1}^{n_C} \frac{\sum_{j=1}^{n_{k-1}^i} w_{k|k-1}^{i,j}\delta_{l^{i,j}}(c^m)}{\sum_{j=1}^{n_{k-1}^i} w_{k|k-1}^{i,j}}\left\langle \sum_{j=1}^{n_{k-1}^i} w_{k|k-1}^{i,m,j}\delta_{\boldsymbol{x}_{k|k-1}^{i,m,j}}(\boldsymbol{x}),p_{D,k}\right\rangle\right)^2}}{\kappa(\widetilde{\boldsymbol{z}})+\sum_{i=1}^{M_{k|k-1}} \frac{r_{k|k-1}^{(i)}\sum_{m=1}^{n_C} \frac{\sum_{j=1}^{n_{k-1}^i} w_{k|k-1}^{i,j}\delta_{l^{i,j}}(c^m)}{\sum_{j=1}^{n_{k-1}^i} w_{k|k-1}^{i,j}}\left\langle \sum_{j=1}^{n_{k-1}^i} w_{k|k-1}^{i,m,j}\delta_{\boldsymbol{x}_{k|k-1}^{i,m,j}}(\boldsymbol{x}_k),g_k^{\mathbf{k}}(\boldsymbol{x}_k)g_k^{\mathrm{c}}(\boldsymbol{x}_k,c^m)p_{D,k}\right\rangle}{1-r_{k|k-1}^{(i)}\sum_{m=1}^{n_C} \frac{\sum_{j=1}^{n_{k-1}^i} w_{k|k-1}^{i,j}\delta_{l^{i,j}}(c^m)}{\sum_{j=1}^{n_{k-1}^i} w_{k|k-1}^{i,j}}\left\langle \sum_{j=1}^{n_{k-1}^i} w_{k|k-1}^{i,m,j}\delta_{\boldsymbol{x}_{k|k-1}^{i,m,j}}(\boldsymbol{x}_k),p_{D,k}\right\rangle}} \\
&= \frac{\sum_{i=1}^{M_{k|k-1}} \frac{r_{k|k-1}^{(i)}(1-r_{k|k-1}^{(i)})\sum_{j=1}^{n_{k-1}^i} w_{k|k-1}^{i,j}g_k^{\mathbf{k}}(\boldsymbol{x}_{k|k-1}^{i,j})g_k^{\mathrm{c}}(\boldsymbol{x}_{k|k-1}^{i,j},c^m)p_{D,k}}{\left(1-r_{k|k-1}^{(i)}\sum_{j=1}^{n_{k-1}^i} w_{k|k-1}^{i,j}p_{D,k}\right)^2}}{\kappa(\widetilde{\boldsymbol{z}})+\sum_{i=1}^{M_{k|k-1}} \frac{r_{k|k-1}^{(i)}\sum_{j=1}^{n_{k-1}^i} w_{k|k-1}^{i,j}g_k^{\mathbf{k}}(\boldsymbol{x}_{k|k-1}^{i,j})g_k^{\mathrm{c}}(\boldsymbol{x}_{k|k-1}^{i,j},c^m)p_{D,k}}{1-r_{k|k-1}^{(i)}\sum_{j=1}^{n_{k-1}^i} w_{k|k-1}^{i,j}p_{D,k}}} \\
&= \frac{\sum_{i=1}^{M_{k|k-1}} \frac{r_{k|k-1}^{(i)}(1-r_{k|k-1}^{(i)})\sum_{j=1}^{n_{k-1}^i} w_{k|k-1}^{i,j}g_k^{\mathbf{k}}(\boldsymbol{x}_{k|k-1}^{i,j})g_k^{\mathrm{c}}(\boldsymbol{x}_{k|k-1}^{i,j})p_{D,k}}{\left(1-r_{k|k-1}^{(i)}\sum_{j=1}^{n_{k-1}^i} w_{k|k-1}^{i,j}p_{D,k}\right)^2}}{\kappa(\widetilde{\boldsymbol{z}})+\sum_{i=1}^{M_{k|k-1}} \frac{r_{k|k-1}^{(i)}\sum_{j=1}^{n_{k-1}^i} w_{k|k-1}^{i,j}g_k^{\mathbf{k}}(\boldsymbol{x}_{k|k-1}^{i,j})g_k^{\mathrm{c}}(\boldsymbol{x}_{k|k-1}^{i,j})p_{D,k}}{1-r_{k|k-1}^{(i)}\sum_{j=1}^{n_{k-1}^i} w_{k|k-1}^{i,j}p_{D,k}}}
\end{aligned} \tag{A12}
$$

$$
\begin{aligned}
p^{(i)}_{U,k}(\mathbf{x}_k|c^m,\mathcal{Z}^k) &= \frac{\sum_{i=1}^{M_{k|k-1}} \frac{r^{(i)}_{k|k-1} p^{(i)}_{k|k-1}(\mathbf{x}_k|c^m,\mathcal{Z}^{k-1}) p^{(i)}_{k|k-1}(c^m|\mathcal{Z}^{k-1}) p_{D,k} g^{\mathbf{k}}_k(\mathbf{x}_k) g^{\mathrm{c}}_k(\mathbf{x}_k,c^m)}{1-r^{(i)}_{k|k-1}}}{\sum_{i=1}^{M_{k|k-1}} \frac{r^{(i)}_{k|k-1} p^{(i)}_{k|k-1}(c^m|\mathcal{Z}^{k-1}) \left\langle p^{(i)}_{k|k-1}(\mathbf{x}_k|c^m,\mathcal{Z}^{k-1}), p_{D,k} g^{\mathbf{k}}_k(\mathbf{x}_k) g^{\mathrm{c}}_k(\mathbf{x}_k,c^m)\right\rangle}{1-r^{(i)}_{k|k-1}}} \\
&= \frac{\sum_{i=1}^{M_{k|k-1}} \frac{r^{(i)}_{k|k-1} \sum_{j=1}^{n^i_{k-1}} w^{i,m,j}_{k|k-1} \delta_{\mathbf{x}^{i,m,j}_{k|k-1}}(\mathbf{x}_k) \frac{\sum_{j=1}^{n^i_{k-1}} w^{i,j}_{k|k-1} \delta_{l^{i,j}}(c^m)}{\sum_{j=1}^{n^i_{k-1}} w^{i,j}_{k|k-1}} p_{D,k} g^{\mathbf{k}}_k(\mathbf{x}_k) g^{\mathrm{c}}_k(\mathbf{x}_k,c^m)}{1-r^{(i)}_{k|k-1}}}{\sum_{i=1}^{M_{k|k-1}} \frac{r^{(i)}_{k|k-1} \frac{\sum_{j=1}^{n^i_{k-1}} w^{i,j}_{k|k-1} \delta_{l^{i,j}}(c^m)}{\sum_{j=1}^{n^i_{k-1}} w^{i,j}_{k|k-1}} \left\langle \sum_{j=1}^{n^i_{k-1}} w^{i,m,j}_{k|k-1} \delta_{\mathbf{x}^{i,m,j}_{k|k-1}}(\mathbf{x}_k), p_{D,k} g^{\mathbf{k}}_k(\mathbf{x}_k) g^{\mathrm{c}}_k(\mathbf{x}_k,c^m)\right\rangle}{1-r^{(i)}_{k|k-1}}} \\
&= \frac{\sum_{i=1}^{M_{k|k-1}} \frac{r^{(i)}_{k|k-1} \sum_{j=1}^{n^i_{k-1}} w^{i,m,j}_{k|k-1} \frac{\sum_{j=1}^{n^i_{k-1}} w^{i,j}_{k|k-1} \delta_{l^{i,j}}(c^m)}{\sum_{j=1}^{n^i_{k-1}} w^{i,j}_{k|k-1}} p_{D,k} g^{\mathbf{k}}_k(\mathbf{x}_k) g^{\mathrm{c}}_k(\mathbf{x}_k,c^m) \delta_{\mathbf{x}^{i,m,j}_{k|k-1}}(\mathbf{x}_k)}{1-r^{(i)}_{k|k-1}}}{\sum_{i=1}^{M_{k|k-1}} \frac{r^{(i)}_{k|k-1} \frac{\sum_{j=1}^{n^i_{k-1}} w^{i,j}_{k|k-1} \delta_{l^{i,j}}(c^m)}{\sum_{j=1}^{n^i_{k-1}} w^{i,j}_{k|k-1}} \sum_{j=1}^{n^i_{k-1}} w^{i,m,j}_{k|k-1} p_{D,k} g_k(\mathbf{x}^{i,m,j}_{k|k-1}) g_k(\mathbf{x}^{i,m,j}_{k|k-1})}{1-r^{(i)}_{k|k-1}}} \\
&= \frac{\sum_{i=1}^{M_{k|k-1}} \frac{\sum_{j=1}^{n^i_{k-1}} r^{(i)}_{k|k-1} w^{i,j}_{k|k-1} \delta_{l^{i,j}}(c^m) p_{D,k} g^{\mathbf{k}}_k(\mathbf{x}_k) g^{\mathrm{c}}_k(\mathbf{x}_k,c^m) \delta_{\mathbf{x}^{i,m,j}_{k|k-1}}(\mathbf{x}_k)}{1-r^{(i)}_{k|k-1}}}{\sum_{i=1}^{M_{k|k-1}} \frac{\sum_{j=1}^{n^i_{k-1}} r^{(i)}_{k|k-1} w^{i,j}_{k|k-1} \delta_{l^{i,j}}(c^m) p_{D,k} g_k(\mathbf{x}^{i,m,j}_{k|k-1}) g_k(\mathbf{x}^{i,m,j}_{k|k-1})}{1-r^{(i)}_{k|k-1}}}
\end{aligned} \tag{A13}
$$

$$
\begin{aligned}
p^{(i)}_{U,k}(c^m|\mathcal{Z}^k) &= \frac{\sum_{i=1}^{M_{k|k-1}} \frac{r^{(i)}_{k|k-1} p^{(i)}_{k|k-1}(c^m|\mathcal{Z}^{k-1}) \left\langle p^{(i)}_{k|k-1}(\mathbf{x}_k|c^m,\mathcal{Z}^{k-1}), p_{D,k} g^{\mathbf{k}}_k(\mathbf{x}_k) g^{\mathrm{c}}_k(\mathbf{x}_k,c^m)\right\rangle}{1-r^{(i)}_{k|k-1}}}{\sum_{i=1}^{M_{k|k-1}} \frac{r^{(i)}_{k|k-1}}{1-r^{(i)}_{k|k-1}} \sum_{m=1}^{n_{\mathrm{c}}} p^{(i)}_{k|k-1}(c^m|\mathcal{Z}^{k-1}) \left\langle p^{(i)}_{k|k-1}(\mathbf{x}_k|c^m,\mathcal{Z}^{k-1}), p_{D,k} g^{\mathbf{k}}_k(\mathbf{x}_k) g^{\mathrm{c}}_k(\mathbf{x}_k,c^m)\right\rangle} \\
&= \frac{\sum_{i=1}^{M_{k|k-1}} \frac{r^{(i)}_{k|k-1} \frac{\sum_{j=1}^{n^i_{k-1}} w^{i,j}_{k|k-1} \delta_{l^{i,j}}(c^m)}{\sum_{j=1}^{n^i_{k-1}} w^{i,j}_{k|k-1}} \left\langle \sum_{j=1}^{n^i_{k-1}} w^{i,m,j}_{k|k-1} \delta_{\mathbf{x}^{i,m,j}_{k|k-1}}(\mathbf{x}_k), p_{D,k} g^{\mathbf{k}}_k(\mathbf{x}_k) g^{\mathrm{c}}_k(\mathbf{x}_k,c^m)\right\rangle}{1-r^{(i)}_{k|k-1}}}{\sum_{i=1}^{M_{k|k-1}} \frac{r^{(i)}_{k|k-1}}{1-r^{(i)}_{k|k-1}} \sum_{m=1}^{n_{\mathrm{c}}} \frac{\sum_{j=1}^{n^i_{k-1}} w^{i,j}_{k|k-1} \delta_{l^{i,j}}(c^m)}{\sum_{j=1}^{n^i_{k-1}} w^{i,j}_{k|k-1}} \left\langle \sum_{j=1}^{n^i_{k-1}} w^{i,m,j}_{k|k-1} \delta_{\mathbf{x}^{i,m,j}_{k|k-1}}(\mathbf{x}_k), p_{D,k} g^{\mathbf{k}}_k(\mathbf{x}_k) g^{\mathrm{c}}_k(\mathbf{x}_k,c^m)\right\rangle} \\
&= \frac{\sum_{i=1}^{M_{k|k-1}} \frac{r^{(i)}_{k|k-1} \sum_{j=1}^{n^i_{k-1}} w^{i,j}_{k|k-1} \delta_{l^{i,j}}(c^m) p_{D,k} g^{\mathbf{k}}_k(\mathbf{x}^{i,m,j}_{k|k-1}) g^{\mathrm{c}}_k(\mathbf{x}^{i,m,j}_{k|k-1})}{1-r^{(i)}_{k|k-1}}}{\sum_{i=1}^{M_{k|k-1}} \frac{r^{(i)}_{k|k-1}}{1-r^{(i)}_{k|k-1}} \sum_{m=1}^{n_{\mathrm{c}}} \sum_{j=1}^{n^i_{k-1}} w^{i,j}_{k|k-1} \delta_{c^{i,j}}(m) p_{D,k} g^{\mathbf{k}}_k(\mathbf{x}^{i,m,j}_{k|k-1}) g^{\mathrm{c}}_k(\mathbf{x}^{i,m,j}_{k|k-1})} \\
&= \frac{\sum_{i=1}^{M_{k|k-1}} \frac{r^{(i)}_{k|k-1} \sum_{j=1}^{n^i_{k-1}} w^{i,j}_{k|k-1} \delta_{l^{i,j}}(c^m) p_{D,k} g^{\mathbf{k}}_k(\mathbf{x}^{i,m,j}_{k|k-1}) g^{\mathrm{c}}_k(\mathbf{x}^{i,m,j}_{k|k-1})}{1-r^{(i)}_{k|k-1}}}{\sum_{i=1}^{M_{k|k-1}} \frac{r^{(i)}_{k|k-1}}{1-r^{(i)}_{k|k-1}} \sum_{j=1}^{n^i_{k-1}} w^{i,j}_{k|k-1} p_{D,k} g^{\mathbf{k}}_k(\mathbf{x}^{i,j}_{k|k-1}) g^{\mathrm{c}}_k(\mathbf{x}^{i,j}_{k|k-1})}
\end{aligned} \tag{A14}
$$

References

1. Ristic, B.; Smets, P. Target classification approach based on the belief function theory. *IEEE Trans. Aerosp. Electron. Syst.* **2005**, *41*, 574–583. [CrossRef]
2. Challa, S.; Pulford, G.W. Joint target tracking and classification using radar and ESM sensors. *IEEE Trans. Aerosp. Electron. Syst.* **2001**, *37*, 1039–1055. [CrossRef]
3. Maskell, S. Joint tracking of manoeuring targets and classification of their manoeurability. *EURASIP J. Appl. Signal Process.* **2014**, *15*, 2339–2350.
4. Farina, A.; Lombardo, P.; Marsella, M. Joint tracking and identification algorithms for multisensor data. *IEE Pro.-Radar Sonar Navig.* **2002**, *149*, 271–280. [CrossRef]
5. Donka, A.; Lyudmila, M. Joint target tracking and classification with particle filtering and mixture Kalman filtering using kinematic radar. *Digit. Signal Process.* **2006**, *16*, 180–204.
6. Magnant, C.; Giremus, A.; Grivel, E.; Ratton, L.; Joseph, B. Joint tracking and classification based on kinematic and target extent measurements. In Proceedings of the 18th International Conference on Information Fusion, Washington, DC, USA, 6–9 July 2015; pp. 1748–1755.
7. Magnant, C.; Kemkemian, S.; Zimmer, L. Joint tracking and classification for extended targets in maritime surveillance. In Proceedings of the IEEE Radar Conference, Oklahoma City, OK, USA, 23–27 April 2018; pp. 1117–1122.
8. Lan, J.; Li, X.R. Joint tracking and classification of extended object using random matrix. In Proceedings of the 16th International Conference on Information Fusion, Istanbul, Turkey, 9–12 July 2013; pp. 1550–1557.
9. Minvielle, P.; Doucet, A.; Marrs, A.; Maskell, S. A Bayesian approach to joint tracking and identification of geometric shapes in video sequences. *Image Vis. Comput.* **2010**, *28*, 111–123. [CrossRef]
10. Gong, J.L.; Fan, G.L.; Yu, L.J.; Havlicek, J.P.; Chen, D.R.; Fan, N.J. Joint view-identity manifold for infrared target tracking and recognition. *Comput. Vis. Image Underst.* **2014**, *118*, 211–224. [CrossRef]
11. Ma, C.H.; Wen, G.J.; Ding, B.Y.; Zhong, J.R.; Yang, X.L. Three-dimensional electromagnetic model–based scattering center matching method for synthetic aperture radar automatic target recognition by combining spatial and attributed information. *J. Appl. Remote Sens.* **2016**, *10*, 016025. [CrossRef]
12. Ding, B.Y.; Wen, G.J. Target reconstruction based on 3-D scattering center model for robust SAR ATR. *IEEE Trans on Geosci. Remote.* **2018**, *56*, 3772–3785. [CrossRef]
13. Hu, J.M.; Wang, W.; Zhai, Q.L.; Ou, J.P.; Zhan, R.H.; Zhang, J. Global scattering center extraction for radar targets using a modified RANSAC method. *IEEE Trans. Antenn. Propag.* **2016**, *64*, 3573–3586. [CrossRef]
14. Zhan, R.H.; Wang, L.P.; Zhang, J. Joint target tracking and classification with scattering center model for radar sensor. In Proceedings of the 2019 International Conference on Control Automation and Information Sciences (ICCAIS), Chengdu, China, 23–26 October 2019; pp. 1–5.
15. Barshalom, Y. *Multitarget-Multisensor Tracking: Applications and Advances*; Artech House: Norwood, MA, USA, 1992.
16. Musicki, D.; Evans, R. Joint integrated probabilistic data association: *JIPDA. IEEE Trans. Aerosp. Electron. Syst.* **2004**, *40*, 1093–1099. [CrossRef]
17. Reid, D. An algorithm for tracking multiple targets. *IEEE Trans. Autom. Control* **1979**, *24*, 843–854. [CrossRef]
18. Mahler, R. *Statiscal Multisource-Multitarget Information Fusion*; Artech House: Norwood, MA, USA, 2007.
19. Mahler, R. *Advances in Statiscal Multisource-Multitarget Information Fusion*; Artech House: Norwood, MA, USA, 2014.
20. Mahler, R. Multitarget Bayes filtering via first-order multitarget moments. *IEEE Trans. Aerosp. Electron. Syst.* **2003**, *39*, 1152–1178. [CrossRef]
21. Mahler, R. PHD filters of higher order in target number. *IEEE Trans. Aerosp. Electron. Syst.* **2007**, *43*, 1523–1543. [CrossRef]
22. Vo, B.T.; Vo, B.N.; Cantoni, A. The cardinality balanced multitarget multi-Bernoulli filter and its implementations. *IEEE Trans. Signal Proces.* **2009**, *57*, 409–423.
23. Reuter, S.; Vo, B.-T.; Vo, B.-N.; Dietmayer, K. The labeled multi-Bernoulli filter. *IEEE Trans. Signal Process.* **2014**, *62*, 3246–3260.
24. Vo, B.N.; Vo, B.T.; Pham, N.T.; Suter, D. Joint detection and estimation of multiple objects from image observations. *IEEE Trans. Signal Process.* **2010**, *58*, 5129–5141. [CrossRef]

25. Fu, Z.Y.; Angelini, F.; Chambers, J.; Naqvi, S.M. Multi-level cooperative fusion of GM-PHD filters for online multiple human tracking. *IEEE Trans. Multimedia.* **2019**, *21*, 2277–2291. [CrossRef]
26. LI, T.C.; Juan, M.C.; Sun, S.D. Partial consensus and conservative fusion of Gaussian mixtures for distributed PHD fusion. *IEEE Trans. Aerosp. Electron. Syst.* **2019**, *55*, 2150–2163. [CrossRef]
27. Vo, B.N.; Vo, B.T.; Beard, M. Multi-sensor multi-object tracking with the generalized labeled multi-Bernoulli filter. *IEEE Trans. Signal Process.* **2019**, *67*, 5952–5967. [CrossRef]
28. Yi, W.; Li, S.Q.; Wang, B.L.; Hoseinnezhad, R.; Kong, L.J. Computationally efficient distributed multi-sensor fusion with multi-Bernoulli filter. *IEEE Trans. Signal Process.* **2020**, *68*, 241–256. [CrossRef]
29. Vo, B.N.; Singh, S.; Doucet, A. Sequential Monte Carlo methods for Bayesian multi-target filtering with random finite sets. *IEEE Trans. Aerosp. Electron. Syst.* **2005**, *41*, 1224–1245.
30. Vo, B.N.; Ma, W.K. The Gaussian mixture probability hypothesis density filter. *IEEE Trans. Signal Proces.* **2006**, *54*, 4091–4104. [CrossRef]
31. Vo, B.T.; Vo, B.N.; Cantoni, A. Analytic implementations of the cardinalized probability hypothesis density filter. *IEEE Trans. Signal Proces.* **2007**, *55*, 3553–3567. [CrossRef]
32. Ristic, B.; Vo, B.T.; Vo, B.N.; Farina, A. A tutorial on Bernoulli filters: theory, implementation and applications. *IEEE Trans. Signal Proces.* **2013**, *61*, 3406–3430. [CrossRef]
33. Arulampalam, S.; Maskell, S.; Gordan, N.; Gordon, N.; Clapp, T. A tutorial on particle filter for on-line nonlinear/non-Gaussian Bayesian tracking. *IEEE Trans. Signal Process.* **2002**, *50*, 174–188. [CrossRef]
34. Schuhmacher, D.; Vo, B.T.; Vo, B.N. A consistent metric for performance evaluation of multi-object filters. *IEEE Trans. Signal Process.* **2008**, *56*, 3447–3457. [CrossRef]

Article

Joint Dwell Time and Bandwidth Optimization for Multi-Target Tracking in Radar Network Based on Low Probability of Intercept

Lintao Ding [1], Chenguang Shi [1,2,*], Wei Qiu [1] and Jianjiang Zhou [1]

1 Key Laboratory of Radar Imaging and Microwave Photonics, Ministry of Education, Nanjing University of Aeronautics and Astronautics, Nanjing 210016, China; dltnuaa@163.com (L.D.); 15250956004@163.com (W.Q.); zjjee@nuaa.edu.cn (J.Z.)

2 Science and Technology on Electro-Optic Control Laboratory, Luoyang 471009, China

* Correspondence: scg_space@163.com; Tel.: +86-151-9589-5178

Received: 6 January 2020; Accepted: 25 February 2020; Published: 26 February 2020

Abstract: Radar network systems have been demonstrated to offer numerous advantages for target tracking. In this paper, a low probability of intercept (LPI)-based joint dwell time and bandwidth optimization strategy is proposed for multi-target tracking in a radar network. Since the Bayesian Cramer–Rao lower bound (BCRLB) provides a lower bound on parameter estimation, it can be utilized as the accuracy metric for target tracking. In this strategy, in order to improve the LPI performance of the radar network, the total dwell time consumption of the underlying system is minimized, while guaranteeing a predetermined tracking accuracy. There are two adaptable parameters in the optimization problem: one for dwell time, and the other for bandwidth allocation. Since the nonlinear programming-based genetic algorithm (NPGA) can solve the nonlinear problem well, we develop a method based upon NPGA to solve the resulting problem. The simulation results demonstrate that the proposed strategy has superiority over traditional algorithms, and can achieve a better LPI performance of this radar network.

Keywords: low probability of intercept (LPI); Bayesian Cramer–Rao lower bound (BCRLB); multi-target tracking; radar network

---

1. Introduction

Recently, radar network systems, such as multiple-input multiple-output (MIMO) radar, have attracted great attention from academic researchers [1–5]. It has been shown that a radar network system has numerous potential advantages over traditional monostatic and bistatic radar, such as waveform diversity [1], multiplexing gain [2], enhanced target tracking, localization performance [6,7], etc. As far as multi-target tracking in a radar network, in order to best utilize the system potential under the limited system resources, the resource allocation is of great importance, and receives more and more attention in recent years [8–21].

An effective radar resource allocation strategy can efficiently optimize system parameters, leading to performance enhancements. Therefore, it is necessary to allocate the total launch resources in the radar networks reasonably. As we all know, power allocation is one crucial factor in the resource management of any radar network [8–11]. Godrich et al. (2011) [9] proposed a power allocation strategy for target localization in distributed MIMO radar systems, whose objective can be divided into two parts. In the first part, the total transmission power is minimized for a given accuracy requirement, while in the latter part, the tracking accuracy is maximized under the constraint of a given power budget. As an extension, Xie et al. [10] extended this work to a more general case of unknown previous position information, which promotes the real-time applications.

A performance-driven power allocation algorithm is proposed by maximizing the achievable tracking accuracy with a given total power budget [11]. The algorithm can be regarded as the response of the cognitive transmitter to the environment, which is observed by the receiver in the radar network.

In addition, the time resource allocation is also critical, such as revisit time and dwell time allocation [12–14]. The concept of radar dwell time optimization for target tracking is studied for the first time [12], under the premise of meeting the predetermined target tracking accuracy requirements, and the total dwell time of the phased array radars is minimized. Narykov et al. [13] employed the Markov decision to manage the time resource for target tracking. Specifically, the dwell time and revisit time are adjusted adaptively to increase the maximum number of tracking targets. Wang et al. [14] proposed a joint revisit and dwell time management strategy for single target tracking, which aims to minimize the total time resource used for target tracking, while meeting a desired tracking accuracy requirement.

However, most of the above researches only focus on the single parameter optimization. On the basis of the research mentioned above, many joint resource management optimization algorithms are proposed. Yan et al. [15] proposed a joint beam selection and power allocation strategy for multiple targets tracking, whose basis is to allocate the limited beam and power resource of the radar network for the purpose of achieving an accurate target state estimation. Xie et al. [16] take two variable parameters into consideration: The number of radar nodes and the transmitted power of radar network, and then propose a joint node selection and power allocation strategy with the objective of tracking multiple targets. A cooperative nodes and transmit waveform scheduling scheme is proposed for multiple targets tracking in a distributed radar network [17], where this scheme aims at minimizing the cost of the allocation of waveforms, while guaranteeing a predefined target tracking accuracy.

Although the above works provide us an opportunity to deal with resource management, they have little regard of the low probability of intercept (LPI) performance in radar network systems. With the development of passive detectors, such as the radar warning receiver (RWR), electronic warfare support (ES), anti-radiation missile (ARM), and so on, a serious threat is posed to the radar network. As a result, the study of LPI optimization for radar network systems has attracted significant interest in recent years [18–23]. She et al. [21] proposed a sensor selection and power allocation algorithm for multi-target tracking, whose basis is to reduce the total transmitted power under the constraint of target tracking accuracy, with the purpose of improving the LPI performance of the radar network. A joint transmitter selection and resource management strategy based upon LPI is proposed by controlling transmitting resources while meeting a specified target-tracking accuracy requirement [22]. Generally, the above literature have put forward the idea of joint resource management for LPI performance in radar network systems, which lays a foundation for future study.

For multi-target tracking in a radar network, the information from each monostatic component must be gathered to the fusion center for fusion and processing. However, the data processing rate is commonly limited. Therefore, in order to process all the measurement data before the next observation time and feed back to the radar transmitter in time, it is necessary to strictly control the total amount of data, which is related to the bandwidth of transmitted waveform. Furthermore, the target tracking accuracy is also related to the bandwidth of the radar-transmitted signal. Garcia et al. [24] take the signal bandwidth into account for the first time, and propose a joint power and bandwidth allocation (JPBA) method, with the purpose of maximizing the localization accuracy of a single target. Yan et al. [25] extend the JPBA strategy to the target tracking scenario, where signal bandwidth is allocated to meet the real-time processing requirements. To conclude, bandwidth allocation is also one of the critical factors which needs to be considered in the resource management of radar transmission.

However, to the best of our knowledge, the problem of dwell time allocation and bandwidth allocation to realize the LPI performance optimization for multi-target tracking in a radar network, which has never been taken into consideration, needs to be analyzed in detail.

In this paper, an LPI-based joint dwell time and bandwidth allocation optimization strategy in a radar network is proposed. The strategy can adaptively adjust the radar selection, dwell time and signal bandwidth allocation according to the target motion characteristics at each observation moment. As the Bayesian Cramer–Rao lower bound (BCRLB) combines the revisit time, dwell time, target RCS, transmission signal bandwidth and some other variables, it offers insight effect into the parameters on the tracking performance. Consequently, we utilized BCRLB as the accuracy metric for target tracking. For a predefined target tracking accuracy threshold, the resulting problem is minimizing the total dwell time by optimizing the radar selection, dwell time and transmit signal bandwidth. Then, an efficient two-step method is proposed to solve this problem. Finally, two different RCS cases is considered in this paper to verify the superiority of the proposed strategy.

The remainder of this paper is organized as follows. The system model is introduced in Section 2. Section 3 presents the joint dwell time and bandwidth optimization strategy. In Section 3.1 we derive the BCRLB as the performance metric of the target tracking accuracy. Then, the LPI performance optimization problem based on BCRLB is formulated in Section 3.2. A nonlinear programming-based genetic algorithm (NPGA)-based method is proposed to solve this problem in Section 3.3. Simulation results are provided in Section 4. Finally, conclusions are given in Section 5.

2. System Model

2.1. Target Dynamic Model

Suppose there are Q scattered targets in a two dimensional space. The qth$(q = 1, 2, \ldots, Q)$ target is initially located at $\left(x_0^q, y_0^q\right)$, with initial velocity $\left(\dot{x}_0^q, \dot{y}_0^q\right)$. Assuming that all of the targets move in a uniform linear line, the dynamic model of the target can be described as:

$$\mathbf{X}_k^q = \mathbf{F}\mathbf{X}_{k-1}^q + \mathbf{W}^q \tag{1}$$

in (1), $\mathbf{X}_k^q = \left[x_k^q, y_k^q, \dot{x}_k^q, \dot{y}_k^q\right]^{\mathrm{T}}$ is the state vector of target q at time index k, where $\left(x_k^q, y_k^q\right)$ and $\left(\dot{x}_k^q, \dot{y}_k^q\right)$ are the position and velocity of target q at time index k, respectively. $\mathbf{F}$ is the target state transition matrix, which can be expressed as:

$$\mathbf{F} = \begin{bmatrix} 1 & 0 & T & 0 \\ 0 & 1 & 0 & T \\ 0 & 0 & 1 & 0 \\ 0 & 0 & 0 & 1 \end{bmatrix} \tag{2}$$

where T denotes the revisit time. The term $\mathbf{W}^q$ is the process noise of target q, which can be assumed as zero-mean Gaussian noise with a known covariance $\mathbf{Q}^q$,

$$\mathbf{Q}^q = \sigma_{q,w}^2 \begin{bmatrix} \frac{T^3}{3} & 0 & \frac{T^2}{2} & 0 \\ 0 & \frac{T^3}{3} & 0 & \frac{T^2}{2} \\ \frac{T^2}{2} & 0 & T & 0 \\ 0 & \frac{T^2}{2} & 0 & T \end{bmatrix} \tag{3}$$

where $\sigma_{q,w}^2$ denotes the process noise intensity of target q.

2.2. Observation Model

Consider a radar network with N two-dimensional phased array radars (PARs) working in space, time and frequency synchronization. In order to simplify the problem, we give some moderate assumptions:

(1) Each radar can only receive its own echo signals;
(2) A single radar tracks at most one target in a revisit period.

The traditional radar network system requires all of the radars in the system to radiate a target at all times. Due to the limitation of spectrum resources, communication resources, energy resources etc., multi-target tracking in a traditional radar network is inefficient. As a result, it is not necessary for all radars to work in a revisit period. Thus, we define a set of binary variables $u_{i,k}^{q} \in \{0,1\}$ to represent the radar selection index:

$$u_{i,k}^{q} = \begin{cases} 1, \text{ if the } q\text{th target is tracked by the } i\text{th radar at time index } k \\ 0, \text{ otherwise} \end{cases} \quad (4)$$

Assuming that all PARs in the radar network are able to extract the distance and angle information from the echo signal, then the measurement equation can be written as:

$$\mathbf{Z}_{i,k}^{q} = \begin{cases} \mathrm{h}_i\left(\mathbf{X}_k^q\right) + \mathbf{V}_{i,k}^{q}, & \text{if } u_{i,k}^{q} = 1 \\ 0, & \text{if } u_{i,k}^{q} = 0 \end{cases} \quad (5)$$

where $\mathbf{Z}_{i,k}^{q}$ represents the measured value, and $\mathrm{h}_i\left(\mathbf{X}_k^q\right)$ is a nonlinear transfer function with the following expression:

$$\mathrm{h}_i\left(\mathbf{X}_k^q\right) = \begin{bmatrix} R_{i,k}^{q} \\ \theta_{i,k}^{q} \end{bmatrix} = \begin{bmatrix} \sqrt{\left(x_k^q - x_i\right)^2 + \left(y_k^q - y_i\right)^2} \\ \arctan\left[\frac{y_k^q - y_i}{x_k^q - x_i}\right] \end{bmatrix} \quad (6)$$

here, (x_i, y_i) denotes the ith radar's position, $R_{i,k}^{q}$ and $\theta_{i,k}^{q}$ are the qth target's distance and azimuth to radar i. In (5), $\mathbf{V}_{i,k}^{q}$ is the measurement noise and can be written as $\mathbf{V}_{i,k}^{q} = \left[\Delta R_{i,k}^{q}, \Delta\theta_{i,k}^{q}\right]^{\mathrm{T}}$, where $\Delta R_{i,k}^{q}$ and $\Delta\theta_{i,k}^{q}$ are the measurement errors of distance and azimuth, respectively. Assuming that $\mathbf{V}_{i,k}^{q}$ is zero-mean Gaussian noise with covariance $\mathbf{G}_{i,k}^{q}$, which can given by:

$$\mathbf{G}_{i,k}^{q} = \begin{bmatrix} \sigma_{R_{i,k}^{q}}^{2} & 0 \\ 0 & \sigma_{\theta_{i,k}^{q}}^{2} \end{bmatrix} \quad (7)$$

herein, $\sigma_{R_{i,k}^{q}}^{2}$ and $\sigma_{\theta_{i,k}^{q}}^{2}$ are the mean square estimation error of distance and azimuth, respectively. Both of them are related to the signal-to-noise ratio (SNR) of the echo at the current moment and can be calculated as [26]:

$$\begin{cases} \sigma_{R_{i,k}^{q}} = \frac{c}{4\pi\beta_{i,k,q}\sqrt{\mathrm{SNR}_{i,k}^{q}}} \\ \sigma_{\theta_{i,k}^{q}} = \frac{\sqrt{3}\lambda}{\pi\gamma\sqrt{\mathrm{SNR}_{i,k}^{q}}} \end{cases} \quad (8)$$

where $\mathrm{SNR}_{i,k}^{q}$ denotes the ith radar' SNR to target q at time index k. The term $c = 3 \times 10^8$ m/s is the speed of light, λ and γ are the transmitted wavelength and antenna aperture, respectively. $\beta_{i,q,k}$ is the effective bandwidth of the ith radar's transmitted waveform to target q.

It can be seen that under the same conditions of other parameters, the higher the $\beta_{i,q,k}$ in (8), the smaller the measurement error of distance. In addition, the amount of radar data samples from the illuminated targets is also related to the transmitted signal bandwidth. Given the oversampling ratio $\rho \geq 1$, the ith radar's sampling frequency on the qth target at time index k is $f_{i,k}^{s} = \rho\beta_{i,q,k}$ [25]. Then, given the observation area V of radar network, the number of the qth target's from ith radar can be calculated as:

$$N_{i,q,k} = u_{i,k}^{q} \frac{\rho\beta_{i,q,k}}{c} VM \quad (9)$$

From Equation (8), we can conclude that the measurement error of distance and azimuth is inversely proportional to the SNR of the echo. According to the radar equation, if the beams are unbiased with target when the ith radar irradiate target q at time index k, the echo SNR of a single pulse, can be expressed as:

$$\mathrm{SNR}^{\mathrm{s}}_{i,q,k} = \frac{P_{\mathrm{t}} G_{\mathrm{t}} G_{\mathrm{r}} \sigma^q \lambda^2 G_{\mathrm{RP}}}{(4\pi)^3 k T_{\mathrm{o}} B_{\mathrm{r}} F_{\mathrm{r}} \left(R^q_i\right)^4} \tag{10}$$

where P_{t} denotes the transmitted power of radar; G_{t} is the transmit antenna gain; G_{r} is the receive antenna gain; σ^q is the radar cross section (RCS) of the target q; G_{RP}, T_{o} and F_{r} are the processing gain, noise temperature and noise coefficient of the radar receiver, respectively; k is the Boltzmann constant; B_{r} is the bandwidth of the radar receiver-matched filter, and R^q_i is the distance from the ith radar to target q.

During the dwell time of a single irradiation to the target, the radar can receive several reflection pulses from the target. Since the radar has known its own emission parameters, all of the target reflections can be accumulated by coherent accumulation technology to improve the SNR of the echo. Suppose $T^d_{i,q,k}$ represents the dwell time of the ith radar's irradiation on target q at time index k, and T_{r} represents the pulse repetition period of radar, then the number of coherent accumulated pulses can be given by:

$$n_{i,q,k} = \frac{T^d_{i,q,k}}{T_r} \tag{11}$$

Assuming that coherent accumulation is ideal, the SNR obtained after $n_{i,q,k}$ pulses can be written as:

$$\mathrm{SNR}^{\mathrm{CI}}_{i,q,k} = n_{i,q,k} \mathrm{SNR}^{\mathrm{s}}_{i,q,k} \tag{12}$$

When there is an angle difference $\widetilde{\alpha}^q_i$ between the true azimuth of target q and the beam pointing of the ith radar, the echo SNR after coherent accumulation can be expressed as:

$$\mathrm{SNR}^q_{i,k} = \mathrm{SNR}^{\mathrm{CI}}_{i,q,k} \exp\left(-4\ln(2)\frac{\left(\widetilde{\alpha}^q_i\right)^2}{\theta^2_{3\mathrm{dB}}}\right) \tag{13}$$

where $\theta_{3\mathrm{dB}}$ denotes 3dB antenna beam width. Substitute Equations (10)–(12) into Equation (13), then we can obtain:

$$\mathrm{SNR}^q_{i,k} = \frac{T^d_{i,q,k}}{T_r} \frac{P_{\mathrm{t}} G_{\mathrm{t}} G_{\mathrm{r}} \sigma^q \lambda^2 G_{\mathrm{RP}}}{(4\pi)^3 k T_{\mathrm{o}} B_{\mathrm{r}} F_{\mathrm{r}} \left(R^q_i\right)^4} \exp\left(-4\ln(2)\frac{\left(\widetilde{\alpha}^q_i\right)^2}{\theta^2_{3\mathrm{dB}}}\right) \tag{14}$$

2.3. Fusion Center

We assume that the radar network adopts an indirect centralized fusion method. Specifically, each radar illuminates the assigned target, extracts the measurement information from the echo signal, and transmits the distance and azimuth information to the fusion center through a radio frequency (RF) stealth data link for processing. In this system, suppose that the fusion center can make full use of the original measurement data without any loss of information, and thus the fusion results are the optimal. Therefore, the measurement information about the target q at time index k can be formulated as:

$$\mathbf{Z}^q_k = \left[[1,1]^{\mathrm{T}} \otimes \mathbf{u}^q_k\right] \odot \left[\left[\left(\mathbf{R}^q_k\right)^{\mathrm{T}}, \left(\boldsymbol{\theta}^q_k\right)^{\mathrm{T}}\right]^{\mathrm{T}} + \left[\left(\Delta\mathbf{R}^q_k\right)^{\mathrm{T}}, \left(\Delta\boldsymbol{\theta}^q_k\right)^{\mathrm{T}}\right]^{\mathrm{T}}\right] \tag{15}$$

where $\mathbf{R}^q_k = \left[R^q_{1,k}, R^q_{2,k}, \ldots, R^q_{N,k}\right]^{\mathrm{T}}$ and $\boldsymbol{\theta}^q_k = \left[\theta^q_{1,k}, \theta^q_{2,k}, \ldots, \theta^q_{N,k}\right]^{\mathrm{T}}$ denotes the sets of the distance and azimuth measurement parameters of target q at time index k, respectively, $\Delta\mathbf{R}^q_k = \left[\Delta R^q_{1,k}, \Delta R^q_{2,k}, \ldots, \Delta R^q_{N,k}\right]^{\mathrm{T}}$ and $\Delta\boldsymbol{\theta}^q_k = \left[\Delta\theta^q_{1,k}, \Delta\theta^q_{2,k}, \ldots, \Delta\theta^q_{N,k}\right]^{\mathrm{T}}$ are the sets of the distance and azimuth measurement parameter errors, respectively. In (15), the term $\mathbf{u}_k^q$ represents the radar allocation index set of target q at time index k, $\otimes$ is the matrix direct product operation, and $\odot$ is the matrix dot product.

It is assumed that the measurement errors of each radar are independent of each other's, so the qth target's measurement noise covariance matrix $\mathbf{G}_k^q$ can be given by:

$$\mathbf{G}_k^q = \text{diag}\left\{u_{1,k}^q\sigma_{R_{i,k}^q}^2, u_{2,k}^q\sigma_{R_{i,k}^q}^2, \ldots, u_{N,k}^q\sigma_{R_{i,k}^q}^2, u_{1,k}^q\sigma_{\theta_{i,k}^q}^2, u_{2,k}^q\sigma_{\theta_{i,k}^q}^2, \ldots, u_{N,k}^q\sigma_{\theta_{i,k}^q}^2\right\} \tag{16}$$

where diag$\{\cdot\}$ denotes diagonal matrix.

Since the fusion center receives the measurement information from all of the radars in the network on each target, the total number of samples that need to be processed can be calculated as follows:

$$N_k = \sum_{q=1}^{Q}\sum_{i=1}^{N} N_{i,q,k} \tag{17}$$

3. Joint Dwell Time and Bandwidth Optimization Strategy

Dwell time allocation is one of the critical problems to address for LPI performance in a radar network. Under the assumption that the radiation interval is fixed, in order to improve the RF stealth performance, we should minimize the total dwell time in the radar network. However, according to the statement in Section 2.2 and (8), we can get: the reduction of the dwell time will reduce the echo SNR, which will lead to the decrease of detection probability and tracking accuracy. As a result, the purpose of our work is to minimize the total dwell time of the radar network, which is constrained by a predefined accuracy requirement for target tracking. Furthermore, when it comes to the bandwidth of the transmitted waveform, transmitting a larger bandwidth signal means that the system has a higher accuracy of target distance. However, it will increase the workload of the fusion center at the same time, and even make the fusion center unable to process all of the target information within the effective time. Therefore, under the premise of meeting the constraints of target tracking accuracy, data processing capacity and the limited radar resources, we propose a joint dwell time and bandwidth optimization strategy for multi-target tracking with the objective of improving the LPI performance in the radar network.

3.1. Performance Metric

The BCRLB provides a lower bound on the mean square error (MSE) of parameter unbiased estimation, and compares to the posterior Cramer–Rao lower bound (PCRLB) [27,28]. In this paper, BCRLB is derived and used as an optimization criterion for the joint dwell time and bandwidth optimization strategy. At time index k, we use the observation vector $\mathbf{Z}_k^q$ to estimate the state of qth target, which can be defined as $\hat{\mathbf{X}}_k^q(\mathbf{Z}_k^q)$, then the MSE of $\hat{\mathbf{X}}_k^q(\mathbf{Z}_k^q)$ satisfies the following equation:

$$\mathbb{E}\left\{\left(\hat{\mathbf{X}}_k^q(\mathbf{Z}_k^q) - \mathbf{X}_k^q\right) \cdot \left(\hat{\mathbf{X}}_k^q(\mathbf{Z}_k^q) - \mathbf{X}_k^q\right)^{\mathrm{T}}\right\} = \mathbf{C}_k^q \geq \mathbf{J}^{-1}(\mathbf{X}_k^q) \tag{18}$$

where $\mathbb{E}\{\bullet\}$ denotes mathematical expectation, $\mathbf{C}_k^q$ is the qth target's BCRLB at time index k, and $\mathbf{J}(\mathbf{X}_k^q)$ is the Bayesian information matrix (BIM), which can be written as:

$$\mathbf{J}(\mathbf{X}_k^q) = -\mathbb{E}_{\mathbf{X}_k^q,\mathbf{Z}_k^q}\left\{\Delta_{\mathbf{X}_k^q}^{\mathbf{X}_k^q} \log p(\mathbf{Z}_k^q, \mathbf{X}_k^q)\right\} \tag{19}$$

where $\Delta_{\mathbf{X}_k^q}^{\mathbf{X}_k^q} = \nabla_{\mathbf{X}_k^q} \nabla_{\mathbf{X}_k^q}^{\mathrm{T}}$, here $\nabla_{\mathbf{X}_k^q}$ denotes the first-order partial derivative vectors. In (19),

$$p\left(\mathbf{Z}_k^q, \mathbf{X}_k^q\right) = p\left(\mathbf{X}_k^q\right) p\left(\mathbf{Z}_k^q \middle| \mathbf{X}_k^q\right) \tag{20}$$

is the joint probability density function (PDF) [11].

The BIM $\mathbf{J}\left(\mathbf{X}_k^q\right)$ can be expressed as the sum of two matrices:

$$\mathbf{J}\left(\mathbf{X}_k^q\right) = \mathbf{J}_{\mathrm{P}}\left(\mathbf{X}_k^q\right) + \mathbf{J}_{\mathrm{D}}\left(\mathbf{X}_k^q\right) \tag{21}$$

where $\mathbf{J}_{\mathrm{P}}\left(\mathbf{X}_k^q\right)$ and $\mathbf{J}_{\mathrm{D}}\left(\mathbf{X}_k^q\right)$ are the Fisher information matrix (FIM) of the priori information and the data, respectively.

$$\mathbf{J}_{\mathrm{P}}\left(\mathbf{X}_k^q\right) = \mathbb{E}_{\mathbf{X}_k^q}\left\{-\Delta_{\mathbf{X}_k^q}^{\mathbf{X}_k^q} \log p\left(\mathbf{X}_k^q\right)\right\} \tag{22}$$

$$\mathbf{J}_{\mathrm{D}}\left(\mathbf{X}_k^q\right) = \mathbb{E}_{\mathbf{X}_k^q, \mathbf{Z}_k^q}\left\{-\Delta_{\mathbf{X}_k^q}^{\mathbf{X}_k^q} \log p\left(\mathbf{Z}_k^q \middle| \mathbf{X}_k^q\right)\right\} \tag{23}$$

Combined with the system model in Section 2, the qth target is tracked by a fixed number of radars at the time index k. Since the radar independently observes the target at the same moment, the BIM of the target state can be simply expressed as:

$$\mathbf{J}\left(\mathbf{X}_k^q\right) = \mathbf{J}_{\mathrm{P}}\left(\mathbf{X}_k^q\right) + \sum_{i=1}^{N} u_{i,k}^q \mathbf{J}_{\mathrm{D}}^{(i)}\left(\mathbf{X}_k^q\right) \tag{24}$$

where $\mathbf{J}_{\mathrm{D}}^{(i)}\left(\mathbf{X}_k^q\right)$ is the FIM of the ith radar's measurement on qth target. In (24), the term $\mathbf{J}_{\mathrm{P}}\left(\mathbf{X}_k^q\right)$ can be calculated iteratively through the following formula:

$$\mathbf{J}_{\mathrm{P}}\left(\mathbf{X}_k^q\right) = \mathbf{D}_{k-1}^{22} - \mathbf{D}_{k-1}^{21}\left(\mathbf{J}\left(\mathbf{X}_{k-1}^q\right) + \mathbf{D}_{k-1}^{11}\right)^{-1} \mathbf{D}_{k-1}^{12} \tag{25}$$

where,

$$\mathbf{D}_{k-1}^{11} = \mathbb{E}_{\mathbf{X}_{k-1}^q, \mathbf{X}_k^q}\left\{-\Delta_{\mathbf{X}_{k-1}^q}^{\mathbf{X}_{k-1}^q} \log p\left(\mathbf{X}_k^q \middle| \mathbf{X}_{k-1}^q\right)\right\} \tag{26}$$

$$\mathbf{D}_{k-1}^{12} = \mathbf{D}_{k-1}^{21} = \mathbb{E}_{\mathbf{X}_{k-1}^q, \mathbf{X}_k^q}\left\{-\Delta_{\mathbf{X}_k^q}^{\mathbf{X}_{k-1}^q} \log p\left(\mathbf{X}_k^q \middle| \mathbf{X}_{k-1}^q\right)\right\} \tag{27}$$

$$\mathbf{D}_{k-1}^{22} = \mathbb{E}_{\mathbf{X}_{k-1}^q, \mathbf{X}_k^q}\left\{-\Delta_{\mathbf{X}_k^q}^{\mathbf{X}_k^q} \log p\left(\mathbf{X}_k^q \middle| \mathbf{X}_{k-1}^q\right)\right\} \tag{28}$$

Combined with the target dynamic model in Section 2.1, $\mathbf{J}_{\mathrm{P}}\left(\mathbf{X}_k^q\right)$ can be written as:

$$\mathbf{J}_{\mathrm{P}}\left(\mathbf{X}_k^q\right) = \left[\mathbf{Q}^q + \mathbf{F}\mathbf{J}^{-1}\left(\mathbf{X}_{k-1}^q\right)\mathbf{F}^{\mathrm{T}}\right]^{-1} \tag{29}$$

For the ith radar, the FIM of the data can be given by:

$$\mathbf{J}_{\mathrm{D}}^{(i)}\left(\mathbf{X}_k^q\right) = \mathbb{E}_{\mathbf{X}_k^q, \mathbf{Z}_{i,k}^q}\left\{-\Delta_{\mathbf{X}_k^q}^{\mathbf{X}_k^q} \log p\left(\mathbf{Z}_{i,k}^q \middle| \mathbf{X}_k^q\right)\right\} = \mathbb{E}_{\mathbf{X}_k^q}\left\{\mathbb{E}_{\mathbf{Z}_{i,k}^q | \mathbf{X}_k^q}\left\{-\Delta_{\mathbf{X}_k^q}^{\mathbf{X}_k^q} \log p\left(\mathbf{Z}_{i,k}^q \middle| \mathbf{X}_k^q\right)\right\}\right\} \tag{30}$$

According to [15], we can get:

$$\mathbf{J}_{\mathrm{D}}^{(i)}\left(\mathbf{X}_k^q\right) = \mathbb{E}_{\mathbf{X}_k^q}\left\{\left(\mathbf{H}_{i,k}^q\right)^{\mathrm{T}}\left(\mathbf{G}_{i,k}^q\right)^{-1}\mathbf{H}_{i,k}^q\right\} \tag{31}$$

where $\mathbf{H}_{i,k}^{q}$ is the Jacobi matrix of $\mathrm{h}_i\left(\mathbf{X}_k^q\right)$ and can be expressed as:

$$\mathbf{H}_{i,k}^{q}=\left[\nabla_{\mathbf{X}_k^q}\left(\mathrm{h}_i\left(\mathbf{X}_k^q\right)\right)^{\mathrm{T}}\right]^{\mathrm{T}}=\left[\nabla_{\mathbf{X}_k^q}R_{i,k}^{q},\nabla_{\mathbf{X}_k^q}\theta_{i,k}^{q}\right] \tag{32}$$

where

$$\nabla_{\mathbf{X}_k^q}R_{i,k}^{q}=\left[\nabla_{x_k^q}R_{i,k}^{q},\nabla_{\dot{x}_k^q}R_{i,k}^{q},\nabla_{y_k^q}R_{i,k}^{q},\nabla_{\dot{y}_k^q}R_{i,k}^{q}\right]^{\mathrm{T}} \tag{33}$$

$$\nabla_{\mathbf{X}_k^q}\theta_{i,k}^{q}=\left[\nabla_{x_k^q}\theta_{i,k}^{q},\nabla_{\dot{x}_k^q}\theta_{i,k}^{q},\nabla_{y_k^q}\theta_{i,k}^{q},\nabla_{\dot{y}_k^q}\theta_{i,k}^{q}\right]^{\mathrm{T}} \tag{34}$$

are the first-order partial derivatives of the target distance and azimuth to the position and velocity, respectively.

Substitute (29) and (31) into (24), we can get the BIM of the target state $\mathbf{X}_k^q$:

$$\mathbf{J}\left(\mathbf{X}_k^q\right)=\left[\mathbf{Q}^q+\mathbf{F}\mathbf{J}^{-1}\left(\mathbf{X}_{k-1}^q\right)\mathbf{F}^{\mathrm{T}}\right]^{-1}+\sum_{i=1}^{N}u_{i,k}^{q}\mathbb{E}_{\mathbf{X}_k^q}\left\{\left(\mathbf{H}_{i,k}^{q}\right)^{\mathrm{T}}\left(\mathbf{G}_{i,k}^{q}\right)^{-1}\mathbf{H}_{i,k}^{q}\right\} \tag{35}$$

The first prior information FIM of $\mathbf{J}\left(\mathbf{X}_k^q\right)$ is only related to the BIM of the target state at the time index $k-1$ and the target dynamic model in Section 2.1. According to (7) and (8), the $\mathbf{G}_{i,k}^{q}$ in the second item is related to the ith radar's bandwidth on the qth target and the radar echo SNR at time index k. Meanwhile, SNR is a function of the dwell time. As a result, $\mathbf{J}\left(\mathbf{X}_k^q\right)$ is related to the bandwidth and the dwell time at time index k, thus laying the foundation for the joint dwell time and the bandwidth optimization strategy. Furthermore, in order to satisfy the demand of real-time, we can approximate (35) as:

$$\mathbf{J}\left(\mathbf{X}_k^q\right)=\left[\mathbf{Q}^q+\mathbf{F}\mathbf{J}^{-1}\left(\mathbf{X}_{k-1}^q\right)\mathbf{F}^{\mathrm{T}}\right]^{-1}+\sum_{i=1}^{N}u_{i,k}^{q}\left(\mathbf{H}_{i,k}^{q}\right)^{\mathrm{T}}\left(\mathbf{G}_{i,k}^{q}\right)^{-1}\mathbf{H}_{i,k}^{q} \tag{36}$$

According to (18), the corresponding BCRLB matrix of the target state estimation error can be calculated as:

$$\mathbf{C}_{\mathrm{BCRLB},k}^{q}=\mathbf{J}^{-1}\left(\mathbf{X}_k^q\right)=\left[\left[\mathbf{Q}^q+\mathbf{F}\mathbf{J}^{-1}\left(\mathbf{X}_{k-1}^q\right)\mathbf{F}^{\mathrm{T}}\right]^{-1}+\sum_{i=1}^{N}u_{i,k}^{q}\left(\mathbf{H}_{i,k}^{q}\right)^{\mathrm{T}}\left(\mathbf{G}_{i,k}^{q}\right)^{-1}\mathbf{H}_{i,k}^{q}\right]^{-1} \tag{37}$$

3.2. Problem Formulation

This part our main task is to formulate the optimization problem, whose objective is minimizing the total dwell time of the radar network with the tracking performance meeting a predefined threshold.

In Section 3.1, we derived the BCRLB of the target tracking error, which can be used to measure the target tracking accuracy. Moreover, given the updated BIM $\mathbf{J}\left(\mathbf{X}_{k-1}^q\right)$ at the time index $k-1$ and the radar radiation parameters, we can now determine the predictive BCRLB of the target q at time index k according to the formula (37):

$$\mathbf{C}_{\mathrm{BCRLB},k|k-1}^{q}=\left[\left[\mathbf{Q}^q+\mathbf{F}\mathbf{J}^{-1}\left(\mathbf{X}_{k-1}^q\right)\mathbf{F}^{\mathrm{T}}\right]^{-1}+\sum_{i=1}^{N}u_{i,k}^{q}\left(\mathbf{H}_{i,k|k-1}^{q}\right)^{\mathrm{T}}\left(\mathbf{G}_{i,k|k-1}^{q}\right)^{-1}\mathbf{H}_{i,k|k-1}^{q}\right]^{-1} \tag{38}$$

where $\mathbf{G}_{i,k|k-1}^{q}$ and $\mathbf{H}_{i,k|k-1}^{q}$ are the predicted values of $\mathbf{G}_{i,k}^{q}$ and $\mathbf{H}_{i,k}^{q}$, respectively. The diagonal element of $\mathbf{C}_{\mathrm{BCRLB},k|k-1}^{q}$ is the lower bound of the estimated MMSE of the target state estimation, which can be extracted as a measurement metric of target tracking accuracy:

$$F_{k|k-1}^{q}=\sqrt{\mathbf{C}_{k|k-1}^{q}(1,1)+\mathbf{C}_{k|k-1}^{q}(3,3)} \tag{39}$$

where $\mathbf{C}^q_{k|k-1}(1,1)$ and $\mathbf{C}^q_{k|k-1}(3,3)$ are the first variable and the third variable on the diagonal $\mathbf{C}^q_{k|k-1}$, respectively.

Since the tracking accuracy meets a predefined threshold $F_{\max}$, the constraint on the accuracy is:

$$F^q_{k|k-1} \leq F_{\max}, \qquad \forall q = 1, 2, \ldots, Q \tag{40}$$

Then, with respect to the total bandwidth budget, if $u^q_{i,k} = 1$, the bandwidth of the ith radar's illumination on the qth target at time index k should satisfy an upper bound $\beta_{\max}$ and a lower bound $\beta_{\min}$:

$$\begin{cases} \beta_{i,q,k} = 0, & u^q_{i,k} = 0 \\ \beta_{\min} \leq \beta_{i,q,k} \leq \beta_{\max}, & u^q_{i,k} = 1 \end{cases} \tag{41}$$

Similarly, the dwell time constraints can be denoted as:

$$\begin{cases} T^{\mathrm{d}}_{i,q,k} = 0, & u^q_{i,k} = 0 \\ T^{\mathrm{d}}_{\min} \leq T^{\mathrm{d}}_{i,q,k} \leq T^{\mathrm{d}}_{\max}, & u^q_{i,k} = 1 \end{cases} \tag{42}$$

We define the data processing rate of fusion center as ε, and the total number of samples in the radar network should satisfy the following constraints:

$$\sum_{q=1}^{Q} N_k = \frac{1}{\varepsilon} \tag{43}$$

By fusing (40), (41), (42) and (43) together, we can formulate the optimization problem for the joint dwell time and bandwidth optimization strategy:

$$\begin{aligned} &\min_{T^d_{m,q,k}\beta_{i,k,q}} \sum_{q=1}^{Q}\sum_{i=1}^{N} T^d_{m,q,k} \\ &s.t. \begin{cases} F^q_{k|k-1} \leq F_{max}, & \forall q = 1, 2, \ldots, Q \\ \begin{cases} \beta_{i,q,k} = 0, & u^q_{i,k} = 0 \\ \beta_{min} \leq \beta_{i,q,k} \leq \beta_{max}, & u^q_{i,k} = 1 \end{cases} \\ \begin{cases} T^d_{i,q,k} = 0, & u^q_{i,k} = 0 \\ T^d_{min} \leq T^d_{i,q,k} \leq T^d_{max}, & u^q_{i,k} = 1 \end{cases} \\ \sum_{q=1}^{Q} N_k = \frac{1}{\varepsilon}, \ \sum_{q=1}^{Q} u^q_{i,k} \leq 1, \ \sum_{m=1}^{N} u^q_{i,k} = M \end{cases} \end{aligned} \tag{44}$$

where $\sum_{q=1}^{Q} u^q_{i,k} \leq 1$ represents that a single radar tracks at most one target in a revisit period. The term $\sum_{m=1}^{N} u^q_{i,k} = M$ represents that each radar is tracked by M radars at time index k.

Since $u^q_{i,k} \in \{0, 1\}$ is a binary variable, the optimization problem described in (44) is a non-convex problem with three parameters: radar selection, dwell time and the transmitted signal's bandwidth. However, for a given $\mathbf{u}^q_k$, assuming that the radar i is assigned to qth target, the unique radar node selection scheme for the qth can be determined. Furthermore, in order to ensure that all targets have enough information, assuming that each target has the same amount of samples which needs to be sent to the fusion center, then the optimization problem can be converted to the following formula, which only has the variables $T^{\mathrm{d}}_{m,q,k}$ and $\beta_{m,q,k}$ $(1 \leq m \leq M)$:

$$\min_{T^{d}_{m,q,k}\beta_{i,k,q}} \sum_{m=1}^{M} T^{\mathrm{d}}_{m,q,k}$$
$$s.t. \begin{cases} F^{q}_{k|k-1} \leq F_{\max} \\ \sum_{m=1}^{M} \beta_{m,q,k} = \frac{c}{Q\rho\varepsilon V} = \beta_{total} \\ \beta_{\min} \leq \beta_{i,q,k} \leq \beta_{\max} \\ T^{\mathrm{d}}_{\min} \leq T^{\mathrm{d}}_{i,q,k} \leq T^{\mathrm{d}}_{\max} \end{cases} \tag{45}$$

where β_{total} is the total bandwidth of the transmitted waveform of all radars that are assigned to the same target.

3.3. Joint Dwell Time and Bandwidth Optimization Problem Solution

The optimization problem proposed in Equation (45) is non-convex, containing two parameters $T^{\mathrm{d}}_{m,q,k}$ and $\beta_{m,q,k}$. We can use the exhaustive method to solve it, which is simple but too inefficient. The genetic algorithm uses selection, cross and mutation operators for searching, which has a great global search ability. However, the local search ability of this genetic algorithm is weak. In contrast, most of the classical nonlinear algorithms adopt the means of the gradient method, which has a strong local search ability, while also possessing a weak global search ability. As a result, we will solve the problem in (45) by NPGA [29], which combines the global search ability of the genetic algorithm and the local search ability of the classical nonlinear programming algorithms. The flowchart of NPGA is shown in Figure 1:

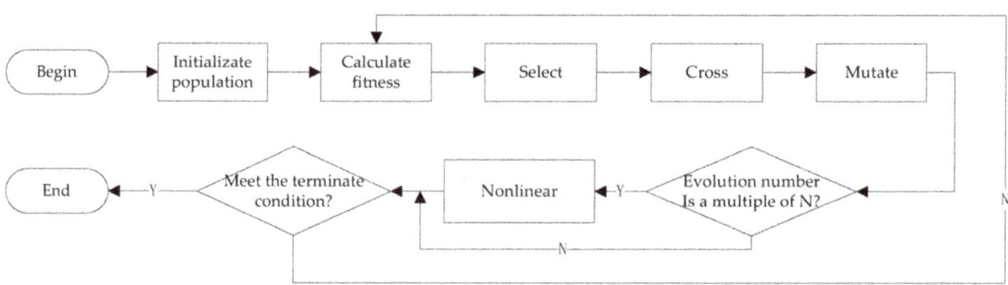

Figure 1. The nonlinear programming-based genetic algorithm (NPGA) flowchart.

By working out the problem (45) for $Q \cdot \mathrm{C}_N^M$ times, we can get all the optimal solutions of the dwell time with respect to different target and radar combinations in the constraint of $\sum_{i=1}^{N} u^q_{i,k} = M$. Then we can use the exhaustive method to obtain the optimal results of the dwell time and radar allocation index in the constraints of $\sum_{q=1}^{Q} u^q_{i,k} \leq 1$. However, the exhaustive method is complex and inefficient. As a result, we propose a radar node selection algorithm with lower computation complexity.

Assuming $M = 2$, which means that each target is fixed to be tracked by two radars at each moment. We define $R_l = \{a,b\}(l = 1,2,\ldots,L)$ as the combinations of the two radars in the radar network, where $L = \mathrm{C}_N^2 = \frac{N!}{(N-2)!2!}$. When the target q is illuminated by the R_l index radars, suppose $S_{l,k,q,\min} = \left(T^{\mathrm{d}}_{a,q,k,\min}\right)^{(l)} + \left(T^{\mathrm{d}}_{b,q,k,\min}\right)^{(l)}$ denotes the minimum dwell time which is solved in (45) through

NPGA, where $\left(T^{\mathrm{d}}_{a,q,k,\min}\right)^{(l)}$ and $\left(T^{\mathrm{d}}_{b,q,k,\min}\right)^{(l)}$ denotes the dwell times of radar a and radar b, respectively. The minimum dwell time matrix $\mathbf{S}_{k,\min}$ which is composed of $S_{l,k,q,\min}$, is shown in Table 1.

Table 1. Minimum dwell time matrix for the fixed radar combination ($M = 2$).

The Minimum Dwell Time of Different Radar Combination		**Target**			
		1	**2**	...	**Q**
Radar Combination	$R_1 = \{1,2\}$	$S_{1,1,k,\min}$	$S_{1,2,k,\min}$	...	$S_{1,Q,k,\min}$
	$R_2 = \{1,3\}$	$S_{2,1,k,\min}$	$S_{2,2,k,\min}$	...	$S_{2,Q,k,\min}$
	$\vdots$	$\vdots$	$\vdots$	$\vdots$	$\vdots$
	$R_L = \{N-1,N\}$	$S_{L,1,k,\min}$	$S_{L,2,k,\min}$	...	$S_{L,Q,k,\min}$

Similar to the term $u^q_{i,k}$, we define a set of binary variables $U^q_{l,k} \in \{0,1\}$ to represent the radar combination selection index.

$$U^q_{l,k} = \begin{cases} 1, \text{ if the } q\text{th radar is tracked by the } l\text{th radar combination at time index } k \\ 0, \text{ otherwise} \end{cases} \tag{46}$$

Then the optimization model of the radar combination allocation index can be described as:

$$\begin{aligned} &\min \sum_{q=1}^{Q} \sum_{l=1}^{L} U^q_{l,k} S_{l,q,k,\min} \\ &s.t. \begin{cases} \sum\limits_{l=1}^{L} U^q_{l,k} = 1 \\ \left(\bigcup\limits_{l=1}^{L} U^r_{l,k} R_l\right) \cap \left(\bigcup\limits_{l=1}^{L} U^m_{l,k} R_l\right) = \varnothing, \forall r \neq m, r,m = 1,2,\ldots,Q \end{cases} \end{aligned} \tag{47}$$

where the first constraints imply that each target is tracked by a fixed radar combination at time index k, while the second one suggests that a single radar tracks at most one target at k. The solution method of (47) can be shown in Algorithm 1.

Algorithm 1. Radar allocation method

Step (1): Working out the problem in (47) $Q \cdot \frac{N!}{(N-2)!2!}$ times, then we can get the minimum dwell time matrix $\mathbf{S}_{k,\min}$ in the constraint of $\sum\limits_{i=1}^{N} u^q_{i,k} = 2$.
Step (2): Sort the columns of matrix $\mathbf{S}_{k,\min}$ in ascending order and assign the target corresponding to the smallest element in the first row to the corresponding radar combination.
Step (3): Remove the column vectors corresponding to the target assigned in Step (2). Remove all the row vectors of the radar which is contained in the radar combination assigned in Step (2).
Step (4): Repeat Step (2) and Step (3) until all the targets are assigned in order to obtain the optimal allocation matrix $\mathbf{U}_{k,\mathrm{opt}}$.

By using the above algorithm, we can obtain the optimal radar allocation results $\mathbf{U}_{k,\mathrm{opt}}$, where $\mathbf{U}_{k,\mathrm{opt}} = \left[\mathbf{U}^1_{k,\mathrm{opt}}, \mathbf{U}^2_{k,\mathrm{opt}}, \ldots, \mathbf{U}^Q_{k,\mathrm{opt}}\right]$, $\mathbf{U}^q_{k,\mathrm{opt}} = \left[U^q_{1,k,\mathrm{opt}}, U^q_{2,k,\mathrm{opt}}, \ldots, U^q_{L,k,\mathrm{opt}}\right]^{\mathrm{T}}$. When $U^q_{l,k,\mathrm{opt}} = 1$, $u^q_{a,k,\mathrm{opt}} = u^q_{b,k,\mathrm{opt}} = 1$, $T^{\mathrm{d}}_{a,q,k,\mathrm{opt}} = \left(T^{\mathrm{d}}_{a,q,k,\min}\right)^{(l)}$, $T^{\mathrm{d}}_{b,q,k,\mathrm{opt}} = \left(T^{\mathrm{d}}_{b,q,k,\min}\right)^{(l)}$. When $U^q_{l,k,\mathrm{opt}} = 0$, $u^q_{a,k,\mathrm{opt}} = u^q_{b,k,\mathrm{opt}} = 0$, $T^{\mathrm{d}}_{a,q,k,\mathrm{opt}} = T^{\mathrm{d}}_{b,q,k,\mathrm{opt}} = 0$. Then we can get $\mathbf{u}_{k,\mathrm{opt}}$ and $\mathbf{T}^{\mathrm{d}}_{k,\mathrm{opt}}$ at time index k, which are the radar allocation index and dwell time optimization results, respectively.

The computational complexity of 0 is $O\left(\frac{Q^2}{2} \times \frac{N!}{(N-2)!2!} \log_2\left(\frac{N!}{(N-2)!2!}\right)\right)$, while the computational complexity of the exhaustive method is $O\left(\left(\frac{N!}{(N-2)!2!}\right)^Q\right)$. Compared with the enumeration method, 0 can greatly reduce the computational complexity and improve the real-time performance.

4. Simulation Results

In this section, some numerical results are provided to illustrate the performance of the proposed LPI-based joint dwell time and bandwidth optimization strategy for multi-target tracking in a radar network. A multi-target tracking scenario with six radars and two targets is considered. In order to simplify the problem, we assume that all the radars in the network systems have the same system parameters. Then we can utilize the default values for the system parameters, as given in Table 2.

Table 2. Radar network system parameters.

Parameter	Value	Parameter	Value
P_t	500 W	σ^q	1 m^2
λ	0.03 m	T_r	5×10^{-4} s
β_{total}	2 MHz	$F_{\max}$	30 m
$\beta_{\min}$	0.1 MHz	$\beta_{\max}$	1.9 MHz
$\theta_{3\text{dB}}$	2°	γ	1 m^2
$T^{\text{d}}_{\min}$	5×10^{-4} s	$T^{\text{d}}_{\max}$	0.1 s

The velocities of target 1 and target 2 are: $(1300, 530)$m/s and $(-1300, -530)$m/s, respectively. It is also assumed that the tracking process lasts 150 s.

Figure 2 depicts the distribution of the radar network, the true trajectories of the two targets and the estimated trajectories of the targets according to the proposed strategy.

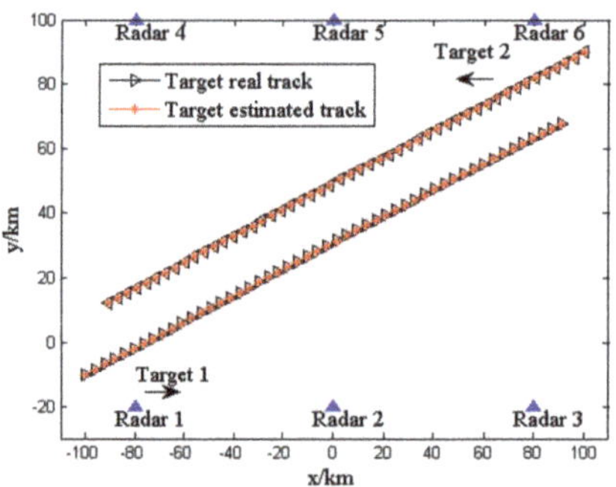

Figure 2. Target trajectory and radar network deployment.

This part first gives the simulation results under the non-undulating RCS model. Assuming that the reflection coefficients of all targets is 1 at any observation time, we define this situation as RCS case 1. In this case, the radar selection and dwell time allocation are only related to the distance and relative position of the target to the radar.

Figure 3 shows the radar selection and bandwidth allocation results of the two targets, while Figure 4 gives the dwell time allocation results. In each figure, on the left side is the radar index, on the right side is the different intensity of the bandwidth and dwell time, which is depicted in different colors. Moreover, the blue areas in each figure indicate that the radar selection variable $u^q_{i,k} = 0$, while the lines in different colors mean that $u^q_{i,k} = 1$, with different colors representing the intensity of the transmitted bandwidth and dwell time. We can conclude that the radar network tends to assign the two radars closest to a specified target for tracking tasks, and more dwell time and bandwidth resources will be allocated to the selected radar, which is farther from the target.

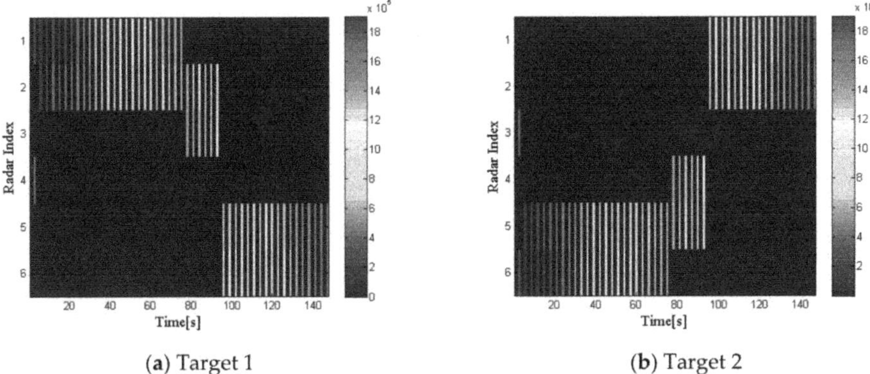

Figure 3. Radar selection and bandwidth allocation in radar cross section (RCS) case 1.

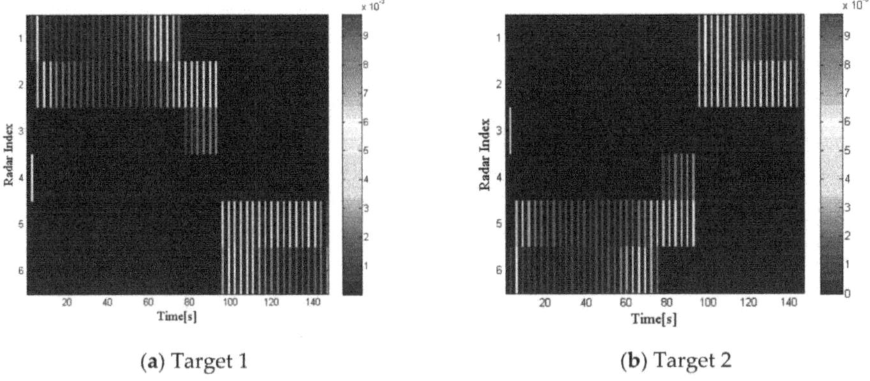

Figure 4. Radar selection and dwell time allocation in RCS case 1.

To show the superiority of the proposed joint dwell time and bandwidth optimization strategy, the optimization algorithm without considering the bandwidth allocation is compared to a benchmark. Figure 5 shows the comparison of total dwell time for two different algorithms. From the result we can see that the proposed strategy can reduce the total dwell time of the radar network compared with the benchmark.

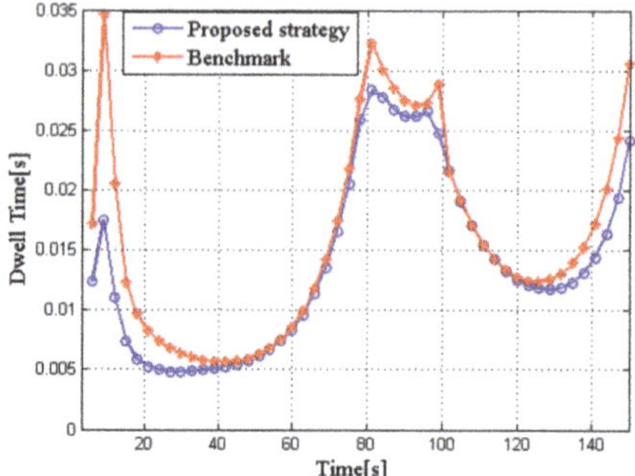

Figure 5. Comparison of total dwell time for different algorithms in RCS case 1.

Define the root mean square error (RMSE) for the tracking accuracy of all targets at time index k as:

$$\text{RMSE}(k) = \sum_{q=1}^{Q} \sqrt{\frac{1}{N_{\text{MC}}} \sum_{n=1}^{N_{\text{MC}}} \left\{ \left[x_k^q - \hat{x}_{n,k|k}^q \right]^2 + \left[y_k^q - \hat{y}_{n,k|k}^q \right]^2 \right\}} \tag{48}$$

where $N_{\text{MC}} = 100$ represents the Monte Carlo experiment number, and $\left(\hat{x}_{n,k|k}^q, \hat{y}_{n,k|k}^q \right)$ is the location estimate at the nth trial.

The RMSE of the proposed strategy and the benchmark are evaluated in Figure 6, respectively. The results prove that the tracking accuracy has not been sacrificed too much after allocating the bandwidth, which is acceptable to our tracking tasks.

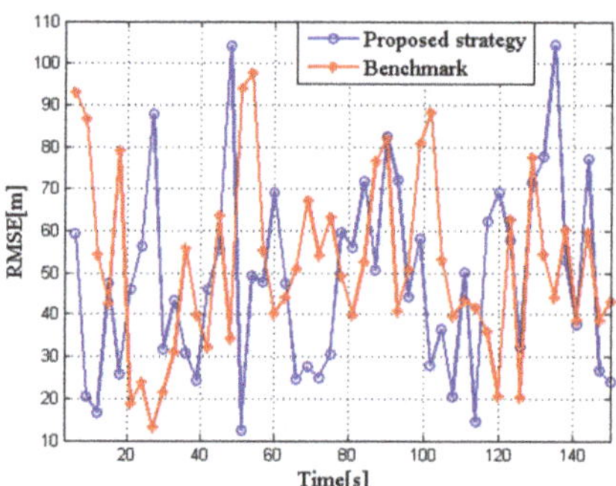

Figure 6. Root mean squared error (RMSE) in two algorithms for target tracking in RCS case 1.

In order to further analyze the impact of the target RCS on radar selection and radar resource allocation results, a second RCS model is also considered, which can be defined as RCS case 2, where it is depicted in Figure 7. In this case, the reflection coefficient of the two targets to radar 3 and radar 4 change with time, while the RCS of the two targets to the other radars remain unchanged at any observation time. In Figure 7, the red and black lines represent the RCS values of target 1 to radar 3 and target 2 to radar 4 at each moment, respectively, which fluctuate around 10.3 m^2. Similarly, the green and blue lines represent the RCS values of target 1 to radar 4 and target 2 to radar 3, respectively, which fluctuate around 2.3 m^2.

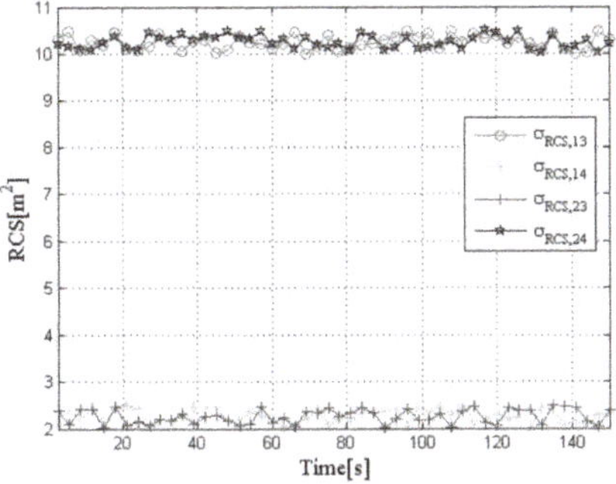

Figure 7. RCS case 2.

Figures 8 and 9 illustrate the optimization results of target 1 and target 2 with the proposed strategy in RCS case 2 at every time index, respectively.

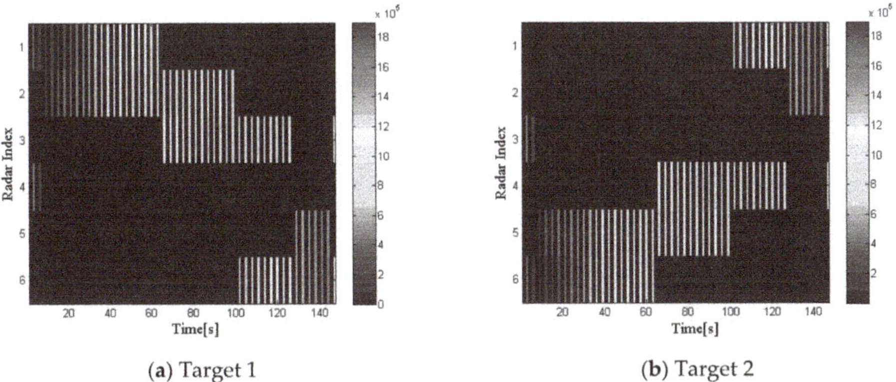

Figure 8. Radar selection and bandwidth allocation in RCS case 2.

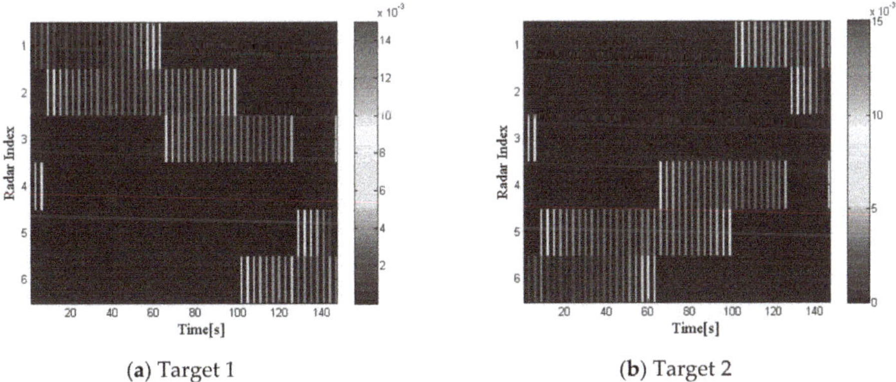

Figure 9. Radar selection and bandwidth allocation in RCS case 2.

Compared with Figures 3 and 4, we can draw the following conclusions. During the whole tracking process, the number of times that radar 3 irradiated target 1 and radar 4 irradiated target 2 increase significantly. In addition, during the period in which radar 2 and radar 3 irradiate target 1 together, radar 2, which is closer to the target, but has a lower reflection coefficient, is allocated more bandwidth and dwell time resources. Similarly, this phenomenon also exists in the resource allocation of target 2. In summary, it can be concluded that the reflection coefficient of the target also affects the radar selection and radar resource allocation results. The radar network system will preferentially select the radar with higher reflection coefficient to irradiate the target. Furthermore, the system tends to allocate more resources to the radar with lower reflection coefficient to the target.

Figures 10 and 11 show the performance comparison of the two algorithms in RCS case 2. Obviously, it is consistent with the conclusions of RCS case 1, thus verifying the stability of the proposed strategy.

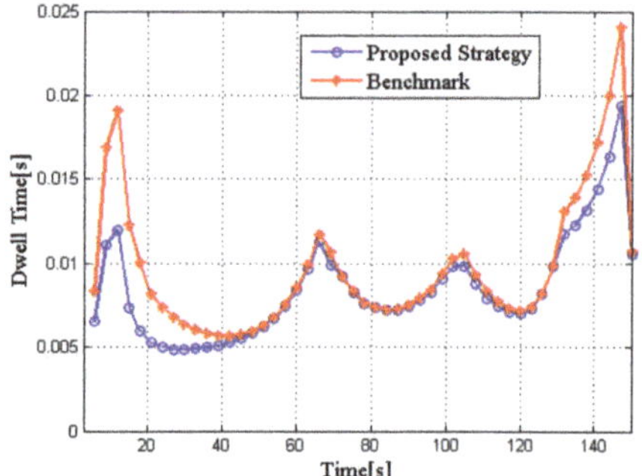

Figure 10. Comparison of total dwell time for different algorithms in RCS case 2.

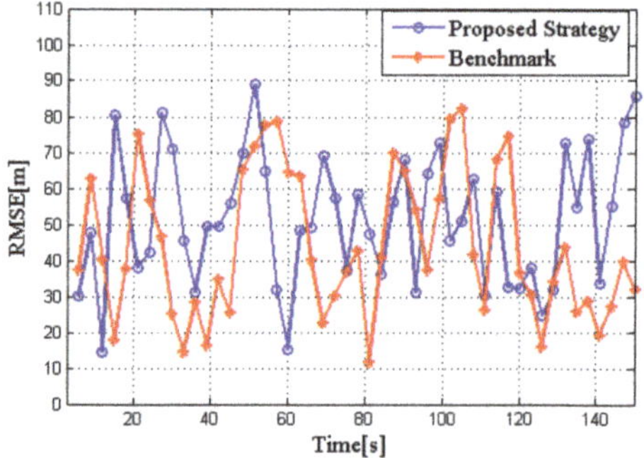

Figure 11. RMSE in two algorithms for target tracking in RCS case 2.

Define the target tracking average root mean square error (ARMSE) as:

$$\text{ARMSE}(k) = \sum_{q=1}^{Q} \sqrt{\frac{1}{N_{\text{MC}}} \sum_{n=1}^{N_{\text{MC}}} \frac{1}{N_k^q(n)} \sum_{k=1}^{N_k^q(n)} \left\{ \left[x_k^q - \hat{x}_{n,k|k}^q \right]^2 + \left[y_k^q - \hat{y}_{n,k|k}^q \right]^2 \right\}} \tag{49}$$

where $N_k^q(n)$ denotes the number of times that the radar network radiated qth target at time index k. Figure 12 shows the ARMSE comparison between the proposed strategy and the benchmark in the two RCS cases. With respect to the target tracking accuracy, the latter is slightly better than the former, but the gap is not large, and is within an acceptable range. In conclusion, the proposed strategy effectively improves the LPI performance of the radar network without sacrificing too much accuracy.

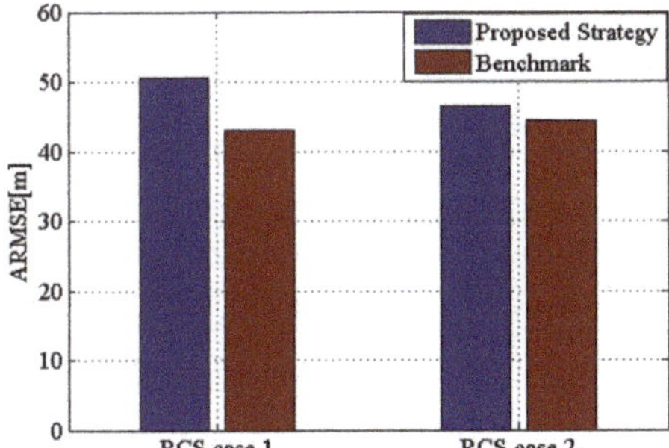

Figure 12. Average root mean square error (ARMSE) comparison of two algorithms for target tracking.

5. Conclusions

An LPI-based joint dwell time and bandwidth allocation strategy is proposed in this paper. The basis of this strategy is to use the optimization technique to control the radars' illumination in the radar network for the purpose of improving the LPI performance. Meanwhile, the tracking accuracy of each target must be guaranteed, which means that the BCRLB meets a predefined threshold. The physical explanation of this strategy can be described as: (1) For each target, select a suitable radar group to complete tracking tasks; (2) Under the premise of tracking tasks requirements, minimize the total dwell time of radar network. The resulting optimization problem contains two adaptable vectors, one for dwell time and the other for bandwidth allocation, which is solved by NPGA, and then a proposed algorithm. Simulation results demonstrate that the proposed strategy can achieve a better LPI performance compared with the benchmark.

In future work, more illumination resources, such as the transmitted power of each radar, will be taken into consideration. Furthermore, the cases of detection probability less than 1 and false alarm probability greater than 0 are of practical importance, which should be taken into account [27,28].

Author Contributions: L.D., C.S. and W.Q. conceived and designed the experiments; L.D., C.S. and W.Q. performed the experiments; L.D. and J.Z. analyzed the data; L.D. wrote the paper; C.S. and J.Z. contributed to data analysis revision; C.S. and J.Z. contributed to English language correction. All authors of article provided substantive comments. All authors have read and agreed to the published version of the manuscript.

Funding: This research has received no external funding.

Acknowledgments: This work is supported in part by the National Natural Science Foundation of China (Grant No. 61801212), in part by the Natural Science Foundation of Jiangsu Province (Grant No. BK20180423), in part by the Fundamental Research Funds for the Central Universities (Grant No. NT2019010), in part by the China Postdoctoral Science Foundation (Grant No. 2019M650113) and in part by Key Laboratory of Radar Imaging and Microwave Photonics (Nanjing Univ. Aeronaut. Astronaut.), Ministry of Education, Nanjing University of Aeronautics and Astronautics, Nanjing 210016, China.

Conflicts of Interest: The authors declare no conflict of interest.

References

1. Li, J.; Stoica, P. MIMO radar with colocated antennas. *IEEE Signal Process. Mag.* **2007**, *24*, 106–114. [CrossRef]
2. Haimovich, A.M.; Blum, R.S.; Cimini, L.J. MIMO radar with widely separated antennas. *IEEE Signal Process. Mag.* **2008**, *25*, 116–129. [CrossRef]
3. Wei, Y.; Meng, H.; Liu, Y.; Wang, X. Extended target recognition in cognitive radar networks. *Sensors* **2010**, *10*, 10181–10197. [CrossRef] [PubMed]
4. Niu, R.; Blum, R.S.; Varshney, P.K.; Drozd, A.L. Target Localization and Tracking in Noncoherent Multiple-Input Multiple-Output radar systems. *IEEE Trans. Aerosp. Electron. Syst.* **2012**, *48*, 1466–1489. [CrossRef]
5. Xie, M.; Yi, W.; Kong, L.; Kirubarajan, T. Receive-Beam Resource Allocation for Multiple Target Tracking with Distributed MIMO radars. *IEEE Trans. Aerosp. Electron.* **2018**, *54*, 2421–2436. [CrossRef]

6. Yan, J.; Pu, W.; Zhou, S.; Liu, H.; Bao, Z. Collaborative detection and power allocation framework for target tracking in multiple radar system. *Inf. Fusion* **2020**, *55*, 173–183. [CrossRef]
7. Godrich, H.; Chiriac, V.M.; Haimovich, A.M.; Blum, R.S. Target tracking in MIMO radar systems: Techniques and performance analysis. In Proceedings of the 2010 IEEE Radar Conference, Washington, DC, USA, 10–14 May 2010; pp. 1111–1116.
8. Alirezaei, G.; Taghizadeh, O.; Mather, R. Optimum power allocation in sensor networks for active applications. *IEEE Trans. Wirel. Commun.* **2015**, *14*, 2854–2867. [CrossRef]
9. Godrich, H.; Petropulu, A.P.; Poor, H.V. Power Allocation Strategies for Target Localization in Distributed Multiple-Radar Architectures. *IEEE Signal Process. Mag.* **2011**, *59*, 3226–3240. [CrossRef]
10. Xie, M.; Yi, W.; Kong, L. Power allocation strategy for target localization in distributed MIMO radar systems without previous position estimation. In Proceedings of the 2016 19th International Conference on Information Fusion, Heidelberg, Germany, 5–8 July 2016; pp. 1661–1665.
11. Yan, J.; Liu, H.; Jiu, B.; Bao, Z. Power Allocation Algorithm for Target tracking in Unmodulated Continuous Wave Radar Network. *IEEE Sens. J.* **2015**, *15*, 1098–1108. [CrossRef]
12. Zwaga, J.H.; Boers, Y.; Driessen, H. On tracking performance constrained MFR parameter control. In Proceedings of the Sixth International Conference of Information Fusion, Cairns, QLD, Australia, 8–11 July 2003; pp. 712–718.
13. Narykov, A.S.; Krasnov, O.A.; Yarovoy, A. Algorithm for resource management of multiple phased array radars for target tracking. In Proceedings of the 16th International Conference on Information Fusion, Istanbul, Turkey, 9–12 July 2013; pp. 1258–1264.
14. Wang, X.; Yi, W.; Xie, M.; Zhai, B.; Kong, L. Time management for target tracking based on the predicted Bayesian Cramer-Rao lower bound in phase array system. In Proceedings of the 2017 20th International Conference on Information Fusion, Xi'an, China, 10–13 July 2017; pp. 1–5.
15. Yan, J.; Liu, H.; Pu, W.; Zhou, S.; Liu, Z.; Bao, Z. Joint beam selection and power allocation for multiple target tracking in netted collocated MIMO radar system. *IEEE Trans. Signal Process.* **2016**, *64*, 6417–6427. [CrossRef]
16. Xie, M.; Yi, W.; Kirubarajan, T.; Kong, L. Joint node selection and power allocation strategy for multitarget tracking in decentralized radar networks. *IEEE Trans. Signal Process.* **2018**, *66*, 729–743. [CrossRef]
17. Lu, Y.; He, Z.; Deng, M.; Liu, S. A Cooperative Node and Waveform Allocation Scheme in Distributed Radar Network for Multiple Targets Tracking. In Proceedings of the 2019 IEEE Radar Conference, Boston, MA, USA, 22–26 April 2019; pp. 1–6.
18. Shi, C.; Wang, F.; Sellathurai, M.; Zhou, J.; Salous, S. Power Minimization-Based Robust OFDM Radar Waveform Design for Radar and Communication Systems in Coexistence. *IEEE Trans. Signal Process.* **2018**, *66*, 1316–1330. [CrossRef]
19. Shi, C.; Wang, F.; Sellathurai, M.; Zhou, J.; Salous, S. Low probability of Intercept-Based Optimal Power Allocation Scheme for an Integrated Multistatic Radar and Communication System. *IEEE Syst. J.* **2019**, 1–12. [CrossRef]
20. Shi, C.; Wang, F.; Salous, S.; Zhou, J. Joint Subcarrier Assignment and Power Allocation Strategy for Integrated Radar and Communications System Based on Power Minimization. *IEEE Sens. J.* **2019**, *19*, 11167–11179. [CrossRef]
21. She, J.; Wang, F.; Zhou, J. A Novel Sensor Selection and Power Allocation Algorithm for Multiple-Target Tracking in an LPI Radar Network. *Sensors* **2016**, *16*, 2193. [CrossRef] [PubMed]
22. Shi, C.G.; Wang, F.; Salous, S.; Zhou, J.J. Joint transmitter selection and resource management strategy based on low probability of intercept optimization for distributed radar networks. *Radio Sci.* **2018**, *53*, 1108–1134. [CrossRef]
23. Zhang, Z.; Salous, S.; Li, H.; Tian, Y. Optimal coordination method of opportunistic array radars for multi-target-tracking-based radio frequency stealth in clutter. *Radio Sci.* **2015**, *50*, 1187–1196. [CrossRef]
24. Garcia, N.; Haimovich, A.M.; Coulon, M.; Lops, M. Resource Allocation in MIMO Radar with Multiple Targets for Non-Coherent Localization. *IEEE Trans. Signal Process.* **2014**, *62*, 2656–2666. [CrossRef]
25. Yan, J.; Pu, W.; Liu, H.; Bao, Z. Joint power and bandwidth allocation for centralized target tracking in multiple radar system. In Proceedings of the 2016 CIE International Conference on Radar, Guangzhou, China, 10–13 October 2016; pp. 1–5. [CrossRef]

26. Sun, H.; Xu, H.; Wang, Y. Asymptotic analysis of sample average approximation for stochastic optimization problems with joint chance constraints via conditional value at risk and difference of convex functions. *J. Optim. Theory Appl.* **2012**, *8*, 1–28. [CrossRef]
27. Hernandez, M.L.; Farina, A.; Ristic, B. PCRLB for tracking in cluttered environments: Measurement sequence conditioning approach. *IEEE TAES* **2006**, *42*, 680–704. [CrossRef]
28. Hernandez, M.L.; Farina, A.; Ristic, B. References. In *The Posterior Cramer-Rao Bounds for Target Tracking*; Klemm, R., Lombardo, P., Eds.; IET: London, UK, 2017; Volume 2, Part II; pp. 263–302.
29. Shi, C.; Zhou, J.; Wang, F. Low probability of intercept optimization for radar network based on mutual information. In *IEEE China Summit & International Conference on Signal & Information Processing*; IEEE: Beijing, China, 2014; pp. 683–687.

Article

Batch Processing through Particle Swarm Optimization for Target Motion Analysis with Bottom Bounce Underwater Acoustic Signals †

Raegeun Oh [1], Taek Lyul Song [2,*] and Jee Woong Choi [1,*]

1 Department of Marine Science & Convergence Engineering, Hanyang University ERICA, Ansan 15588, Korea; rgoh@hanyang.ac.kr
2 Department of Electronic Systems Engineering, Hanyang University ERICA, Ansan 15588, Korea
* Correspondence: tsong@hanyang.ac.kr (T.L.S.); choijw@hanyang.ac.kr (J.W.C.); Tel.: +82-31-400-4156 (T.L.S.); +82-31-400-5531 (J.W.C.)

† This paper is an extended version of our paper published in Oh, R.; Song, T.L; Choi, J.W. Bearings-Only Target Motion Analysis using Ray Tracing. In Proceedings of ICCAIS 2019 Conference, Chengdu, China, 23–26 October 2019.

Received: 29 January 2020; Accepted: 21 February 2020; Published: 24 February 2020

Abstract: A target angular information in 3-dimensional space consists of an elevation angle and azimuth angle. Acoustic signals propagating along multiple paths in underwater environments usually have different elevation angles. Target motion analysis (TMA) uses the underwater acoustic signals received by a passive horizontal line array to track an underwater target. The target angle measured by the horizontal line array is, in fact, a conical angle that indicates the direction of the signal arriving at the line array sonar system. Accordingly, bottom bounce paths produce inaccurate target locations if they are interpreted as azimuth angles in the horizontal plane, as is commonly assumed in existing TMA technologies. Therefore, it is necessary to consider the effect of the conical angle on bearings-only TMA (BO-TMA). In this paper, a target conical angle causing angular ambiguity will be simulated using a ray tracing method in an underwater environment. A BO-TMA method using particle swarm optimization (PSO) is proposed for batch processing to solve the angular ambiguity problem.

Keywords: target motion analysis; bottom bounce path; ray tracing; particle swarm optimization

1. Introduction

Acoustic signals are used to indirectly obtain information about objects located underwater. Most passive sonar systems use multiple hydrophones in an array for enhanced performance. A horizontal line array (HLA), used for detecting the azimuth angle of an underwater target, receives acoustic signals with a high signal to noise ratio from designated directions using a beamforming technique. If the target signal intensity is high enough in a designated direction, the target direction is detected. The estimated target direction is represented as a conical angle that indicates the direction of the incoming signal measured by the HLA. Unfortunately, it is impossible to distinguish between up and down or right and left from the conical angle. This is called the cone of ambiguity [1].

Sequential processing and batch processing algorithms are used to estimate the target's state, including position and velocity, through bearings-only target motion analysis (BO-TMA). There exist several conventional sequential processing algorithms, including the extended Kalman filter [2], the pseudo-measurement filter [3], and the modified gain extended Kalman filter [4]. In addition to these filters, particle filter approaches [5] and random finite set approaches [6–8] have been recently introduced. If sufficient computational performance is achieved, sequential processing is suitable for implementation in real time systems. However, good sequential estimation results require small errors in the initial state estimates, and a batch processing algorithm is used for this purpose. Batch processing delivers stable initial values, even though it is not designed to operate in real time because it requires a batch of stored measurements. Robust target localization performance is expected if both types of algorithms are employed properly [9].

In most of the previous studies on sonar systems [2,3,6–9], it is assumed that the received signal arrives at the HLA through the horizontal plane when the distance between the observer and target is large, or when the observer and the target are located at equal depths. Therefore, the cone of ambiguity of the HLA is simplified to left/right ambiguity, which can be easily addressed through a ship maneuver. However, eigenray tracing results show that the received signal can arrive at the HLA with a high elevation angle, especially along a bottom bounce path [10]. The studies in [11,12] consider the elevation angles in BO-TMA for different sensor depths between the observer and the target. They treat only direct paths without considering the reflection of the ray from the waveguide boundaries (i.e., sea surface and bottom) or the refraction of the ray from the vertical sound speed profile. However, bottom bounce paths, which are generated from the reflection of acoustic waves at the ocean bottom, can produce inaccurate target bearings [13] that affect BO-TMA results.

The ray tracing method [14] is used to calculate the elevation angle due to the refraction and reflection of sound waves in underwater waveguides. This method describes the path of each ray as sound waves propagating through the underwater waveguide. In particular, it is possible to calculate the eigenray [15], which represents the path of a ray that propagates from the source to the receiver. The elevation angle of the target signal can be simulated through eigenray tracing, and the conical angle can be calculated using the azimuth angle and the elevation angle.

In this paper, a study is based on the published conference paper [16] and it is conducted to confirm the observability of TMA using the conical angle including the elevation angle of the path reflected from the bottom interface for a given scenario. A discrete target dynamic equation is established with the target state vector, and the conical angle measurement is obtained from the relative geometry of the observer and the target using the ray tracing method in Section 2. Section 3 presents a method of converting the conical angle into a bearing line in Cartesian coordinates using knowledge of the ocean environment (i.e., bottom bathymetry and a sound speed profile). Additionally, a BO-TMA using the particle swarm optimization (PSO) algorithm is proposed. In Section 4, simulation results for the BO-TMA are analyzed using ray tracing. Finally, a summary and conclusion are given in Section 5.

2. Problem Formulation

2.1. Dynamic Model

The target state vector at the discrete time instance k, $1 \leq k \leq K$, is defined as:

$$X_s(k) = \left[p_{xs}(k),\ p_{ys}(k),\ v_{xs}(k),\ v_{ys}(k)\right], \tag{1}$$

$$U_s(k) = \left[u_{xs}(k),\ u_{ys}(k)\right], \tag{2}$$

where $p_{xs}(k)$ and $p_{ys}(k)$ are the target locations in Cartesian coordinates. Here, the x-axis indicates East and the y-axis indicates North. Additionally, $v_{xs}(k)$ and $v_{ys}(k)$ are the target velocities for each direction, and $u_{xs}(k)$ and $u_{ys}(k)$ are the target accelerations. The observer state vector is similarly defined as:

$$X_o(k) = \left[p_{xo}(k),\ p_{yo}(k),\ v_{xo}(k),\ v_{yo}(k)\right], \tag{3}$$

$$U_o(k) = \left[u_{xo}(k),\ u_{yo}(k)\right], \tag{4}$$

where the subscript o indicates the observer. Then, the discrete-time system state equation can be described by:

$$X_i(k+1) = FX_i^T(k) + GU_i^T(k), \tag{5}$$

where X_i and U_i are the state vectors of the target (when $i = s$) and the observer (when $i = o$), and control input, respectively. The superscript T denotes a transpose. The state transition matrix F and input coefficient matrix G are defined, respectively, as:

$$F = \begin{bmatrix} I_2 & \Delta t I_2 \\ 0_2 & I_2 \end{bmatrix}, \quad G = \begin{bmatrix} \Delta t^2/2 I_2 \\ \Delta t I_2 \end{bmatrix}, \tag{6}$$

where I_2 is the 2-dimensional identity matrix, 0_2 is the 2×2 zero matrix, and Δt is the time interval. For system observability, we assume that the sensor outmaneuvers the target while the target is moving with a constant velocity [17].

The horizontal plane trajectories of the target and the observer located at equal depths of 200 m are shown in Figure 1. The total simulation time is 580 s with a sampling period of 20 s so that the total number of scans is 30. The initial state vector of the target, $X_s(1)$, is $[0\ \text{m},\ 2500\ \text{m},\ 0\ \text{m/s},\ -3\ \text{m/s}]$ with zero acceleration over the simulation time. The initial state vector of the observer, $X_o(1)$, is $[2000\ \text{m},\ -7000\ \text{m},\ 2.6\ \text{m/s},\ 1.5\ \text{m/s}]$. To ensure system observability, the course of the observer is changed once from 60 to 0° via lateral acceleration starting at 200 s. The bearing change rate is 0.6° per second. The distance between the observer and the target is decreased from a maximum distance of 9.7 km to a minimum distance of 6.9 km.

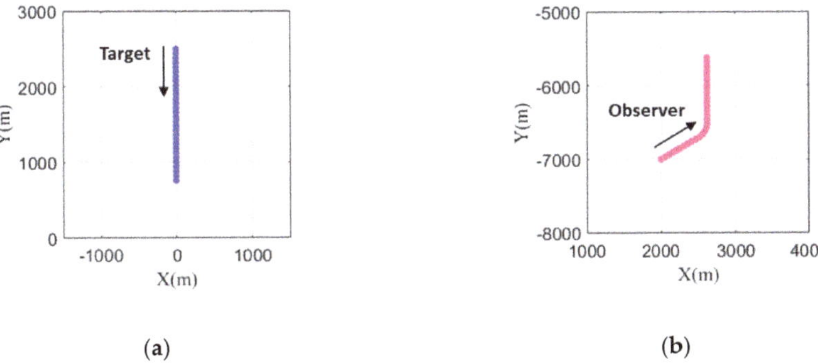

Figure 1. Trajectories of (**a**) the target and (**b**) the observer in the horizontal plane.

2.2. Measurement Model

Conventional TMA assumes that the target information obtained from passive line array sonar consists of only the azimuth angle in the horizontal plane, neglecting bottom bounce signals to avoid conical angle ambiguity. In this case, the azimuth angle measured from the north axis, $\varphi_n(k)$, is expressed as:

$$\varphi_n(k) = atan2\big(p_{xs}(k) - p_{xo}(k),\ p_{ys}(k) - p_{yo}(k)\big), \tag{7}$$

where $atan2(x,\ y)$ denotes a four-quadrant arctangent function that describes the angle between the position of the target and the north axis (positive y-axis). The azimuth angle from the north axis, $\varphi_n(k)$, is converted to the azimuth angle from the direction of the HLA, $\varphi_l(k)$, by subtracting the heading angle of the HLA, $c_o(k)$, at each scan time k:

$$\varphi_l(k) = \varphi_n(k) - c_o(k). \tag{8}$$

In this paper, BO-TMA along with a ray tracing method is used to achieve accurate estimation of target localization in environments with conical angle ambiguity. The conical angle, $\theta(k)$, is expressed as:

$$\theta(k) = cos^{-1}(cos(\varphi_l(k)) \times cos(\mu(k))) + v(k), \tag{9}$$

where $\mu(k)$ is the elevation angle in the vertical plane, and $v(k)$ is the measurement noise modeled as zero mean Gaussian noise with standard deviation σ_m. The sign of $\theta(k)$ is unknown from Equation (9), and the conical angle indicates the magnitude of the angle measured from the heading direction of the line array. Thus, the inability to know the exact direction of the arriving signal is known as left/right ambiguity. Various angles used in this paper are shown in Figure 2. φ_l and φ_n are the azimuth angles from due north and the direction of the HLA, respectively. c_o, θ, and μ are heading angle of the HLA, conical angle, and elevation angle in the vertical plane, respectively.

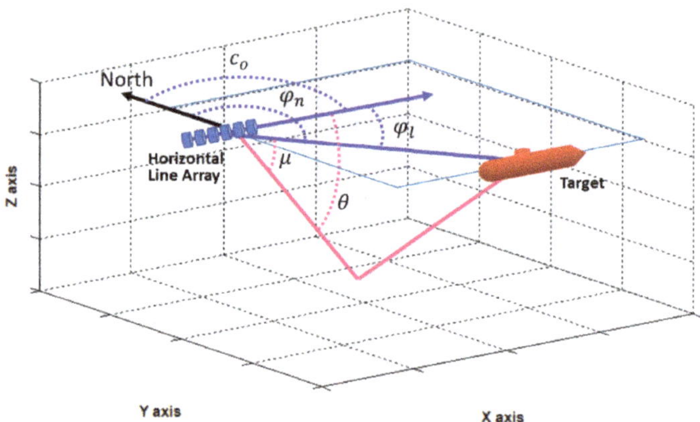

Figure 2. Geometry between observer and target. φ_l and φ_n are the azimuth angles from due north and the direction of the HLA, respectively. c_o, θ, and μ are heading angle of the HLA, conical angle, and elevation angle in the vertical plane, respectively.

A ray tracing method is used to estimate the elevation angle of the target signal in Equation (9). In the ocean, propagation paths of acoustic rays are strongly affected by sound speed profile and bottom bathymetry. These environmental data can be obtained through measurements, from a database, or from an ocean prediction model. In this study, a scenario is constructed that assumes a simple environment. The bathymetry is assumed to be flat with a depth of 2000 m. The sound speed profile $C(z)$ in water is assumed to follow Munk's sound speed profile and is given by [18]:

$$C(z) = C_0[1.0 + \epsilon\{e^{-\eta} - (1-\eta)\}], \tag{10}$$

where z is depth, and C_0 is a reference sound speed equal to 1500 m/s as the sound speed at the depth of channel axis z_C ($z_C = 400$ m), $\eta = 2(z - z_C)/z_C$ is a dimensionless depth relative to the channel axis, and the perturbation coefficient ϵ is equal to 7.4×10^{-3}.

The ray paths predicted by the ray tracing method using Munk's sound speed profile are shown in Figure 3. Although ray tracing was conducted based on observer position, the ray tracing results obtained for opposite directions are the same due to the reciprocity of ray diagrams [14]. In addition, the ray tracing results for all azimuth angles are the same because it is assumed that acoustical ocean parameters are independent of azimuth angle. It is shown in this scenario that only bottom reflected paths exist between the target and the observer, and a direct path from the target does not exist. The elevation angle of the bottom bounce path was calculated by ray tracing to be between 23 and 29° at a target distance of 9.7—6.9 km.

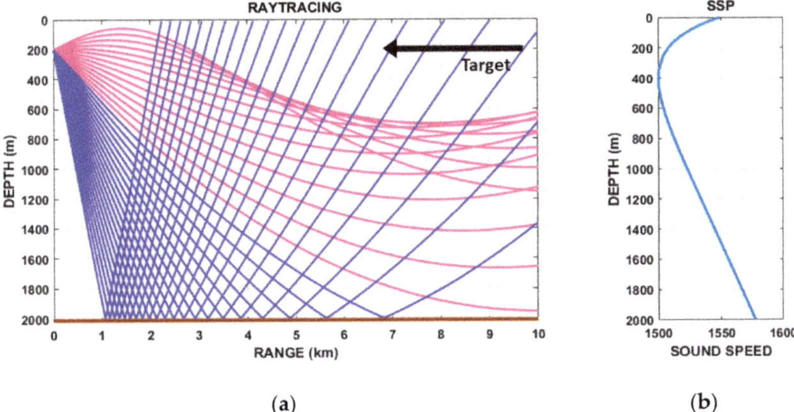

(a) (b)

Figure 3. (**a**) Ray paths predicted by ray tracing method based on (**b**) Munk's sound speed profile. Direct and bottom bounce paths are plotted with magenta and blue lines, respectively.

Figure 4 shows the simulation results of the bearing measurements from 30 scans over same time period, which is known as BTR (Bearing-Time Record). The red dashed line that represents the azimuth angle from the north axis $\varphi_n(k)$ was plotted as additional information for assessing the bearing error compared to the conical angle of the bottom bounce path. The bearing error is defined as the difference between $\varphi_l(k)$ and $+\theta(k)$ or its mirror angle $-\theta(k)$ due to conical angle ambiguity. Figure 4 contains the time histories of $c_o(k)$ (the observer heading angle), $\varphi_n(k)$ (the true target azimuth angle), $c_o(k) + |\theta(k)|$, and $c_o(k) - |\theta(k)|$ (two possible bearing angles for TMA that stem from the bottom bounce path). The right/left ambiguity in the horizontal plane is shown in Figure 4 and can be resolved by comparing the histories of $c_o(k) + |\theta(k)|$ and $c_o(k) - |\theta(k)|$. The history of $c_o(k) - |\theta(k)|$ has smaller variations than that of $c_o(k) + |\theta(k)|$. Note that these two angle histories correspond to the history of the true azimuth angle $\varphi_n(k)$. The history of $\varphi_n(k)$ in Figure 4 shows small variations for the entire period that includes the times before and after the observer maneuver, which implies that $c_o(k) - |\theta(k)|$ rather than $c_o(k) + |\theta(k)|$ should be applied as the bearing history for this scenario in TMA. From the selection process, the correct sign of $\theta(k)$ in Equation (9) for this scenario is negative. However, even after choosing the bearing history with the correct sign of $\theta(k)$, $c_o(k) - |\theta(k)|$ still contains bearing error when compared to the true azimuth angle history $\varphi_n(k)$. Figure 4 shows that this error is ~1° before the observer maneuver and ~13° after the maneuver. This discrepancy is due to $\mu(k)$, the elevation angle of the bottom bounce path. Conventional TMA methods for target localization cannot avoid localization errors resulting from bearing errors. Therefore, a new TMA method that accounts for the bottom bounce path is needed.

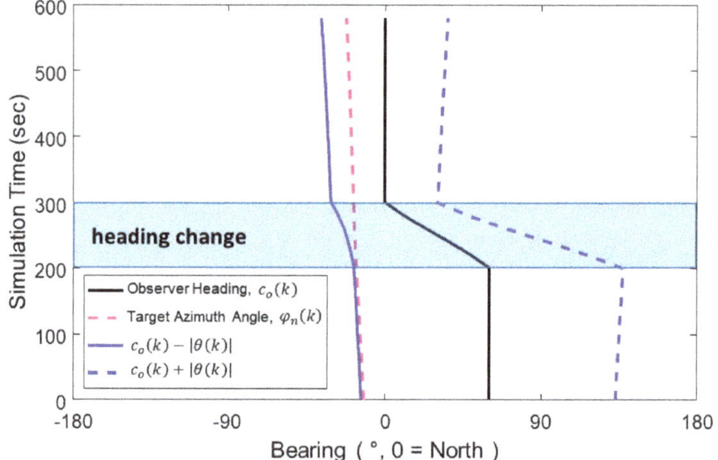

Figure 4. Bearing-time records (BTRs) of the scenario.

3. Target Motion Analysis with Bottom Bounce Path

3.1. Bearing Lines of Bottom Bounce Path

Bearing error is due to the elevation angle $\mu(k)$ of the bottom bounce path, which is unknown even after selection of the correct sign of $\theta(k)$. In this study, the bearing line in Cartesian coordinates is introduced. Define the i-th expected azimuth angle $\hat{\varphi}_l(k, i)$ for $1 \leq i \leq I$, which represents a possible target azimuth angle relative to the heading direction of the HLA. According to Equation (9), $\hat{\varphi}_l(k, i)$ must lie within the range between zero and the conical angle $\theta(k)$, and then the elevation angle $\hat{\mu}(k, i)$ can be estimated as:

$$\hat{\mu}(k, i) = cos^{-1}\left(\frac{cos(\theta(k))}{cos(\hat{\varphi}_l(k, i))}\right). \tag{11}$$

The sign of $\hat{\varphi}_l(k, i)$ is equal to the sign of $\theta(k)$. For each $\hat{\varphi}_l(k, i)$, ray tracing for the ray launched at an angle of $\hat{\mu}(k, i)$ from the observer position is conducted to find the range $\hat{r}(k, i)$ of target location if it exists in the direction of $\hat{\varphi}_l(k, i)$ (Figure 5a). Since the target depth was assumed to be 200 m, the distance at which the ray arrives at a water depth of 200 m after bottom reflection becomes the target range in the $\hat{\varphi}_l(k, i)$ direction. This process is repeated $i = I$ times (Figure 5b). In this study, the expected azimuth angle was varied every 0.5°. Accordingly, $|\theta(k)|$ divided by 0.5° was used to determine the value of I for each scan k.

For the k-th scan, I possible target positions in the horizontal plane corresponding to every $\hat{\varphi}_l(k, i)$ are connected in a line, which is defined as a *bearing line* in this paper. The possible target position vector in Cartesian coordinates with $\hat{\varphi}_l(k, i)$ and $\hat{r}(k, i)$ is denoted as:

$$\hat{L}(k, i) = \left[\hat{p}_{xl}(k, i), \hat{p}_{yl}(k, i)\right]. \tag{12}$$

Figure 6 is drawn in the horizontal plane and it shows the bearing lines corresponding to $k = 1$ and $k = K$. The lines (denoted by line of conical angle) indicating the measured conical angle $\theta(k)$ in the horizontal plane for $k = 1$ and $k = K$. If the elevation angle is not considered, as in previous studies, the bearing line is displayed as a straight line. However, the bearing line $\hat{L}(k, i)$ is displayed as a curved line when the elevation angle is considered. Conventional batch estimation methods for TMA utilize the conical angles to determine the initial target states, while the proposed TMA method utilizes the bearing lines. The objective of the proposed TMA method is to find the optimal initial position and velocity of the target based on the bearing lines in Cartesian coordinates using the PSO algorithm to minimize the objective function.

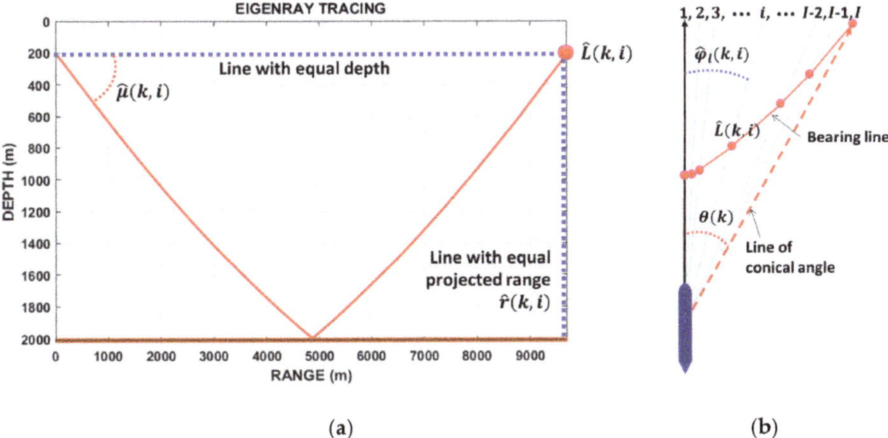

(a) (b)

Figure 5. (**a**) Eigenray tracing result conducted to determine the expected target range. The distance at which the ray arrives at an expected target depth after bottom reflection becomes the estimated target range in $\hat{\varphi}_I(k,\ i)$ direction. (**b**) Top-view illustration showing the line of conical angle and bearing line. For k-th scan, the line connecting I possible target positions estimated using the eigenray tracing is a bearing line (red line in figure).

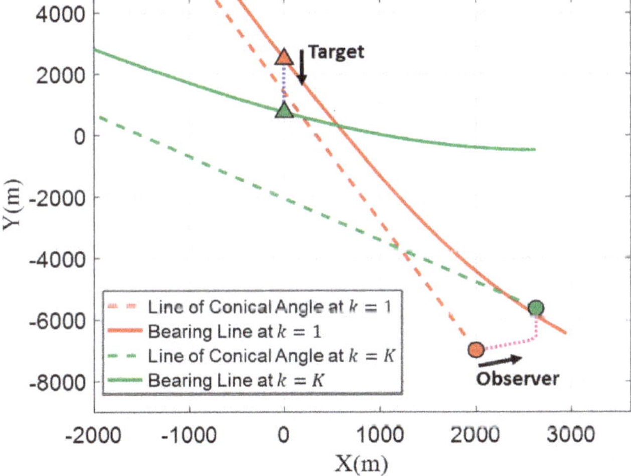

Figure 6. Bearing lines (solid lines) and lines of conical angles (dashed lines) at $k = 1$ and $k = K$.

3.2. Particle Swarm Optimization

The PSO algorithm is a stochastic optimization algorithm used to find the optimal positions of particles and is based on the social behavior of animals moving in flocks [19,20]. In BO-TMA studies, each particle representing an estimated initial target state vector consists of four elements: the positions and velocities in the x and y directions. First, at $k = 1$, the particles are uniform, randomly spread along the bearing line within the target-observer distance from 1 to 30 km. A specific velocity vector, which is randomly selected in the range of $0|\hat{v}|$ 10 m/s, where $|\hat{v}| = \sqrt{|\hat{v}_x|^2 + |\hat{v}_y|^2}$, is assigned to each particle. Then, the position of each particle at the next scan time ($k = 2$) is calculated using the dynamic model shown in Section 2.1 from the position at $k = 1$. In this manner, a total of K positions are determined for each particle, which forms a particle trajectory. After that, the shortest distance between each particle position and the bearing line corresponding to the same scan time number is calculated. This distance is then normalized by the distance between the observer and particle position at each scan time to avoid excessive convergence to local optima, which happens because distance error increases as the distance between the observer and the particle increases. Finally, the normalized distance errors for all K particle positions are summed to obtain an objective function J^m for the m-th particle, which is expressed as:

$$J^m = \sum_{k=1}^{K} \frac{\underset{i}{min} \sqrt{(\hat{p}_{xm}(k) - \hat{p}_{xl}(k,\ i))^2 + \left(\hat{p}_{ym}(k) - \hat{p}_{yl}(k,\ i)\right)^2}}{\sqrt{(\hat{p}_{xm}(k) - p_{xo}(k))^2 + \left(\hat{p}_{ym}(k) - p_{yo}(k)\right)^2}}, \tag{13}$$

where $\hat{p}_{xm}(k)$ and $\hat{p}_{ym}(k)$, respectively, are the positions in the x and y directions of the m-th particle at scan time k; and $p_{xo}(k)$ and $p_{yo}(k)$ are the observer positions in the x and y directions at scan time k. The total particle number used here was 200 (Table 1). Since each particle is considered a candidate for the target, the next step is to find the initial state vector of the particle that produces the minimum value of J^m. In this study, the PSO algorithm was used as an optimization technique to find the optimal target trajectory. In each generation, the best values for the state vectors consisting of the positions and velocities of the particles are evaluated by comparison with state vectors selected during previous generations, and the state change rates of the particles are adjusted based on the experiences of the particles and their companions. The state vectors in the next generation are updated with the sum of the present state vectors and the adjusted state change rates of the particles [20]. The process is expressed as [19]:

$$v_p(n+1,\ m,\ d) = c_1 v_p(n,m,d) + v_l(n,m,d) + v_s(n,m,d), \tag{14}$$

$$v_l(n,m,d) = c_2 r_1 \left\{x_{pl}(m,d) - x_p(n,m,d)\right\}, \tag{15}$$

$$v_s(n,m,d) = c_3 r_2 \left\{x_{ps}(d) - x_p(n,m,d)\right\}, \tag{16}$$

$$\text{and } x_p(n+1,m,d) = x_p(n,m,d) + v_p(n+1,\ m,\ d), \tag{17}$$

where $x_p(n,m,\ d)$ represents the state vector of the m-th particle for the n-th generation with dimension d. Dimension d is one of 1, 2, 3, and 4 corresponding to the positions and velocities of the particles at $k = 1$, that is, $\hat{p}_{xm}(1)$, $\hat{p}_{ym}(1)$, $\hat{v}_{xm}(1)$, and $\hat{v}_{ym}(1)$, respectively. In addition, $v_p(n,m,\ d)$ represents the state change rates of the particles for $x_p(n,m,\ d)$. Finally, $v_l(n,m,\ d)$ and $v_s(n,m,\ d)$ are the local state change rate and the social state change rate for the m-th particle, respectively. The local state vector $x_{pl}(m,\ d)$ is the best state vector of the m-th particle obtained from the first generation to the n-th generation, and the social state vector $x_{ps}(d)$ is the best state vector of the particle with the smallest J^m of all particles up to the n-th generation. In the above equations, c_1, c_2, and c_3 are acceleration weight constants determined empirically through many trial runs to be 0.73, 0.1, and 0.2, respectively. Random numbers r_1 and r_2 are selected in the range between 0 and 1. The process is iterated until the state vector of each particle converges to the best state vector that satisfies the minimum position errors. In our case, the generation is terminated when the standard deviations of the positions, σ_p, and velocities, σ_v, of 200 particles converge to values less than 100 m and 0.2 m/s, respectively. Finally, the trajectory of the particle with the best state vector is selected as the target trajectory.

Table 1. Particle swarm optimization parameters used to find the optimal initial position and velocity of target.

Parameter	Symbol	Value
Number of particles	m	200
Number of dimensions	d	4
Number of generations	n	$\sigma_p < 100$ m and $\sigma_v < 0.2$ m/s
Acceleration weight constants c_1	c_1	0.73
Acceleration weight constants c_2	c_2	0.1
Acceleration weight constants c_3	c_3	0.2
Random number r_1	r_1	0—1
Random number r_1	r_2	0—1

4. Simulation Result

For the observability test, it was assumed that the water depth was 2000 m and the bottom topography was flat. The conical angle was calculated using the azimuth and elevation angle of the acoustic ray path between the target and the observer. Munk's sound speed profile was used for ray tracing to calculate the elevation angle. To test the applicability of batch processing using the PSO algorithm proposed in this paper, it was assumed that Gaussian noise with zero mean and standard deviation σ_m was included in the conical angle measurements. Three values of σ_m (0.2, 0.4, and 0.6°) were considered for comparison purposes. For this scenario, the conical angle was estimated to change at a rate of approximately 0.5°/scan except during the period in which the observer heading changed. Figure 7 shows the histories of conical angles with measurement errors corresponding to three different standard deviations.

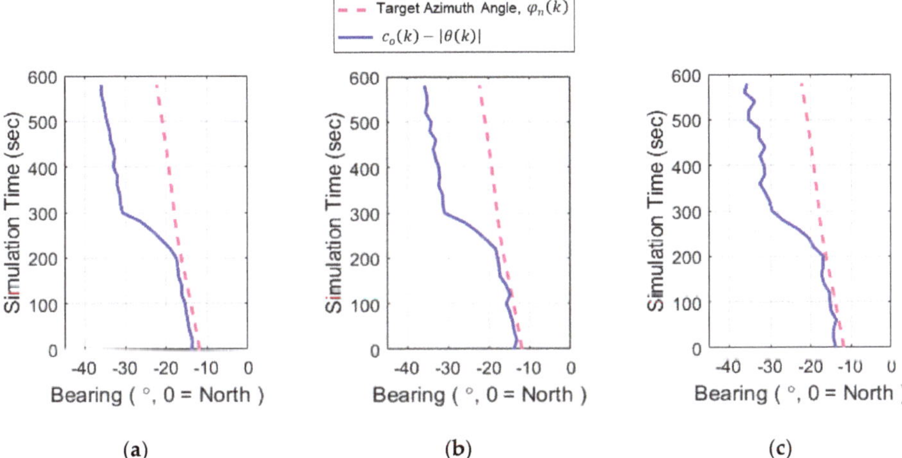

Figure 7. The BTRs for conical angle measurements including Gaussian measurement error with zero mean and standard deviation of (**a**) 0.2, (**b**) 0.4, and (**c**) 0.6°.

One thousand random runs were generated for each of the three standard deviations of the conical angle measurement errors, and TMA was carried out for each run. The results are shown in Figure 8, in which the left column shows the scatter plots for the estimates of initial target position for the 1000 runs, and the right column shows the scatter plots of target velocity. For the different standard deviations, the mean values of the estimated initial state vectors and their variances are listed in Table 2. The results show that, as the standard deviation of the measurement error increases, the distribution of the initial state vector obtained from the proposed BO-TMA becomes wider. In particular, as the measurement error increases, the estimated positions of the target tend to spread wider along the bearing line at $k = 1$, which is reasonable because the particles were spread along the bearing line at $k = 1$. The mean value of the initial state vector estimated for the standard deviation of 0.2° (marked by a yellow triangle in the figure) has the best agreement with the true initial target state vector (marked by red circle), and as the standard deviation increases, the difference increases slightly. However, the mean values for the three cases are still in good agreement with the true values.

Table 2. The means and variances of the estimated initial state vector $[\hat{p}_{xm}(1), \hat{p}_{ym}(1), \hat{v}_{xm}(1), \hat{v}_{ym}(1)]$ for three values of standard deviation of measurement error.

Standard Deviation of Measurement Noise, σ_m	Mean of Initial Target State Vector, $\hat{X}_s$, [m, m, m/s, m/s]	Variance of Initial Target State Vector, $\hat{\sigma}_s^2$, [m^2, m^2, m/s^2, m/s^2]
0.2°	[18, 2524, −0.2, −2.8]	$[71^2, 178^2, 0.5^2, 0.5^2]$
0.4°	[5, 2607, −0.4, −2.9]	$[105^2, 340^2, 0.6^2, 0.7^2]$
0.6°	[−6, 2693, −0.5, −2.8]	$[138^2, 473^2, 0.7^2, 0.9^2]$

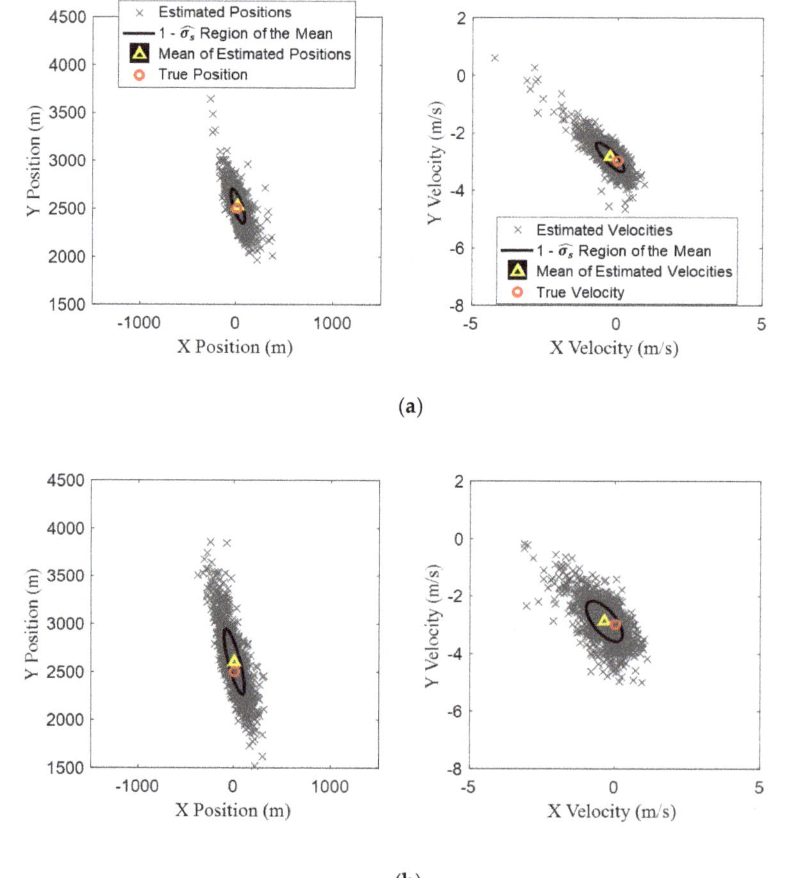

(a)

(b)

Figure 8. *Cont.*

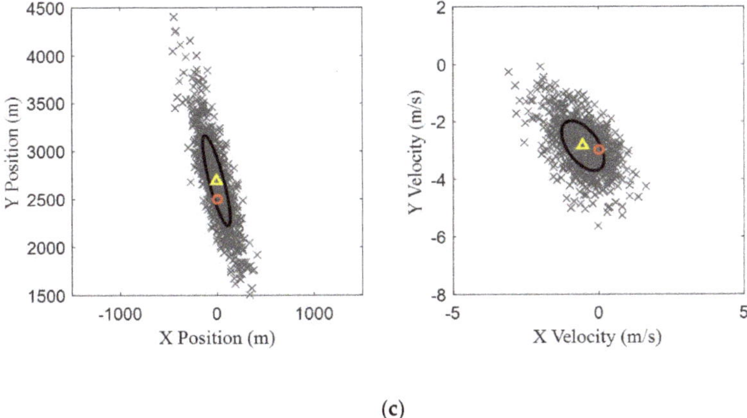

(c)

Figure 8. The distribution of initial states estimated using TMA for 1000 random runs for standard deviations of zero mean Gaussian measurement errors of (**a**) 0.2, (**b**) 0.4, and (**c**) 0.6°. The true initial state vector of the target is [0 m, 2500 m, 0 m/s, −3 m/s]. The left column shows the initial target position estimates, and the right shows target velocity estimates. The true initial state vector of the target and the mean of estimated state vectors are indicated by red circles and yellow triangles, respectively. The regions within one standard deviation of the mean are indicated by black ellipses.

To investigate the accuracy of the TMA results with increasing the number of scans k, the processes were repeated with the scan numbers of 15, 30, and 60 which correspond to the sampling periods of 40, 20, and 10 s, respectively. The standard deviation of the conical angle measurements were assumed to be 0.4°. The estimation results of the initial target state vector with the three scan numbers are shown in Figure 9, and the resulting mean values and variances are listed in Table 3. Figure 9 and Table 3 indicate that more frequent collection of conical angle measurements achieves more accurate TMA results with increased expense of computational resources.

Table 3. The means and variances of the estimated initial state vector $[\hat{p}_{xm}(1), \hat{p}_{ym}(1), \hat{v}_{xm}(1), \hat{v}_{ym}(1)]$ for three different measurement numbers.

Number of Measurements, k	**Mean of Initial Target State Vector, $\hat{X}_s$, [m, m, m/s, m/s]**	**Variance of Initial Target State Vector, $\hat{\sigma}_s{}^2$, [m^2, m^2, m/s^2, m/s^2]**
15	[−18, 2655, −0.2, −3.0]	$[138^2, 509^2, 0.6^2, 0.9^2]$
30	[5, 2607, −0.4, −2.9]	$[105^2, 340^2, 0.6^2, 0.7^2]$
60	[11, 2590, −0.4, −2.9]	$[82^2, 236^2, 0.5^2, 0.6^2]$

As shown in Figure 7, the bearing errors due to elevation angle after the observer maneuver are larger than 10°. Conventional TMA methods produce large localization errors in environments dominated by acoustic rays being strongly reflected or refracted up and down. However, the proposed BO-TMA method using ray tracing shows good localization performance in such environments, which implies that the proposed TMA method is a more effective tool for increasing solution accuracy in real underwater applications, especially in waveguide environments where bottom bounce paths are dominant.

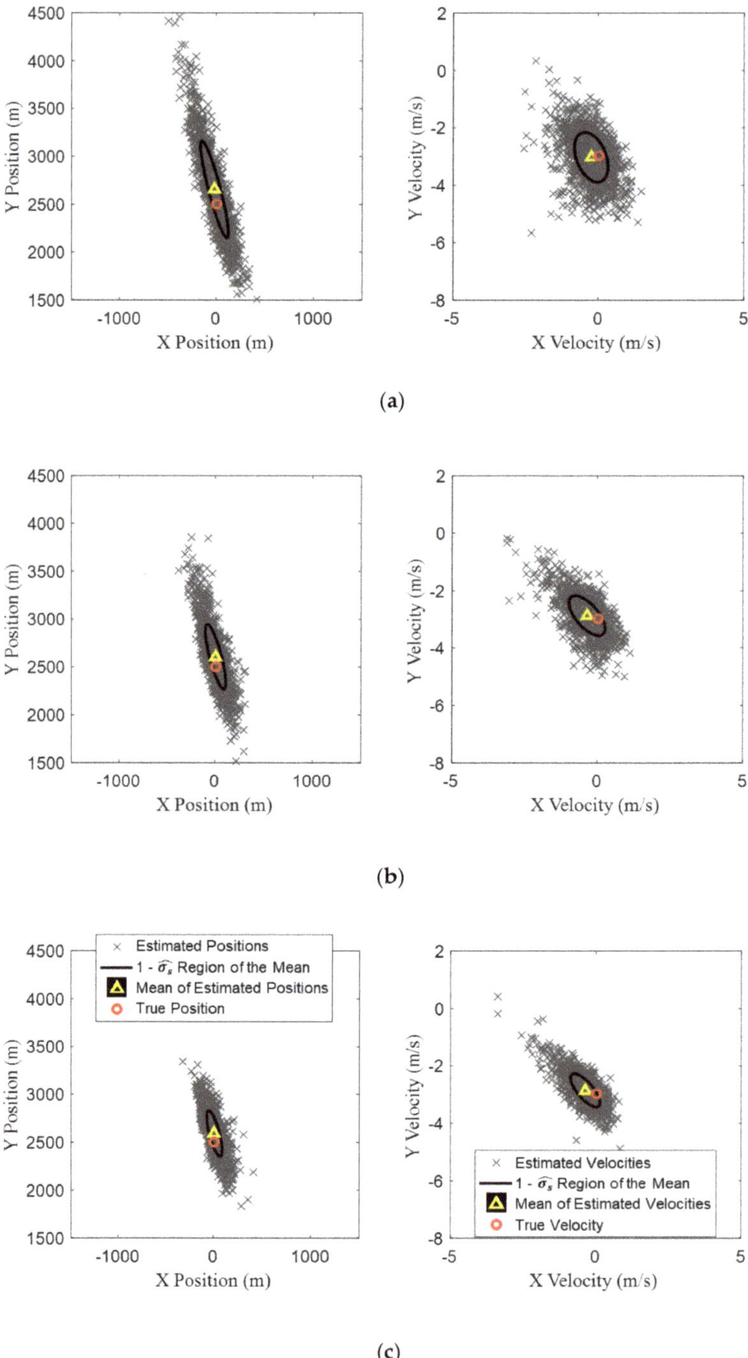

Figure 9. The distribution of initial states estimated using TMA for 1000 random runs with standard deviations of zero mean Gaussian measurement error of 0.4° with the measurement numbers of (**a**) 15, (**b**) 30, and (**c**) 60.

5. Summary and Conclusion

In this paper, a BO-TMA algorithm using a ray tracing method is proposed to accurately consider the conical angles generated by bottom bounce paths. The 3-dimensional conical angle information was converted to bearing lines in a 2-dimensional plane using a ray tracing method. Then, the PSO algorithm was carried out based on the constructed bearing lines to find optimal target state vectors.

The BO-TMA method using ray tracing and the PSO algorithm proposed in this paper is summarized below.

(1) Convert the conical angles of the bottom bounce path to a bearing line using the ray tracing technique. Set the generation number $n = 1$.

(2) Initialize particles with the bearing line at $k = 1$. Uniform, randomly spread particles on the bearing line and assign velocities randomly selected in the range $0 \leq |\hat{v}| \leq 10$ m/s.

(3) For each particle with a four-element state vector, calculate the objective function J^m using the particle trajectories and the bearing lines corresponding to $k = 1, \cdots, K$.

(4) Find the particle that produces the minimum value of J^m.

(5) Generate the next generation particle group by applying the PSO algorithm.

(6) Go to Step (3), and then iterate the process.

(7) Terminate the iteration when the state vectors of the particles reach the termination condition.

In this paper, a ray tracing technique was used to calculate the elevation angle. The conical angle of the target was then calculated based on the estimated elevation angle. Characteristics of the oceanic environment are known, allowing for accurate estimation of elevation angles. However, since the oceanic environment fluctuates temporally and spatially, errors can arise from uncertainty in environmental information. In addition, we assumed that target depth is the same as observer depth. Uncertainty in target depth may also result in target distance errors. Therefore, further research into various target-observer geometries and various ocean environments is required to generalize the results shown in this paper.

Author Contributions: Conceptualization, R.O. and J.W.C.; methodology, R.O. and T.L.S.; software, R.O.; validation, R.O., T.L.S. and J.W.C.; formal analysis, R.O. and T.L.S.; investigation, R.O.; resources, R.O.; data curation, R.O.; writing—original draft preparation, R.O.; writing—review and editing, T.L.S. and J.W.C.; visualization, R.O.; supervision, T.L.S. and J.W.C.; project administration, J.W.C.; funding acquisition, J.W.C. All authors have read and agreed to the published version of the manuscript.

Funding: This research was a part of the project titled "Development of Distributed Underwater Monitoring and Control Networks", funded by the Ministry of Oceans and Fisheries, Korea.

Conflicts of Interest: The authors declare no conflict of interest.

References

1. Abraham, D.A. *Underwater Acoustic Signal Processing: Modeling, Detection, and Estimation, Modern Acoustics and Signal Processing*; Springer Nature Switzerland AG: Cham, Switzerland, 2019.
2. Aidala, V.J.; Hammel, S.E. Utilization of Modified Polar Coordinates for Bearings-Only Tracking. *IEEE Trans. Autom. Control* **1983**, *AC-28*, 283–294. [CrossRef]
3. Aidala, V.J.; Nardone, S.C. Biased Estimation Properties of the Pseudolinear Tracking Filter. *IEEE Trans. Aerosp. Electron. Syst.* **1982**, *AES-18*, 432–441. [CrossRef]
4. Song, T.L.; Speyer, J.L. A stochastic analysis of a modified gain extended Kalman filter with applications to estimation with bearings only measurements. *IEEE Trans. Autom. Control* **1985**, *AC-30*, 940–949. [CrossRef]
5. Arulampalam, S.; Maskell, S.; Gordon, N.; Clapp, T. A tutorial on particle filters for online nonlinear/non-Gaussian Bayesian tracking. *IEEE Trans. Signal process.* **2002**, *50*, 174–188. [CrossRef]
6. Beard, M.; Vo, B.T.; Vo, B.N.; Arulampalam, S. A partially uniform target birth model for gaussian mixture PHD/CPHD Filtering. *IEEE Trans. Aerosp. Electron. Syst.* **2013**, *49*, 2835–2844. [CrossRef]
7. Zhang, Q.; Song, T.L. Improved bearings-only multi-target tracking with GM-PHD filtering. *Sensors* **2016**, *16*, 1469. [CrossRef] [PubMed]

8. Xie, Y.; Song, T.L. Bearings-only multi-target tracking using an improved labeled multi-Bernoulli filter. *Signal Process.* **2018**, *151*, 32–44. [CrossRef]
9. Kirubarajan, T.; Shalom, Y.B.; Lerro, D. Bearings-Only Tracking of Maneuvering Targets Using a Batch-Recursive Estimator. *IEEE Trans. Aerosp. Electron. Syst.* **2001**, *37*, 770–780. [CrossRef]
10. Oh, R.; Gu, B.S.; Kim, S.; Song, T.L.; Choi, J.W.; Son, S.U.; Kim, W.K.; Bae, H.S. Estimation of bearing error of line array sonar system caused by bottom bounced path. *J. Acoust. Soc. Korea* **2018**, *37*, 412–421.
11. Jauffret, C.; Pillon, D. Observability in Passive Target Motion Analysis. *IEEE Trans. Aerosp. Electron. Syst.* **1996**, *32*, 1290–1300. [CrossRef]
12. Hammel, S.E.; Aidala, V.J. Observability Requirements for Three-Dimensional Tracking via Angle Measurements. *IEEE Trans. Aerosp. Electron. Syst.* **1985**, *AES-21*, 200–207. [CrossRef]
13. Heaney, K.D.; Campbell, R.L.; Murray, J.J. Deep water towed array measurements at close range. *J. Acoust. Soc. Am.* **2013**, *134*, 3230–3241. [CrossRef] [PubMed]
14. Jensen, F.B.; Kuperman, W.A.; Porter, M.B.; Schmidt, H. *Computational Ocean Acoustics, AIP Series in Modern Acoustics and Signal Processing*; American Institute of Physics: New York, NY, USA, 1994.
15. Porter, M.B.; Bucker, H.P. Gaussian beam tracing for computing ocean acoustic fields. *J. Acoust. Soc. Am.* **1987**, *82*, 1349–1359. [CrossRef]
16. Oh, R.; Song, T.L.; Choi, J.W. Bearings-Only Target Motion Analysis using Ray Tracing. In Proceedings of the 8th International Conference on Control Automation & Information Sciences, Chengdu, China, 23–26 October 2019.
17. Song, T.L. Observability of target tracking with bearings-only measurements. *IEEE Trans. Aerosp. Electron. Syst.* **1996**, *32*, 1468–1472. [CrossRef]
18. Munk, W.H. Sound channel in an exponentially stratified ocean with application to SOFAR. *J. Acoust. Soc. Am.* **1974**, *55*, 220–226. [CrossRef]
19. Kennedy, J. Particle Swarm Optimization. In *Encyclopedia of Machine Learning*; Springer: New York, NY, USA, 2011; pp. 760–766.
20. Kim, S.; Choi, J.W. Optimal Deployment of Sensor Nodes Based on Performance Surface of Underwater Acoustic Communication. *Sensors* **2017**, *17*, 2389–2403.

Article

Adaptive Estimation of Spatial Clutter Measurement Density Using Clutter Measurement Probability for Enhanced Multi-Target Tracking

Seung Hyo Park, Sa Yong Chong, Hyung June Kim and Taek Lyul Song *

Department of Electronic Systems Engineering, Hanyang University, Ansan 15588, Korea; gyeonwoo4@naver.com (S.H.P.); syong0329@hanmail.net (S.Y.C.); lovesunday88@naver.com (H.J.K.)
* Correspondence: tsong@hanyang.ac.kr; Tel.: +82-31-400-4156

Received: 10 December 2019; Accepted: 21 December 2019; Published: 23 December 2019

Abstract: The point detections obtained from radars or sonars in surveillance environments include clutter measurements, as well as target measurements. Target tracking with these data requires data association, which distinguishes the detections from targets and clutter. Various algorithms have been proposed for clutter measurement density estimation to achieve accurate and robust target tracking with the point detections. Among them, the spatial clutter measurement density estimator (SCMDE) computes the sparsity of clutter measurement, which is the reciprocal of the clutter measurement density. The SCMDE considers all adjacent measurements only as clutter, so the estimated clutter measurement density is biased for multi-target tracking applications, which may result in degraded target tracking performance. Through the study in this paper, a major source of tracking performance degradation with the existing SCMDE for multi-target tracking is analyzed, and the use of the clutter measurement probability is proposed as a remedy. It is also found that the expansion of the volume of the hyper-sphere for each sparsity order reduces the bias of clutter measurement density estimates. Based on the analysis, we propose a new adaptive clutter measurement density estimation method called SCMDE for multi-target tracking (MTT-SCMDE). The proposed method is applied to multi-target tracking, and the improvement of multi-target tracking performance is shown by a series of Monte Carlo simulation runs and a real radar data test. The clutter measurement density estimation performance and target tracking performance are also analyzed for various sparsity orders.

Keywords: data association; clutter measurement density; spatial clutter measurement density estimator; multi-target tracking

1. Introduction

Signals with strength higher than the detection threshold of the sensor are used as measurements for track initiation and track state update of target tracking. These measurements include not only the target measurements, but also clutter measurements due to environmental factors. Since the source of the measurements in the tracking system is not known in advance, target tracking performance may be significantly degraded if measurements generated by clutter are used when the track state is updated. It is essential to use a tracking algorithm based on data association that can statistically distinguish target and clutter measurements in a cluttered environment [1–4].

Since the number of targets existing in the surveillance region and information on the appearance and disappearance of the target cannot be known in advance, it is important to have a means for determining whether the target is being tracked by a tracking algorithm. For target tracking with track management, integrated probabilistic data association (IPDA) [5,6] and integrated track splitting (ITS) [7–9] have been proposed as data association algorithms for single target tracking, which include a track management method that utilizes the target existence probability of each track for controlling the track status and track number or track label. Linear multitarget-IPDA (LM-IPDA) [10], joint IPDA (JIPDA) [11], and iterative JIPDA (iJIPDA) [12] have been proposed for multiple target tracking by extending IPDA and ITS.

In the aforementioned data association algorithms, it is assumed that the number of clutter measurements is Poisson distributed with a parameter called the clutter measurement density, and the clutter measurements are assumed to be uniformly distributed in the surveillance space. The clutter measurement density is defined as the mean number of clutter measurements per unit volume of the surveillance space. The clutter measurement density is an important parameter used to calculate the data association probability and the target existence probability in the data association algorithms.

If the clutter measurement density is fixed to a design value for target tracking in heterogeneous clutter environments, the error in the clutter measurement density deteriorates not only the target state estimation performance, but also the false track discrimination (FTD) performance because prior information about the clutter measurement is unknown in actual target tracking environments. For accurate and robust target tracking in these environments, it is required to estimate the clutter measurement density adaptively. Clutter measurement density estimation methods are divided into track based estimation methods and measurement based estimation methods. In addition, they are divided into single scan estimation methods and multiple scan estimation methods depending on whether the memory is used in the calculation.

The clutter map method [13,14] is a multi-scan estimation method that uses the measurements from previous scans to calculate the clutter measurement density in the current scan. It divides the surveillance region into a finite number of cells and then estimates the clutter measurement density in each cell by statistically counting the number of existing measurements in the cell during a pre-determined multiple scan period. The clutter map can reduce the influence of bias caused by the target measurements, but estimation performance is sensitive to the parameters such as the cell size and the length of multiple scan period. It is difficult to apply the clutter map when the number of measurements and the spatial probability distribution are time varying.

In [15,16], the clutter measurement density estimation method based on the probability hypothesis density (PHD) filter [17] was handled in conjunction with a target tracking algorithm based on data association. It was designed as a feedback structure that used the intensity of clutter estimated through PHD. However, since the clutter generator is assumed to be a Gaussian function with unknown mean and unknown covariance, it is difficult to use in practical implementations due to heavy computational loads. The work in [18] proposed an interactive clutter measurement density estimator (ICMDE) based on a Gaussian mixture PHD (GM-PHD) filter [19] to estimate the clutter measurement density adaptively in environments where the clutter measurement densities are nonuniform and time varying. In [18], the Gaussian model for the clutter generator was assumed to have a known covariance for reducing the computational loads required to calculate the updated state PHD. By dividing the entire surveillance area, the clutter generator for each partition is represented as a component with the Gaussian model. These processes are performed for multiple scans to generate a reliable clutter map of the surveillance area. In [20], a method of forming a clutter map as proposed by using the histogram probabilistic multi-hypothesis tracker (H-PMHT) based on expectation maximization for image target tracking with each scene composed of millions of pixels. This method forms a clutter map through many iterations until local convergence is guaranteed.

The track based and the measurement based clutter measurement density estimation methods are classified as single scan estimation methods in which the clutter measurement density of the previous scan does not affect the clutter measurement density of the current scan. The track based clutter measurement density estimation method uses the validation gate of the track and the validated measurements existing in this gate. There exist several methods such as the conditional mean estimator based on the target perceivability [21] and the maximum likelihood estimator based on the assumption of unknown, but non-random clutter measurement density [22]. The conditional mean estimator [22] requires prior knowledge of clutter measurement density so that the maximum likelihood estimator may be used as an auxiliary estimator. For the track based clutter measurement density estimation methods, different clutter measurement densities are produced for the same measurement shared by the two tracks as the size of the validation gate of each track is different. This is a drawback of the track based clutter measurement density estimators.

The spatial clutter measurement density estimator (SCMDE) [23] is a measurement based clutter measurement density estimation method that calculates the sparsity as the reciprocal of the clutter measurement density by evaluating the volume of the hyper-sphere centered at the measurement of interest and counting the number of measurements inside the volume. The number of measurements and the hyper-sphere volume are determined by the sparsity order. Unlike the track based clutter measurement density estimation methods, it produces a unique sparsity for each measurement regardless of the validation gate size of the track involved.

It was pointed out in [23] that the existing SCMDE generates the unbiased estimates of clutter measurement density when the point of interest is the target detection for single target tracking environments. It produces biased and bigger clutter measurement density estimates than the actual ones when the point of interest is a clutter detection, which results in improved target tracking performance as the data association probabilities become smaller for the clutter detection. However, when the existing SCMDE is used for multi-target tracking environments, biased clutter measurement density estimation is expected from the nature of SCMDE that all the adjacent measurements to the point of interest are considered to be clutter detections. Through the study in this paper, a major source of biased clutter measurement density estimation of the existing SCMDE for multi-target tracking environments is analyzed, and remedies to reduce the biases are proposed. The new adaptive SCMDE for multi-target tracking (MTT-SCMDE) utilizes the clutter measurement probability to take into account only the clutter measurements for improved accuracy by reducing the biases in the clutter measurement density estimation. Through the analysis, an expansion of the volume of the hyper-sphere corresponding to each sparsity order from that of the existing SCMDE is proposed for more accurate clutter measurement density estimation.

A method that takes into account clutter-originated measurements in the clutter measurement density calculation was proposed in [24]. The performance of the SCMDE algorithm for multi-target tracking was presented in [24]. In this paper, we elaborated the theoretical development by analyzing the source of biases in the MTT-SCMDE algorithm for multi-target tracking, and refined its performance by increasing the hyper-sphere volume for the measurement of interest. The improvement was based on strict analysis presented in this paper. To verify the performance of the proposed clutter measurement density estimation method, a series of simulation runs was executed in heterogeneous clutter environments, and the results were analyzed by performance comparison to check how closely the estimated clutter measurement densities followed the true clutter measurement densities for multiple targets. In addition, the clutter measurement density estimation performance and the target tracking performance were tested for various sparsity orders and various numbers of targets involved. The proposed MTT-SCMDE was also applied to a set of real radar data for performance evaluation.

The remainder of this paper is organized as follows. The stochastic models in the target tracking algorithm are described in Section 2. Section 3 derives the LM-IPDA algorithm for multi-target tracking in a cluttered environment. The SCMDE method is briefly described in Section 4. Section 5 describes the proposed clutter measurement density estimation method in detail. The clutter measurement density estimation performance and multiple target tracking performance of the proposed method are analyzed through a series of Monte Carlo simulation runs in various tracking environments, as well as a set of real radar data in Section 6, followed by the Conclusions. Performance analysis of the existing SCMDE used in multi-target tracking environments is presented in the Appendix A.

2. Models

The following assumptions are applied for using multi-target tracking algorithms in a cluttered environment.

- The sensor has infinite resolution; each measurement is generated only from one source; and its the origin can be either a target or clutter.
- Each target generates at most one measurement at each scan according to the target detection probability P_D.

Superscript τ denotes a target, or the index of a track that follows the target. Target τ's trajectory state $\mathbf{x}_k^\tau$ is an $n_x \times 1$ state vector. In this paper, the dynamics of the targets from scan k to scan $k+1$ are assumed to follow a linear dynamic model in a two-dimensional (2D) plane, such as:

$$\mathbf{x}_{k+1}^\tau = \mathbf{\Phi}_k \mathbf{x}_k^\tau + \mathbf{\Gamma}_k w_k, \tag{1}$$

where $\mathbf{x}_k$ is a 6×1 state vector consisting of the target position, velocity, and acceleration in a 2D plane, $\mathbf{\Phi}_k$ is the state propagation matrix, and $\mathbf{\Gamma}_k$ is the coefficient matrix of w_k, which is a white Gaussian process noise with zero-mean and covariance matrix $diag(q,q)$. The last term of (1) is white Gaussian with zero-mean and covariance matrix $\mathbf{Q}_k = q\mathbf{\Gamma}_k(\mathbf{\Gamma}_k)^\top$, and its distribution is denoted by $N(0, \mathbf{Q}_k)$. The state propagation matrix $\mathbf{\Phi}_k$ and the coefficient matrix for the process noise $\mathbf{\Gamma}_k$ follow a nearly constant velocity (NCV) model or constant turn rate (CTR) model [25].

For the NCV model, the state propagation matrix and the coefficient matrix for the process noise in (1) become:

$$\mathbf{\Phi}_k^{NCV} = \begin{bmatrix} \mathbf{I}_2 & T\mathbf{I}_2 & \mathbf{O}_2 \\ \mathbf{O}_2 & \mathbf{I}_2 & \mathbf{O}_2 \\ \mathbf{O}_2 & \mathbf{O}_2 & \mathbf{O}_2 \end{bmatrix}, \tag{2}$$

$$\mathbf{\Gamma}_k^{NCV} = \begin{bmatrix} \frac{T^2}{2}\mathbf{I}_2 \\ T\mathbf{I}_2 \\ \mathbf{O}_2 \end{bmatrix}, \tag{3}$$

where T is the sampling time of a discrete time interval, $\mathbf{I}_2$ is a 2×2 identity matrix, $\mathbf{O}_2$ is a 2×2 null matrix, and the variance of w_k is set to be $\sigma_a^2\mathbf{I}_2$ such that w_k of the NCV model represents the acceleration uncertainty; this implies $\mathbf{Q}_k^{NCV} = \sigma_a^2\mathbf{\Gamma}_k^{NCV}(\mathbf{\Gamma}_k^{NCV})^\top$. For the NCV model, the acceleration components of $\mathbf{x}_k$ are set to be zero.

For the CTR model, the state propagation matrix and the coefficient matrix for the process noise in (1) become:

$$\mathbf{\Phi}_k^{CTR} = \begin{bmatrix} \mathbf{I}_2 & \frac{\sin(\Omega_k T)}{\Omega_k}\mathbf{I}_2 & \frac{1-\cos(\Omega_k T)}{\Omega_k^2}\mathbf{I}_2 \\ \mathbf{O}_2 & \cos(\Omega_k T)\mathbf{I}_2 & \frac{\sin(\Omega_k T)}{\Omega_k}\mathbf{I}_2 \\ \mathbf{O}_2 & -\Omega_k \sin(\Omega_k T)\mathbf{I}_2 & \cos(\Omega_k T)\mathbf{I}_2 \end{bmatrix}, \tag{4}$$

$$\mathbf{\Gamma}_k^{CTR} = \begin{bmatrix} \frac{\Omega_k T - \sin(\Omega_k T)}{\Omega_k^3}\mathbf{I}_2 \\ \frac{1-\cos(\Omega_k T)}{\Omega_k^2}\mathbf{I}_2 \\ \frac{\sin(\Omega_k T)}{\Omega_k}\mathbf{I}_2 \end{bmatrix}. \tag{5}$$

w_k of the CTR model represents the uncertainty in target jerk, and its variance is $\sigma_j^2\mathbf{I}_2$; this implies $\mathbf{Q}_k^{CTR} = \sigma_j^2\mathbf{\Gamma}_k^{CTR}(\mathbf{\Gamma}_k^{CTR})^\top$. The turn rate Ω_k is adaptively estimated using the target acceleration and velocity estimates while tracking.

Target measurement model $\mathbf{z}_k$ is an $n_z \times 1$ vector, and it is expressed as:

$$\mathbf{z}_k = \mathbf{H}\mathbf{x}_k + v_k, \tag{6}$$

where $\mathbf{H}$ is the measurement matrix denoted by:

$$\mathbf{H} = \begin{bmatrix} \mathbf{I}_2 & \mathbf{O}_2 & \mathbf{O}_2 \end{bmatrix} . \tag{7}$$

In (6), v_k is a white Gaussian measurement noise of the sensor with zero-mean and covariance matrix $\mathbf{R}_k$.

The sensor obtains a set of measurements $\mathbf{Z}_k$ at each scan k. $\mathbf{z}_{k,i}$ is the ith measurement of $\mathbf{Z}_k$, and the measurement vector of $\mathbf{z}_{k,i}$ can be expressed by:

$$\mathbf{z}_{k,i} = \begin{bmatrix} z^x_{k,i} & z^y_{k,i} \end{bmatrix}^\top , \tag{8}$$

where $z^x_{k,i}$ and $z^y_{k,i}$ represent the x and y positions in the 2D Cartesian coordinate system, respectively.

3. LM-IPDA Algorithm for Multi-Target Tracking

In a cluttered environment, multi-target tracking algorithms with data association such as global nearest neighbor (GNN) [26,27] and joint probabilistic data association (JPDA) [28–30] have been widely used. However, these algorithms in general do not include an FTD procedure that can distinguish the true tracks generated by the target measurements from the false tracks generated by the clutter measurements. JIPDA and LM-IPDA are multi-target tracking algorithms with FTD functions for autonomous track management. JIPDA has optimal target tracking performance for single scan data association since it probabilistically takes into account all possible events between measurements and tracks in the cluster for each scan. However, it has heavy computational loads as the number of feasible joint events to be considered increases combinatorially depending on the number of measurements and the number of tracks. In this paper, LM-IPDA instead of JIPDA is used for multi-target tracking as the computation time increases linearly commensurate with the number of targets. In LM-IPDA, the state of track τ is represented as a hybrid state that consists of the target existence event (discrete event) and the trajectory state (continuous variable) such as:

$$p[\mathbf{x}^\tau_{k-1}, \chi^\tau_{k-1} | \mathbf{Z}^{k-1}] = P\left\{\chi^\tau_{k-1} | \mathbf{Z}^{k-1}\right\} p(\mathbf{x}^\tau_{k-1} | \chi^\tau_{k-1}, \mathbf{Z}^{k-1}), \tag{9}$$

where χ^τ_{k-1} represents the existence event of target τ at scan $k-1$, and the probability density function of the target state at scan k satisfies:

$$p(\mathbf{x}^\tau_{k-1} | \chi^\tau_{k-1}, \mathbf{Z}^{k-1}) = N(\mathbf{x}^\tau_{k-1}; \hat{\mathbf{x}}^\tau_{k-1|k-1}, \mathbf{P}^\tau_{k-1|k-1}). \tag{10}$$

The track recursion is composed of the following steps:

- prediction of track state and existence probability,
- selection of validated measurements,
- calculation of modulated clutter measurement density,
- update of track state and existence probability.

3.1. Prediction of Track State and Existence Probability

The existence event of a target in the surveillance region at scan k is denoted by χ^τ_k as a random event, and $\bar{\chi}^\tau_k$ is the complement of χ^τ_k. The existence of a target propagates by the Markov chain one model [5,6]:

$$P\left\{\chi^\tau_{k-1} | \mathbf{Z}^k\right\} = p_{11} P\left\{\chi^\tau_{k-1} | \mathbf{Z}^{k-1}\right\}, \tag{11}$$

where p_{11} is the transition probability of target existence.

The trajectory state of each track τ is propagated using the prediction step of the Kalman filter:

$$\hat{\mathbf{x}}^{\tau}_{k|k-1} = \mathbf{\Phi}_k \hat{\mathbf{x}}^{\tau}_{k-1|k-1} \tag{12}$$

$$\mathbf{P}^{\tau}_{k|k-1} = \mathbf{\Phi}_k \hat{\mathbf{P}}^{\tau}_{k-1|k-1} (\mathbf{\Phi}_k)^{\top} + \mathbf{Q}_k. \tag{13}$$

3.2. Selection of Validated Measurements

Since it is computationally inefficient to use all measurements in the entire surveillance region, the validation gate [2,3] is generated around the track predicted position, and the track is updated using only the measurements that exist inside the gate. If the measurement $\mathbf{z}_{k,i}$ satisfies the following equation for track τ, $\mathbf{z}_{k,i}$ is regarded as a validated measurement. Otherwise, it is not used for updating the states of track τ.

$$(\mathbf{z}_{k,i} - \mathbf{H}\hat{\mathbf{x}}^{\tau}_{k|k-1})^{\top} (\mathbf{S}^{\tau}_{k|k-1})^{-1} (\mathbf{z}_{k,i} - \mathbf{H}\hat{\mathbf{x}}^{\tau}_{k|k-1}) < \tau_G, \tag{14}$$

with:

$$\mathbf{S}^{\tau}_{k|k-1} = \mathbf{H}\mathbf{P}^{\tau}_{k|k-1}\mathbf{H}^{\top} + \mathbf{R}_k, \tag{15}$$

where $\sqrt{\tau_G}$ is the size of the validation gate. The set of validated measurements selected by track τ and the number of validated measurements are denoted by $\mathbf{Z}^{\tau}_k$ and m^{τ}_k, respectively.

3.3. Calculation of Modulated Clutter Measurement Density

The calculating process of the modulated clutter measurement density is a crucial part of the LM approach, which can significantly reduce the amount of computation of JIPDA, which evaluates the probabilities of all the feasible joint events that can occur in multiple target tracking for each scan.

The modulated clutter measurement density of measurement $\mathbf{z}_{k,i}$ can be obtained by adding the influence of other tracks to $\mathbf{z}_{k,i}$ by utilizing the probability that measurement $\mathbf{z}_{k,i}$ is generated from other targets to the pure clutter measurement density $\rho_{k,i}$ at the position of measurement $\mathbf{z}_{k,i}$. Let $\tilde{\rho}^{\tau}_{k,i}$ denote the modulated clutter measurement density, then:

$$\tilde{\rho}^{\tau}_{k,i} = \rho_{k,i} + \sum_{\substack{\sigma \in \mathbf{T}_k \\ \sigma \neq \tau}} \frac{P^{\sigma}_{k,i}}{1 - P^{\sigma}_{k,i}} p^{\sigma}_{k,i}, \tag{16}$$

where $P^{\sigma}_{k,i}$ is a measure of the influence of track σ on $\mathbf{z}_{k,i}$, and it is represented by the prior probability that $\mathbf{z}_{k,i}$ is generated from target σ such as:

$$P^{\sigma}_{k,i} = P_D P_G P\left\{\chi^{\tau}_k | \mathbf{Z}^{k-1}\right\} \frac{p^{\sigma}_{k,i}}{\rho_{k,i}} / \sum_{l=1}^{m^{\sigma}_k} \frac{p^{\sigma}_{k,l}}{\rho_{k,l}}, \tag{17}$$

where $p^{\sigma}_{k,i}$ is a likelihood function of measurement $\mathbf{z}_{k,i}$ with respect to track σ such that:

$$p^{\sigma}_{k,i} = \frac{1}{P_G} N(\mathbf{z}_{k,i}; \mathbf{H}\hat{\mathbf{x}}^{\sigma}_{k|k-1}, \mathbf{S}^{\sigma}_{k|k-1}), \tag{18}$$

where P_G is the gate probability [2].

The modulated clutter measurement density of measurement $\tilde{\rho}_{k,i}$ is used for calculating the data association probabilities in the update step of track trajectory state, as well as the update step of the target existence probability.

3.4. Update of Track State and Existence Probability

Using the validated measurements selected by track τ, the posterior trajectory state of track τ and the posterior target existence probability are calculated.

Let $\beta_{k,i}^{\tau}$ denote the data association probability that is conditioned on the target existence event χ_k^{τ}. The data association probability $\beta_{k,i}^{\tau}$ is expressed for the event $\chi_{k,i}^{\tau}$, which indicates that the ith validated measurement of track τ is a target measurement, and the event $\chi_{k,0}^{\tau}$, which indicates that all the validated measurements of track τ are regarded as clutter measurements.

$$\beta_{k,i}^{\tau} = P\left\{\chi_{k,i}^{\tau}|\chi_k^{\tau}, \mathbf{Z}^k\right\} = \frac{P_D P_G}{\Lambda_k^{\tau}} \frac{p_{k,i}^{\tau}}{\tilde{\rho}_{k,i}^{\tau}}, \tag{19}$$

$$\beta_{k,0}^{\tau} = P\left\{\chi_{k,0}^{\tau}|\chi_k^{\tau}, \mathbf{Z}^k\right\} = \frac{1 - P_D P_G}{\Lambda_k^{\tau}}, \tag{20}$$

where Λ_k^{τ} is the measurement likelihood ratio of track τ such that:

$$\Lambda_k^{\tau} = 1 - P_D P_G + P_D P_G \sum_{i=1}^{m_k^{\tau}} \frac{p_{k,i}^{\tau}}{\tilde{\rho}_{k,i}^{\tau}}. \tag{21}$$

The posterior trajectory state of track τ is calculated by the total probability theorem [31] such as:

$$p(\mathbf{x}_k^{\tau}|\chi_k^{\tau}, \mathbf{Z}^k) = \sum_{i=0}^{m_k^{\tau}} p(\mathbf{x}_k^{\tau}|\chi_k^{\tau}, \chi_{k,i}^{\tau}, \mathbf{Z}^k) P\left\{\chi_{k,i}^{\tau}|\chi_k^{\tau}, \mathbf{Z}^k\right\}, \tag{22}$$

where $p(\mathbf{x}_k^{\tau}|\chi_k^{\tau}, \chi_{k,i}^{\tau}, \mathbf{Z}^k)$ is a single Gaussian distribution as a posterior probability density function for the target trajectory state conditioned on the facts that target τ exists and $\mathbf{z}_{k,i}$ is the target measurement.

$$p(\mathbf{x}_k^{\tau}|\chi_k^{\tau}, \chi_{k,i}^{\tau}, \mathbf{Z}^k) = N(\mathbf{x}_k^{\tau}; \hat{\mathbf{x}}_{k|k,i}^{\tau}, \mathbf{P}_{k|k,i}^{\tau}), \tag{23}$$

where the conditional mean $\hat{\mathbf{x}}_{k|k,i}^{\tau}$ and covariance $\mathbf{P}_{k|k,i}^{\tau}$ satisfy:

$$\hat{\mathbf{x}}_{k|k,i}^{\tau} = \begin{cases} \hat{\mathbf{x}}_{k|k-1}^{\tau} + \mathbf{K}_{k|k-1}^{\tau}(\mathbf{z}_{k,i} - \mathbf{H}\hat{\mathbf{x}}_{k|k-1}^{\tau}) & i > 0 \\ \hat{\mathbf{x}}_{k|k-1}^{\tau} & i = 0 \end{cases} \tag{24}$$

$$\mathbf{P}_{k|k,i}^{\tau} = \begin{cases} (\mathbf{I}_6 - \mathbf{K}_{k|k-1}^{\tau}\mathbf{H})\mathbf{P}_{k|k-1}^{\tau} & i > 0 \\ \mathbf{P}_{k|k-1}^{\tau} & i = 0. \end{cases} \tag{25}$$

where $\mathbf{I}_n$ denote an $n \times n$ identity matrix, and the Kalman gain $\mathbf{K}_{k|k-1}^{\tau}$ is expressed by:

$$\mathbf{K}_{k|k-1}^{\tau} = \mathbf{P}_{k|k-1}^{\tau}\mathbf{H}^{\top}(\mathbf{S}_{k|k-1}^{\tau})^{-1}. \tag{26}$$

Using the data association probabilities for the validated measurements, the updated track state estimates are obtained in the form of a Gaussian mixture such as:

$$\hat{\mathbf{x}}_{k|k}^{\tau} = \sum_{i=0}^{m_k^{\tau}} \beta_{k,i}^{\tau}\hat{\mathbf{x}}_{k|k,i}^{\tau} \tag{27}$$

$$\mathbf{P}^{\tau}_{k|k} = \sum_{i=0}^{m^{\tau}_k} \beta^{\tau}_{k,i} \left(\mathbf{P}^{\tau}_{k|k,i} + \hat{\mathbf{x}}^{\tau}_{k|k,i} (\hat{\mathbf{x}}^{\tau}_{k|k,i})^{\top} \right) - \hat{\mathbf{x}}^{\tau}_{k|k} (\hat{\mathbf{x}}^{\tau}_{k|k})^{\top}. \tag{28}$$

The posterior target existence probability is used as a track score for track management including confirmation and termination. It is obtained by using the prior target existence probability and the measurement likelihood ratio such as [10]:

$$P\left\{\chi^{\tau}_k | \mathbf{Z}^k\right\} = \frac{P\left\{\chi^{\tau}_k | \mathbf{Z}^{k-1}\right\} \Lambda^{\tau}_k}{1 - (1 - \Lambda^{\tau}_k) P\left\{\chi^{\tau}_k | \mathbf{Z}^{k-1}\right\}}. \tag{29}$$

4. The Existing Spatial Clutter Measurement Density Estimator for Single Target Tracking

The clutter measurement density is an important parameter to calculate the data association probability and the posterior target existence probability for track maintenance. In particular, when the LM approach is used in a situation where multiple targets are located in the vicinity, the merging and switching phenomena of the tracks are reduced by utilizing the modulated clutter density with moderate computational loads [10]. Therefore, it is crucial to estimate the clutter measurement density properly.

The clutter measurement density is defined as the average number of measurements that exist within a unit volume. For calculating the sparsity of $\mathbf{z}_{k,i}$, the measurements are aligned in the ascending order of distance from $\mathbf{z}_{k,i}$. If $\mathbf{Y}^i_k$ denotes the set of the aligned measurements such as:

$$\mathbf{Y}^i_k = \bigcup_{l=1} \mathbf{z}^{(l)}_{k,i}, \tag{30}$$

where $\mathbf{z}^{(l)}_{k,i}$ is the lth nearest measurement from $\mathbf{z}_{k,i}$. Let $r^{(n)}_{k,i}$ denote the distance from $\mathbf{z}_{k,i}$ to the nth nearest measurement, $\mathbf{z}^{(n)}_{k,i}$ in $\mathbf{Y}^i_k$. Then, $r^{(n)}_{k,i}$ becomes:

$$r^{(n)}_{k,i} = \left\| \mathbf{z}^{(n)}_{k,i} - \mathbf{z}_{k,i} \right\|. \tag{31}$$

The SCMDE estimates the sparsity of the measurements, which is the reciprocal of the clutter measurement density. The sparsity of $\mathbf{z}_{k,i}$ is obtained from:

$$\hat{\gamma}^{(n)}_{k,i} = \frac{1}{\rho^{(n)}_{k,i}} = \frac{V\left(r^{(n)}_{k,i}\right)}{n}, \tag{32}$$

where n and $V\left(r^{(n)}_{k,i}\right)$ denote the sparsity order and the volume of the hyper-sphere with radius $r^{(n)}_{k,i}$ for $\mathbf{z}_{k,i}$, respectively. $V\left(r^{(n)}_{k,i}\right)$ is expressed by:

$$V\left(r^{(n)}_{k,i}\right) = C_{n_z} \left(r^{(n)}_{k,i}\right)^{n_z}, \tag{33}$$

where n_z represents the dimension of the measurement space and $C_{n_1} = 2$, $C_{n_2} = \pi$, and $C_{n_3} = \frac{4\pi}{3}$. Figure 1 schematically illustrates the hyper-spheres with various sparsity orders for $\mathbf{z}_{k,i}$ in a 2D measurement space.

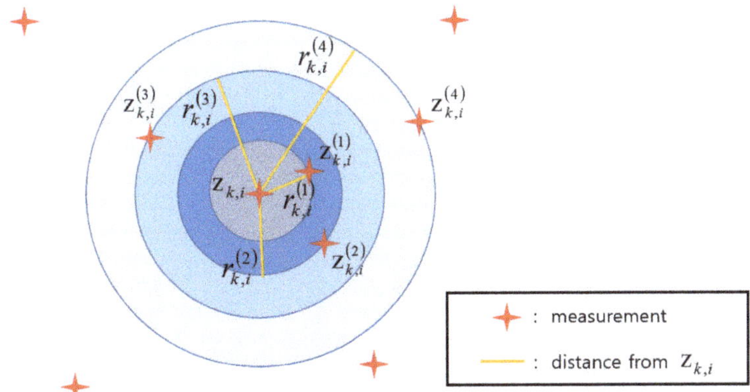

Figure 1. Hyper-spheres for the existing spatial clutter measurement density estimator (SCMDE) in a 2D space.

In the process of deriving the sparsity, track information is not used. A measurement shared by two or more tracks has a unique clutter measurement density regardless of the track states.

5. SCMDE for Multi-Target Tracking Environments

5.1. Drawbacks of the Existing SCMDE for Multi-Target Tracking

It was introduced in [23] that the SCMDE for single target tracking environments yields an accurate clutter measurement density when the point of interest is the target detection. When the point of interest is a clutter detection, the SCMDE generates a biased and smaller sparsity than the actual value, which implies a bigger clutter measurement density. This phenomenon gives benefits to single target tracking as the bigger clutter measurement density decreases the data association probability for the clutter detection in the probabilistic data association (PDA) algorithm. It was also introduced in [23] that these benefits are reduced as the sparsity order increases. Therefore, the existing SCMDE improves target tracking performance for single target tracking.

In this subsection, the performance of the existing SCMDE is analyzed for multi-target tracking in homogeneous clutter environments. The detailed derivations are given in Appendix A of this paper. When the point of interest is a target detection for a two-target case, the average value of the sparsity estimate for the point of interest $\mathbf{z}_{k,i}$ becomes:

$$E\left\{\hat{\gamma}_{k,i}^{(n)}\right\} = \begin{cases} \frac{1}{\rho}(1 - \frac{1-e^{-\rho V^{(1)}}}{\rho V^{(1)}}), & n = 1 \\ \frac{1}{\rho}(1 - \frac{1-e^{-\rho V^{(2)}}}{\rho V^{(2)}} + \frac{e^{-\rho V^{(2)}}}{2}), & n = 2 \end{cases} \tag{34}$$

where n is the sparsity order, ρ is the clutter measurement density of the homogeneous clutter environment, and $V^{(n)}$ is the volume of the hyper-sphere used for the sparsity estimation. If (34) is compared to the true sparsity, $\frac{1}{\rho}$, which can be obtained from the existing SCMDE for single target tracking as shown in (35) of [23], the sparsity estimates are smaller than the true ones and biased. The bias becomes reduced as n increases and $V^{(n)}$ becomes bigger. In contrast to single target tracking environments, the SCMDE generates bigger clutter measurement density estimates when the point of interest is a target detection, which results in a reduced data association probability for the target detection and deteriorated target tracking performance for multi-target tracking environments.

When the point of interest is a clutter detection for two-target cases, the average value of the sparsity estimates for the point of interest, $\mathbf{z}_{k,i}$, becomes:

$$E\left\{\hat{\gamma}_{k,i}^{(n)}\right\} = \begin{cases} \frac{1}{\rho} - \frac{1}{\rho^2 V^{(1)}}(1 - e^{-\rho V^{(1)}}), & n = 1 \\ \frac{1}{\rho} - \frac{1}{2\rho^2 V^{(2)}}(1 - e^{-\rho V^{(2)}}), & n = 2 \end{cases} \tag{35}$$

The average sparsity estimates in (35) are smaller than the actual $\frac{1}{\rho}$, and this fact results in bigger clutter measurement density estimates. The average sparsity estimate for $n = 1$ in (35) is the same as $n = 1$ for single target tracking environments specified in (23) of [23]. When $n = 2$, the average sparsity estimate becomes bigger than $n = 1$, and this indicates that the clutter measurement density estimates become less biased for $n = 2$. This indicates that more accurate clutter measurement density estimation is possible with the PDA algorithm as n increases. From the above analysis, the existing SCMDE has two incompatible aspects in tracking performance for multi-target tracking environments. One aspect is that tracking performance becomes deteriorated as it generates smaller data association probabilities than the actual ones for true target detections. Another aspect is that tracking performance is improved as it generates smaller data association probabilities than the actual one for clutter measurements. These incompatible aspects are due to the biased and reduced sparsity estimates described in (34) and (35).

In order to improve tracking performance for multi-target tracking environments, it is more important to have the improved data association results with less biased clutter measurement density estimates. This can be done by evaluating the clutter measurement probability of each validated measurement for counting only the number of clutter measurements (excluding the number of target measurements), inside the volume of the hyper-sphere $V(r_{k,i}^{(n)})$ specified in (33). The clutter measurement probability is the probability that the measurement is a clutter detection not from a target. If the clutter measurement probability is used for the sparsity estimates, enhanced tracking is expected due to less biased clutter measurement density estimates. This has a more significant effect in performance improvement when the point of interest is a target detection rather than a clutter detection. From the analysis in this section, the magnitude of bias of the sparsity estimate of (34) and (35) becomes smaller as n increases and the volume of the hyper-sphere $V^{(n)}$ increases. In the next subsection, the adaptive SCMDE algorithm for multi-target tracking (MTT-SCMDE) is proposed to take into account the clutter measurement probability and increased hyper-sphere volume for each sparsity order n to achieve enhanced tracking performances.

5.2. MTT-SCMDE

To estimate the clutter measurement density accurately for these multi-target tracking environments, we propose a method to calculate the probability that adjacent measurements are generated from clutter and use this probability to estimate the clutter measurement density.

To derive the clutter measurement probability, two events are defined.

1. $\chi_{k,j}^{0}$: an event that the measurement $\mathbf{z}_{k,j}$ is not a target measurement for any of the tracks at scan k.
2. $\chi_{k,j}^{\tau}$: an event that the measurement $\mathbf{z}_{k,j}$ is a target measurement originated from the target τ at scan k

Let $\mathbf{T}_k$ denote the set of the cluster targets at scan k, and H_k represent the event that $\mathbf{T}_k$ exists at scan k such as:

$$H_k = \bigcup_{\sigma \in \mathbf{T}_k} \chi_k^{\sigma}. \tag{36}$$

The probability that $\mathbf{z}_{k,j}$ is generated from clutter under H_k, the mutual exclusiveness of $\mathbf{z}_{k,j}$ sources, becomes:

$$P(\chi^0_{k,j}|\mathbf{Z}^{k-1}, H_k) = \alpha_k \prod_{\sigma \in \mathbf{T}_k} \left(1 - P^{\sigma}_{k,j}\right), \tag{37}$$

where $P^{\sigma}_{k,j}$ is the prior probability that $\mathbf{z}_{k,j}$ is generated from target σ introduced in (17) and where α_k is a normalization constant.

The probability that $\mathbf{z}_{k,j}$ is generated from target $\tau \in \mathbf{T}_k$ becomes:

$$\begin{aligned} P(\chi^{\tau}_{k,j}|\mathbf{Z}^{k-1}, H_k) &= \alpha_k P^{\tau}_{k,j} \prod_{\substack{\eta \in \mathbf{T}_k \\ \eta \neq \tau}} \left(1 - P^{\eta}_{k,j}\right) && (38) \\ &= \alpha_k \frac{P^{\tau}_{k,j}}{1 - P^{\tau}_{k,j}} \prod_{\eta \in \mathbf{T}_k} \left(1 - P^{\eta}_{k,j}\right). && (39) \end{aligned}$$

From (37) and (39), α_k can be obtained from the mutual exclusiveness of $\mathbf{z}_{k,j}$ sources such as:

$$P(\chi^0_{k,j}|\mathbf{Z}^{k-1}, H_k) + \sum_{\tau \in \mathbf{T}_k} P(\chi^{\tau}_{k,j}|\mathbf{Z}^{k-1}, H_k) = 1. \tag{40}$$

Then, α_k is obtained as:

$$\alpha_k = \frac{1}{\prod_{\sigma \in \mathbf{T}_k} \left(1 - P^{\sigma}_{k,j}\right) \left(1 + \sum_{\eta \in \mathbf{T}_k} \frac{P^{\eta}_{k,j}}{1 - P^{\eta}_{k,j}}\right)}. \tag{41}$$

Therefore, the clutter measurement probability $P(\chi^0_{k,j}|\mathbf{Z}^{k-1}, H_k)$ in (37) can be expressed as:

$$P(\chi^0_{k,j}|\mathbf{Z}^{k-1}, H_k) = \frac{1}{1 + \sum_{\eta \in \mathbf{T}_k} \left(\frac{P^{\eta}_{k,j}}{1 - P^{\eta}_{k,j}}\right)}. \tag{42}$$

The proposed MTT-SCMDE utilizes the clutter measurement probabilities of the element of $\mathbf{Y}^i_k$ defined in (30). Let $C^{(l)}_{k,i}$ be the clutter measurement probability of $\mathbf{z}^{(l)}_{k,i}$, the lth nearest measurement from $\mathbf{z}_{k,i}$. If the cumulative sum $C^{(l)}_{k,i}$ from $l = 1$ is bigger than the predetermined sparsity order n, which is the expected number of clutter measurements, the summation is stopped for the sparsity calculation. If:

$$\sum_{l=1}^{m-1} C^{(l)}_{k,i} < n \leq \sum_{l=1}^{m} C^{(l)}_{k,i}, \tag{43}$$

then the radius $r^{(n)}_{k,i}$ for the hyper-sphere volume calculation becomes:

$$r^{(n)}_{k,i} = \left\| \mathbf{z}^{(m+1)}_{k,i} - \mathbf{z}_{k,i} \right\|, \tag{44}$$

where $\mathbf{z}^{(m+1)}_{k,i} \in \mathbf{Y}^i_k$.

In this paper, $\mathbf{z}^{(n+1)}_{k,i}$ is used to calculate $r^{(n)}_{k,i}$ instead of $\mathbf{z}^{(n)}_{k,i}$ used in [23] for single target tracking to produce less biased estimates of the clutter measurement density for multi-target tracking environments.

The estimated sparsity of order n becomes:

$$\hat{\gamma}_{k,i}^{(n)} = \frac{V\left(r_{k,i}^{(n)}\right)}{\sum_{l=1}^{m} C_{k,i}^{(l)}}, \quad (45)$$

where $V\left(r_{k,i}^{(n)}\right)$ is the volume of the hyper-sphere with radius $r_{k,i}^{(n)}$ defined in (44).

When estimating the sparsity, the existing SCMDE utilizes the number of measurements in the hyper-sphere, while the proposed MTT-clutter measurement density estimation method utilizes the mean number of clutter measurements with the clutter measurement probability to reduce biases in the clutter measurement density estimates in multi-target tracking applications.

Figure 2 shows an expansion of the volume of the hyper-sphere of the MTT-SCMDE if it is used for data association in single target environments. Compared to Figure 1, the volume of the hyper-sphere for each sparsity order is increased.

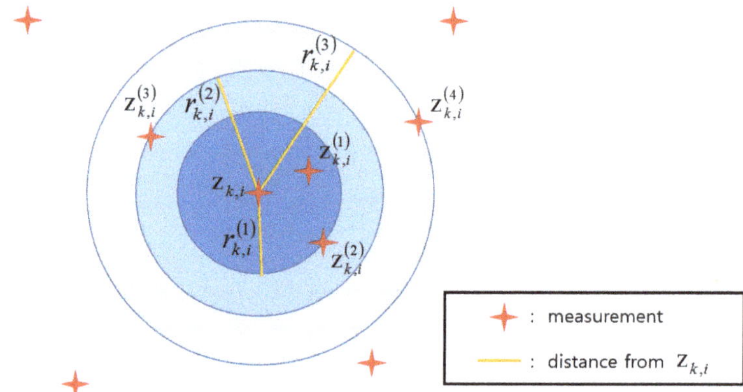

Figure 2. Hyper-spheres for the multi-target tracking (MTT)-SCMDE used for single target tracking in a 2D space.

6. Performance Tests

6.1. Simulation Experiments

The performance test was done to compare the results with respect to FTD performance and the accuracy of estimated clutter measurement density for the cases, which utilized:

- True clutter measurement density (true CMD),
- SCMDE with various sparsity orders,
- MTT-SCMDE with various sparsity orders.

The sets of simulation experiments are presented for multi-target tracking in a heterogeneous environment with varying the number of targets.

The sampling time of sensor T was 1 s, and the measurement noise covariance was $\mathbf{R}_k = 25\mathbf{I}_2 m^2$. The target detection probability P_D and the gate probability P_G were 0.8 and 0.99, respectively. One simulation run consisted of 50 scans, and the total number of Monte Carlo simulation runs was 500. To initialize the track, the two point differencing method [14] was employed if the calculated velocity obtained from two consecutive scans was smaller than the predetermined maximum velocity constraint $V_{max} = 25$ m/s.

In these simulations, the following four evaluation indices were calculated:

- confirmed true track rate (CTTR) [1],
- position root mean squared error (RMSE),
- track retention test statistics [10],
- clutter measurement density estimation performance.

The confirmed track was defined as one whose posterior target existence probability calculated by (29) was bigger than a predetermined confirmed threshold. Among confirmed tracks, the tracks satisfying the following equation were classified as the confirmed true tracks for tracking performance evaluation purposes:

$$(\mathbf{x}_k - \hat{\mathbf{x}}^{\tau}_{k|k})^{\top}(\mathbf{P}^{\tau}_{0|0})^{-1}(\mathbf{x}_k - \hat{\mathbf{x}}^{\tau}_{k|k}) < \gamma_{true}, \tag{46}$$

where $\mathbf{x}_k$ and $\mathbf{P}^{\tau}_{0|0}$ are the state vector of the true target and the initial error covariance matrix of the confirmed track τ, respectively. Conversely, each confirmed track met the following test, and it became a confirmed false track.

$$(\mathbf{x}_k - \hat{\mathbf{x}}^{\tau}_{k|k})^{\top}(\mathbf{P}^{\tau}_{0|0})^{-1}(\mathbf{x}_k - \hat{\mathbf{x}}^{\tau}_{k|k}) > \gamma_{false}, \tag{47}$$

The confirmed true track rate (CTTR) is an evaluation index showing the statistical ratio of confirmed true tracks over time. The position root mean squared error is the distance error between the confirmed true track and the true target, and it was obtained only for the confirmed true tracks over time. The track retention test statistic is an evaluation index of the multi-target tracking algorithms and accumulates statistics on how much the confirmed true tracks are retained or lost between retention test start time (RST) and retention test end time (RET). In these simulations, RST and RET were designated to be 15 and 35, respectively. The retention test was to check the following items, and they were used to indicate the statistical ratio representing the robustness of each algorithm for the period in which the targets were located in the immediate vicinity:

- nCase: the total number of CTTs at RST,
- nOk: the percentage of nCase CTTs that still followed the original target at RET,
- nSwitch: the percentage of nCase CTTs that did not follow the original target at RET,
- nMerge: the percentage of merging two or more nCase CTTs during the retention test,
- nLost: the percentage of nCase CTTs that were terminated during the retention test.

The clutter measurement density estimation performance is an evaluation index of how closely the estimated clutter measurement density follows the true clutter measurement density using the clutter measurement density estimation method. The performance of the proposed algorithm was tested in comparison with the existing SCMDE algorithm for multi-target tracking by varying the number of targets in a heterogeneous clutter environment. Simulations were performed for 3, 5, and 7 targets in three scenarios, and the performance of clutter measurement density estimation was analyzed as the number of targets increased. In addition, the effectiveness and the robustness of the tracking performance were verified through a test with real radar data.

6.1.1. The Number of Targets: 3

The simulation considered the 2D surveillance region depicted in Figure 3. The targets maneuvered slightly to form curved trajectories within the surveillance region. To track maneuvering targets, the LM-IPDA-interacting multiple model (LM-IPDA-IMM) [32] was employed. The LM-IPDA-IMM algorithm used in this study utilized the NCV model and the CTR model introduced in Section 2. The targets were located apart at the beginning of the scenario, then they were located in the immediate vicinity at Scan 25, and then moved away from each other.

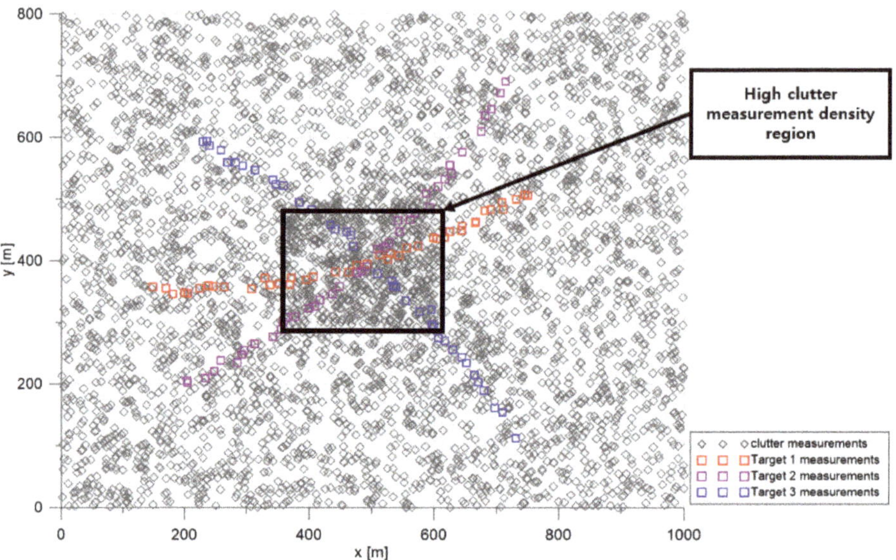

Figure 3. Simulation scenario with three targets.

The base clutter measurement density was 1×10^{-4} scans/m^2, and it increased to 3×10^{-4} scans/m^2 in the high clutter measurement density region; clutter measurements were spatially distributed with a uniform distribution inside each cluttered region for every scan. In Figure 3, the squares represent the measurements of each target. The gray symbols represent the clutter measurements generated during a single simulation run.

Figure 4 represents the CTTR for three targets in 500 Monte Carlo runs, and the position RMSE for Target 1 and the estimated clutter measurement density for Target 1 over time are listed in Figures 5 and 6, respectively. For fair comparisons, the number of confirmed false tracks of each case was made to be almost 40 for all 500 Monte Carlo simulation runs by adjusting the initial target existence probability while the confirmation threshold was equal for all the algorithms in comparison. Using the true clutter measurement density showed that the CTTR had the fastest build-up. Even if the same sparsity order was applied, the proposed clutter measurement density estimation method provided better tracking results than the SCMDE. The closer to the true clutter measurement density the estimated clutter measurement density was, the better the performance was. At around Scan 25, when the targets were located in the immediate vicinity, the SCMDE estimated the clutter measurement density of the target measurement, which appeared to be bigger than the actual. This resulted in a slow build-up of the CTTR. By comparing the CTTR results for the sparsity order of $n = 1$ and $n = 5$ for the same clutter measurement density estimation methods, one could find that higher sparsity order resulted in better tracking performance because the higher the sparsity order was, the more accurate the estimated clutter measurement density was, as shown in Figure 6. The position RMSEs shown in Figure 5 were calculated for only the confirmed true tracks, which satisfied (46) such that the RMSEs looked similar in the order of magnitudes for all the algorithms in comparison as the confirmed true tracks passed the condition of (46). However, the number of samples involved in the RMSE calculation was quite different for each algorithms, as shown by the CTTR of Figure 4, which implied high reliability in RMSE for the algorithms with high CTTR and low reliability in RMSE for the algorithms with low CTTR. Figure 7 shows the true states and the estimated states of Target 1 over time for the position, velocity, and acceleration elements of each coordinate axis. Only the averaged state estimates of the confirmed tracks are shown in Figure 7. The existing SCMDE with the sparsity order of $n = 1$

showed the worst estimation performance among the algorithms in comparison. The target tracking algorithm using the proposed MTT-SCMDE with the sparsity order of $n = 5$ showed similar estimation performance to the one using the true clutter measurement density, and its state estimates were close to the true target states. This implied that the proposed MTT-SCMDE produced more reliable and accurate estimates for multi-target tracking than the existing SCMDE.

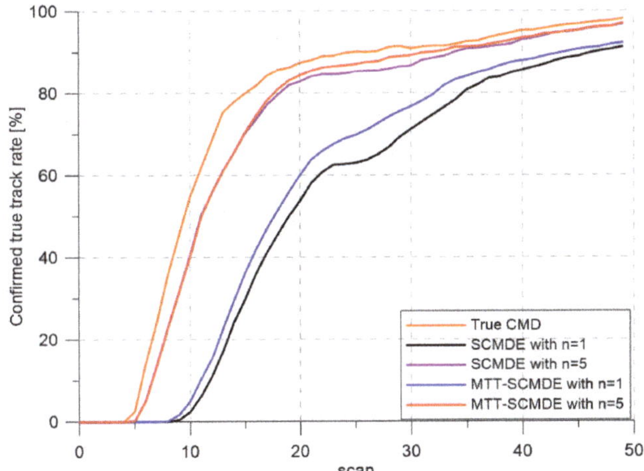

Figure 4. Confirmed true track rate. CMD, clutter measurement density.

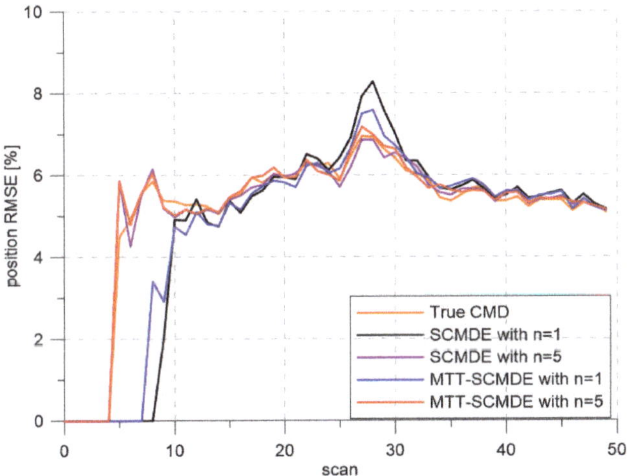

Figure 5. Position RMSE.

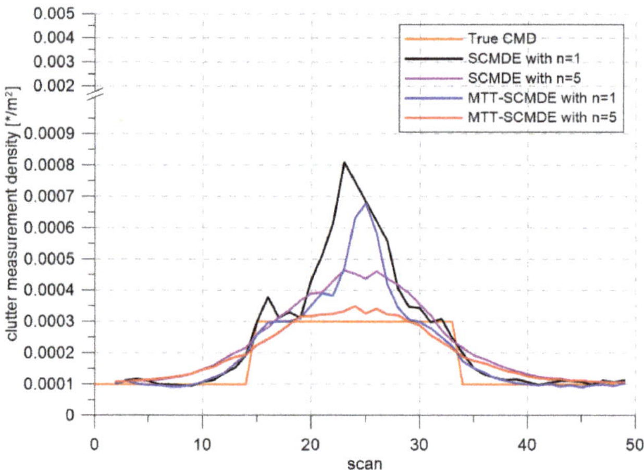

Figure 6. True clutter measurement density and estimated clutter measurement density.

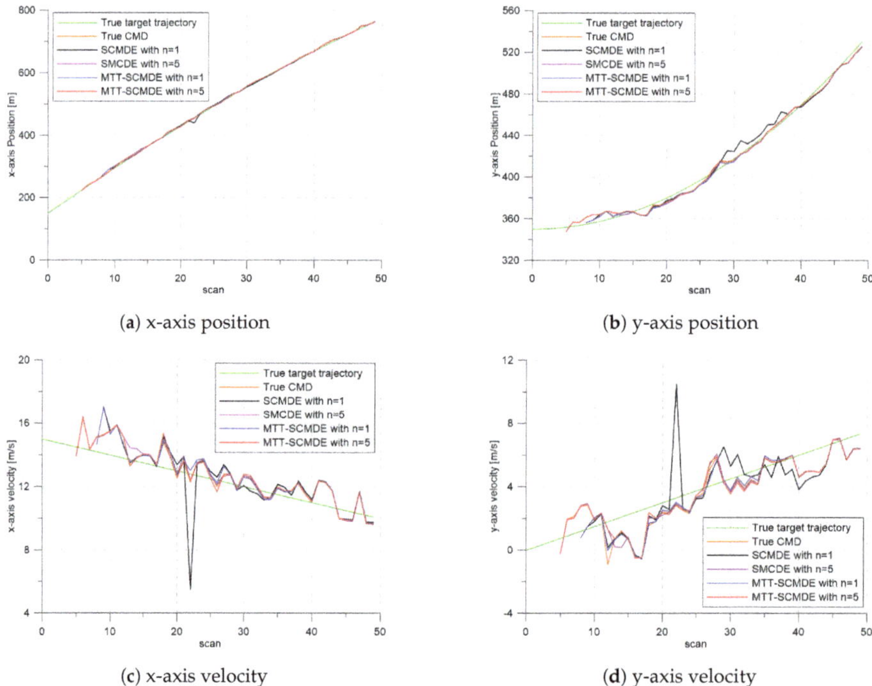

(**a**) x-axis position

(**b**) y-axis position

(**c**) x-axis velocity

(**d**) y-axis velocity

Figure 7. *Cont.*

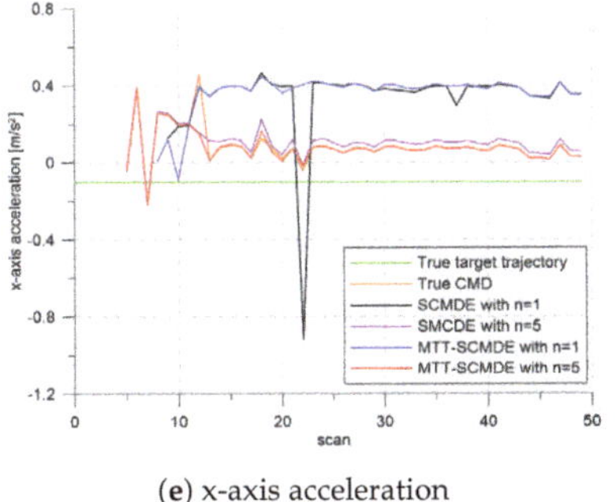

(**e**) x-axis acceleration

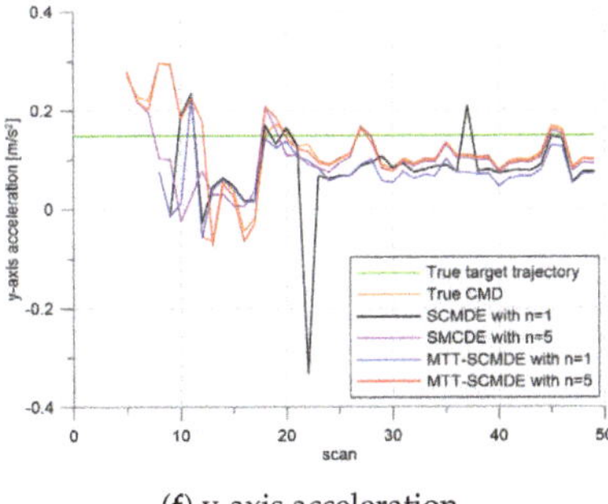

(**f**) y-axis acceleration

Figure 7. The true states and the estimated states of Target 1 over time.

Although the clutter measurement density was estimated close to the actual for the proposed method with sparsity order of $n = 5$, the tracking performance was slightly worse than using the true clutter measurement density. It produced the best tracking performance among the methods in comparison. Therefore, the proposed method with a high sparsity order was a viable solution for this environment.

Table 1 shows the statistics of the track retention test. The proposed clutter measurement density estimation method had a higher track maintenance performance in terms of true track confirmation and track losses including switch and merge than the SCMDE method with the same sparsity order.

Table 1. Track retention statistics for Monte Carlo simulation.

	True CMD	SCMDE with $n = 1$	SCMDE with $n = 5$	MTT-SCMDE with $n = 1$	MTT-SCMDE with $n = 5$
nCase	1199	446	1053	673	1056
nOk (%)	95.91	93.05	95.35	94.11	95.45
nSwitch (%)	1.08	2.91	1.33	2.03	1.42
nMerge (%)	0.59	0.67	0.85	1.10	0.76
nLost (%)	2.42	3.37	2.47	2.76	2.37

6.1.2. The Number of Targets: 5

In this scenario, we analyzed the clutter measurement density estimation performance by increasing the number of targets to five, as shown in Figure 8. The parameters except the number of targets were the same as in the previous scenario. The number of confirmed false tracks was made almost equal as in the previous scenario by adjusting the initial target existence probability.

Figures 9–11 represent CTTR, position RMSE for Target 1, and the estimated clutter measurement density for Target 1 over time for the scenario, respectively. All the algorithms had the same trend in estimation performance as in the previous scenario. The proposed clutter measurement density estimation method with the sparsity order of $n = 5$ showed the best tracking performance among the methods in comparison because it estimated the clutter measurement density similar to the true clutter measurement density even if the number of closely located targets increased. As shown in Table 2, nCase and nOk for the MTT-SCMDE with $n = 5$ represented the best tracking performance among the adaptive estimation methods in comparison.

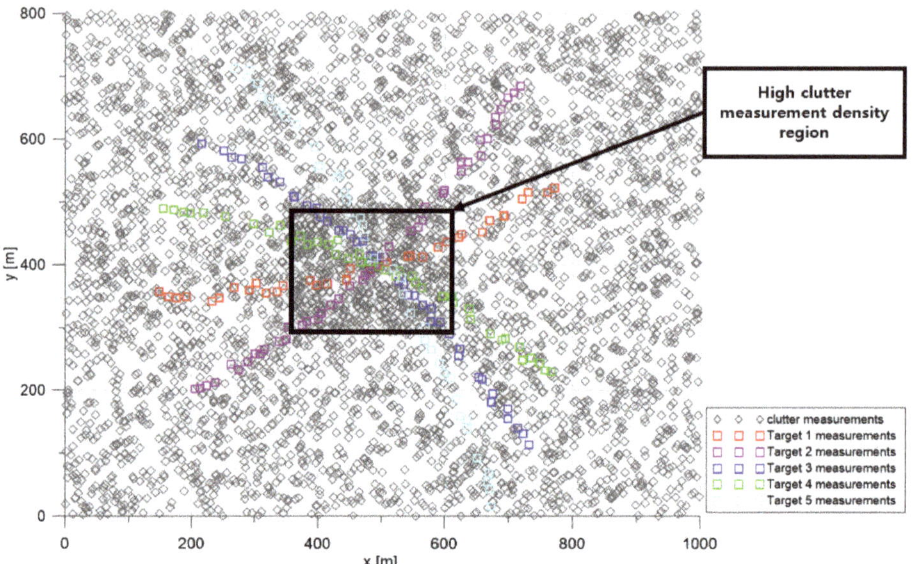

Figure 8. Simulation scenario with five targets.

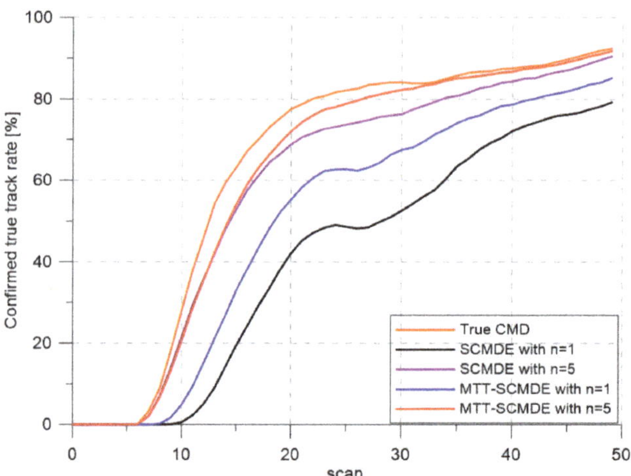

Figure 9. Confirmed true track rate.

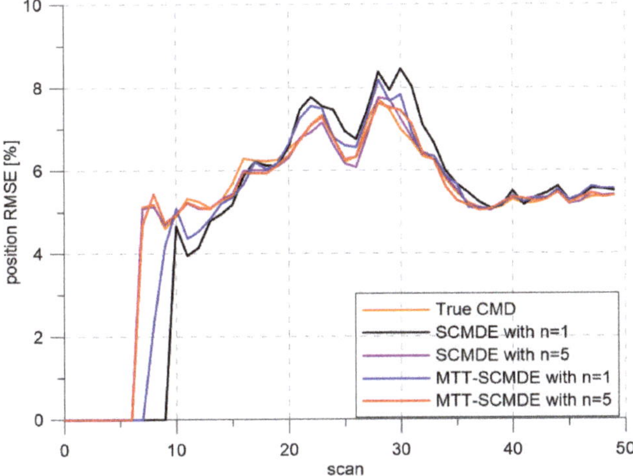

Figure 10. Position RMSE.

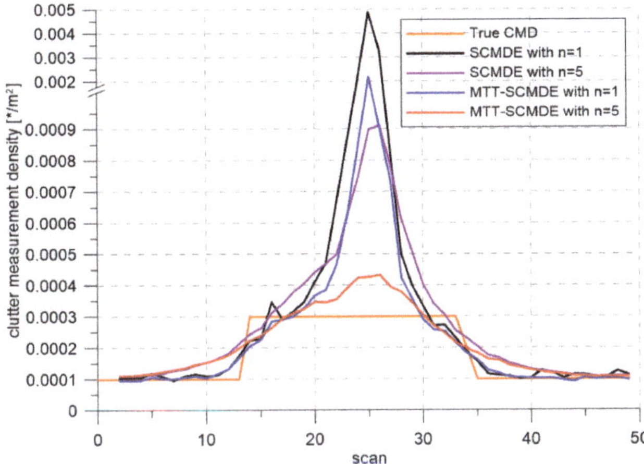

Figure 11. True clutter measurement density and estimated clutter measurement density.

Table 2. Track retention statistics for Monte Carlo simulation.

	True CMD	SCMDE with $n = 1$	SCMDE with $n = 5$	MTT-SCMDE with $n = 1$	MTT-SCMDE with $n = 5$
nCase	1581	485	1324	825	1354
nOk (%)	87.86	74.64	86.25	81.94	87.59
nSwitch (%)	5.76	13.20	6.42	7.52	6.28
nMerge (%)	4.87	8.45	5.66	6.91	4.36
nLost (%)	1.51	3.71	1.67	3.63	1.77

6.1.3. The Number of Targets: 7

The measurement histories of the seven closely located targets are shown in Figure 12. In this scenario, considering that the number of targets was seven, the simulation was performed by extending the sparsity order to 7 in addition to the 1 and 5 used in the previous scenarios. As in the previous scenarios, multiple targets were gathered in the high density clutter region.

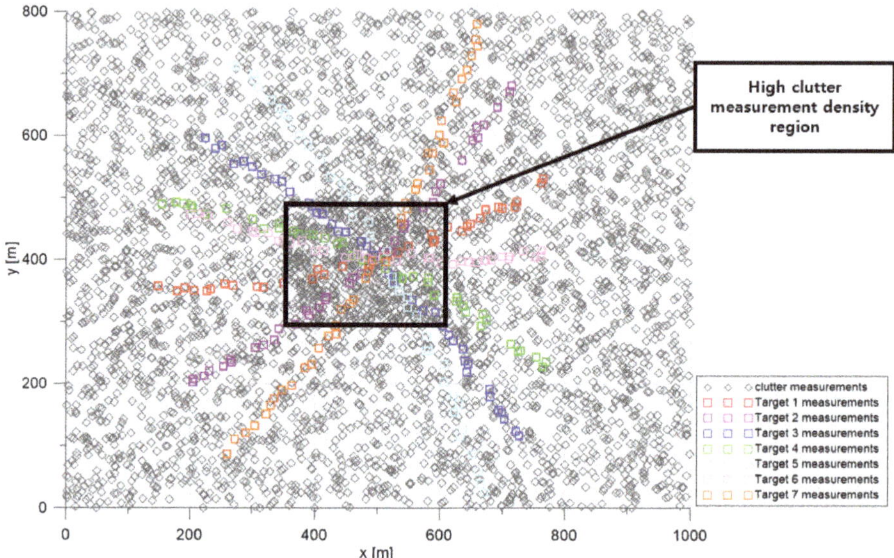

Figure 12. Simulation scenario with seven targets.

Figure 13 shows the CTTR over time. As shown in Figure 14, the estimation errors of with the sparsity order $n = 7$ were similar to the result using the true clutter measurement density. As the number of targets increased, increasing the sparsity order implied that better tracking results could be obtained, and the proposed MTT-SCMDE had better tracking performance compared to the existing SCMDE with the same sparsity order. Figure 15 represents the estimated clutter measurement density over time and shows that even with a large number of closely located targets, the proposed method had the best performance of estimating the clutter measurement density. In Table 3, the MTT-SCMDE with $n = 7$ showed more than 80% track retention performance, similar to the case with true clutter measurement density. It showed the best tracking performance among the adaptive estimation algorithms in comparison.

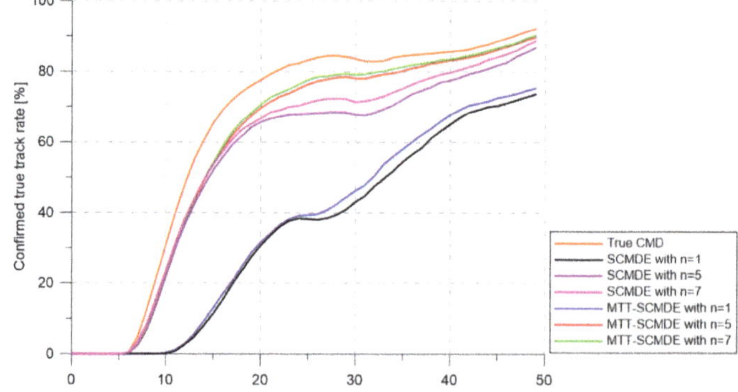

Figure 13. Confirmed true track rate.

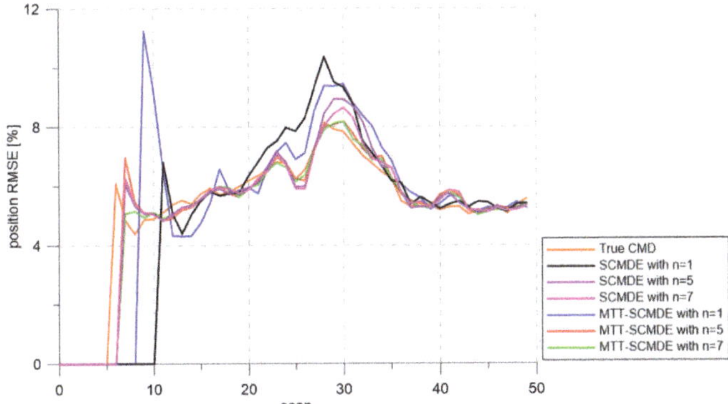

Figure 14. Position RMSE.

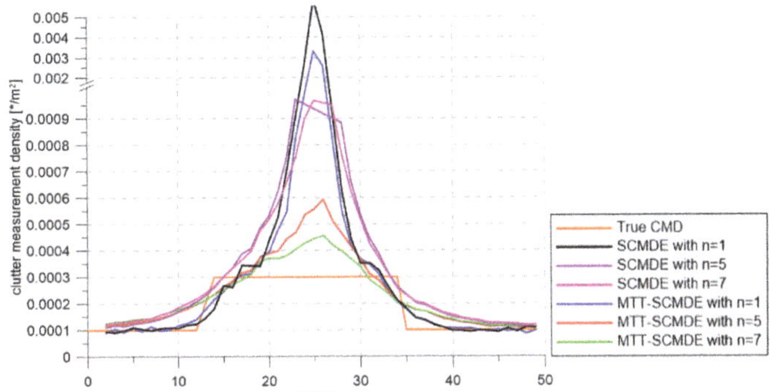

Figure 15. True clutter measurement density and estimated clutter measurement density.

Table 3. Track retention statistics for Monte Carlo simulation.

	True CMD	SCMDE with $n = 1$	SCMDE with $n = 5$	SCMDE with $n = 7$	MTT-SCMDE with $n = 1$	MTT-SCMDE with $n = 5$	MTT-SCMDE with $n = 7$
nCase	2291	398	1876	1881	457	1894	1900
nOk (%)	83.72	64.57	79.26	80.33	67.83	81.73	82.32
nSwitch (%)	8.07	18.59	10.13	10.10	19.26	10.09	9.74
nMerge (%)	7.03	13.82	9.22	7.55	8.97	6.02	5.84
nLost (%)	1.18	3.02	1.39	2.02	3.94	2.16	2.10

6.2. Test with Real Radar Data

In this section, a set of measurements obtained from a surveillance radar system is utilized for performance analysis of the proposed algorithm. The main focus of the analysis was to verify the robustness of the algorithm for tracking in clutter without track loss and switching, especially in the region where the multiple targets were located in the vicinity. For this data gathering experiment, there were no other reference sensors to measure the exact locations of the target. Therefore, it was not possible to analyze the accuracy of the target tracking, so we focused on the maintenance performance for the confirmed tracks and the discrimination performance for the false tracks caused by clutter.

The radar acquired measurements every one second. The 2D radar measurements consisted of distance and azimuth information. The measurements of the distance and azimuth information were converted to the x, y positions in the Cartesian coordinate system for the tracking algorithms. LM-IPDA with the NCV model in Section 2 was used for tracking in this performance test, and the results of target tracking were compared for three cases, which employed a fixed value (1×10^{-7} scan/m^2) for the clutter measurement density, adaptive clutter measurement density estimation with the existing SCMDE, and the proposed MTT-SCMDE.

The initial target existence probability of the track was set to be 0.1. When the target existence probability of track was smaller than $\frac{1}{10}$ of the initial value, the track would be terminated, and if the target existence probability was bigger than 0.95, it was classified as a confirmed track.

Figure 16 contains the measurement dataset for the entire period of 92 s. As shown in Figure 16, the radar detection range was 90 km, and the radar measurements were used from $-90°$ to $0°$ from the north. The gray symbols represent the measurements obtained from the radar.

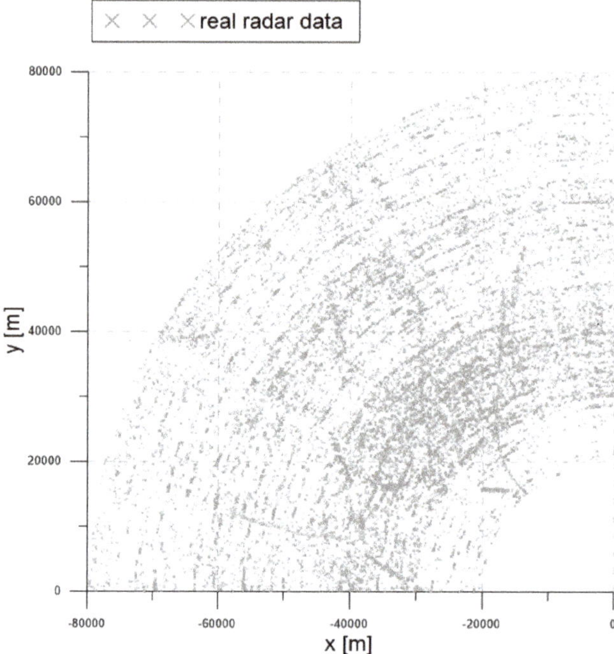

Figure 16. Real radar measurements within the surveillance region.

Figures 17–19 show the trajectories of the confirmed tracks estimated by the LM-IPDA algorithm with the NCV model, which utilized fixed clutter measurement density, the SCMDE, and the MTT-SCMDE, respectively. The sparsity order $n = 5$ was used for the SCMDE and the MTT-SCMDE. The main difference in the tracking results of the three cases was shown for the two targets in a formation flight in the high clutter measurement density region, which was specified by a green circle of each figure. In the case of using the fixed clutter measurement density, no confirmed track was generated for the left of the two targets in a formation flight. When the SCDME was used, the tracks for both targets were confirmed in the beginning, but one of the confirmed tracks was lost as the distance between the two targets became smaller. As the SCMDE did not distinguish the nature of adjacent measurements when estimating the clutter measurement density, a bias in the clutter measurement density estimates was included for the closely located targets, and this bias decreased the data association probability of the true target measurement. This resulted in the loss of the confirmed track. However, in the case of the proposed MTT-SCMDE, it can be seen from Figure 19 that the tracks for both targets were confirmed without loss of tracks. This demonstrated the robustness of the proposed MTT-SCMDE algorithm in practical applications.

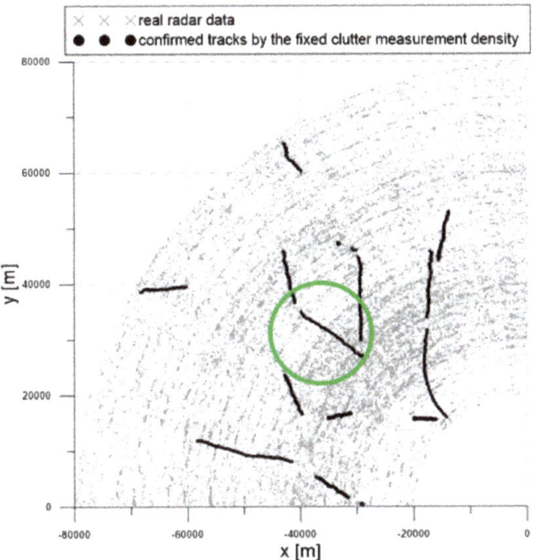

Figure 17. The trajectories of the confirmed tracks by using the fixed clutter measurement density.

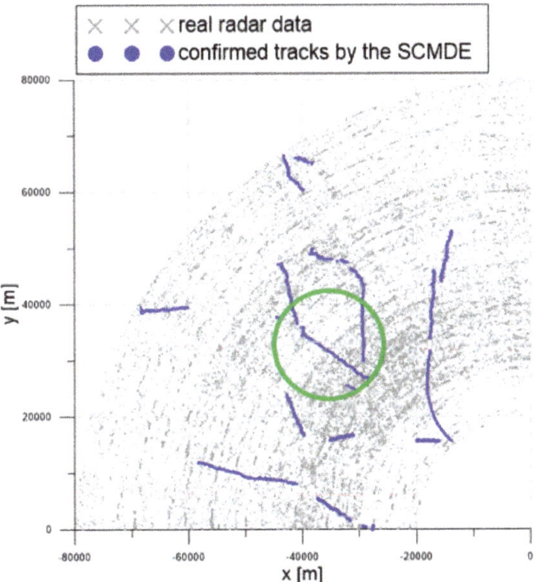

Figure 18. The trajectories of the confirmed tracks by using the SCMDE.

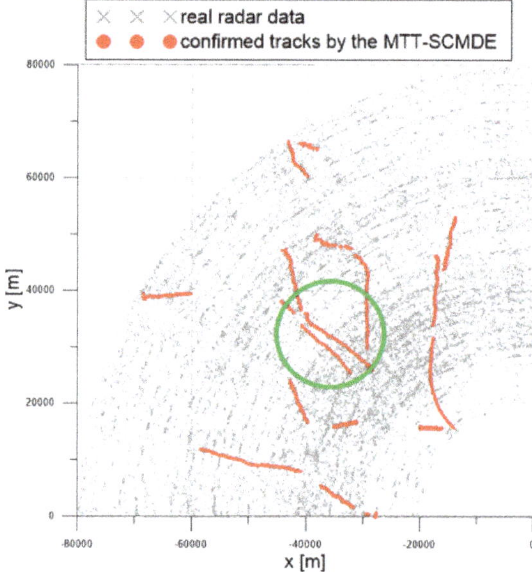

Figure 19. The trajectories of the confirmed tracks by using the proposed MTT-SCMDE.

7. Conclusions

The clutter measurement density is a parameter required to calculate the data association probability of the measurement and target existence probability of a track and has a large impact on target tracking performance even with small changes. This paper presented the SCMDE with clutter measurement probability to estimate the clutter measurement density adaptively for non-parametric multi-target tracking in environments where there is no prior information about clutter distribution. The algorithm was developed by analyzing the causes of estimation performance deterioration of the existing SCMDE. The proposed clutter measurement density estimation method calculated the sparsity of the measurements by probabilistically classifying adjacent measurements as a target measurement or as a clutter measurement. We demonstrated the effectiveness of the proposed clutter measurement density estimation method, which was designed to achieve more accurate and robust clutter measurement density estimation by showing the performance improvement for multi-target tracking through simulation studies in various environments and a test with real radar data.

Author Contributions: Conceptualization, S.H.P. and T.L.S.; methodology, S.H.P.; software, S.H.P. and S.Y.C.; validation, S.H.P. and T.L.S.; formal analysis, H.J.K. and T.L.S.; investigation, T.L.S.; resources, S.H.P.; data curation, S.H.P. and S.Y.C.; writing, original draft preparation, S.H.P. and H.J.K.; writing, review and editing, S.H.P.; visualization, S.H.P. and H.J.K.; supervision, T.L.S.; project administration, T.L.S.; funding acquisition, T.L.S.

Funding: This work was supported by Hanwha Research and Development Center.

Conflicts of Interest: The authors declare no conflict of interest.

Appendix A. Performance Analysis of the Existing SCMDE Used in Multi-Target Tracking Environments

In this analysis, clutter is distributed with a Gamma probability density function (pdf) with the number of clutter measurements, n, inside the hyper-sphere volume of V_n as [23],

$$p(V_n) = \frac{\rho}{(n-1)!}(\rho V_n)^{n-1}e^{-\rho V_n}, \tag{A1}$$

where ρ is the clutter measurement density. In this case, the number of clutter measurements is Poisson distributed inside the volume of a hyper-sphere, V, such as:

$$P(m) = \frac{(\rho V)^m}{m!} e^{-\rho V}. \tag{A2}$$

When the point of interest $\mathbf{z}_{k,i}$ is a target detection, the sparsity estimate that the existing SCMDE generates is described in (30) of Section 4.

For the target cardinality $|T| = 2$ case under the assumption that the position of another target is known, the conditional pdf of the sparsity estimates for $n = 1$ becomes:

$$p\left(\hat{\Upsilon}_{k,i}^{(1)} | D\right) = e^{-\rho D} \delta\left(\hat{\Upsilon}_{k,i}^{(1)} - D\right) + \rho e^{-\rho \hat{\Upsilon}_{k,i}^{(1)}} h\left(\hat{\Upsilon}_{k,i}^{(1)} - D\right), \tag{A3}$$

where $D = V(r_D)$ in (31) with r_D is the distance to the position of another target detection from the point of interest for $|T| = 2$. In (A3), δ is the delta function, and h is the Heaviside unit step function. For $n = 2$, the sparsity estimate of the existing SCMDE satisfies the following conditional pdf.

$$\begin{aligned} p\left(\hat{\Upsilon}_{k,i}^{(2)} | D\right) &= \rho D e^{-\rho D} \delta\left(\hat{\Upsilon}_{k,i}^{(2)} - \frac{D}{2}\right) + 2\rho e^{-2\rho \hat{\Upsilon}_{k,i}^{(2)}} h\left(\hat{\Upsilon}_{k,i}^{(2)} - \frac{D}{2}\right) \\ &+ 4\rho^2 \hat{\Upsilon}_{k,i}^{(2)} e^{-2\rho \hat{\Upsilon}_{k,i}^{(2)}} \left(h\left(\hat{\Upsilon}_{k,i}^{(2)}\right) - h\left(\hat{\Upsilon}_{k,i}^{(2)} - \frac{D}{2}\right)\right). \end{aligned} \tag{A4}$$

The conditional sparsity can be calculated from (A3) and (A4), and it becomes:

$$E\left\{\hat{\Upsilon}_{k,i}^{(n)} | D\right\} = \begin{cases} \frac{1}{\rho}(1 - e^{-\rho D}), & n = 1 \\ \frac{1}{\rho} - \frac{1}{\rho}\frac{(1+\rho D)}{2}, & n = 2 \end{cases}. \tag{A5}$$

As the target detection in (A3) and (A4) can be uniformly distributed in a hyper-sphere volume of $V^{(n)}$ with $0 \le D \le V^{(n)}$, the pdf of D becomes:

$$p(D) = \frac{1}{V^{(n)}}. \tag{A6}$$

Then, the average sparsity estimate is calculated, and it results in:

$$E\left\{\hat{\Upsilon}_{k,i}^{(n)}\right\} = \int_V E\left\{\hat{\Upsilon}_{k,i}^{(n)}\right\} p(D) dD = \begin{cases} \frac{1}{\rho}\left(1 - \frac{1 - e^{-\rho V^{(1)}}}{\rho V^{(1)}}\right), & n = 1 \\ \frac{1}{\rho}\left(1 - \frac{1 - e^{-\rho V^{(2)}}}{\rho V^{(2)}} + \frac{e^{-\rho V^{(2)}}}{2}\right), & n = 2 \end{cases}. \tag{A7}$$

Note that when the point of interest is the target detection for single target tracking environments, the original SCMDE generates the average sparsity such as:

$$E\left[\hat{\Upsilon}_{k,i}^{(n)}\right] = \frac{1}{\rho} \quad for\ all\ n. \tag{A8}$$

Similarly, the conditional pdf of the sparsity estimate under the known D assumption when the point of interest is a clutter detection for $|T| = 2$ can be derived as:

$$p\left(\hat{\Upsilon}_{k,i}^{(1)} | D\right) = e^{-\rho D} \delta\left(\hat{\Upsilon}_{k,i}^{(1)} - D\right) + \rho e^{-\rho \hat{\Upsilon}_{k,i}^{(1)}} \left(h\left(\hat{\Upsilon}_{k,i}^{(1)}\right) - h\left(\hat{\Upsilon}_{k,i}^{(1)} - D\right)\right) for\ n = 1, \tag{A9}$$

and:

$$p\left(\hat{\gamma}_{k,i}^{(2)}|D\right) = e^{-\rho D}\delta\left(\hat{\gamma}_{k,i}^{(2)} - \frac{D}{2}\right) + \rho D e^{-\rho D}\delta\left(\hat{\gamma}_{k,i}^{(2)} - \frac{D}{2}\right) + 2\rho e^{-\rho\hat{\gamma}_{k,i}^{(2)}} h\left(\hat{\gamma}_{k,i}^{(2)} - \frac{D}{2}\right)$$
$$+4\rho^2\hat{\gamma}_{k,i}^{(2)} e^{-2\rho\hat{\gamma}_{k,i}^{(2)}}\left(h\left(\hat{\gamma}_{k,i}^{(2)}\right) - h\left(\hat{\gamma}_{k,i}^{(2)} - \frac{D}{2}\right)\right) for\ n = 2. \quad \text{(A10)}$$

Then, the average sparsity estimate for (A9) and (A10) can be obtained as:

$$E\left[\hat{\gamma}_{k,i}^{(n)}\right] = \begin{cases} \frac{1}{\rho} - \frac{1}{\rho^2 V^{(1)}}\left(1 - e^{-\rho V^{(1)}}\right) & n = 1 \\ \frac{1}{\rho} - \frac{1}{2\rho^2 V^{(2)}}\left(1 - e^{-\rho V^{(2)}}\right) & n = 2 \end{cases}. \quad \text{(A11)}$$

Note that for $n = 1$ in (A11), the average sparsity estimate for $|T| = 2$ is the same as the one for single target tracking as shown in (23) of [23]. As the sparsity order n increases, the bias in the sparsity estimate reduces.

References

1. Challa, S.; Evans, R.; Morelande, M.; Mušicki, D. *Fundamentals of Object Tracking*; Cambridge University Press: Cambridge, UK, 2011.
2. Bar-Shalom, Y.; Fortmann, T.E. *Tracking and Data Association*; Academic Press: New York, NY, USA, 1988.
3. Bar-Shalom, Y.; Li, X.R. *Estimation with Application to Tracking and Navigation*; Wiley-Interscience: Hoboken, NJ, USA, 2001.
4. Blackman, S.S.; Popoli, R. *Design and Analysis of Modern Tracking Systems*; Artech House: Boston, MA, USA, 1999.
5. Mušicki, D. Integrated Probabilistic Data Association. In Proceedings of the 31st Conference on Decision and Control, Tucson, AZ, USA, 16–18 December 1992; pp. 3796–3798.
6. Mušicki, D.; Evans, R.; Stankovic, S. Integrated Probabilistic Data Association. *IEEE Trans. Autom. Control* **1994**, *39*, 1237–1241. [CrossRef]
7. Mušicki, D.; Evans, R.; Scala, B.L. Integrated Track Splitting Filter-Efficient Multi-Scan Single Target Tracking in Clutter. *IEEE Trans. Aerosp. Electron. Syst.* **2007**, *43*, 1409–1425.
8. Mušicki, D.; Evans, R. Multi-Scan Multi-Target Tracking in Clutter with Integrated Track Splitting Filter. *IEEE Trans. Aerosp. Electron. Syst.* **2007**, *45*, 1432–1447. [CrossRef]
9. Mušicki, D.; Evans, R.; Scala, B.L. Integrated Track Splitting Suite of Target Tracking Filters. In Proceedings of the 6th International Conference of Information Fusion, Cairns, Australia, 8–11 July 2003; pp. 1039–1046.
10. Mušicki, D.; Evans, R.; Scala, B.L. Multi-Target Tracking in Clutter Without Measurement Assignment. *IEEE Trans. Aerosp. Electron. Syst.* **2008**, *44*, 877–896. [CrossRef]
11. Mušicki, D.; Evans, R. Joint Integrated Probabilistic Data Association: JIPDA. *IEEE Trans. Aerosp. Electron. Syst.* **2004**, *40*, 1093–1099. [CrossRef]
12. Song, T.L.; Mušicki, D.; Kim, H.W. Iterative Joint Integrated Probabilistic Data Association for Multitarget Tracking. *IEEE Trans. Aerosp. Electron. Syst.* **2015**, *51*, 642–653. [CrossRef]
13. Mušicki, D.; Suvorova, S.; Morelande, M.; Moran, B. Clutter Map and Target Tracking. In Proceedings of the 7th International Conference of Information Fusion, Philadelphia, PA, USA, 25–28 July 2005; pp. 69–76.
14. Mušicki, D.; Evans, R. Clutter Map Information for Data Association and Track Initialization. *IEEE Trans. Aerosp. Electron. Syst.* **2004**, *40*, 387–398. [CrossRef]
15. Chen, X.; Tharmarasa, R.; Pelletier, M.; Kirubarajan, T. Integrated Clutter Estimation and Target Tracking using Poisson Point Processes. *IEEE Trans. Aerosp. Electron. Syst.* **2012**, *48*, 1210–1235. [CrossRef]
16. Chen, X.; Tharmarasa, R.; Pelletier, M.; Kirubarajan, T. Integrated Bayesian Clutter Estimation with JIPDA/MHT Trackers. *IEEE Trans. Aerosp. Electron. Syst.* **2013**, *49*, 395–414. [CrossRef]
17. Mahler, R.P.S. Bayes filtering via first-order multitarget moments. *IEEE Trans. Aerosp. Electron. Syst.* **2003**, *39*, 1152–1178. [CrossRef]
18. Kim, W.C.; Song, T.L. Interactive clutter measurement density estimator for multitarget data association. *IET Radar Sonar Navigation.* **2017**, *11*, 125–132. [CrossRef]

19. Vo, B.N.; Ma, W.K. The Gaussian Mixture Probability Hypothesis Density Filter. *IEEE Trans. Signal Process.* **2006**, *54*, 4091–4104. [CrossRef]
20. Davey, S.J.; Vu, H.X.; Arulampalam, S.; Fletcher, F.; Lim, C.C. Clutter mapping for Histogram PMHT. In Proceedings of the 2014 IEEE Workshop on Statistical Signal Processing, Gold Coast, Australia, 29 June–2 July 2014; pp. 153–156.
21. Li, N.; Li, X.R. Target perceivability and its applications. *IEEE Trans. Signal Process.* **2001**, *49*, 2588–2604.
22. Li, X.R.; Li, N. Integrated Real-time Estimation of Clutter Density for Tracking. *IEEE Trans. Signal Process.* **2000**, *48*, 2797–2805. [CrossRef]
23. Song, T.L.; Mušicki, D. Adaptive Clutter Measurement Density Estimation for Improved Target Tracking. *IEEE Trans. Aerosp. Electron. Syst.* **2011**, *47*, 1457–1466. [CrossRef]
24. Park, S.H.; Xie, Y.; Han, D.H.; Song, T.L. Spatial Clutter Measurement Density Estimation with the Clutter Probability for Improving Multi-Target Tracking Performance in Cluttered Environments. In Proceedings of the 2018 International Conference on Control, Automation and Information Sciences (ICCAIS), Hangzhou, China, 24–27 October 2018; pp. 90–95.
25. Li, X.R.; Jilokv, V.P. Survey of maneuvering target tracking. Part I. Dynamic models. *IEEE Trans. Aerosp. Electron. Syst.* **2003**, *39*, 1333–1364.
26. Konstantinova, P.; Udvarev, A.; Semerdjiev, T. A Study of a Target Tracking Algorithm Using Global Nearest Neighbor Approach. In Proceedings of the 4th international conference on Computer systems and technologies, Rousse, Bulgaria, 19–20 June 2003; pp. 290–295.
27. Sinha, A.; Ding, Z.; Kirubarajan, T.; Farooq, M. Track Quality Based Multitarget Tracking Approach for Global Nearest-Neighbor Association. *IEEE Trans. Aerosp. Electron. Syst.* **2012**, *48*, 1179–1191. [CrossRef]
28. Fortmann, T.E.; Bar-Shalom, Y.; Scheffe, M. Multi-target Tracking Using Joint Probabilistic Data Association. In Proceedings of the 19th IEEE Conference on Decision and Control including the Symposium on Adaptive Processes, Albuquerque, NM, USA, 10–12 December 1980; pp. 807–812.
29. Fortmann, T.E.; Bar-Shalom, Y.; Scheffe, M. Sonar Tracking of Multiple Targets Using Joint Probabilistic Data Association. *IEEE J. Ocean. Eng.* **1983**, *38*, 173–184. [CrossRef]
30. Chang, K.C.; Chong, C.Y.; Bar-Shalom, Y. Joint Probabilistic Data Association in Distributed Sensor Networks. *IEEE Trans. Autom. Control* **1986**, *31*, 889–897. [CrossRef]
31. Papoulis, A.; Pillai, U. *Probability, Random Variables and Stochastic Processes;* MC Graw Hill: New York, NY, USA, 2002.
32. Mušicki, D.; Suvorova, S.; Challa, S. Multi target tracking of ground targets in clutter with LMIPDA-IMM. In Proceedings of the 7th International Conference on Information Fusion, Stockholm, Sweden, 28 June–1 July 2004; pp. 1104–1110.

Article

Multi-Target Localization and Tracking Using TDOA and AOA Measurements Based on Gibbs-GLMB Filtering †

Zhengwang Tian, Weifeng Liu * and Xinfeng Ru

Xiasha Higher Education Zone, Hangzhou 310018, Zhejiang, China; 17764590774@163.com (Z.T.); a840064210@hdu.edu.cn (X.R.)

* Correspondence: dashan_liu@163.com; Tel.: +86-0571-86915193

† This paper is an extended version of our paper published in Z. Tian, W. Liu, and X. Ru. "Multi-acoustic Array Localization and Tracking Method based on Gibbs-GLMB". In Proceedings of ICCAIS 2019 Conference, Chengdu, China, 23–26 October 2019.

Received: 31 October 2019; Accepted: 6 December 2019; Published: 10 December 2019

Abstract: This paper deals with mobile multi-target detection and tracking. In the traditional method, there are uncertainties such as misdetection and false alarm in the measurement data, and it will be inevitable having to deal with the data association. To solve the target trajectory and state estimation problem under a cluttered environment, this paper proposes a non-concurrent multi-target acoustic localization tracking method based on the Gibbs-generalized labelled multi-Bernoulli (Gibbs-GLMB) filter and considers an acoustic array of a fixed arrangement for the tracking of targets by joint time difference of arrival (TDOA) and angle of arrival (AOA) measurements. Firstly, the TDOAs are calculated by using the generalized cross-correlation algorithm (GCC) and the AOAs are derived from the received signal directions. Secondly, we assume the independence of the targets and fuse the measurements which are used to track the multiple targets via the Gibbs-GLMB filter. Finally, the effectiveness of the method is verified by Monte Carlo simulation experiments.

Keywords: passive localization; time difference of arrival; angle of arrival; random finite sets; Gibbs sampling; GLMB filter; multi-target tracking

1. Introduction

Passive detection, such as multi-sensor array localization of acoustic sources, plays an important role in the field of target tracking [1]. Localization through acoustic signals has broad applications in both civil and military fields, for example, the detection of unknown objects in the airport, detection of illegal traffic whistles/horns, localization of submarines or marine animals and the localization of explosion sources. Existing passive localization and tracking techniques include Time Difference of Arrival (TDOA) [2], and Angle of arrival (AOA) [3]. TDOA requires multiple observation devices to be used at the same time to ensure that the clock time of the sensor is consistent among the multiple groups of sensors [4].

In multi-target scenarios, it is difficult to determine the number and states of targets due to clutter effects, miss detection and data association uncertainty. Reference [5] proposed the route-based dynamic modeling to improve data association performance. For the complex and non-linear acoustic signals, traditional signal processing techniques such as the Fang algorithm, the music algorithm and the Taylor series expansion method are used [2]. To address the nonlinearity of the measurement process, the extended Kalman filter, one of the common methods in target tracking, can be used to provide better information for multi-sensor information fusion. In Reference [6], they performed data fusion that combines the active detection and the passive interception using

Maximum Likelihood Estimation (MLE). In Reference [7], MLE based on compressed sensing is proposed for the TDOA method. In Reference [8], the authors developed MLE for the proposed model through the Gauss-Newton iteration and semidefinite relaxation. An extended Kalman particle filtering (EKPF) approach for non-concurrent multiple acoustic tracking (NMAT) has been studied in Reference [9]; however, this paper only considers single scan tracking.

Currently, The Joint Probabilistic Data Association Filter (JPDAF), Multiple Hypothesis Tracking (MHT) and Random Finite Set (RFS) [10] are three mainstream methods for multi-target tracking. Traditional approaches such as JPDAF and MHT are extended to solve multi-target tracking based on single-target tracking, mostly in an ad-hoc manner. As the number of targets increases, the calculation amount of these methods will increase exponentially. The RFS approach pioneered by Mahler [11–14] provides a top-down state-space model formulation for multiple object system based on fundamental concepts in estimation theory, such as multi-target estimation error [15] and Bayes optimality [13,14]. Due to its mathematically rigorous foundation, the RFS theory has received worldwide attention in recent years and is considered to be a way to solve multi-target tracking.

Many well-known multi-target filters have been developed from the RFS framework, for example, the Probability Hypothesis Density filter [11,16,17], cardinality-banlance multi-target multi-Bernoulli filter [18], Cardinalized PHD filter [12,19]. However, these filters can only obtain the scatter set estimation of the target but cannot form the target trajectories, though several heuristics have been proposed to join state estimates from different times steps to form trajectories. In spite of this, these filters have been widely used in many fields, for example, computer vision [20–22]; sensor scheduling [23,24]; multi-sensor fusion [25]. Reference [26] propose to solve the multi-target sensor management by using the random set method in the POMDP] framework; References [27–29] use Cauchy-Schwarz divergence and Rényi divergence as information gains, respectively, provide a new sensor scheduling; robotics [30,31] and group target tracking [32].

The RFS multi-target trackers are formulated via labeled RFS [13,14,33]. In Reference [33], Mahler has shown that labeled RFS is the only principled approach that can provide target trajectories from the filtering density. The recent breakthrough in multi-target tracking is a filter labeled RFS called the Generalized Labeled Multi-Bernoulli (GLMB) or the Vo-Vo filter [34,35]. This filter is the first analytic method to Bayes filter with the multi-target, which provides estimates of target trajectories with linear complexity, and can be efficiently implemented by jointing the update step and the prediction step, for more effective multi-target tracking [36]. GLMB filtering has been demonstrated to track more than one million targets in heavy clutter, misdetections and data association uncertainty [37]. Another advantage of labeled RFS over unlabeled RFS is that it can provide ancestry or lineage information in problems that involve spawning targets [38]. Such capability is not possible without labels.

GLMB RFS has been applied in many fields, such as tracking with merged measurements [39], extended targets [40], computer vision [41–43], cell tracking [44,45], track-before-detect [46,47], sensor scheduling [48,49], field robotics [50–52], distributed tracking [53,54] and cell microscopy [55]. The GLMB solution has also been applied to the multi-sensor case [56] and the multi-scan case [57]. The multi-sensor GLMB filter [56] is the first multi-sensor solution with linear complexity in the sum of number of measurements. The multi-scan GLMB filter [57] is the first solution that is demonstrated to handle as much as 100 scans as well as providing posterior statistics about the set of target trajectories.

The work in this paper is based on Reference [1] and the published conference papers [58]. The measurement data is calculated from the real sound source signal, and the positioning and tracking of multiple target states via the GLMB filter under the RFS framework. To implement the GLMB filter more effectively, Vo et al. proposed joint the update step and the prediction step, eliminating the inefficiency caused by the primal two process redundancy and adopting the Gibbs sampling method in the truncation process [36], which provide a more effective solution for the ranked assignment problem(the data association problems). The Gibbs-GLMB has been proven to provide faster and more accurate results based on the GLMB filter.

Multi-acoustic array localization is typical of multi-sensor, passive localization and nonlinear problems. In the case of traditional algorithms dealing with multi-sensor measurement uncertainty and target uncertainty, the target tracking is based on the method of associating data and there is no effective way to estimate the number of targets. The Gibbs-GLMB filtering can effectively solve these problems. We change from the original active tracking to passive tracking based on the original Gibbs-GLMB filter [36]. TDOA and AOA measurements are generated by calculating the true signal correlation, which is a nonlinear estimation problem. The TDOAs and the AOAs are computed by the generalized cross-correlation (GCC) [4] and receiving direction of acoustic signal, respectively.

For the reasons above, we use the Gibbs-GLMB filter. First, we assume that the target obeys a cv motion model and the observation information includes TDOA and AOA. Secondly, target tracking in a multi-sensor array is done using the Gibbs sampling implementation of the GLMB filter (Gibbs-GLMB filter), to reduce the computational complexity of the algorithm without sacrificing accuracy. The effectiveness of the algorithm is verified by three pairs of acoustic array sensors are deployed to track three targets.

2. Background

2.1. NOTATION

- Single-target state is expressed by a small letter, (e.g., x).
- Multi-target states are represented by an italic capital letter, (e.g., X).
- The labeled states and distribution are bolded, (e.g., $\boldsymbol{x}, \boldsymbol{X}, \boldsymbol{\pi}$).
- The spaces are represented by blackboard bold (e.g., the state space $\mathbb{X}$ and measurement space $\mathbb{Z}$).
- $\mathcal{F}(\mathbb{X})$ is the all finite subsets of $\mathbb{X}$.
- The inner product symbol is abbreviated as:$\langle f, g\rangle \triangleq \int f(x)\, g(x)\, dx$.
- The following multi-target exponential notation $h^X \triangleq \prod_{x\in X} h(x)$, where h is a real-valued function, with $h^{\emptyset} = 1$.
- The generalization of the Kronecker delta for sets, vectors and integers:

$$\delta_Y(X) = \begin{cases} 1, if X = Y \\ 0, otherwise, \end{cases}$$

where inclusion function is denoted as:

$$1_Y(X) = \begin{cases} 1, if X \subseteq Y \\ 0, otherwise. \end{cases}$$

- $X_{m:n}$ is shorthand for the list of variables $X_m, X_{m+1}, \cdots, X_n$.

2.2. Random Finite Set

In the multi-target environment based on RFS framework in time k, the states of multiple targets can be denoted by a set as $X_k = \left\{x_{k,1}, \cdots, x_{k,N(k)}\right\} \in \mathcal{F}(\mathcal{X})$ [13], where $\mathcal{F}(\mathcal{X})$ is the all of finite subsets of state space and $N(k)$ is the number of surviving targets. In the similar way, the observation about TDOAs and AOAs of the qth sensor pair can be described as $Z_k^{[q]} = \left\{z_{k,1}^{[q]}, \cdots, z_{k,M_q(k)}^{[q]}\right\} \in \mathcal{F}(\mathcal{Z})$, where $\mathcal{F}(\mathcal{Z})$ is the space of finite subsets of observation space $\mathcal{Z}$ and $M_q(k)$ is the number of observed measurement.

In the location tracking area, there are many uncertainties in the number and measurement of the detection process, such as birth, death, derivation, false alarm and missed detection. Consequently, The multiple targets state in time k can be defined as [59,60]:

$$X_k = \left[\bigcup_{x \in X_{k-1}} S_{k|k-1}(x) \right] \cup \left[\bigcup_{x \in X_{k-1}} B_{k|k-1}(x) \right] \bigcup \Gamma_k, \tag{1}$$

where $S_{k|k-1}(x)$, $B_{k|k-1}(x)$ and Γ_k are the RFS of survival target at time $k-1$, the RFS of the target spawn at time k from the survival target at time $k-1$ and the RFS of target new-born, respectively. There are clutter or false alarms in the tracking area, which can be expressed as:

$$Z_k^{[q]} = \left[\bigcup_{x \in X_k} \Theta_k^{[q]}(x) \right] \bigcup \mathcal{K}_k^{[q]}, \tag{2}$$

where $\Theta_k^{[q]}(x)$ is the measurements with the RFS which produced by targets in the tracking area:

$$\Theta_k^{[q]}(x_k) = \begin{cases} \phi, & H_{miss} \\ \left\{ z_k^{[q]} \right\}, & \bar{H}_{miss} \end{cases} \tag{3}$$

here, the $\bar{H}_{miss}$ and H_{miss} are the hypotheses of detection and miss detection, more generally, whether the sensor has received the signal generated by targets. Moreover, $\mathcal{K}_k^{[q]}$ is the measurement set of alarms or clutter false which follows a poisson distribution with a uniform density $\mathcal{U}(z)$ on the observation area and is given by:

$$\mathcal{K}_k(z_k) = \frac{\lambda_c}{\int \mathcal{U}(z)\, dz} \mathcal{U}(z). \tag{4}$$

In RFS framework, the probability density function that the state of multi-target makes a transition from state X_{k-1} to X_k can be described as:

$$f_{k|k-1}(X_k | X_{k-1}) = \sum_{W \in X_k} \pi_{T,k|k-1}(W | X_{k-1}) \times \pi_{\Gamma,k}(X_k - W), \tag{5}$$

where $\pi_{T,k|k-1}(\cdot|\cdot)$ is the probability density of spontaneous target birth and $\pi_{\Gamma,k}(\cdot)$ is the probability density of target new-born.

2.3. Multi-Bernoulli RFS

The state of the target and the measurements are random variables. The Bernoulli distribution can be used to describe a single target $X \in \mathbb{X}$. Hence, a singleton target probability X is r which satisfies the spatial distribution of a probability density $p(x)$ and the probability that the target does not exist is $1-r$. The probability density distribution of the Bernoulli RFS is written as follows:

$$\pi(x) = \begin{cases} 1-r & X = \emptyset \\ r \cdot p(x) & X = \{x\} \\ 0 & \text{otherwise.} \end{cases} \tag{6}$$

The Multi-Bernoulli RFS is given by combining of the M independent Bernoulli RFS $X^{(i)} \in \mathbb{X}, i = 1, \cdots, M$(satisfying $X = \bigcup_{i=1}^{M} X^{(i)}$) with existence probability $r^{(i)} \in (0,1)$, which is described by $\left\{ \left(r^{(i)}, p^{(i)} \right) \right\}_{i=1}^{M}$. $\sum_{i=1}^{M} r^{(i)}$ is the mean cardinality of the Multi-Bernoulli RFS. Therefore, the probability density distribution of multi-Bernoulli is expressed by [12,18]:

$$\pi(\{x_1, \ldots, x_n\}) = \prod_{j=1}^{M} \left(1 - r^{(j)}\right) \sum_{1 \leqslant i_1 \neq \cdots \neq i_n \leqslant M} \prod_{j=1}^{n} \frac{r^{(i_j)} p^{(i_j)}(x_j)}{1 - r^{(i_j)}}. \tag{7}$$

2.4. Labeled Multi-Bernoulli RFS

A labeled-random finite set (L-RFS) [34,35] means that each state of the RFS has a unique tag. This means we attach a unique label $l \in \mathbb{L} = \{\alpha_i : i \in \mathbb{N}\}$ to each state $x \in \mathbb{X}$ where $\mathbb{L}$ is discrete countable space and $\mathbb{N}$ is the positive integer set space. The single target state is expressed as:

$$\mathrm{x}_{k,N(k)} = \left(x_{k,N(k)}, l_{k,N(k)}\right) \in \mathbb{X} \times \mathbb{L}. \tag{8}$$

The labels of the set $\mathrm{X} \subset \mathbb{X} \times \mathbb{L}$ can be represented by $\mathcal{L}(\mathrm{X}) = \{\mathcal{L}(\mathrm{x}) : \mathrm{x} \in \mathrm{X}\}$, where $\mathcal{L} : \mathbb{X} \times \mathbb{L} \rightarrow \mathbb{L}$ is defined by $\mathcal{L}((x,l)) = l$. The distinct label indicator is defined by $\Delta(\mathrm{X}) = \delta_{|\mathrm{X}|}(|\mathcal{L}(\mathrm{X})|)$.

The parameter of a labeled multi-Bernoulli(LMB) RFS can be described as a set $\{(r^{(\zeta)}, p^{(\zeta)}) : \zeta \in \Psi\}$ with index set Ψ. We extend the problem on space$\mathbb{X}$ to space$\mathbb{X} \times \mathbb{L}$, thus, the probability density distribution of labeled LMB-RFS is given by [34]:

$$\pi(\{(x_1, l_1), \ldots (x_n, l_n)\}) = \delta_n(|\{l_1, \ldots l_n\}|) \prod_{\zeta \in \Psi} \left(1 - r^{(\zeta)}\right) \prod_{j=1}^{n} \frac{1_{\alpha(\Psi)}(l_j)\, r^{(\alpha^{-1}(l_j))} p^{(\alpha^{-1}(l_j))}(x_j)}{1 - r^{(\alpha^{-1}(l_j).)}} \tag{9}$$

The following simplified alternative form of the LMB can be simplified as:

$$\pi(\mathrm{X}) = \Delta(\mathrm{X})\, 1_{\alpha(\Psi)}(\mathcal{L}(\mathrm{X}))\, [p(\cdot)]^{\mathrm{X}} \tag{10}$$

2.5. GLMB RFS

A generalized label multi-Bernoulli RFS under the state space $\mathbb{X}$ and the label space $\mathbb{L}$ has the following distribution [34]:

$$\pi(\mathrm{X}) = \Delta(\mathrm{X}) \sum_{\xi \in \Xi} \omega^{(\xi)}(\mathcal{L}(\mathrm{X})) \left[p^{(\xi)}\right]^{\mathrm{X}}, \tag{11}$$

where $\xi = (\theta_{1:k}) \in \Theta$ is a historical association maps. The non-negative $\omega^{(\xi)}(L)$ and a probability density $p^{(\xi)}$ satisfy:

$$\sum_{L \in \mathbb{L}} \sum_{\xi \in \Xi} \omega^{(\xi)}(L) = 1 \tag{12}$$

$$\int p^{(\xi)}(x,l)\, dx = 1. \tag{13}$$

3. Problem Formulation

3.1. Model Environment

There are multiple sets of acoustic sensor arrays in the detection range, denoted as $S_{1:Q} = \{s_1, \cdots, s_q, \cdots, s_Q\}, q \in \{1, 2, \cdots, Q\}$, where $s_q = \{s_{q,1}, \cdots, s_{q,j}, \cdots, s_{q,N}\}, j \in \{1, \cdots, N\}$. Each $s_{q,j}$ can be defined as $s_{q,j} = (x_{q,j}, y_{q,j})$, in 2-dimensional space.

Assuming a single target state of position is $\mathrm{x}_i = (x_i, y_i)$, each set of sensors consists of two acoustic sensors.

3.1.1. Time Difference of Arrival

The time difference τ is expressed as:

$$TDOA_q = \left| \frac{\|\mathrm{x}_i - s_{q,1}\| - \|\mathrm{x}_i - s_{q,2}\|}{v} \right|, \tag{14}$$

where both x_i and $s_{q,j}$ are in Cartesian coordinates, $\|\cdot\|$ is the Euclidean-norm and v is the velocity of sound.

3.1.2. Angle of Arrival

For sensor 1, AOA can be expressed as:

$$AOA_{q,1} = \arctan\left(\frac{y_i - y_{q,1}}{x_i - x_{q,1}}\right). \tag{15}$$

For ease of understanding, the TDOA and AOA are illustrated in the positioning system of Figure 1, where: $TDOA = \left|\frac{MN}{velocity\ of\ sound}\right|$, $AOA = \alpha$.

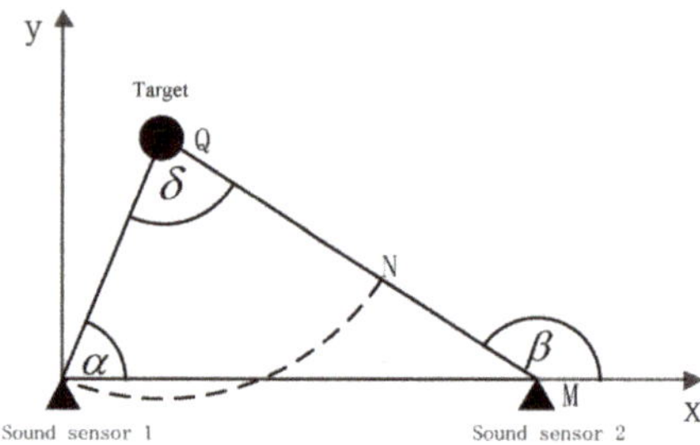

Figure 1. A pair of sensor array model diagrams.

3.2. Measurement

3.2.1. TDOA Measurement

The signals observed by a pair of sensors can be mathematically described as:

$$z_1(t) = \alpha_1 s(t) + n_1(t) \tag{16}$$

$$z_2(t) = \alpha_2 s(t - \tau_{1:2}) + n_2(t), \tag{17}$$

where $z_1(t)$ and $z_2(t)$ are the signals received by the pair of sensor array, $s(t)$ is the true signal, $n_1(t)$ and $n_2(t)$ are noise signals, $\tau_{1:2}$ is the time difference between two sensors detecting the signal, α_1 and α_2 are signal amplitudes [8].

The time difference can be estimated by the generalized cross correlation (GCC) method [4]:

$$R_{GCC}(\tau_q) = \int_{-\infty}^{\infty} \psi_{12}(\omega) Z_1(\omega) Z_2^*(\omega) e^{-j\omega\tau_q} d\omega \tag{18}$$

$$\hat{\tau}_q = \arg\max R_{GCC}(\tau_q) \tag{19}$$

$$\psi_{1,2}(\omega) = \frac{1}{|G_{x_1x_2}(\omega)|} = \frac{1}{|Z_1(\omega) Z_2^*(\omega)|}. \tag{20}$$

Here, $R_{GCC}(\tau_q)$ is the GCC, where $Z_1(\omega)$ is the Fourier transforms of $z_1(t)$ and $Z_2^*(\omega)$ is the Fourier transform conjugate of $z_2(t)$. $\psi_{1,2}(\omega)$ is the weight function of GCC. To reduce environmental noise and reverberation interference, we choose the phase transform (PHAT) as our weight function, the formula is given by $\psi_{1,2}(\omega) = \frac{1}{|G_{x_1x_2}(\omega)|}$.

3.2.2. AOA Measurement

The AOA is calculated by the positional relationship between the sensor and the target position. The difference in the AOAs of the sensor array is calculated by combining the measurements of the two sensors:

$$\delta_q = \left| \arctan\left(\frac{y_i - y_{q,1}}{x_i - x_{q,1}}\right) - \arctan\left(\frac{y_i - y_{q,2}}{x_i - x_{q,2}}\right) \right|. \tag{21}$$

3.3. Motion Model

We take the CV model as an example of a linear model, also known as a non-maneuver model:

$$\begin{bmatrix} \dot{x} \\ \ddot{x} \end{bmatrix} = \begin{bmatrix} 0 & 1 \\ 0 & 0 \end{bmatrix} \begin{bmatrix} x \\ \dot{x} \end{bmatrix} + \begin{bmatrix} 0 \\ 1 \end{bmatrix} w(t), \tag{22}$$

where x is the location of the target, $\dot{x}$ is the velocity of the target, $\ddot{x}$ is the acceleration of the target, $w(t)$ is zero mean white noise. Let T denotes the sampling interval, then the discrete-time model is given by:

$$\begin{bmatrix} x_{k+1} \\ \dot{x}_{k+1} \end{bmatrix} = \begin{bmatrix} 1 & \mathrm{T} \\ 0 & 1 \end{bmatrix} \begin{bmatrix} x_k \\ \dot{x}_k \end{bmatrix} + \begin{bmatrix} \mathrm{T}^2/2 \\ \mathrm{T} \end{bmatrix} w(t). \tag{23}$$

4. TDOA Localization Algorithm Based on SMC-GLMB Filtering

4.1. Target State Estimation

The purpose of multi-target tracking is to jointly estimate the target cardinality and target states based on the observations. Multi-target tracking can be transformed into the recursive Bayesian estimation problem by modeling the state and measurement of multi-target using RFS. The RFS approach can effectively deal with the uncertainty of data association between the target and the measurement and the state probability density function of the set of targets. We use $\pi_k(\cdot|Z_{1:k})$ to indicate the RFS posterior probability density of multi-target state; $f_{k|k-1}(\cdot|\cdot)$ to represent the multi-target transition density; $g_k(\cdot|\cdot)$ to represent the likelihood function. The posterior probability density of multiple targets is recursively calculated by the following prediction and update steps [11–14]:

$$\pi_{k|k-1}(X_k|Z_{1:k-1}) = \int f_{k|k-1}(X_k|X_{k-1})\pi_{k-1}(X_{k-1}|Z_{1:k-1})\,\delta X_{k-1}$$

$$\pi_k(X_k|Z_{1:k}) = \frac{g_k(Z_k|X_k)\,\pi_{k|k-1}(X_k|Z_{1:k-1})}{\int g_k(Z_k|X_k)\,\pi_{k|k-1}(X_k|Z_{1:k-1})\delta X_k},$$

where the set integral on $\mathcal{F}(\mathbb{X} \times \mathbb{L}) \to \mathbb{R}$ is defined as:

$$\int f(\mathrm{X})\delta\mathrm{X} = \sum_{i=0}^{\infty} \frac{1}{i!} \int f(\{\mathrm{x}_1, \cdots, \mathrm{x}_i\}) d(\mathrm{x}_1, \cdots, \mathrm{x}_i).$$

All information about the multiple targets states are included in the multi-target posterior, for example, the number and state of the target at the current time.

We experience with q pairs of sensors, therefore the above update step and prediction step can be rewritten as [59–61]:

$$\pi_{k|k-1}\left(X_k \middle| Z_{1:k-1}^{[1:Q]}\right) = \int f_{k|k-1}(X_k|X_{k-1})\pi_{k-1}\left(X_{k-1}\middle|Z_{1:k-1}^{[1:Q]}\right)\delta X_{k-1} \tag{24}$$

$$\pi_k\left(X_k\middle|Z_{1:k-1}^{[1:Q]}\right) = \frac{\prod_{q=1}^{Q} g_k\left(Z_k^{[q]}|X_k\right)\pi_{k|k-1}\left(X_k\middle|Z_{1:k-1}^{[1:Q]}\right)}{\int \prod_{q=1}^{Q} g_k\left(Z_k^{[q]}|X_k\right)\pi_{k|k-1}\left(X_k\middle|Z_{1:k-1}^{[1:Q]}\right)\delta X_k}. \tag{25}$$

The recursive process is not easy to deal with exactly due to the non-linear of the observation equation. The sequential Monte Carlo(SMC) methods are a viable approach to approximate the integrals of interest using random samples.

4.2. Particle Filter Implementation

Since the multi-target posterior probability density recursion requires the calculation of multi-set integral (24) and (25), its computational complexity is much larger than that of the single-target filtering process [16,61,62]. By the SMC method, the weighted particles can be estimated by recursive approximation to estimate the posterior probability density.

At the current time k, the particles are sampled by SMC, obtained from the spatial distribution of the target.

$$\tilde{X}_k^{(i)} \sim p\left(\cdot \left| X_{k-1}^{(i)}, Z_k \right.\right) \tag{26}$$

$\left\{\omega_{k-1}^{(i)}, X_{k-1}^{(i)}\right\}_{i=1}^{N}$ represents the set of importance weighted particles at time $k-1$ and the multi-target posterior probability density can be expressed as:

$$\pi_{k-1|k-1}\left(X_k \left| Z_{1:k-1}\right.\right) \approx \sum_{i=1}^{N} \omega_{k-1}^{(i)} \delta_{X_{k-1}^{(i)}}\left(X_{k-1}\right) \tag{27}$$

Algorithm 1 Particle Filter

1: **for** $k = 1, \cdots, T$ **do**

2: **for** $i = 1, \cdots, N$ **do**

3: Sample $\tilde{X}_k^{(i)} \sim p\left(\cdot \left| X_{k-1}^{(i)}, Z_k \right.\right)$;

4: Set $\tilde{\omega}_k^{(i)} = \omega_{k-1}^{(i)} \frac{g_k\left(Z_k \left| \tilde{X}_k^{(i)}\right.\right) f_{k|k-1}\left(\tilde{X}_k^{(i)} \left| X_{k-1}^{(i)}\right.\right)}{p_{k|k-1}\left(\tilde{X}_k^{(i)} \left| X_{1:k-1}^{(i)}, Z_{1:k}\right.\right)}$;

5: Normalise weights $\omega_k^{(i)} = \frac{\tilde{\omega}_k^{(i)}}{\sum_{j=1}^{N} \tilde{\omega}_k^{(j)}}$, here $\sum_{i=1}^{N} \omega_k^{(i)} = 1$;

6: **end for**

7: **end for**

8: Resample $\left\{\omega_k^{(i)}, X_k^{(i)}\right\}_{i=1}^{N}$ and get $\left\{\tilde{\omega}_k^{(i)}, \tilde{X}_k^{(i)}\right\}_{i=1}^{N}$;

9: Set $\hat{\pi}_k = \sum_{i=1}^{N} \omega_k^{(i)} \delta_{X_k^{(i)}}$ as the estimated posterior probability density;

4.3. The Multi-Sensor Likelihood

Given the multi-target state X, each $(x, l) \in X$ is either detected with probability $p_{D,m}(x, l)$ and generates observation z with likelihood function $g(z|x, l)$. For S sensors, the multi-sensor and multi-target mapping [56] is defined by $\theta^{(m)} : \mathbb{L} \to \left\{0, 1 \cdots, \left|Z^{(m)}\right|\right\}$, $m = 1, ..., S$. The set Θ represents the space of vector maps $\theta = (\theta^{(1)}, ..., \theta^{(S)})$. Assuming that the target and clutter generation are independent and the multi-sensor likelihood function is given by [34]:

$$g(Z|X) \propto \sum_{\theta \in \Theta(\mathcal{L}(X))} \left[\psi_{Z_1}\left(\cdot; \theta^{(1)}\right)\right]^X \cdots \left[\psi_{Z_m}\left(\cdot; \theta^{(S)}\right)\right]^X$$

$$\psi_{Z_m}(x,l;\theta) = \begin{cases} \frac{p_{D,m}(x,l) g\left(z_{\theta(l,m)}|x,l\right)}{\mathcal{K}_m\left(z_{\theta(l,m)}\right)}, & \text{if } \theta(l,m) > 0 \\ 1 - p_{D,m}(x,l), & \text{if } \theta(l,m) = 0 \end{cases},$$

where $p_{D,m}(x,l)$ is the probability detection, $\mathcal{K}_m$ is Poisson clutter, for sensor m.

4.4. GLMB Filter

The GLMB filter is a Bayesian recursion from the multi-Bernoulli distribution, which satisfies the following formula [34]:

$$\begin{gathered} C = \mathcal{F}(\mathbb{L}) \times \Xi \\ \omega^{(c)}(L) = \omega^{(I,\xi)}(L) = \omega^{(I,\xi)} \delta_I(L) \\ p^{(c)} = p^{(I,\xi)} = p^{(\xi)}. \end{gathered} \tag{28}$$

The forward propagation expression of GLMB Filter is as follows:

$$\pi(\mathrm{X}) = \Delta(\mathrm{X}) \sum_{(I,\xi) \in \mathcal{F}(\mathbb{L}) \times \Xi} \omega^{(I,\xi)} \delta_I(\mathcal{L}(\mathrm{X})) \left[p^{(\xi)}\right]^{\mathrm{X}}. \tag{29}$$

The distribution of multi-target prior probability is given by the Equation (29), thus, the multi-target prediction is still the multi-Bernoulli distribution and the prediction step can be expressed as:

$$\pi_+(\mathrm{X}_+) = \Delta(\mathrm{X}_+) \sum_{(I_+,\xi) \in \mathcal{F}(\mathbb{L}) \times \Xi} \omega_+^{(I_+,\xi)} \delta_{I_+}(\mathcal{L}(\mathrm{X}_+)) \left[p_+^{(\xi)}\right]^{\mathrm{X}^+} \tag{30}$$

where

$$\omega_+^{(I_+,\xi)} = \omega_s^{(\xi)}(I_+ \cap \mathbb{L})\, \omega_B(I_+ \cap \mathbb{B}) \tag{31}$$

$$p_+^{(\xi)}(x,l) = 1_{\mathbb{L}}(l)\, p_s^{(\xi)}(x,l) + (1 - 1_{\mathbb{L}}(l))\, p_{\mathbb{B}}(x,l) \tag{32}$$

$$p_s^{(\xi)}(x,l) = \frac{\left\langle p_s(\cdot,l) f(x|\cdot,l), p^{(\xi)}(\cdot,l)\right\rangle}{\eta_s^{(\xi)}(l)} \tag{33}$$

$$\eta_s^{(\xi)}(l) = \int \left\langle p_s(\cdot,l) f(x|\cdot,l), p^{(\xi)}(\cdot,l)\right\rangle dx \tag{34}$$

$$\omega_s^{(\xi)}(L) = \left[\eta_s^{(\xi)}\right]^L \sum_{I \in L} 1_I(L) \left[q_s^{(\xi)}\right]^{I-L} \omega^{(I,\xi)} \tag{35}$$

$$q_s^{(\xi)}(l) = \left\langle q_s(\cdot,l), p_s^{(\xi)}(\cdot,l)\right\rangle \tag{36}$$

Here, the $\omega_B(I_+ \cap \mathbb{B})$ and $\omega_s^{(\xi)}(I_+ \cap \mathbb{L})$ are weights of the birth labels $(I_+ \cap \mathbb{B})$ and surviving labels $(I_+ \cap \mathbb{L})$, respectively. $p_{\mathbb{B}}(x,l)$ is density of a new-born target, $p_s^{(\xi)}(x,l)$ is the probability density of the surviving target obtained from the prior probability $p^{(\xi)}(\cdot,l)$. $f(x|\cdot,l)$ means density weighted by the probability of survival $p_s(\cdot,l)$.

Given the predicted density as Equation (29), the update step can be expressed in the form of a truncated estimate:

$$\pi(\mathrm{X}|\mathrm{Z}) \approx \Delta(\mathrm{X}) \sum_{(I,\xi) \in \mathcal{F}(\mathbb{L}) \times \Xi} \sum_{\theta \in \Theta^{(M)}} \tilde{\omega}^{(I,\xi,\theta)}(\mathrm{Z}) \delta_I(\mathcal{L}(\mathrm{X})) \left[p^{(\xi,\theta)}(\cdot|\mathrm{Z})\right]^{\mathrm{X}}, \tag{37}$$

where (I,ξ) is fixed parameter. For M elements set $\Theta^{(M)} = \left\{\xi^{(1)}, \cdots, \xi^{(M)}\right\}$ there is the highest weight $\omega^{(I,\xi,\theta)}$, $\tilde{\omega}^{(I,\xi,\theta)}$ is the re-normalized weight after truncation and

$$\tilde{\omega}^{(I,\xi,\theta)}(Z) = \frac{\delta_{\theta^{-1}(\{0:|Z\})}(I)\,\omega^{(I,\xi)}\left[\eta_Z^{(\xi,\theta)}\right]^I}{\sum\limits_{(I,\xi)\in\mathcal{F}(\mathbb{L})\times\Xi}\sum\limits_{\theta\in\Theta}\delta_{\theta^{-1}(\{0:|Z\})}(I)\,\omega^{(I,\xi)}\left[\eta_Z^{(\xi,\theta)}\right]^I} \tag{38}$$

$$p^{(\xi,\theta)}(x,l\,|Z) = \frac{p^{(\xi)}(x,l)\,\psi_Z(x,l;\theta)}{\eta_Z^{(\xi,\theta)}(l)} \tag{39}$$

$$\eta_Z^{(\xi,\theta)}(l) = \left\langle p^{(\xi)}(\cdot,l)\,\psi_Z(\cdot,l;\theta)\right\rangle \tag{40}$$

$$\psi_Z(x,l;\theta) = \delta_0(\theta(l))\,q_D(x,l) + (1-\delta_0(\theta(l)))\,\frac{p_D(x,l)\,g\left(z_{\theta(l)}|x,l\right)}{\mathcal{K}\left(z_{\theta(l)}\right)}. \tag{41}$$

4.5. Gibbs-GLMB Filter

Gibbs sampling is a special case of continuous Markov Chain Monte Carlo (MCMC), which can transform sampling from high-dimensional space to low-dimensional one [63]. Assuming that the target state is $X_k = \left\{x_{k,1},\cdots,x_{k,N(k)}\right\}$, which obeys the probability distribution π, the probability distribution of the target state is $\pi\left(\left\{x_{k,1},\cdots,x_{k,N(k)}\right\}\right)$.

Algorithm 2 Gibbs sampling

1: **for** $k = 1 : T$ **do**
2:   **for** $n = 1 : N(k)$ **do**
3:     $x_{k,n} \sim \pi_n\left(\cdot\,\middle|x_{k,1:n-1},x_{k-1,n+1:N(k)}\right)$;
4:   **end for**
5:   $X_k = \left\{x_{k,1},\cdots,x_{k,N(k)}\right\}$;
6: **end for**

In Algorithm 2, $x_{k,1:n-1}$ is the samples $x_{k,1},\cdots,x_{k,n1}$ that have generated at current time, $x_{k-1,n+1:N(k)}$ is associations $x_{k,n+1},\cdots,x_{k,N(k)}$ at previous time. The Gibbs sampling algorithm reduces the joint density estimation problem to conditional probability to reduce the sampling difficulty and finally updates all parameters by the iterative process of each parameter.

In the calculation process of the update step and the prediction step of the GLMB filter, the number of weights and data quantities of the update and the prediction step are exponentially increasing. By using optimal assignment implementation and the kth shortest path, the complexity of the measurement quantity is cubic [35]. The GLMB filter is truncated by Gibbs sampling, thereby joint prediction and update reduce the complexity of the measurements to linear.

Given the GLMB distribution (11) at the current time, the GLMB distribution at the next time is [36]:

$$\pi_{Z_+}(\mathbf{X}) \propto \Delta(\mathbf{X}) \sum_{I,\xi,I_+,\theta_+} \omega^{(I,\xi)}\omega_{Z_+}^{(I,\xi,I_+,\theta_+)}\delta_{I_+}(\mathcal{L}(\mathbf{X}))\left[p_{Z_+}^{(\xi,\theta_+)}\right]^{\mathbf{X}} \tag{42}$$

where $I \in \mathcal{F}(\mathbb{L})$, $\xi \in \Xi$, $I_+ \in \mathcal{F}(\mathbb{L}_+)$, $\theta_+ \in \Theta_+$ and

$$\omega_{Z_+}^{(I,\xi,I_+,\theta_+)} = 1_{\Theta_+(I_+)}(\theta_+)\left[1-\bar{P}_S^{(\xi)}\right]^{I-I_+}\left[\bar{P}_S^{(\xi)}\right]^{I\cap I_+}[1-r_{B,+}]^{\mathbb{B}_+-I_+}r_{B,+}^{\mathbb{B}_+\cap I_+}\left[\bar{\psi}_{Z_+}^{(\xi,\theta_+)}\right]^{I_+} \tag{43}$$

$$\bar{P}_S^{(\xi)}(l) = \left\langle p^{(\xi)}(\cdot,l),P_S(\cdot,l)\right\rangle \tag{44}$$

$$\bar{\psi}_{Z_+}^{(\xi,\theta_+)}(l_+) = \left\langle \bar{p}_+^{(\xi)}(\cdot,l_+), \psi_{Z_+}^{(\theta_+(l_+))}(\cdot,l_+) \right\rangle \tag{45}$$

$$\bar{p}_+^{(\xi)}(x_+,l_+) = 1_{\mathbb{L}}(l_+) \frac{\left\langle P_S(\cdot,l_+) f_+(x_+|\cdot,l_+), p^{(\xi)}(\cdot,l_+) \right\rangle}{\bar{P}_S^{(\xi)}(l_+)} + 1_{\mathbb{B}_+}(l_+) p_{B,+}(x_+,l_+) \tag{46}$$

$$p_{Z_+}^{(\xi,\theta_+)}(x_+,l_+) = \frac{\bar{p}_+^{(\xi)}(x_+,l_+) \psi_{Z_+}^{(\theta_+(l_+))}(x_+,l_+)}{\bar{\psi}_{Z_+}^{(\xi,\theta_+)}(l_+)}. \tag{47}$$

Note that $r_{B,+}(l_+)$ is the birth probability of the target with label l_+, $p_{B,+}(x_+,l_+)$ is its kinematic state distribution and $f_+(x_+|\cdot,l_+)$ is the Markov state transition function.

5. Experiment

5.1. Simulation Environment Settings

We use six sensors consisted of three arrays to observe three acoustic targets as shown in Figure 2. The sensors are at (100 m, 95 m), (95 m, 100 m), (5 m, 100 m), (0 m, 95 m), (0 m, 5 m) and (5 m, 0 m), in $[0, 100] \times [0, 100]$ m^2.

Three pairs of sensors track the target, each sensor's observation distance is 150 m, the simulated sound velocity is 344 m/s, the surviving probability is $P_S = 0.99$ and the clutter intensity of Poisson distribution is $\lambda_c = 2$. The scenario last 100 s, maximum number of targets is 3. The motion model is a linear state space equation(CV motion model) and the state of target is expressed as:

$$X_k = AX_{k-1} + B\omega_k \tag{48}$$

$$A = \begin{bmatrix} 1 & \mathrm{T} & 0 & 0 \\ 0 & 1 & 0 & 0 \\ 0 & 0 & 1 & \mathrm{T} \\ 0 & 0 & 0 & 1 \end{bmatrix} \quad B = \begin{bmatrix} \mathrm{T}^2/2 & 0 \\ \mathrm{T} & 0 \\ 0 & \mathrm{T}^2/2 \\ 0 & \mathrm{T} \end{bmatrix} \omega_k(t), \tag{49}$$

where A is the target state transition matrix, B is the noise matrix and ω_k is process noise and follows a standard Gaussian distribution. The sampling period is T = 1.

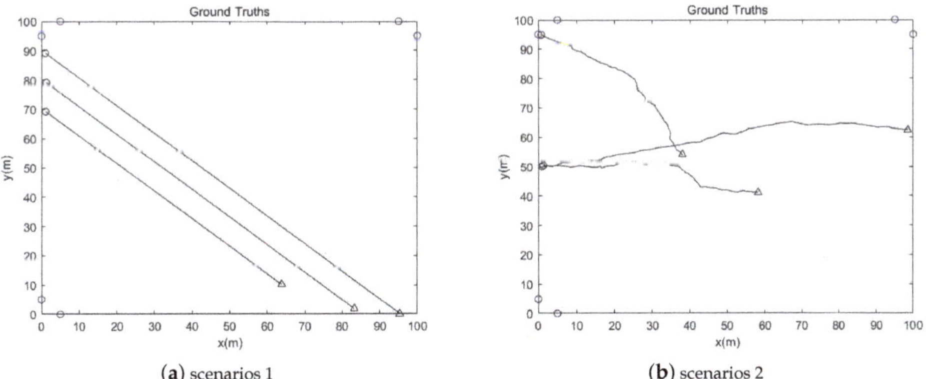

(**a**) scenarios 1 (**b**) scenarios 2

Figure 2. Detection model diagram.

In Figure 2, the sensors are displayed by the blue circle, the black circle represents the starting point and the triangle represents the end position. The target location is unknown and two scenarios were compared in this section. The position and velocity vector of target and sensor are represented as

$x_k = \left[p_{k,x}, \dot{p}_{k,x}, p_{k,y}, \dot{p}_{k,y}\right]^{\mathrm{T}}$ and $s_k^i = \left[q_{k,x}^i, \dot{q}_{k,x}^i, q_{k,y}^i, \dot{q}_{k,y}^i\right]^{\mathrm{T}}$, respectively. The range dependent detection probability is defined as:

$$P_{D,i}(x_k) = P_{D,\max} \exp\left(-\frac{\left[x_k - s_k^i\right]^{\mathrm{T}} C^{\mathrm{T}} \Sigma_D^{-1} C \left[x_k - s_k^i\right]}{2}\right), \tag{50}$$

where $P_{D,max} = 0.98$, $\Sigma_D = \mathrm{diag}(500, 500)^2$ and $C = \begin{bmatrix} 1 & 0 & 0 & 0 \\ 0 & 0 & 1 & 0 \end{bmatrix}$.

In the scenario 1, We build a parallel model and the survival period of three targets are 1 s–100 s, 10 s–90 s and 20 s–80 s, respectively. The initial states of the three targets are an LMB-RFS with parameters $\{r_{B,k}(l_i), p_{B,k}(l_i)\}_{i=1}^3$, where $l_i = (k,i)$, $r_{B,k}(l_i) = 0.02$ and $p_B(x_{0,i}, l_i) = \mathcal{N}\left(x_{0,i}; \mu_B^{(i)}; P_B\right)$ with:

$$\begin{aligned} \mu_B^{(1)} &= [0\ \mathrm{m}, 1\ \mathrm{m/s}, 90\ \mathrm{m}, -1\ \mathrm{m/s}]^{\mathrm{T}} \\ \mu_B^{(2)} &= [0\ \mathrm{m}, 1\ \mathrm{m/s}, 80\ \mathrm{m}, -1\ \mathrm{m/s}]^{\mathrm{T}} \\ \mu_B^{(3)} &= [0\ \mathrm{m}, 1\ \mathrm{m/s}, 70\ \mathrm{m}, -1\ \mathrm{m/s}]^{\mathrm{T}} \\ P_B &= diag\left([0.2, 0.08, 0.2, 0.1]^{\mathrm{T}}\right)^2. \end{aligned} \tag{51}$$

In the scenario 2, the survival period of three targets are 1 s–90 s, 1 s–80 s and 30 s–100 s, respectively. Target 1 and target 2 are born in the same position at the same time. The initial parameters $\{r_{B,k}(l_i), p_{B,k}(l_i)\}_{i=1}^2$, $r_{B,k}(l_i) = 0.02$ and $p_B(x_{0,i}, l_i) = \mathcal{N}\left(x_{0,i}; \mu_B^{(i)}; P_B\right)$ with:

$$\begin{aligned} \mu_B^{(1)} &= [0\ \mathrm{m}, 1\ \mathrm{m/s}, 50\ \mathrm{m}, 0\ \mathrm{m/s}]^{\mathrm{T}} \\ \mu_B^{(2)} &= [0\ \mathrm{m}, 0.8\ \mathrm{m/s}, 95\ \mathrm{m}, -0.5\ \mathrm{m/s}]^{\mathrm{T}} \\ P_B &= diag\left([0.2, 0.08, 0.2, 0.1]^{\mathrm{T}}\right)^2. \end{aligned} \tag{52}$$

The experiment uses the three Matlab audio files sample1.wav, sample2.wav and sample3.wav as the acoustic signals of the three targets in the Figure 3.

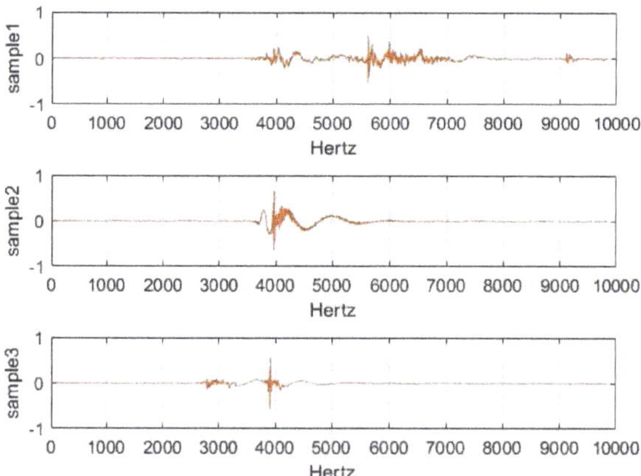

Figure 3. Acoustic signals of the three experimental targets

Taking the acoustic time difference as $\tau = 0.02$ s as an example, the simulation results of the three signals through the cross-correlation algorithm are as shown in the Figure 4.

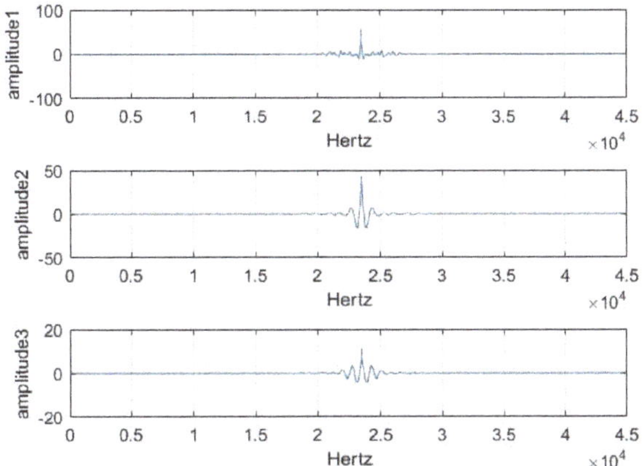

Figure 4. Cross-correlation waveform with a time difference of 0.02 s.

The time difference of the received signals of sensor arrays are calculated according to the GCC function and the angle difference of each group of sensors is calculated according to the signal receiving direction. The observation equation of the target is defined as:

$$\mathrm{z}_k^{[q]} = \begin{bmatrix} \hat{\tau}_q \\ \delta_q \end{bmatrix} + \begin{bmatrix} \sigma_\tau \\ \sigma_\delta \end{bmatrix}, q = 1, \ldots, Q \tag{53}$$

$$\hat{\tau}_q = \arg\max \int_{-\infty}^{+\infty} \psi_{12}(\omega)\, Z_1(\omega)\, Z_2^*(\omega) e^{-j\omega\tau_q} d\omega \tag{54}$$

$$\delta_q = \left| \arctan\left(\frac{p_{k,y} - q_{k,y}^1}{p_{k,x} - q_{k,x}^1} \right) - \arctan\left(\frac{p_{k,y} - q_{k,y}^2}{p_{k,x} - q_{k,x}^2} \right) \right|, \tag{55}$$

where, $\mathrm{z}_k^{[q]}$ is nonlinear. At time k, $\hat{\tau}_q$ is the time difference observed by a pair of sensors, δ_q is the angle difference between a pair of sensors receiving signals, $\sigma_\tau = 0.001$ s and $\sigma_\delta = (\pi/720)$ rad are the standard deviations of the Gaussian distributed measurement noise. In the scenario 1, three pairs of sensor arrays detected the measurements data of target 1 as show in the Figure 5.

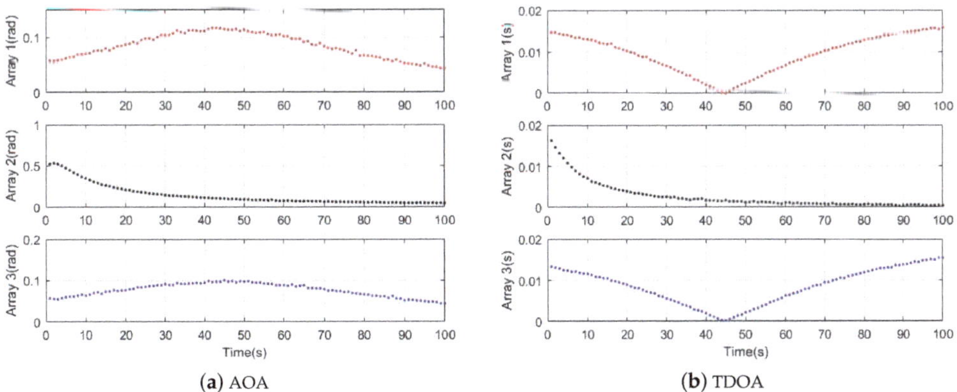

(a) AOA **(b)** TDOA

Figure 5. Measurement data.

5.2. Algorithm Estimation Analysis

5.2.1. Scenario 1

The simulation time is 100 s. The black line in the Figure 6 is the real trajectory of the target and the Red circle and the Color points are the estimated target location. It can be seen from the two pictures that the target tracks obtained from the Gibbs-GLMB filtering and the GLMB filtering are basically consistent with the true trajectory of the targets.

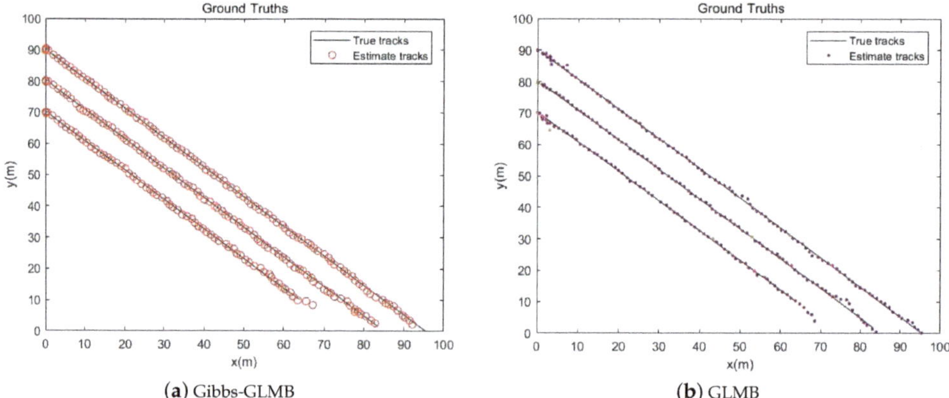

(a) Gibbs-GLMB (b) GLMB

Figure 6. Track estimation.

From the simulation results in Figures 6–8, it can be seen that the tracking performance of both algorithms is better. In Figures 7 and 8, the cross points are all measurements in the simulation of 1 s–100 s and the points generated outside the target track are false alarms caused by clutter interference. When a target is born, the random clutter may cause false alarms at the position. Nevtheless, in the subsequent tracking and localization, most of these false alarms will be eliminated. Through 100 times Monte Carlo(MC) simulations, the number of targets is estimated as shown in the Figure 9. The red line is algorithm Gibbs-GLMB and the black dotted line is algorithm GLMB. We can see from the comparison of the two algorithms that the number of Gibbs-GLMB estimates is more accurate overall but the estimated number of targets has a large deviation when the actual number of targets changes. The cardinality estimates of targets based on GLMB Filter is always a slightly higher than the true comparison.

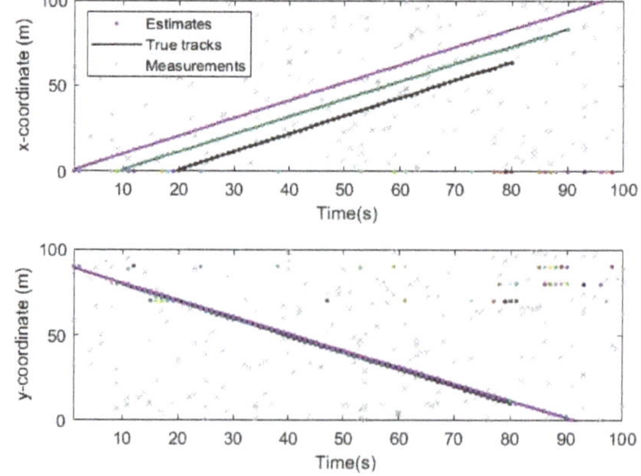

Figure 7. Track result on x and y coordinates by Gibbs-GLMB.

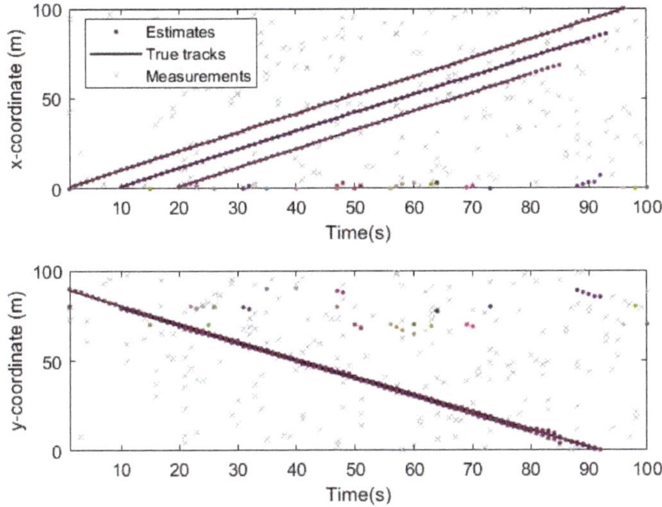

Figure 8. Track result on x and y coordinates by GLMB.

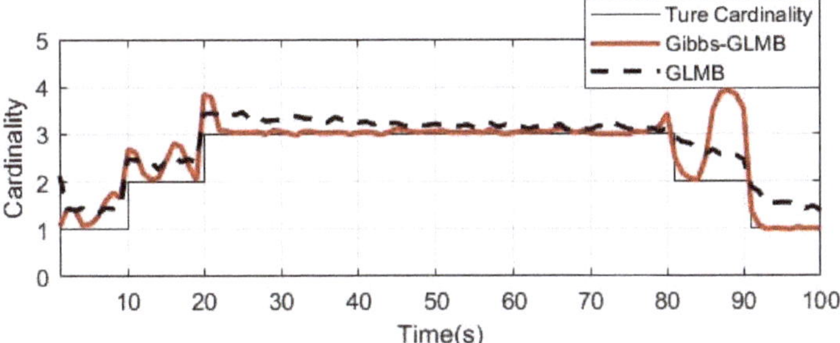

Figure 9. The cardinality estimates (100 times MC).

We use the Optimal Subpattern Assignment (OSPA) distance [64] ($c = 100, p = 1$) to analyze tracking performance. Figure 10 shows the simulation result over 100 MC runs. We can see that OSPA-Loc of two algorithms are very small in the whole process, indicating good estimation performance of the tracker. As shown in the results, the number of targets increases in 0 s, 10 s and 20 s, the OSPA of the GLMB fluctuated, however, the Gibbs-GLMB fluctuated more strongly than GLMB. When the number of targets decreases in 80 s and 90 s, the Gibbs-GLMB results have a large fluctuation, because the three parallel targets are relatively close, the target 3 is false detection; In 87 s–90 s, false cardinality estimates occurs due to the symmetrical geometric relationship between the sensors and the targets. while the GLMB stays a little high but remains relatively stable.

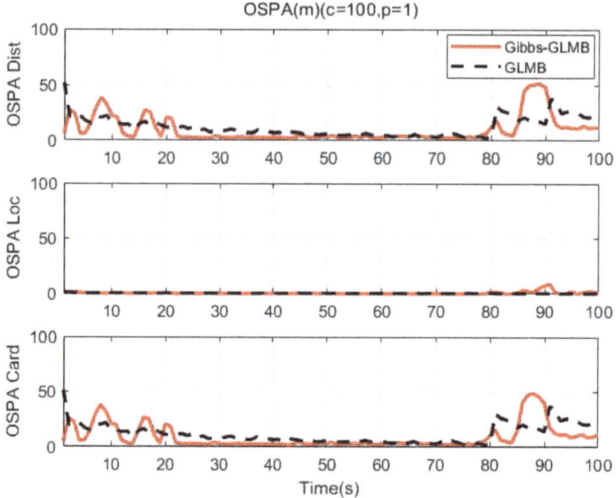

Figure 10. OSPA distance (100 times MC).

5.2.2. Scenario 2

The target simulation time is 100 s in the Figure 11. The legend in the pictures is the same as that in scenario 1. It also can be seen from the two pictures that the target tracks obtained from the Gibbs-GLMB filtering and the GLMB filtering are basically consistent with true trajectory of the targets.

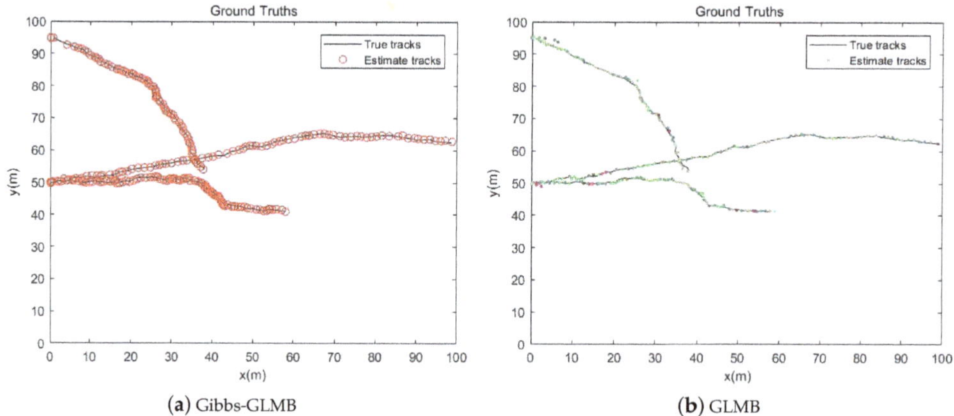

(**a**) Gibbs-GLMB (**b**) GLMB

Figure 11. Track estimation.

In the Figure 12, the blue box and the red circle are the estimated position of the particle point and the true position of the target point, respectively. We can see that the targets can be effectively detected through the particles after some steps even the two targets overlap in the same position at beginning.

From the simulation results in Figures 11–14, it can be seen that Gibbs-GLMB and GLMB are able to track the target by acoustic and quickly detect the new-born target. But target $3(x_3 = [0\ \text{m},\ 0.8\ \text{m/s},\ 95\ \text{m},\ -0.5\ \text{m/s}]^T)$ with a special starting position which on a sensor position $(0\ \text{m}, 95\ \text{m})$, there will be a measurement error and the tracking will be missing detection. Through 100 times Monte Carlo(MC) simulations, the cardinality of targets is estimated as shown in Figure 15. Both algorithms can accurately estimate the cardinality of targets but the cardinality estimates has a large error at the beginning of Gibbs-GLMB. In this scenario, there is no fluctuation when the cardinality of targets changes.

Figure 12. The estimated position of the particle point.

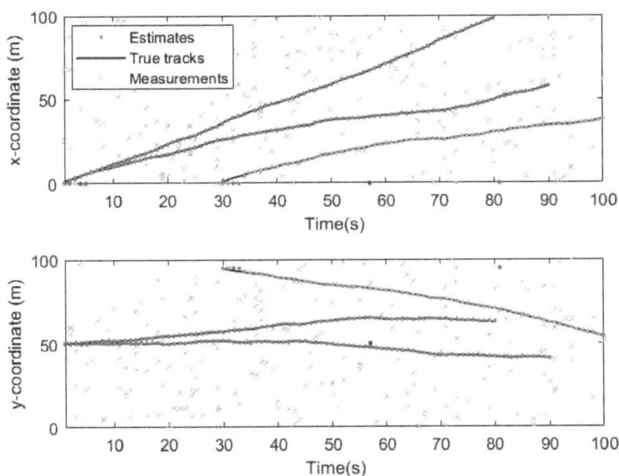

Figure 13. Track result on x and y coordinates by Gibbs-GLMB.

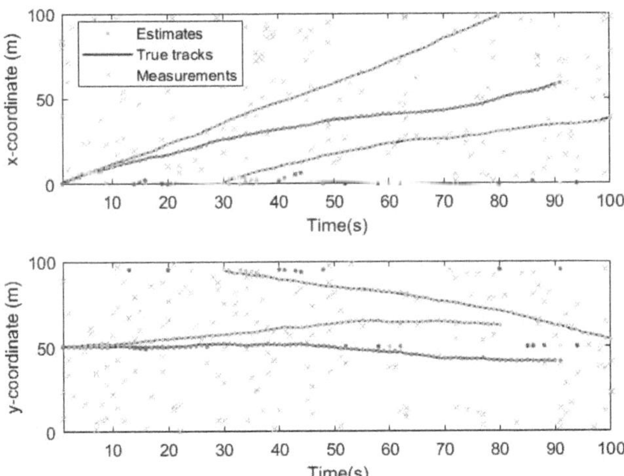

Figure 14. Track result on x and y coordinates by GLMB.

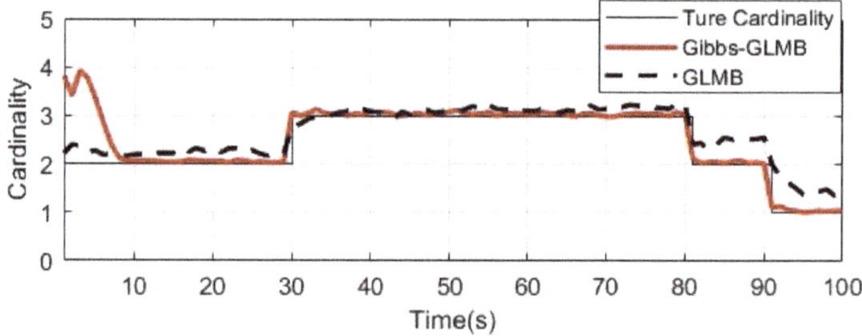

Figure 15. The cardinality estimates (100 times MC).

Figure 16 is the OSPA distance over 100 MC runs. The statistical results can further illustrate that the proposed Gibbs-GLMB is more accurate than the GLMB throughout the tracking process, although the error is larger at the beginning of the experiment.

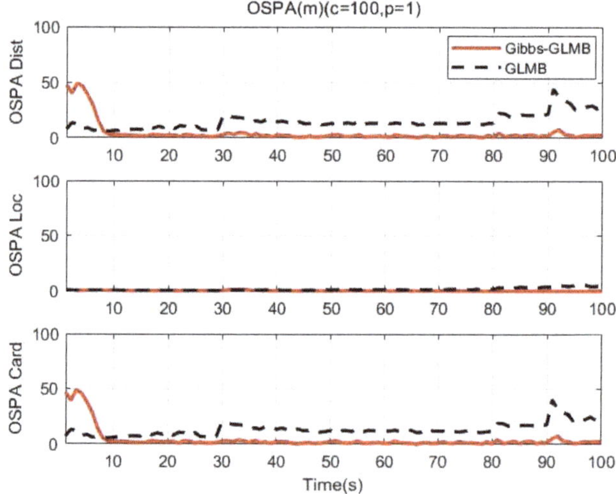

Figure 16. OSPA distance (100 times MC).

5.3. Performance

The filtering algorithms are run in the same PC. The configuration is as follows: CPU: Inter Core MLi5-4200H@2.80 GHz, RAM: 8 GB, using the software matlab2017b, the computation load and accuracy is as described in the following table:

As shown in Table 1, the proposed Gibbs-GLMB filtering is significantly reduced in time-consuming and the cardinality estimates is more accurate compared to the GLMB filtering. The Gibbs-GLMB filtering improves the speed of operation greatly and reduces the complexity of the procedure by integrating updating and prediction of the GLMB filter into one step and by combining the Gibbs sampling algorithm to evaluate the updated target combination, eliminating the target combination with smaller possibility and retaining the target combination with larger weight so the tracking accuracy will be better, and the computational complexity is also significantly reduced due to the reduction of the target combination.

Table 1. Performance comparison.

Method	Scenario 1		Scenario 2	
	Running Time (s/step)	The Cardinality Accuracy	Running Time (s/step)	The Cardinality Accuracy
Gibbs-GLMB	0.5790	76.86%	0.5950	88.61%
GLMB	1.4188	67.64%	2.1950	61.60%

6. Conclusions

By introducing the multi-sensor acoustic array and signal detection model, we proposes to use TDOA and AOA measurements, combined with the Gibbs-GLMB filter to track multiple acoustic sources. In this paper, we use RFS theory which can solve the loss of correspondence between set elements with labels and use PHAT combined with the GCC algorithm which improves the result of TDOA calculation through real acoustic signals. The feasibility of the method is verified by tracking multiple nonlinear moving targets. The experimental simulation results show that the Gibbs-GLMB filter can effectively track multi-target but the sensor position will affect the results of the tracking. Compared with the GLMB filter, Gibbs-GLMB filter runs faster and the results are more accurate. The method proposed in this paper is only implemented under ideal simulation conditions. In the future, we will consider applying it to real audio experiments and design an effective sensor array distribution.

Author Contributions: For conceptualization, W.L.; methodology, W.L.; software, Z.T.; validation, W.L. and Z.T.; formal analysis, W.L. and Z.T.; investigation, W.L.; writing—original draft preparation, Z.T.; writing—review and editing, W.L and X.R.

Funding: This work is supported by the National Natural Science Foundation of China (61771177).

Conflicts of Interest: The authors declare no conflicts of interest. The funders had no role in the design of the study; in the collection, analyses, or interpretation of data; in the writing of the manuscript, or in the decision to publish the results.

References

1. Wang, X.; Liu, W.; Chen, Y. Passive localization based on multi-sensor GLMB filter Using a TDOA Approach. In Proceedings of the 2017 36th Chinese Control Conference (CCC), Dalian, China, 26–28 July 2017; pp. 5230–5235.
2. Brandstein, M.S.; Silverman, H.F. A practical methodology for speech source localization with microphone arrays. *Comput. Speech Lang.* **1997**, *11*, 91–126. [CrossRef]
3. Schmidt, R. Multiple emitter location and signal parameter estimation. *IEEE Trans. Antennas Propag.* **1986**, *34*, 276–280. [CrossRef]
4. Knapp, C.; Carter, G. The generalized correlation method for estimation of time delay. *Signal Process.* **1976**, *24*, 320–327. [CrossRef]
5. Xu, L.; Liang, Y.; Duan, Z.; Zhou, G. Route-based dynamics modeling and tracking with application to air traffic surveillance. *IEEE Trans. Intell. Transp. Syst.* **2019**. [CrossRef]
6. Yang, J.; Zhang, M. Research on Multi Source Heterogeneous Data Fusion Location Algorithm Based on Active Detection and Passive Acquisition. *Acta Electron. Sin.* **2018**, *46*, 165–169. [CrossRef]
7. Cao, H.; Chan, Y.T.; So, H.C. Maximum Likelihood TDOA Estimation from Compressed Sensing Samples without Reconstruction. *IEEE Signal Process. Lett.* **2017**, *24*, 564–568. [CrossRef]
8. Wang, Y.; Ho, K.C. TDOA Positioning Irrespective of Source Range. *IEEE Trans. Signal Process.* **2017**, *65*, 1447–1460. [CrossRef]
9. Zhong, X.; Hopgood, J.R. Particle filtering for TDOA based acoustic source tracking: Nonconcurrent Multiple Talkers. *Signal Process.* **2014**, *96*, 382–394. [CrossRef]
10. Vo, B.-N.; Mallick, M.; Bar-Shalom, Y.; Coraluppi, S.; Osborne, R., III; Mahler, R.; Vo, B.-T. *Multitarget Tracking*; Wiley Encyclopedia of Electrical and Electronics Engineering: New York, NY, USA, September 2015.

11. Mahler, R. Multitarget Bayes filtering via first-order multitarget moments. *IEEE Trans. Aerosp. Electron. Syst.* **2003**, *39*, 1152–1178. [CrossRef]
12. Mahler, R. PHD filters of higher order in target number. *IEEE Trans. Aerosp. Electron. Syst.* **2007**, *43*, 1523–1543. [CrossRef]
13. Mahler, R. *Statistical Multisource-Multitarget Information Fusion*; Artech House: Norwood, MA, USA, 2007.
14. Mahler, R. *Advances in Statistical Multisource-Multitarget Information Fusion*; Artech House: Norwood, MA, USA, 2014.
15. Vo, B.-N.; Vo, B.-T.; Pham, N.-T.; Suter, D. Joint Detection and Estimation of Multiple Objects from Image Observations. *IEEE Trans. Signal Process.* **2010**, *58*, 5129–5241. [CrossRef]
16. Vo, B.-N.; Singh, S.; Doucet, A. Sequential Monte Carlo methods for multi-target filtering with random finite sets. *IEEE Trans. Aerosp. Electron. Syst.* **2005**, *41*, 1224–1245.
17. Vo, B.-N.; Ma, W.K. The Gaussian mixture Probability Hypothesis Density Filter. *IEEE Trans. Signal Process.* **2006**, *54*, 4091–4104. [CrossRef]
18. Vo, B.-T.; Vo, B.-N.; Cantoni, A. The Cardinality Balanced Multi-target Multi-Bernoulli filter and its implementations. *IEEE Trans. Signal Process.* **2009**, *57*, 409–423.
19. Vo, B.-T.; Vo, B.-N.; Cantoni, A. Analytic implementations of the Cardinalized Probability Hypothesis Density Filter. *IEEE Trans. Sig. Proc.* **2007**, *55 Pt 2*, 3553–3567. [CrossRef]
20. Maggio, E.; Taj, M.; Cavallaro, A. Efficient multitarget visual tracking using random finite sets. *IEEE Trans. Circuits Syst. Video Technol.* **2008**, *18*, 1016–1027. [CrossRef]
21. Hoseinnezhad, R.; Vo, B.-N.; Vo, B.T.; Suter, D. Visual tracking of numerous targets via multi-bernoulli filtering of image data. *Pattern Recognit.* **2012**, *45*, 3625–3635. [CrossRef]
22. Hoseinnezhad, R.; Vo, B.-N.; Vo, B.-T. Visual tracking in background subtracted image sequences via multi-bernoulli filtering. *IEEE Trans. Signal Process.* **2013**, *61*, 392–397. [CrossRef]
23. Ristic, B.; Vo, B.-N. Sensor Control for Multi-Object State-Space Estimation Using Random Finite Sets. *Automatica* **2010**, *46*, 1812–1818. [CrossRef]
24. Ristic, B.; Vo, B.-N.; Clark, D. A note on the reward function for PHD filters with sensor control. *IEEE Trans. Aerosp. Electron. Syst.* **2011**, *47*, 1521–1529. [CrossRef]
25. Yi, W.; Li, S.; Wang, B.; Hoseinnezhad, R.; Kong, L. Computationally Efficient Distributed Multi-sensor Fusion with Multi-Bernoulli Filter. *IEEE Trans. Signal Process.* **2020**. [CrossRef]
26. Hoang, H.G.; Vo, B.T. Sensor management for multi-target tracking via multi-Bernoulli filtering. *Automatica* **2014**, *50*, 1135–1142. [CrossRef]
27. Hoang, H.; Vo, B.-N.; Vo, B.-T.; Mahler, R. The Cauchy-Schwarz divergence for Poisson point processes. *IEEE Trans. Inf. Theory* **2015**, *61*, 4475–4485. [CrossRef]
28. Gostar, A.K.; Hoseinnezhad, R.; Bab-Hadiashar, A. Multi-Bernoulli sensor control using Cauchy-Schwarz divergence. In Proceedings of the 2016 19th International Conference on Information Fusion (FUSION), Heidelberg, Germany, 5–8 July 2016; pp. 651–657.
29. Gostar, A.K.; Hoseinnezhad, R.; Bab-Hadiashar, A. Multi-Bernoulli sensor-selection for multi-target tracking with unknown clutter and detection profiles. *Signal Process.* **2016**, *119*, 28–42. [CrossRef]
30. Mullane, J.; Vo, B.-N.; Adams, M.; Vo, B.-T. A random-finite-set approach to Bayesian SLAM. *IEEE Trans. Robot.* **2011**, *27*, 268–282. [CrossRef]
31. Lundquist, C.; Hammarstrand, L.; Gustafsson, F. Road intensity based mapping using radar measurements with a Probability Hypothesis Density filter. *IEEE Trans. Signal Process.* **2011**, *59*, 1397–1408. [CrossRef]
32. Liu, W.; Zhu, S.; Wen, C.L.; Yu, Y. Structure modeling and estimation of multiple resolvable group targets via graph theory and multi-Bernoulli filter. *Automatica* **2018**, *89*, 274–289. [CrossRef]
33. Mahler, R. Exact Closed-Form Multitarget Bayes Filters. *Sensors* **2019**, *19*, 2818. [CrossRef]
34. Vo, B.-T.; Vo, B.-N. Labeled random finite sets and multi-object conjugate priors. *IEEE Trans. Signal Process.* **2013**, *61*, 3460–3475. [CrossRef]
35. Vo, B.-N.; Vo, B.-T.; Phung, D. Labeled Random Finite Sets and the Bayes Multi-Target Tracking Filter. *IEEE Trans. Signal Process.* **2014**, *62*, 6554–6567. [CrossRef]
36. Vo, B.-N.; Vo, B.-T.; Hoang, H.G. An Efficient Implementation of the Generalized Labeled Multi-Bernoulli Filter. *IEEE Trans. Signal Process.* **2017**, *65*, 1975–1987. [CrossRef]
37. Beard, M.; Vo, B.-T.; Vo, B.-N. A Solution for Large-Scale Multi-Object Tracking. *arXiv* **2018**, arXiv:1804.06622.

38. Bryant, D.; Vo, B.-T.; Vo, B.-N.; Jones, B. A Generalized Labeled Multi-Bernoulli Filter with Object Spawning. *IEEE Trans. Signal Process.* **2018**, *66*, 6177–6189. [CrossRef]
39. Beard, M.; Vo, B.-T.; Vo, B.-N. Bayesian multi-target tracking with merged measurements using labelled random finite sets. *IEEE Trans. Signal Process.* **2015**, *63*, 1433–1447. [CrossRef]
40. Beard, M.; Reuter, S.; Granström, K.; Vo, B.-T.; Vo, B.-N.; Scheel, A. Multiple extended target tracking with labeled random finite sets. *IEEE Trans. Signal Process.* **2016**, *64*, 1638–1653. [CrossRef]
41. Rathnayake, T.; Gostar, A.K.; Hoseinnezhad, R.; Bab-Hadiashar, A. Labeled multi-Bernoulli track-before-detect for multi-target tracking in video. In Proceedings of the 2015 18th International Conference on Information Fusion (Fusion), Washington, DC, USA, 6–9 July 2015; pp. 1353–1358.
42. Punchihewa, Y.; Vo, B.-T.; Vo, B.-N.; Kim, D.Y. Multiple Object Tracking in Unknown Backgrounds with Labeled Random Finite Sets. *IEEE Trans. Signal Process.* **2018**, *66*, 3040–3055. [CrossRef]
43. Kim, D.Y.; Vo, B.-N.; Vo, B.-T.; Jeon, M. A Labeled Random Finite Set Online Multi-Object Tracker for Video Data. *Pattern Recognit.* **2019**, *90*, 377–389. [CrossRef]
44. Xu, B.; Lu, M.; Cong, J.; Nener, B. An Ant Colony Inspired Multi-Bernoulli Filter for Cell Tracking in Time-Lapse Microscopy Sequences. *IEEE J. Biomed. Health Inform.* **2019**. [CrossRef]
45. Xu, B.; Shi, J.; Lu, M.; Cong, J.; Wang, L.; Nener, B. An Automated Cell Tracking Approach with Multi-Bernoulli Filtering and Ant Colony Labor Division. *IEEE/ACM Trans. Comput. Biol. Bioinform.* **2019**. [CrossRef]
46. Papi, F.; Vo, B.-N.; Vo, B.-T.; Fantacci, C.; Beard, M. Generalized Labeled Multi-Bernoulli Approximation of Multi-Object Densities. *IEEE Trans. Signal Process.* **2015**, *63*, 5487–5497. [CrossRef]
47. Papi, F.; Kim, D.Y. A particle multi-target tracker for superpositional measurements using labeled random finite sets. *IEEE Trans. Signal Process.* **2015**, *63*, 4348–4358. [CrossRef]
48. Beard, M.; Vo, B.-T.; Vo, B.-N.; Arulampalam, S. Void probabilities and Cauchy-Schwarz divergence for Generalized Labeled Multi-Bernoulli Models. *IEEE Trans. Signal Process.* **2017**, *65*, 5047–5061. [CrossRef]
49. Gostar, A.K.; Hoseinnezhad, R.; Bab-Hadiashar, A. Sensor control for multi-object tracking using labeled multi-Bernoulli filter. In Proceedings of the 17th International Conference on Information Fusion (FUSION), Salamanca, Spain, 7–10 July 2014; pp. 1–8.
50. Deusch, H.; Reuter, S.; Dietmayer, K. The labeled multi-Bernoulli SLAM filter. *IEEE Signal Process. Lett.* **2015**, *22*, 1561–1565 . [CrossRef]
51. Moratuwage, D.; Adams, M.; Inostroza, F. δ-Generalised Labelled Multi-Bernoulli Simultaneous Localisation and Mapping. In Proceedings of the 2018 International Conference on Control, Automation and Information Sciences (ICCAIS), Hangzhou, China, 24–27 October 2018; pp. 175–182.
52. Moratuwage, D.; Adams, M.; Inostroza, F. δ-Generalized Labeled Multi-Bernoulli Simultaneous Localization and Mapping with an Optimal Kernel-Based Particle Filtering Approach. *Sensors* **2019**, *19*, 2290. [CrossRef]
53. Fantacci, C.; Vo, B.-N.; Vo, B.-T.; Battistelli, G.; Chisci, L. Robust fusion for multisensor multiobject tracking. *IEEE Signal Proc. Lett.* **2018**, *25*, 640–644. [CrossRef]
54. Li, S.; Battistelli, G.; Chisci, L.; Yi, W.; Wang, B.; Kong, L. Computationally Efficient Multi-Agent Multi-Object Tracking with Labeled Random Finite Sets. *IEEE Trans. Sig. Process.* **2019**, *67*, 260–275. [CrossRef]
55. Hadden, W.J.; Young, J.L.; Holle, A.W.; McFetridge, M.L.; Kim, D.Y.; Wijesinghe, P.; Taylor-Weiner, H.; Wen, J.H.; Lee, A.R., Bieback, K.; et al. Stem cell migration and mechanotransduction on linear stiffness gradient hydrogels. *Proc. Natl. Acad. Sci. USA* **2017**, *114*, 5647–5652. [CrossRef]
56. Vo, B.; Vo, B.; Beard, M. Multi-Sensor Multi-Object Tracking With the Generalized Labeled Multi-Bernoulli Filter. *IEEE Trans. Signal Process.* **2019**, *67*, 5952–5967. [CrossRef]
57. Vo, B.-N.; Vo, B.-T. A Multi-Scan Labeled Random Finite Set Model for Multi-object State Estimation. *IEEE Trans. Signal Process.* **2019**, *67*, 4948–4963. [CrossRef]
58. Tian, Z.; Liu, W.; Ru, X. Multi-acoustic Array Localization and Tracking Method based on Gibbs-GLMB. In Proceedings of the IEEE International Conference on Computer Applications & Information Security (ICCAIS), Chengdu, China, 23–26 October 2019; pp. 1–6.
59. Vo, B.-N.; Ma, W.-K.; Singh, S. Localizing an unknown time-varying number of speakers: A Bayesian random finite set approach. In Proceedings of the IEEE International Conference on Acoustics, Speech, and Signal Processing (ICASSP '05), Philadelphia, PA, USA, 23 March 2005; Volume 4, pp. iv/1073–iv/1076.
60. Ma, W.-K.; Vo, B.-N.; Singh, S.S.; Baddeley, A. Tracking an unknown time-varying number of speakers using TDOA measurements: A random finite set approach. *IEEE Trans. Signal Process.* **2006**, *54*, 3291–3304.

61. Arulampalam, M.S.; Maskell, S.; Gordon, N.; Clapp, T. A tutorial on particle filters for online nonlinear/non-Gaussian Bayesian tracking. *IEEE Trans. Signal Process.* **2002**, *50*, 174–188. [CrossRef]
62. Doucet, A.; Godsill, S.; Andrieu, C. On sequential Monte Carlo sampling methods for Bayesian filtering. *Stat. Comput.* **2000**, *10*, 197–208. [CrossRef]
63. Geman, S.; Geman, D. Stochastic relaxation, Gibbs distributions, and the Bayesian restoration of images. *IEEE Trans. Pattern Anal. Mach. Intell.* **1984**, *6*, 721–741. [CrossRef] [PubMed]
64. Schuhmacher, D.; Vo, B.-T.; Vo, B.-N. A Consistent Metric for Performance Evaluation of Multi-Object Filters. *IEEE Trans. Signal Process.* **2008**, *56*, 3447–3457. [CrossRef]

Article

Anti-Clutter Gaussian Inverse Wishart PHD Filter for Extended Target Tracking

Yuan Huang, Liping Wang, Xueying Wang * and Wei An

College of Electronic Science, National University of Defense Technology, Changsha 410073, China; huangyuan09@nudt.edu.cn (Y.H.); wangliping17@nudt.edu.cn (L.W.); anwei@nudt.edu.cn (W.A.)
* Correspondence: wang_xueying87@126.com; Tel.: +1-354-894-9042

Received: 11 October 2019; Accepted: 22 November 2019; Published: 23 November 2019

Abstract: The extended target Gaussian inverse Wishart probability hypothesis density (ET-GIW-PHD) filter overestimates the number of targets under high clutter density. The reason for this is that the source of measurements cannot be determined correctly if only the number of measurements is used. To address this problem, we proposed an anti-clutter filter with hypothesis testing, we take into account the number of measurements in cells, the target state and spatial distribution of clutter to decide whether the measurements in cell are clutter. Specifically, the hypothesis testing method is adopted to determine the origination of the measurements. Then, the likelihood functions of targets and clutter are deduced based on the information mentioned above, resulting in the likelihood ratio test statistic. Next, the likelihood ratio test statistic is proved to be subject to a chi-square distribution and a threshold corresponding to the confidence coefficient is introduced and the measurements below this threshold are considered as clutter. Then the correction step of ET-GIW-PHD is revised based on hypothesis testing results. Extensive experiments have demonstrated the significant performance improvement of our proposed method.

Keywords: extended target; target tracking; PHD filter; high clutter density

---

1. Introduction

Extended target tracking (ETT) draws lots of attention in recent years because of its wide range of applications in traffic control [1], autonomous driving [2–4], person tracking [5,6] and etc. [7–11]. Since one extended target generates more than one measurement per time step, its shape information can be obtained. Using this information, the kinematic state and extent of the target can be estimated simultaneously. The extent of the target including the size, shape, and orientation can be further used for target identification.

The difference between point target tracking and extended target tracking lies in the measurement model and hypotheses. Point target generates at most one measurement per time step, while the extended target generates multiple measurements. Many algorithms were proposed to track point target based on point target hypothesis, such as probability hypothesis density (PHD) filter [12] and cardinalized PHD (CPHD) filter [13]. Since the extended target violates one measurement hypothesis, the number of targets will be overestimated if point target tracking algorithms are directly used for tracking extended targets. To address this problem, Mahler proposed an extended target tracking algorithm based on the inhomogeneous Poisson Point Process model (PPP model) [14] in random finite sets (RFSs) frame, namely extended target probability hypothesis density (ET-PHD) [15], Jiang et al. [16] proposed a novel time-matching ET-PHD filter, a Gaussian mixture implementation of the ET-PHD, called the extended target Gaussian inverse Wishart probability hypothesis density (ET-GIW-PHD) filter, has been presented in [17]. However, ET-PHD only estimates the kinematic state of the target (such as position, velocity) and does not estimate the extent of the target. Therefore, this method cannot extract the shape of the target. Nevertheless, the estimation of the target extent is important because it can be used to classify target and improve tracking accuracy [18–20].

Measurement partition is an important step in ETT. In ETT, measurements are partitioned into several non-empty subsets, each subset contains measurements that are all from the same source, either a single target or a clutter source, the subset is defined as cell. In ETT, the increase of measurements gives rise to the quick increase of the set partitions, thus the partition algorithm should be designed to achieve tractable computational complexity. Distance partition [17] is the most widely used method. Modified Bayesian adaptive resonance theory (MB-ART) [21] can also achieve good performance. For more details about other partition algorithms, please see [22–24].

One of the most important works in extended target tracking is how to model the target extent. To address this problem, the stick model is used for bicycle and pedestrian tracking [25,26]. The object extension is represented by a symmetric positive definite (SPD) random matrix [27], namely a random matrix (RM) model. Feldmann et al. [28] adapted the RM model for the case when the sensor error cannot be ignored. Lan et al. [29] took into account time variation and distortion of target extension in RM frame. In order to handle irregular shapes, a random hypersurface model (RHM) is introduced in [30–32]. Gaussian Processes (GP) was used to represent the target shape and achieved good performance [33–36]. Since shape estimation is similar to curving fitting, Kaulbersch et al. [37] applied a curve fitting method for shape estimation. Granström et al. [38] proposed an extension model for specific sensor. Granström et al. [39] proposed extended target Gaussian inverse Wishart PHD (ET-GIW-PHD) filter to incorporate widely used RM model into PHD filter and approximate the estimated PHD with an unnormalized mixture of Gaussian inverse Wishart (GIW) distributions. Later, Granström et al. [40] proposed extended target Gamma Gaussian inverse Wishart PHD (ET-GGIW-PHD) filter to estimate the measurement rate and target state simultaneously. The combination of several RM model was used to model nonelliptic targets in [41,42]. As mentioned in [39], more experiments that test ET-GIW-PHD filters are needed, e.g., for data that contains more clutter than typical laser data does, this provides the general motivation for this paper. We found that the number of targets will be overestimated which degrades the final performance when severe clutters are partitioned into one cell in ET-GIW-PHD. More analyses are presented in Section 3. In this paper, we proposed an anti-clutter ET-GIW-PHD filter for better cardinality estimation performance.

The main contributions of this paper are twofold. First, the reason why ET-GIW-PHD overestimates the number of targets is discussed detailedly, and the probability of the measurement generated by clutter against different scenario parameters is presented. Second, in order to deal with the cardinality overestimation in ET-GIW-PHD, we proposed an anti-clutter ET-GIW-PHD filter which revises the correction step of ET-GIW-PHD with hypothesis testing. Hypothesis testing is introduced to determine the source of measurements in the cell, hypothesis testing results are integrated into the correction step in ET-GIW-PHD. In order to deal with the source of measurements correctly, the essential differences between the measurements of targets and clutter should be recognized. Since the variation of target state over time follows certain rules (motion model and shape transition model), the state of targets could be predicted while clutter could not. Then, the likelihood functions of targets and clutter are deduced. The likelihood functions are built based on not only the number of measurements but also the target state and spatial distribution of clutter. Since the likelihood ratio test statistic is proved to be subject to chi-square distribution, a threshold corresponding to the confidence coefficient is introduced, this threshold is used to determine the source of measurements in the cell. It worth note that the perfect sensor resolution is advocated as a theoretical hypothesis in this paper. In reality, the results in the Section 5 will be affected by the limited sensor resolution. Future work will tackle the sensor's limited resolution.

The rest of the paper is outlined as follows. Section 2 reviews the ET-GIW-PHD filter. Section 3 discusses the reason why ET-GIW-PHD overestimates the number of targets. Our anti-clutter ET-GIW-PHD is presented detailedly in Section 4. We conduct experiments in different simulation scenarios to demonstrate the effectiveness of our proposed approach in Section 5; Conclusion is drawn in Section 6.

2. ET-GIW-PHD Review

In ET-GIW-PHD, both predicted PHD and corrected PHD can be approximated as an unnormalized mixture of Gaussian inverse Wishart distributions. Let $\zeta_k = \{x_k, \mathbf{X}_k\}$ be the sufficient statistics of the GIW components at time which contains kinematical state x_k and extension state $\mathbf{X}_k$ which is mathematically described by a symmetric and positively definite (SPD) random matrix. The iterative formulae for ξ_k are obtained in [39]. More implementation details, such as pruning and merging, can also be found in [39].

Prediction:

$$D_{k+1|k}(\xi_{k+1}) = \int p_s(\xi_k)p_{k+1|k}(\xi_{k+1}|\xi_k) \times D_{k|k}(\xi_k)d\xi_k + D^b_{k+1}(\xi_{k+1}), \tag{1}$$

where $p_s(\cdot)$ is the probability of survival, $p_{k+1|k}(\xi_{k+1}|\xi_k)$ is the state transition density, $D^b_{k+1}(\cdot)$ is the birth PHD, new target spawning is omitted [39].

Correction:

The corrected PHD $D_{k|k}(\xi_k)$ can be summarized as:

$$D_{k|k}(\xi_k) = D^{ND}_{k|k}(\xi_k) + \sum_{\mathrm{p}\angle\mathbf{Z}_k} \sum_{W\in p} D^D_{k|k}(\xi_k, W), \tag{2}$$

where $\mathrm{p}\angle\mathbf{Z}_k$ means that the measurement sets $\mathbf{Z}_k$ are partitioned into non-empty cells, $W \in \mathrm{p}$ means that the cell W is in the partition p.

$D^{ND}_{k|k}(\xi_k)$ handles the undetected target case, because $D_{k+1|k}(\xi_{k+1})$ is approximated as an unnormalized mixture of Gaussian inverse Wishart distributions, it is given by

$$D^{ND}_{k|k}(\xi_k) = \sum_{j=1}^{J_{k|k-1}} w^j_{k|k}\mathcal{N}(x_k; m^{(j)}_{k|k}, \mathbf{P}^{(j)}_{k|k} \otimes \mathbf{X}_k)\mathcal{IW}(\mathbf{X}_k; v^{(j)}_{k|k}, \mathbf{V}^{(j)}_{k|k}), \tag{3}$$

where $J_{k|k-1}$ is the number of components of predicted PHD, $w^{(j)}_{k|k}$ is the weight of GIW component. $\mathcal{N}(x; m, \mathbf{P})$ means that a vector x is subject to Gaussian distribution with mean m and covariance $\mathbf{P}$, $\mathcal{IW}(\mathbf{X}; v, \mathbf{V})$ means that a matrix is subject to inverse Wishart distribution with degree of freedom v and inverse scale matrix $\mathbf{V}$. $\otimes$ is the Kronecker product.

$D^D_{k|k}(\xi_k, W)$ handles the detected target case, which is given by

$$D^D_{k|k}(\xi_k, W) = \sum_{j=1}^{J_{k|k-1}} w^{(j,W)}_{k|k}\mathcal{N}(x_k; m^{(j,W)}_{k|k}, \mathbf{P}^{(j,W)}_{k|k} \otimes \mathbf{X}_k)\mathcal{IW}(\mathbf{X}_k; v^{(j,W)}_{k|k}, \mathbf{V}^{(j,W)}_{k|k}). \tag{4}$$

$w^{(j,W)}_{k|k}$ can be obtained by

$$w^{(j,W)}_{k|k} = \frac{\omega_p}{d_W} e^{-\gamma^{(j)}} \left(\frac{\gamma^{(j)}}{\lambda_c c_k}\right)^{|W|} p^{(j)}_D \Lambda^{(j,W)}_k w^{(j)}_{k|k-1}, \tag{5}$$

where

$$d_W = \delta_{|W|,1} + \sum_{j=1}^{J_{k|k-1}} e^{-\gamma^{(j)}} \left(\frac{\gamma^{(j)}}{\lambda_c c_k}\right)^{|W|} p^{(j)}_D \Lambda^{(j,W)}_k w^{(j)}_{k|k-1} \tag{6}$$

$$\Lambda_k^{(j,W)} = \frac{1}{(\pi^{|W|}|W|\mathbf{S}_{k|k-1}^{(j,W)})^{\frac{d}{2}}} \frac{|\mathbf{V}_{k|k-1}^{(j)}|^{\frac{v_{k|k-1}^{(j)}}{2}} \Gamma_d(\frac{v_{k|k}^{(j,W)}}{2})}{|\mathbf{V}_{k|k-1}^{(j,W)}|^{\frac{v_{k|k}^{(j,W)}}{2}} \Gamma_d(\frac{v_{k|k-1}^{(j)}}{2})} \tag{7}$$

$$\omega_p = \frac{\prod_{W\in P} d_W}{\sum_{P'\angle Z_k} \prod_{W'\in P'} d_{W'}}. \tag{8}$$

$\Lambda_k^{(j,W)}$ presents the likelihood of the jth GIW component given the measurements of the Wth cell, ω_p is the weight of pth partition, $p_D^{(j)}$ is the detection probability of jth GIW component, $\gamma^{(j)}$ is the expected number of measurements generated by jth GIW component, λ_c is the mean number of clutter measurements, c_k is the spatial distribution of the clutter over the surveillance volume, $\delta_{i,j}$ is the Kronecker delta, $|W|$ is the the number of measurements in the Wth cell, $\mathbf{S}_{k|k-1}^{(j,W)}$ is innovation factor, $\Gamma_d(\cdot)$ is the multivariate Gamma function.

3. Analysis of ET-GIW-PHD

In ET-GIW-PHD, the calculation of $w_{k|k}^{(j,W)}$ is important. If the measurements in Wth cell are generated by clutter, $w_{k|k}^{(j,W)}$ is expected to be smaller than the pruning threshold, then the corresponding component will be eliminated and the clutter will be eliminated.

In Equation (5), $w_{k|k}^{(j,W)}$ contains two parts, one is the weight of the pth partition, denoted by ω_p, the other is the weight of Wth cell in partition. Without loss of generality, only one partition is considered for clarity, therefore $\omega_p = 1$. Substituting Equation (6) into Equation (5), we arrive at

$$w_{k|k}^{(j,W)} = \frac{e^{-\gamma^{(j)}}\left(\frac{\gamma^{(j)}}{\lambda_c c_k}\right)^{|W|} p_D^{(j)} \Lambda_k^{(j,W)} w_{k|k-1}^{(j)}}{\delta_{|W|,1} + \sum_{l=1}^{J_{k|k-1}} e^{-\gamma(l)}\left(\frac{\gamma^{(l)}}{\lambda_c c_k}\right)^{|W|} p_D^{(l)} \Lambda_k^{(l,W)} w_{k|k-1}^{(l)}}. \tag{9}$$

From Equation (9), the numerator is a part of denominator, the measurements of Wth cell is used to correct each GIW component, then $\Lambda_k^{(j,W)}$ can be obtained, $w_{k|k}^{(j,W)}$ can be given based on some prior parameters, such as p_D, γ, λ_c and c_k (for brevity, the subscript and superscript are omitted here).

If the measurements in the cell are generated by clutter, the likelihood $\Lambda_k^{(j,W)}$ of each GIW component will be very small since clutter does not obey the kinematic and extent model of target. If the number of clutter measurements in the cell is equal to one, then $|W| = 1$, $\delta_{|W|,1} = 1$, $\sum_{l=1}^{J_{k|k-1}} e^{-\gamma(l)}\left(\frac{\gamma^{(l)}}{\lambda_c c_k}\right)^{|W|} p_D^{(l)} \Lambda_k^{(l,W)} w_{k|k-1}^{(l)}$ will be much smaller than 1 because the likelihood $\Lambda_k^{(j,W)}$ achieves a small value mentioned above and other parameters can be considered as constants, the value of $w_{k|k}^{(j,W)}$ will be close to 0 and is smaller than the pruning threshold, then the corresponding component will be eliminated and the clutter is eliminated. However, if the number of clutter measurements in the cell is more than one, then $|W| \neq 1$, $\delta_{|W|,1} = 0$, Equation (9) is the normalization process. Although $\Lambda_k^{(j,W)}$ is close to zero, $w_{k|k}^{(j,W)}$ can still take a large value. In this case, ghost targets will emerge and the number of targets will be overestimated. Further details on numerical implementation can be found in Section 5.

According to the analysis above, if the measurement in the cell is clutter, $\sum_{l=1}^{J_{k|k-1}} e^{-\gamma(l)}\left(\frac{\gamma^{(l)}}{\lambda_c c_k}\right)^{|W|} p_D^{(l)} \Lambda_k^{(l,W)} w_{k|k-1}^{(l)}$ (denoted by $\sum_{l=1}^{J_{k|k-1}} \psi_{l,W}$) in Equation (9) should be added by 1. Otherwise, it should be added by 0 and the clutter can be eliminated. However, from Equation (9), if the cell contains only one measurement, $\sum_{l=1}^{J_{k|k-1}} \psi_{l,W}$ is added by 1, it means that the cell contains only one measurement is considered as clutter in ET-GIW-PHD. Otherwise, it is considered as a target if the cell contains more than one measurement. In fact, this assumption can be violated under strong clutter. The criterion whether measurements in the cell are generated by clutter based on only the number of measurements can be erroneous. A simple numerical calculation is shown below to illustrate this point.

In ET-GIW-PHD, the probability of the measurements of the cell generated by clutter is obtained based on the Bayesian theorem, see Equation (10). Note that, only the number of measurement is considered in this calculation.

$$\begin{aligned} P(\mathbf{Z}_W \subset \mathrm{C}|n_W = 1) &= 1 - P(\mathbf{Z}_W \subset \mathrm{T}|n_W = 1) \\ &= \frac{P(n_W = 1|\mathbf{Z}_W \subset \mathrm{C})P(\mathbf{Z}_W \subset \mathrm{C})}{P(n_W = 1|\mathbf{Z}_W \subset \mathrm{T})P(\mathbf{Z}_W \subset \mathrm{T}) + P(n_W = 1|\mathbf{Z}_W \subset \mathrm{C})P(\mathbf{Z}_W \subset \mathrm{C})}, \end{aligned} \tag{10}$$

where $\mathbf{Z}_W$ presents the measurements in cell, n_W is the number of measurements in cell, C and T mean clutter and target respectively, $P(\mathbf{Z}_W \subset \mathrm{C})$ and $P(\mathbf{Z}_W \subset \mathrm{T})$ are the prior information.

The number of measurements generated by the target is subject to Poisson distribution with Poisson rate γ, the detection probability is p_d, then

$$\begin{aligned} P(n_W = 1|\mathbf{Z}_w \subset \mathrm{T}) &= \sum_{i=1}^{\infty} p_d(1-p_d)^{i-1} C_i^1 \frac{\gamma^i}{i!} e^{-\gamma} \\ &= \sum_{j=0}^{\infty} p_d(1-p_d)^j (j+1) \frac{\gamma^{j+1}}{(j+1)!} e^{-\gamma} \\ &= p_d \gamma e^{-p_d \gamma} \sum_{j=0}^{\infty} \frac{((1-p_d)\gamma)^j}{j!} e^{-(1-P_d)\gamma} \\ &= p_d \gamma e^{-p_d \gamma}. \end{aligned} \tag{11}$$

where $C_m^n = \frac{n!}{m!(m-n)!}$ denotes the combinatorial number of the events that m out of n.

Remark: p_d is not equal to p_D in Equation (6). p_d is the probability that one measurement generated by target or clutter is detected, while p_D is the probability that an extended target will generate a measurement set [15]. p_D can be derived if p_d is already known.

The clutter measurements are assumed to be uniformly distributed over the surveillance area, and the number of clutter is subject to Poisson distribution with Poisson rate λ_c. So we have

$$\begin{aligned} P(n_W = 1|\mathbf{Z}_w \subset \mathrm{C}) &= \sum_{i=1}^{\infty} p_d(1-p_d)^{i-1} C_i^1 \frac{{\lambda_c}^i}{i!} e^{-\lambda_c} \\ &= p_d \lambda_c e^{-p_d \lambda_c}. \end{aligned} \tag{12}$$

When the number of measurements is 1, the probability of the measurement in the cell generated by clutter is shown in Table 1 with different γ and λ_c. In this simulation, the prior information is set to 0.5, then $P(\mathbf{Z}_W \subset \mathrm{C}) = P(\mathbf{Z}_W \subset \mathrm{T}) = 0.5$.

Table 1. The probability of the measurement generated by clutter.

	$\lambda_c = 5$ $\gamma = 5$	$\lambda_c = 5$ $\gamma = 10$	$\lambda_c = 25$ $\gamma = 10$	$\lambda_c = 35$ $\gamma = 10$
$p_d = 0.6$	1	1	0.944	0.055
$p_d = 0.7$	1	1	0.965	0.034
$p_d = 0.8$	1	1	0.979	0.021
$p_d = 0.9$	1	1	0.987	0.013

From Table 1 we can see that when $\gamma = 10$, $\lambda_c = 35$ and $p_d = 0.9$, $P(\mathbf{Z}_W \subset \mathrm{C}|n_W = 1) = 0.013$. Although the cell contains only one measurement, the probability of the measurement in the cell generated by clutter is close to 0. Consequently, the criterion of ET-GIW-PHD does not work well in this case. when $\gamma = 10$ and $\lambda_c = 5$, $P(\mathbf{Z}_W \subset \mathrm{C}|n_W = 1) = 1$. In this case, the clutter is distinguished correctly based on the criterion of ET-GIW-PHD. In summary, the determination whether measurements are generated by clutter based on only the number of measurements can be erroneous.

4. Anti-Clutter ET-GIW-PHD

The difference between ET-GIW-PHD and anti-clutter ET-GIW-PHD is how to determine the source of measurements in the cell, specifically, the difference is how to obtain d_W in Equation (6). Using only the number of measurements in ET-GIW-PHD to determine whether the measurements in the cell is the target or not may be erroneous. In contrast, our anti-clutter ET-GIW-PHD uses hypothesis testing to deal with this problem. The number of measurements, the kinematic state and extent state of target and clutter spatial distribution are taken into account to obtain the likelihood ratio test statistic.

There are two hypotheses:

$$H_0 : \mathbf{Z}_W \subset \mathrm{C}, \tag{13}$$

$$H_1 : \mathbf{Z}_W \subset \mathrm{T}, \tag{14}$$

where $\mathbf{Z}_W = \{z_1, z_2, ..., z_{n_W}\}$ is the measurements of Wth cell, n_W is the number of the measurements, C and T represent clutter and target respectively.

The likelihood ratio test statistic for hypothese is given by

$$\eta = \frac{L(\mathbf{Z}_W|\mathrm{T})}{L(\mathbf{Z}_W|\mathrm{C})}, \tag{15}$$

where $L(\mathbf{Z}_W|\mathrm{T})$ and $L(\mathbf{Z}_W|\mathrm{C})$ are the likelihood to measure the set $\mathbf{Z}_W$ given $\mathbf{Z}_W \subset \mathrm{T}$ and $\mathbf{Z}_W \subset \mathrm{C}$ respectively and $L(\mathbf{Z}_W|\mathrm{T})$ and $L(\mathbf{Z}_W|\mathrm{C})$ will be presented later.

If $\log(\cdot)$ is applied to Equation (15), $\log(\eta)$ can be obtained.

$$\log \eta = \log L(\mathbf{Z}_W|\mathrm{T}) - \log L(\mathbf{Z}_W|\mathrm{C}). \tag{16}$$

Because $\log(\cdot)$ is monotony increase, $\log(\eta)$ is also the test statistic for hypothesis H_0 versus H_1. If these measurements are generated by the target, $L(\mathbf{Z}_W|\mathrm{T})$ will achieve a large value and $L(\mathbf{Z}_W|\mathrm{C})$ will be small. Consequently, the statistics $\log(\eta)$ will grow to a large value. Using a threshold, we can distinguish between targets and clutter. Specifically, if $\log(\eta)$ is greater than the threshold, the measurements in the cell is considered to be generated by targets. Otherwise, these measurements are considered to be clutter. The expression of $L(\mathbf{Z}_W|\mathrm{T})$ and $L(\mathbf{Z}_W|\mathrm{C})$ are given below, the setting of the threshold is discussed.

If the measurements are generated by a target, different extent models lead to a different expression of $L(\mathbf{Z}_W|\mathrm{T})$. In this paper, $L(\mathbf{Z}_W|\mathrm{T})$ is deduced based on the model in [27],

$$\begin{aligned} L(\mathbf{Z}_W|\mathrm{T}) &= \sum_{j=n_W}^{\infty} p_d^{n_W}(1-p_d)^{j-n_W} C_j^{n_W} \frac{\gamma^j}{j!} e^{-\gamma} \cdot \prod_{i=1}^{n_W} \mathcal{N}(z_i; (\mathbf{H}_k \otimes \mathbf{I}_d) x_k, \mathbf{X}_k) \\ &= \sum_{j=n_W}^{\infty} p_d^{n_W}(1-p_d)^{j-n_W} \frac{j!}{n_W!(j-n_W)!} \frac{\gamma^j}{j!} e^{-\gamma} \cdot \prod_{i=1}^{n_W} \mathcal{N}(z_i; (\mathbf{H}_k \otimes \mathbf{I}_d) x_k, \mathbf{X}_k) \\ &= \frac{p_d^{n_W}\gamma^{n_W}}{n_W!} e^{-p_d\gamma} \sum_{j=n_W}^{\infty} \frac{((1-p_d)\gamma)^{j-n_W}}{(j-n_W)!} e^{-(1-p_d)\gamma} \cdot \prod_{i=1}^{n_W} \mathcal{N}(z_i; (\mathbf{H}_k \otimes \mathbf{I}_d) x_k, \mathbf{X}_k) \\ &= \frac{p_d^{n_W}\gamma^{n_W}}{n_W!} e^{-p_d\gamma} \prod_{i=1}^{n_W} \mathcal{N}(z_i; (\mathbf{H}_k \otimes \mathbf{I}_d) x_k, \mathbf{X}_k), \end{aligned} \tag{17}$$

where $\mathbf{I}_d$ is an unit matrix with d dimension, $\mathbf{X}_k$ is the extension of target at time k, $\mathbf{H}_k$ is the 1D observation matrix.

The clutters are assumed to be uniformly distributed over the surveillance area [27], then

$$\begin{aligned}
L(\mathbf{Z}_W|\mathrm{C}) &= \beta_{FA}^{n_W} \sum_{j=n_W}^{\infty} \frac{\lambda_c^j}{j!} e^{-\lambda_c} C_j^{n_W} p_d^{n_W} (1-p_d)^{j-n_W} \\
&= \beta_{FA}^{n_W} \sum_{j=n_W}^{\infty} \frac{\lambda_c^j}{j!} e^{-\lambda_c} \frac{j!}{n_W!(j-n_W)!} p_d^{n_W} (1-p_d)^{j-n_W} \\
&= \beta_{FA}^{n_W} p_d^{n_W} \frac{\lambda^{n_W}}{n_W!} e^{-p_d\lambda_c} \sum_{j=n_W}^{\infty} \frac{((1-p_d)\lambda_c)^{j-n_W}}{(j-n_W)!} e^{-(1-p_d)\lambda_c} \\
&= \beta_{FA}^{n_W} p_d^{n_W} \frac{\lambda_c^{n_W}}{n_W!} e^{-p_d\lambda_c} \sum_{i=0}^{\infty} \frac{((1-p_d)\lambda_c)^{i}}{i!} e^{-(1-p_d)\lambda_c} \\
&= \beta_{FA}^{n_W} \frac{p_d^{n_W}\lambda_c^{n_W}}{n_W!} e^{-p_d\lambda_c},
\end{aligned} \tag{18}$$

where $\beta_{FA} = \lambda_c c_k$, λ_c is the mean number of clutter measurements, c_k is the spatial distribution of the clutter over the surveillance volume.

Substitute Equations (17) and (18) into Equation (16), we have

$$\begin{aligned}
\log(\eta) &= \log L(\mathbf{Z}_W|\mathrm{T}) - \log L(\mathbf{Z}_W|\mathrm{C}) \\
&= \sum_{j=1}^{n_W} \log(\mathcal{N}(z_j; (\mathbf{H}_k \otimes \mathbf{I}_d)x_k, \mathbf{X}_k)) + \log(\frac{p_d^{n_W}\gamma^{n_W}}{n_W!} e^{-p_d\gamma}) - \log(\beta_{FA}^{n_W} \frac{p_d^{n_W}\lambda_c^{n_W}}{n_W!} e^{-p_d\lambda_c}) \\
&= \sum_{j=1}^{n_W} \{-0.5\log 2\pi - 0.5\log|\mathbf{X}_k|\} - \sum_{j=1}^{n_W} \{0.5(z_j - (\mathbf{H}_k \otimes \mathbf{I}_d)x_k)\mathbf{X}_k^{-1}(z_j - (\mathbf{H}_k \otimes \mathbf{I}_d)x_k)^{\mathrm{T}}\} \\
&\quad + (-p_d\gamma + n_W \log p_d\gamma - \sum_{i=1}^{n_W} \log i) - (-p_d\lambda_c + n_W \log p_d\lambda_c - \sum_{i=1}^{n_W} \log i) - n_W \log \beta_{FA} \\
&= -0.5 \sum_{j=1}^{n_W} \{(z_j - (\mathbf{H}_k \otimes \mathbf{I}_d)x_k)\mathbf{X}_k^{-1}(z_j - (\mathbf{H}_k \otimes \mathbf{I}_d)x_k)^{\mathrm{T}}\} - 0.5 n_W \log 2\pi - 0.5 n_W \log|\mathbf{X}_k| \\
&\quad - p_d(\gamma - \lambda_c) + n_W \log(\frac{\gamma}{\lambda_c}) - n_W \log \beta_{FA} \\
&= -0.5G + D,
\end{aligned} \tag{19}$$

where

$$G = \sum_{j=1}^{n_W} (z_j - (\mathbf{H}_k \otimes \mathbf{I}_d)x_k)\mathbf{X}_k^{-1}(z_j - (\mathbf{H}_k \otimes \mathbf{I}_d)x_k)^{\mathrm{T}} \tag{20}$$

$$\begin{aligned}
D = &-0.5 n_W \log 2\pi - 0.5 n_W \log|\mathbf{X}_k| - p_d(\gamma - \lambda_c) \\
&+ n_W \log(\frac{\gamma}{\lambda_c}) - n_W \log \beta_{FA}.
\end{aligned} \tag{21}$$

Because z_j is subject to Gaussian distribution with mean $(\mathbf{H}_k \otimes \mathbf{I}_d)x_k$ and covariance $\mathbf{X}_k$, $z_j \propto \mathcal{N}((\mathbf{H}_k \otimes \mathbf{I}_d)x_k, \mathbf{X}_k)$, thus

$$G = \sum_{j=1}^{n_W} (z_j - (\mathbf{H}_k \otimes \mathbf{I}_d)x_k)\mathbf{X}_k^{-1}(z_j - (\mathbf{H}_k \otimes \mathbf{I}_d)x_k)^{\mathrm{T}} \propto \mathcal{X}^2(n_W), \tag{22}$$

where G is subject to chi-square distribution with degree of freedom n_W. In Equation (21), γ, λ_c and β_{FA} are priori known, the volume of the target extension is proportional to $|\mathbf{X}_k|$, the size of the target could be assumed to be unchanged, then D could be considered as a constant.

The confidence coefficient is set to α and a threshold is introduced (denoted by g), suppose hypothesi H_1 is true, then

$$P(\log\eta > g) = P(-0.5G + D > g) = P\{G < 2(D-g)\} = \alpha, \tag{23}$$

then

$$2(D-g) = \mathcal{X}^2_{1-\alpha} \tag{24}$$

$$g = D - 0.5\mathcal{X}^2_{1-\alpha}, \tag{25}$$

where

$$\alpha = \int_{\mathcal{X}^2_\alpha(n_W)}^{\infty} \mathcal{X}^2(n_W)dx. \tag{26}$$

From Equation (23), the probability of $log(\eta) < g$ is $1-\alpha$. Generally, α is set to be a value close to 1 and $log(\eta) < g$ is a small probability event. If $log(\eta) < g$ is satisfied, hypothesi H_1 should be rejected. Finally, we have

If $log(\eta) < g$, the measurements are generated by clutter, then

$$d_W = 1 + \sum_{l=1}^{J_{k|k-1}} e^{-\gamma(l)} \left(\frac{\gamma^{(l)}}{\beta_{FA}}\right)^{|W|} p_D^{(l)} \Lambda_k^{(l,W)} w_{k|k-1}^{(l)}. \tag{27}$$

If $\log\eta \geq g$,the measurements are generated by targets, then

$$d_W = \sum_{l=1}^{J_{k|k-1}} e^{-\gamma(l)} \left(\frac{\gamma^{(l)}}{\beta_{FA}}\right)^{|W|} p_D^{(l)} \Lambda_k^{(l,W)} w_{k|k-1}^{(l)}. \tag{28}$$

The pseudo-code for anti-clutter ET-GIW-PHD is illustrated in Table 2.

Table 2. Pseudo-code for anti-clutter extended target Gaussian inverse Wishart probability hypothesis density (ET-GIW-PHD) filter.

```
1: Input: Sequence of measurement sets
2: Initialize: parameter initialization
3: for k = 1 : K (K is totally time steps)
4:   Measurements partition
5:   Prediction
6:   Correction, see Table 3.
7:   Prune and merge
8:   Extract target state
9: end for
10: Output: Sequence of estimated targets.
```

The difference between ET-GIW-PHD and anti-clutter ET-GIW-PHD lies in correction step. Pseudo-code for anti-clutter ET-GIW-PHD filter correction is shown in Table 3, pseudo-code for other steps (prediction, prune and merge etc) can be found in [39].

Table 3. Pseudo-code for anti-clutter ET-GIW-PHD filter correction.

Input: GIW components $\{w^j_{k|k-1}, \xi^j_{k|k-1}\}_{j=1}^{J_{k|k-1}}$, measurements partitions $\{p_\rho\}_{\rho=1}^{n}$

Undetected target case:

for $j = 1 : J_{k|k-1}$

$w^j_{k|k} \leftarrow (1 - (1 - e^{-\gamma})p_D)\, w^j_{k|k-1} \quad \xi^j_{k|k} \leftarrow \xi^j_{k|k-1}$

end for

Detected target case:

$l = 0$

for $\rho = 1 : n$

  for $W = 1 : |p_\rho|$

    $l = l + 1$

    for $j = 1 : J_{k|k-1}$

    update $\xi^j_{k|k-1}$ using Kalman filter, see details in [39], $\xi^{j+l\cdot J_{k|k-1}}_{k|k} \xleftarrow{update} \xi^j_{k|k-1}$

    $w^{(j+l\cdot J_{k|k-1},W)}_{k|k} \leftarrow e^{-\gamma^{(j)}} \left(\frac{\gamma^{(j)}}{\lambda_c c_k}\right)^{|W|} p_D^{(j)} \Lambda_k^{(j,W)} w^{(j)}_{k|k-1}$

    $G^j = \sum_{j=1}^{n_w} (z_j - (\mathbf{H}_k \otimes \mathbf{I}_d)x^j_{k|k-1})(\mathbf{X}^j_{k|k-1})^{-1}(z_j - (\mathbf{H}_k \otimes \mathbf{I}_d)x^j_{k|k-1})^{\mathrm{T}}$

    end for

  $D = -0.5n_W \log 2\pi - 0.5n_W \log|\mathbf{X}_{k|k-1}| - p_d(\gamma - \lambda_c) + n_W \log(\frac{\gamma}{\lambda_c}) - n_W \log \beta_{FA}.$

  $\log(\eta) = -0.5 \arg\min_j G^j + D$

  $g = D - 0.5\mathcal{X}^2_{1-\alpha}$,

  $$d_W^{(\rho,W)} = \begin{cases} 1 + \sum_{l=1}^{J_{k|k-1}} e^{-\gamma(l)} \left(\frac{\gamma^{(l)}}{\beta_{FA,k}}\right)^{|W|} p_D^{(l)} \Lambda_k^{(l,W)} w^{(l)}_{k|k-1} & \log\eta < g \\ \sum_{l=1}^{J_{k|k-1}} e^{-\gamma(l)} \left(\frac{\gamma^{(l)}}{\beta_{FA,k}}\right)^{|W|} p_D^{(l)} \Lambda_k^{(l,W)} w^{(l)}_{k|k-1} & \log\eta \geq g \end{cases}$$

  $w^{(j+l\cdot J_{k|k-1},W)}_{k|k} \leftarrow \frac{w^{(j+l\cdot J_{k|k-1},W)}_{k|k}}{d_W}$

  end for

$\omega_{p_\rho} \leftarrow \Pi_{W=1}^{|p_\rho|} d_W^{(\rho,W)}$

end for

$\omega_{p_\rho} \leftarrow \frac{\omega_{p_\rho}}{\sum_{\rho=1}^{n} \omega_{p_\rho}}$ for $\rho = 1 : n$

$J_{k|k} \leftarrow J_{k|k-1}(l+1)\ J_{tmp} = J_{k|k-1}$

for $\rho = 1 : n$

  for $j = 1 : J_{k|k-1}|p_\rho|$

  $w^{(j+J_{tmp})}_{k|k} \leftarrow w^{(j+J_{tmp})}_{k|k} \omega_{p_\rho}$

  end for

$J_{tmp} \leftarrow J_{tmp} + J_{k|k-1}|p_\rho|$

end for

Output: GIW components $\{w^j_{k|k}, \xi^j_{k|k}\}_{j=1}^{J_{k|k}}$

5. Simulation

In this section, the effectiveness of ET-GIW-PHD and anti-clutter ET-GIW-PHD were tested. Two scenarios with multiple targets were established. The surveillance area was set as $[-1000\text{ m}, 1000\text{ m}] \times [-1000\text{ m}, 1000\text{ m}]$, then c_k is 2.5×10^{-7} under the assumption that clutter is uniformly distributed over the surveillance area. We set totally 100 time steps and the sampling time is 1 s.

In the first scenario, four targets moving along different lines were generated:

$$\begin{aligned} x_0^{(1)} &= [-1000\text{ m}, 1000\text{ m}, 25\text{ m/s}, -25\text{ m/s}], t_s^{(1)} = 5\text{ s}, t_e^{(1)} = 45\text{ s};\\ x_0^{(2)} &= [-1000\text{ m}, -1000\text{ m}, 25\text{ m/s}, 25\text{ m/s}], t_s^{(2)} = 15\text{ s}, t_e^{(2)} = 55\text{ s};\\ x_0^{(3)} &= [1000\text{ m}, -1000\text{ m}, -25\text{ m/s}, 25\text{ m/s}], t_s^{(3)} = 25\text{ s}, t_e^{(3)} = 65\text{ s};\\ x_0^{(4)} &= [1000\text{ m}, 1000\text{ m}, -25\text{ m/s}, -25\text{ m/s}], t_s^{(4)} = 35\text{ s}, t_e^{(4)} = 75\text{ s}; \end{aligned} \tag{29}$$

where $x_0^{(j)}$ is the initial state of jth target, $t_s^{(j)}$ is the born time of jth target, $t_e^{(j)}$ is the end time of jth target. The birth intensity in the first scenario is

$$D_b(\xi) = \sum_{j=1}^{4} w_b \mathcal{N}(x; x_0^{(j)}, \mathbf{P}_b \otimes \mathbf{X}_k)\mathcal{IW}(\mathbf{X}_k; v_b, \mathbf{V}_b), \tag{30}$$

where $w_b = 0.03$, $\mathbf{P}_b = \text{diag}([100, 100])$, $v_b = 10$, $\mathbf{V}_b = \text{diag}([100, 100])$.

In the second scenario, two targets were born at (−1000 m, 300 m) and (−1000 m, −300 m), respectively at $k = 0$ (k is time step). Next, they moved close gradually and then moved in parallel before separating. The birth intensity in the second scenario is

$$D_b(\xi) = \sum_{j=1}^{2} w_b \mathcal{N}(x; m_j, \mathbf{P}_b \otimes \mathbf{X}_k)\mathcal{IW}(\mathbf{X}_k; v_b, \mathbf{V}_b), \tag{31}$$

where $m_1 = [-1000, 300, 25, -25]$, $m_2 = [-1000, -300, 25, 25]$, $w_b = 0.03$, $\mathbf{P}_b = \text{diag}([100, 100])$, $v_b = 10$, $\mathbf{V}_b = \text{diag}([100, 100])$. The true trajectories of two scenario are shown in Figure 1.

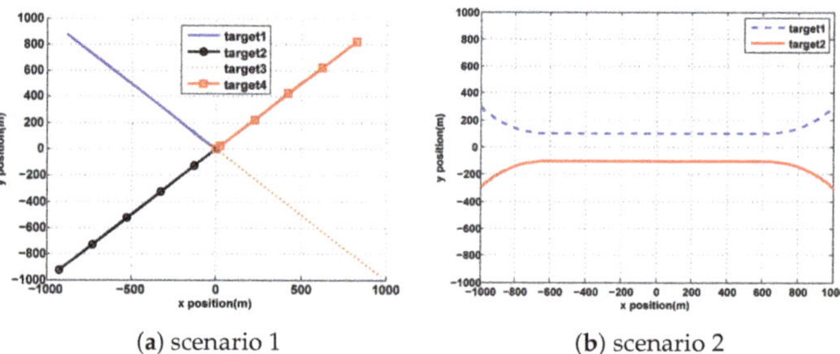

(**a**) scenario 1 (**b**) scenario 2

Figure 1. True trajectories of two scenarios: (**a**) four targets move along different lines in scenario 1. (**b**) Two targets move closeer gradually and then move in parallel before separating in scenario 2.

The dynamic and measurement model are shown below. The target kinematic state is denoted as $x = [r_x, r_y, \dot{r}_x, \dot{r}_y]$, where r_x and $\dot{r}_x$ is the position and velocity in the x direction, likewise of y direction. The time evolution of kinematic state given by

$$x_k^{(j)} = \mathbf{F}_k^{(j)} x_{k-1}^{(j)} + w_k^{(j)}, \tag{32}$$

where $x_k^{(j)}$ is the target state of jth target at time k, w_k is the process noise of jth target and is the Gaussian white noise with zero mean and covariance $\mathbf{Q}_k^{(j)}$, $\mathbf{F}_k^{(j)}$ is the kinematic state transition matrix of jth target, given by

$$\mathbf{F}_k{}^{(j)} = \begin{bmatrix} 1 & 0 & t & 0 \\ 0 & 1 & 0 & t \\ 0 & 0 & 1 & 0 \\ 0 & 0 & 0 & 1 \end{bmatrix}, \mathbf{Q}_k{}^{(j)} = \Omega^2 \begin{bmatrix} \frac{t^2}{2} \\ t \end{bmatrix} \begin{bmatrix} \frac{t^2}{2} & t \end{bmatrix}, \tag{33}$$

where t is the sampling time and Ω represents the acceleration error, $t = 1\,\mathrm{s}$ and $\Omega = 5\,\mathrm{m/s^2}$ in this simulation.

In this simulation, the major and minor axes are 20 m and 15 m respectively for all targets. The major axis was aligned with the direction of motion of the target and the extent of these targets remained unchanged.

The measurement model can be expressed as

$$z_k^{(j)} = (\mathbf{H}_k \otimes \mathbf{I}_d) x_k^{(j)} + q_k, \tag{34}$$

where $z_k^{(j)}$ is the measurements generated by the jth target at time k, q_k is the measurement noise and is the Gaussian white noise with zero mean and covariance $\mathbf{R}_k$, $\mathbf{H}_k \otimes \mathbf{I}_d$ is the observation matrix, given by

$$\mathbf{H}_k \otimes \mathbf{I}_d = \begin{bmatrix} 1 & 0 & 0 & 0 \\ 0 & 1 & 0 & 0 \end{bmatrix}, \mathbf{R}_k = \begin{bmatrix} 1 & 0 \\ 0 & 1 \end{bmatrix}. \tag{35}$$

where $\mathbf{H}_k = [1, 0]$, $\mathbf{I}_d = \begin{bmatrix} 1 & 0 \\ 0 & 1 \end{bmatrix}$.

In our experiment, the confidence coefficient α of anti-clutter ET-GIW-PHD is set to 0.99. A distance partition algorithm [17] is used for both filters, a measurement partition that contains several cells can be obtained for a given distance threshold. Clutter Poisson rate λ_c is set to 35, then clutter density $\lambda_c c_k$ is 8.75×10^{-6} (the clutter density in this paper is higher than that of related references, such as [18,21]). The expected number of measurements generated by targets γ is set to 15. The probability of survival p_s and the detection probability p_D are assumed to be state independent and set to 0.99 and 0.98, respectively. The probability p_d is set to 0.99.

Tracking results are evaluated using the optimal subpattern assignment metric (OSPA) [43], which is widely used to evaluate multiple-target tracking performance [39–42].

The OSPA distance is defined by

$$d_p^c(\varkappa_k, \hat{\varkappa}_k) = \left(\frac{1}{n} \left(\min_{\pi \in \Pi_n} \sum_{i=1}^{m} d^c(x_i, \hat{x}_{\pi(i)})^p + C^p(n-m) \right) \right)^{1/p}, \tag{36}$$

where $m < n$, $\varkappa_k = \{x_k^{(1)}, x_k^{(2)}, ..., x_k^{(m)}\}$is the true RFS at time k, $\hat{\varkappa} = \{\hat{x}_k^{(1)}, \hat{x}_k^{(2)}, ..., \hat{x}_k^{(n)}\}$ is the estimated RFS, $\prod_n$ is the assignment results which assign $\varkappa$ to $\hat{\varkappa}$, p means $p - norm$, c is the penalty cost for cardinality mismatch. In this simulation, $c = 60$ and $p = 2$.

ET-GIW-PHD and anti-clutter ET-GIW-PHD are applied to two scenarios mentioned above for performance evaluation. The trajectories generated by these two methods are presented in Figures 2 and 3.

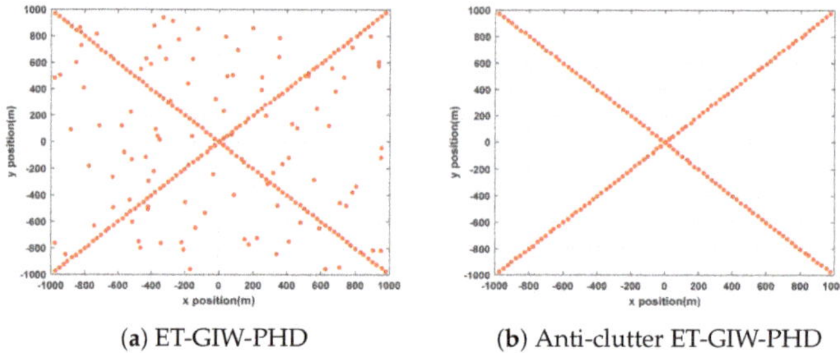

(a) ET-GIW-PHD (b) Anti-clutter ET-GIW-PHD

Figure 2. The obtained trajectories of ET-GIW-PHD and anti-clutter ET-GIW-PHD in scenario 1: (a) ET-GIW-PHD. (b) Anti-clutter ET-GIW-PHD.

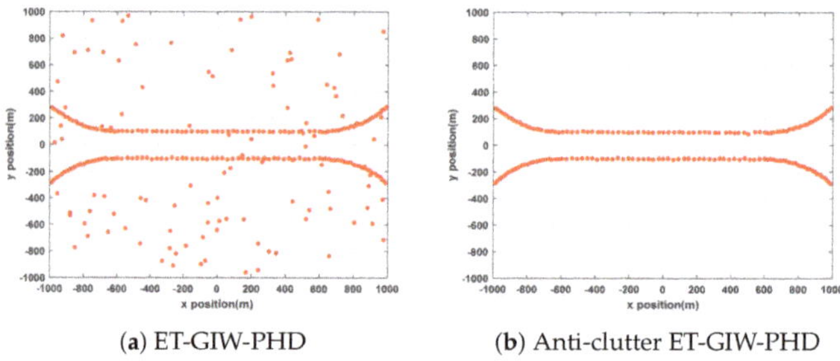

(a) ET-GIW-PHD (b) Anti-clutter ET-GIW-PHD

Figure 3. The obtained trajectories of ET-GIW-PHD and anti-clutter ET-GIW-PHD in scenario 2: (a) ET-GIW-PHD. (b) Anti-clutter ET-GIW-PHD.

From Figures 2 and 3 we can see that the trajectories of anti-clutter ET-GIW-PHD are almost identical to the true trajectories. Note that, in the results of ET-GIW-PHD, some peices of clutter are incorrectly considered as targets. However, our anti-clutter ET-GIW-PHD can deal with the clutter more correctly and achieves better performance.

To further verify the analysis in Section 3, the calculation of $w_{k|k}^{(j,W)}$ in Equation (9) at $k = 40$ (k is time step) in scenario 1 is shown below. The partition result at $k = 40$ is given firstly in Figure 4.

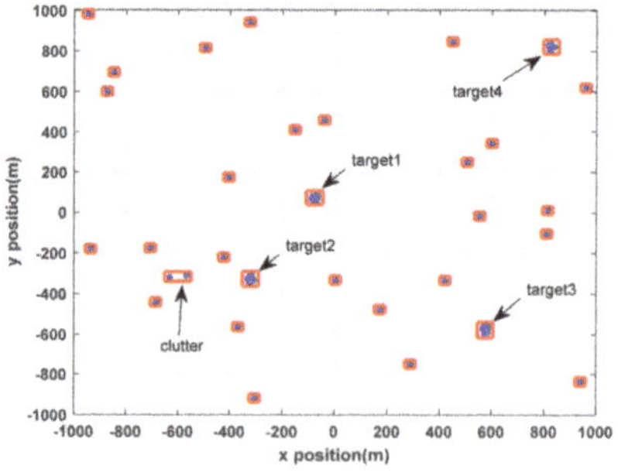

Figure 4. The partition result at $k = 40$ in scenario 1.

From Figure 4 we can see that the measurements of four targets are correctly clustered, and two clutter (marked with arrows in Figure 4) are incorrectly partitioned into one cell.

$e^{-\gamma^{(j)}}\left(\frac{\gamma^{(j)}}{\beta_{FA,k}}\right)^{|W|} p_D^{(j)} \Lambda_k^{(j,W)} w_{k|k-1}^{(j)}$ is denoted as $\psi_{j,W}$ for jth GIW component in the Wth cell, then

$$w_{k|k}^{(j,W)} = \frac{\psi_{j,W}}{\delta_{|W|,1} + \sum_{l=1}^{J_{k|k-1}} \psi_{l,W}}. \tag{37}$$

From simulation results, the number of components of predicted PHD is 14 at $k = 40$, then $J_{k|k-1} = 14$, $\psi_{j,W}$ of the clutter cell (marked with arrows in Figure 4) is obtained and shown in Table 4.

The likelihood $\Lambda_k^{(j,W)}$ of each GIW component in this cell is very small since clutter does not obey the kinematic and extent model of target, therefore $\psi_{j,W}$ achieve small value as shown in Table 4.

Table 4. The $\psi_{j,W}$ of the clutter cell.

j	1	2	3	4	5	6	7
$\psi_{j,W}$	4.7×10^{-21}	2.5×10^{-24}	1.4×10^{-19}	2.7×10^{-15}	2.9×10^{-10}	3.7×10^{-19}	3.5×10^{-70}
j	8	9	10	11	12	13	14
$\psi_{j,W}$	3.8×10^{-50}	4.4×10^{-57}	6.7×10^{-8}	2.6×10^{-30}	2.9×10^{-17}	4.7×10^{-14}	1.1×10^{-55}

Because the number of measurement in this cell is two, Equation (37) is represent as

$$w_{k|k}^{(j,W)} = \frac{\psi_{j,W}}{\sum_{l=1}^{J_{k|k-1}} \psi_{l,W}}. \tag{38}$$

Equation (38) is a normalization process, $w_{k|k}^{(j,W)}$ of the clutter cell is shown in Table 5.

Table 5. $w_{k|k}^{(j,W)}$ of the clutter cell.

j	1	2	3	4	5	6	7
$w_{k|k}^{(j,W)}$	6.9×10^{-14}	3.7×10^{-17}	2.1×10^{-12}	4.1×10^{-8}	4.3×10^{-3}	5.5×10^{-12}	5.1×10^{-63}
j	8	9	10	11	12	13	14
$w_{k|k}^{(j,W)}$	5.6×10^{-43}	6.5×10^{-50}	0.99	3.9×10^{-23}	4.3×10^{-10}	6.9×10^{-7}	1.7×10^{-48}

Although $\psi_{j,W}$ is small, $w_{k|k}^{(j,W)}$ may achieve a large value ($w_{k|k}^{(10,W)} = 0.99$) and results in a ghost target. At $k = 40$, the estimated number of targets was 5 while true number is 4. That is, the number of targets was overestimated.

To test the influence of the clutter density on tracking performance, ET-GIW-PHD and anti-clutter ET-GIW-PHD were tested under different numbers of clutter modeled as Poisson distribution with Poisson rate λ_c. The clutter measurements are assumed to be uniformly distributed over the surveillance area. The OSPA distance of these two filters under different Poisson rate λ_c is shown in Figures 5 and 6.

As we can see from Figures 5 and 6, when λ_c is small, ET-GIW-PHD achieves good performance. However, as λ_c increases, the performance of ET-GIW-PHD degrades significantly. In contrast, our anti-clutter ET-GIW-PHD achieves superior performance with varying λ_c, which demonstrates that anti-clutter ET-GIW-PHD is more robust to clutter than ET-GIW-PHD. The results of cardinality estimation are shown in Figures 7 and 8.

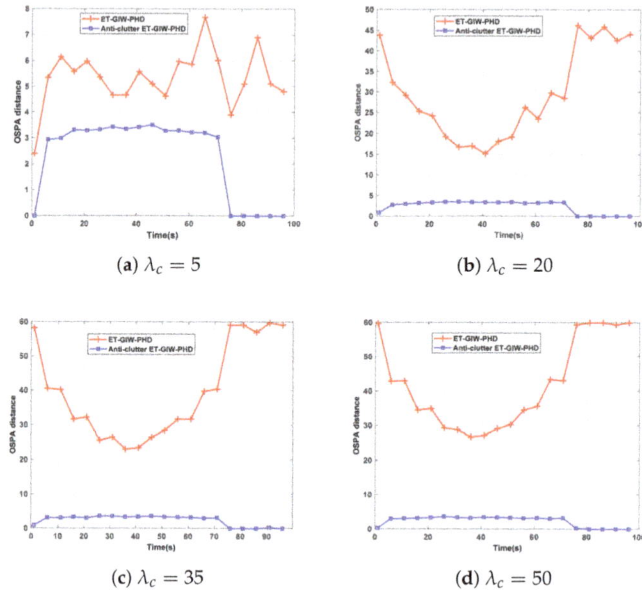

Figure 5. The optimal subpattern assignment metric (OSPA) distance of ET-GIW-PHD and anti-clutter ET-GIW-PHD under different Poisson rate of clutter in scenario 1: (**a**) Poisson rate $\lambda_c = 5$. (**b**) Poisson rate $\lambda_c = 20$. (**c**) Poisson rate $\lambda_c = 35$. (**d**) Poisson rate $\lambda_c = 50$.

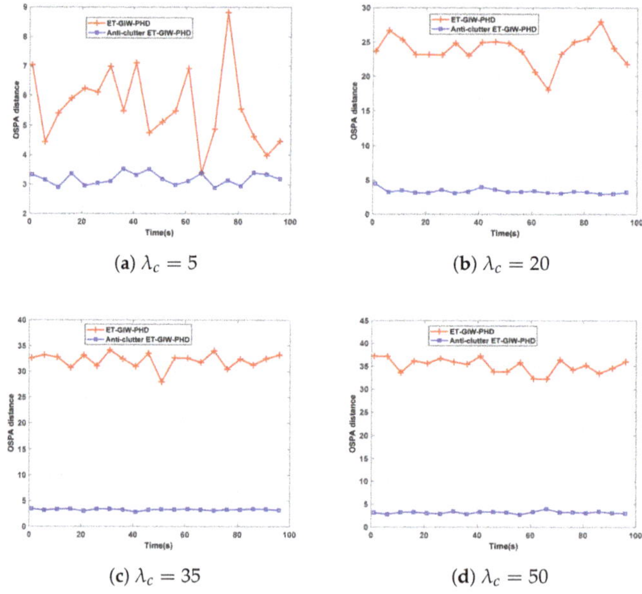

Figure 6. The OSPA distance of ET-GIW-PHD and anti-clutter ET-GIW-PHD under different Poisson rate of clutter in scenario 2: (**a**) Poisson rate $\lambda_c = 5$. (**b**) Poisson rate $\lambda_c = 20$. (**c**) Poisson rate $\lambda_c = 35$. (**d**) Poisson rate $\lambda_c = 50$.

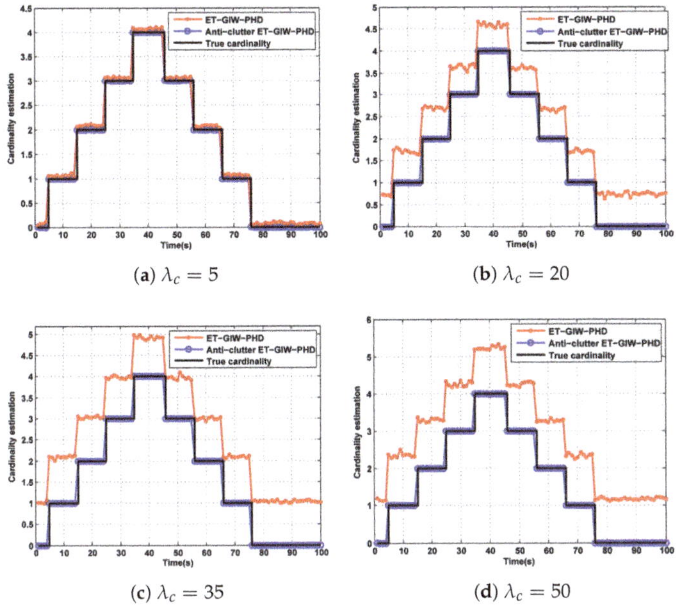

Figure 7. The cardinality estimation of ET-GIW-PHD and anti-clutter ET-GIW-PHD under different Poisson rate of clutter in scenario 1: (**a**) Poisson rate $\lambda_c = 5$. (**b**) Poisson rate $\lambda_c = 20$. (**c**) Poisson rate $\lambda_c = 35$. (**d**) Poisson rate $\lambda_c = 50$.

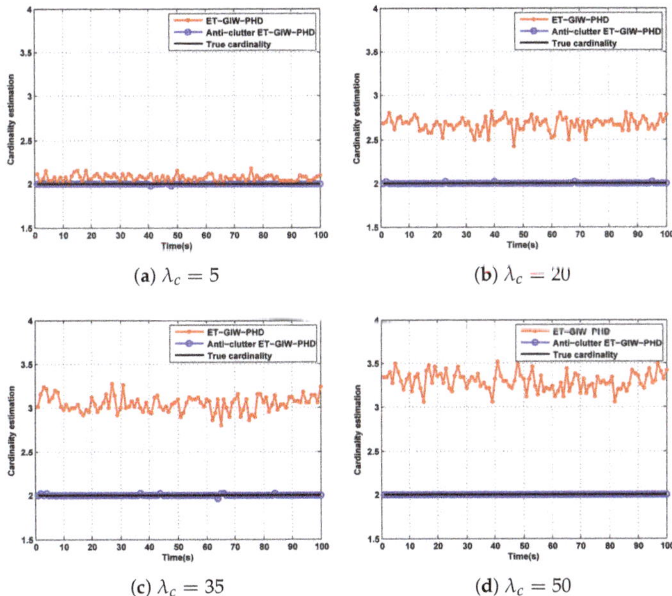

Figure 8. The cardinality estimation of ET-GIW-PHD and anti-clutter ET-GIW-PHD under different Poisson rate of clutter in scenario 2: (**a**) Poisson rate $\lambda_c = 5$. (**b**) Poisson rate $\lambda_c = 20$. (**c**) Poisson rate $\lambda_c = 35$. (**d**) Poisson rate $\lambda_c = 50$.

From Figures 7 and 8 we can see that the cardinality estimation error of ET-GIW-PHD increases as the λ_c grows. That is, ET-GIW-PHD cannot avoid the overestimation of cardinality under high clutter density. When clutter density is small, the clutter spreads apart. Thus, it is unlikely to partition more than one clutter into one cell. In the presence of severe clutter, the probability that multiple clutter being partitioned into one cell increases, and thus ET-GIW-PHD could overestimate the cardinality. However, our anti-clutter ET-GIW-PHD uses not only the number of measurement, but also target state and spatial distribution of clutter for better cardinality estimation performance. Using hypothesis testing, the measurements can be distinguished more correctly. Therefore, a better tracking performance can be achieved. Extensive experiments have demonstrated the effectiveness of anti-clutter ET-GIW-PHD.

6. Conclusions

In this paper, we propose an anti-clutter ET-GIW-PHD filter which revises the correction step of ET-GIW-PHD with hypothesis testing for better tracking performance under severe clutter. Our anti-clutter ET-GIW-PHD adopts a hypothesis testing method to distinguish between measurements from targets and clutter, hypothesis testing results are incorporated into the correction step. Specifically, likelihood functions are built to incorporate the number of measurements, the target state, and clutter spatial distribution in anti-clutter ET-GIW-PHD, the source of measurements in the cell is determined more correctly. Compared with ET-GIW-PHD, our method improves the cardinality estimation accuracy and achieves better tracking performance. The effectiveness of our method has been demonstrated by extensive experiments.

Author Contributions: Y.H. conceived the idea of the study and conducted the research. Y.H. and L.W. analyzed most of the data and wrote the initial draft of the paper. X.W. and W.A. contributed to finalizing this paper.

Funding: This work was supported by the National Natural Science Foundation of China (Nos.61605242).

Conflicts of Interest: The authors declare no conflict of interest. The funders had no role in the design of the study; in the collection, analyses, or interpretation of data; in the writing of the manuscript, or in the decision to publish the results.

References

1. Granström, K.; Reuter, S.; Meissner, D.; Scheel, A. A multiple model PHD approach to tracking of cars under an assumed rectangular shape. In Proceedings of the IEEE International Conference on Information Fusion (FUSION), Salamanca, Spain, 7–10 July 2014; pp. 1–8.
2. Cao, X.; Lan, J.; Li, X.R.; Liu, Y. Extended Object Tracking Using Automotive Radar. In Proceedings of the IEEE International Conference on Information Fusion (FUSION), Cambridge, UK, 10–13 July 2018; pp. 1–5.
3. Haag, S.; Duraisamy, B.; Koch, W.; Dickmann, J. Radar and Lidar Target Signatures of Various Object Types and Evaluation of Extended Object Tracking Methods for Autonomous Driving Applications. In Proceedings of the IEEE International Conference on Information Fusion (FUSION), Cambridge, UK, 10–13 July 2018; pp. 1746–1755.
4. Hammarstrand, L.; Lundgren, M.; Svensson, L. Adaptive Radar Sensor Model for Tracking Structured Extended Objects. *IEEE Trans. Aerosp. Electron. Syst.* **2012**, *48*, 1975–1995. [CrossRef]
5. Wieneke, M.; Davey, S.J. Histogram PMHT with target extent estimates based on random matrices. In Proceedings of the IEEE International Conference on Information Fusion (FUSION), Chicago, IL, USA, 5–8 July 2011; pp. 322–329.
6. Wieneke, M.; Koch, W. A PMHT Approach for Extended Objects and Object Groups. *IEEE Trans. Aerosp. Electron. Syst.* **2012**, *48*, 2349—2370. [CrossRef]
7. Gilholm, K.; Godsill, S.; Maskell, S.; Salmond, D. Poisson models for extended target and group tracking. In Proceedings of the SPIE Signal and Data Processing of Small Targets, San Diego, CA, USA, 31 July–4 August 2005; pp. 230–241.
8. Bar-Shalom, Y.; Willett, P.K.; Tian, X. *Tracking and Data Fusion*; YBS Publishing: Storrs, CT, USA, 2011.
9. Granstrom, K.; Lundquist, C.; Gustafsson, F.; Orguner, U. Random set methods: Estimation of multiple extended objects. *IEEE Robot. Autom. Mag.* **2014**, *21*, 73–82. [CrossRef]

10. Koch, W. *Tracking and Sensor Data Fusion*; Springer: Berlin/Heidelberg, Germany, 2016.
11. Mahler, R.P. *Advances in Statistical Multisource-Multitarget Information Fusion*; Artech House: Norwood, MA, USA, 2014.
12. Vo, B.N.; Ma, W.K. The Gaussian mixture probability hypothesis density filter. *IEEE Trans. Signal Process.* **2006**, *54*, 4091–4104. [CrossRef]
13. Mahler, R. PHD filters of higher order in target number. *IEEE Trans. Aerosp. Electron. Syst.* **2007**, *43*, 1523–1543. [CrossRef]
14. Gilholm, K.; Salmond, D. Spatial distribution model for tracking extended objects. *IEE Proc. Radar Sonar Navig.* **2005**, *152*, 364–371. [CrossRef]
15. Mahler, R. PHD filters for nonstandard targets, I: Extended targets. In Proceedings of the IEEE International Conference on Information Fusion (FUSION), Seattle, WA, USA, 6–9 July 2009; pp. 915–921.
16. Jiang, D.; Liu, M.; Gao, Y.; Gao, Y. Time-matching extended target probability hypothesis density filter for multi-target tracking of high resolution radar. *Signal Process.* **2019**, *157*, 151–160. [CrossRef]
17. Granstrom, K.; Lundquist, C.; Orguner, O. Extended target tracking using a Gaussian-mixture PHD filter. *IEEE Trans. Aerosp. Electron. Syst.* **2012**, *48*, 3268–3286. [CrossRef]
18. Hu, Q.; Ji, H.; Zhang, Y. A standard PHD filter for joint tracking and classification of maneuvering extended targets using random matrix. *Signal Process.* **2018**, *144*, 352–363. [CrossRef]
19. Lan, J.; Li, X.R. Joint tracking and classification of extended object using random matrix. In Proceedings of the IEEE International Conference on Information Fusion (FUSION), Istanbul, Turkey, 9–12 July 2013; pp. 1550–1557.
20. Sun, L.; Lan, J.; Li, X.R. Joint tracking and classification of extended object based on support functions. In Proceedings of the IEEE International Conference on Information Fusion (FUSION), Salamanca, Spain, 7–10 July 2014; pp. 1–8.
21. Zhang, Y.; Ji, H. A robust and fast partitioning algorithm for extended target tracking using a gaussian inverse wishart phd filter. *Knowl.-Based Syst.* **2016**, *95*, 125–141. [CrossRef]
22. Li, P.; Ge, H.w.; Yang, J.l.; Zhang, H. Shape selection partitioning algorithm for Gaussian inverse Wishart probability hypothesis density filter for extended target tracking. *IET Signal Process.* **2016**, *10*, 1041–1051. [CrossRef]
23. Li, P.; Ge, H.; Yang, J. Adaptive Measurement Partitioning Algorithm for a Gaussian Inverse Wishart PHD Filter that Tracks Closely Spaced Extended Targets. *Radioengineering* **2017**, *26*, 573–580. [CrossRef]
24. Zhao, Z.; Xiang, W.; Tao, W. A Novel Measurement Data Classification Algorithm Based on SVM for Tracking Closely Spaced Targets. *IEEE Trans. Instrum. Meas.* **2018**, *68*, 1089–1100. [CrossRef]
25. Baum, M.; Faion, F.; Hanebeck, U.D. Modeling the target extent with multiplicative noise. In Proceedings of the IEEE International Conference on Information Fusion (FUSION), Singapore, 9–12 July 2012; pp. 2406–2412.
26. Granström, K.; Lundquist, C. On the use of multiple measurement models for extended target tracking. In Proceedings of the IEEE International Conference on Information Fusion (FUSION), Istanbul, Turkey, 9–12 July 2013; pp. 1534–1541.
27. Koch, J.W. Bayesian approach to extended object and cluster tracking using random matrices. *IEEE Trans. Aerosp. Electron. Syst.* **2008**, *44*, 1042–1059. [CrossRef]
28. Feldmann, M.; Franken, D.; Koch, W. Tracking of extended objects and group targets using random matrices. *IEEE Trans. Signal Process.* **2011**, *59*, 1409–1420. [CrossRef]
29. Lan, J.; Li, X.R. Tracking of extended object or target group using random matrix—Part I: New model and approach. In Proceedings of the IEEE International Conference on Information Fusion (FUSION), Singapore, 9–12 July 2012; pp. 2177–2184.
30. Baum, M.; Noack, B.; Hanebeck, U.D. Extended object and group tracking with elliptic random hypersurface models. In Proceedings of the IEEE International Conference on Information Fusion (FUSION), Edinburgh, UK, 26–29 July 2010; pp. 1–8.
31. Baum, M.; Hanebeck, U.D. Shape tracking of extended objects and group targets with star-convex RHMs. In Proceedings of the IEEE International Conference on Information Fusion (FUSION), Chicago, IL, USA, 5–8 July 2011; pp. 1–8.

32. Baum, M.; Noack, B.; Hanebeck, U.D. Mixture random hypersurface models for tracking multiple extended objects. In Proceedings of the IEEE Decision and Control and European Control Conference (CDC-ECC), Orlando, FL, USA, 12–15 December 2011; pp. 3166–3171.
33. Wahlström, N.; Özkan, E. Extended target tracking using Gaussian processes. *IEEE Trans. Signal Process.* **2015**, *63*, 4165–4178. [CrossRef]
34. Thormann, K.; Baum, M.; Honer, J. Extended Target Tracking Using Gaussian Processes with High-Resolution Automotive Radar. In Proceedings of the IEEE International Conference on Information Fusion (FUSION), Cambridge, UK, 10–13 July 2018; pp. 1764–1770.
35. Guo, Y.; Li, Y.; Tharmarasa, R.; Kirubarajan, T.; Efe, M.; Sarikaya, B. GP-PDA Filter for Extended Target Tracking with Measurement Origin Uncertainty. *IEEE Trans. Aerosp. Electron. Syst.* **2018**, *55*, 1725–1742. [CrossRef]
36. Aftab, W.; Hostettler, R.; Freitas, A.D.; Arvaneh, M.; Mihaylova, L. Spatio-Temporal Gaussian Process Models for Extended and Group Object Tracking With Irregular Shapes. *IEEE Trans. Veh. Technol.* **2019**, *68*, 2137–2151. [CrossRef]
37. Kaulbersch, H.; Honer, J.; Baum, M. A Cartesian B-Spline Vehicle Model for Extended Object Tracking. In Proceedings of the IEEE International Conference on Information Fusion (FUSION), Cambridge, UK, 10–13 July 2018; pp. 1–5.
38. Granström, K.; Lundquist, C.; Orguner, U. Tracking rectangular and elliptical extended targets using laser measurements. In Proceedings of the IEEE International Conference on Information Fusion (FUSION), Chicago, IL, USA, 5–8 July 2011; pp. 592–599.
39. Granstrom, K.; Orguner, U. A PHD filter for tracking multiple extended targets using random matrices. *IEEE Trans. Signal Process.* **2012**, *60*, 5657–5671. [CrossRef]
40. Granström, K.; Orguner, U. Estimation and maintenance of measurement rates for multiple extended target tracking. In Proceedings of the IEEE International Conference on Information Fusion (FUSION), Singapore, 9–12 July 2012; pp. 2170–2176.
41. Granström, K.; Willett, P.; Bar-Shalom, Y. An extended target tracking model with multiple random matrices and unified kinematics. In Proceedings of the IEEE International Conference on Information Fusion (FUSION), Washington, DC, USA, 6–9 July 2015; pp. 1007–1014.
42. Lan, J.; Li, X.R. Tracking of Maneuvering Non-Ellipsoidal Extended Object or Target Group Using Random Matrix. *IEEE Trans. Signal Process.* **2014**, *62*, 2450–2463. [CrossRef]
43. Schuhmacher, D.; Vo, B.T.; Vo, B.N. A consistent metric for performance evaluation of multi-object filters. *IEEE Trans. Signal Process.* **2008**, *56*, 3447–3457. [CrossRef]

Article

Tracking Multiple Marine Ships via Multiple Sensors with Unknown Backgrounds

Cong-Thanh Do [1,*,†], Tran Thien Dat Nguyen [1] and Weifeng Liu [2]

1 School of Electrical Engineering, Computing, and Mathematical Sciences, Curtin University, Bentley, WA 6102, Australia; t.nguyen172@postgrad.curtin.edu.au

2 School of Automation, Hangzhou Dianzi University, Hangzhou 310018, China; dashan_liu@163.com

* Correspondence: thanh.docong@student.curtin.edu.au or thanhdc@tnu.edu.vn

† Current address: ThaiNguyen University of Technology, ThaiNguyen University, ThaiNguyen 251810, Vietnam.

Received: 23 September 2019; Accepted: 15 November 2019; Published: 18 November 2019

Abstract: In multitarget tracking, knowledge of the backgrounds plays a crucial role in the accuracy of the tracker. Clutter and detection probability are the two essential background parameters which are usually assumed to be known constants although they are, in fact, unknown and time varying. Incorrect values of these parameters lead to a degraded or biased performance of the tracking algorithms. This paper proposes a method for online tracking multiple targets using multiple sensors which jointly adapts to the unknown clutter rate and the probability of detection. An effective filter is developed from parallel estimation of these parameters and then feeding them into the state-of-the-art generalized labeled multi-Bernoulli filter. Provided that the fluctuation of these unknown backgrounds is slowly-varying in comparison to the rate of measurement-update data, the validity of the proposed method is demonstrated via numerical study using multistatic Doppler data.

Keywords: random finite sets; unknown background; bootstrapping method; GLMB filter; multisensor multitarget tracking; Murty's algorithm

1. Introduction

In a multitarget scenario, the targets set cardinality and their dynamic states randomly vary with time. The objective of tracking multiple targets is to estimate the number of targets and their trajectories using the data collected from sensor(s) in a joint manner [1–4]. Currently, there are three major paradigms for this field of study, namely Joint Probability Data Association (JPDA) [1], Multiple Hypotheses Tracking (MHT) [2] and Random Finite Set (RFS) [3,4]. While the first two formers involve modifying single target tracking filters to accommodate the problem of multitarget tracking, the latter applies estimation theory focusing on Bayesian optimality and provide a top-down formulation for solving the multitarget estimation problem [3,4].

Using RFS leads to the development of a series of multitarget estimation algorithms. Several RFS-based filters has been proposed in both the literature and practical applications, such as the Probability Hypothesis Density (PHD) [5], Cardinalized PHD (CPHD) [6,7], and the multi-Bernoulli filters [8]. While these filters and their extensions can give good estimates of the current target states, they do not produce target trajectories without using heuristics [9,10]. A theoretically rigorous and systematic consideration of the multitarget trajectory estimation based on RFS approach was proposed in [11]. This work also derives an exact closed-form solution to the multitarget tracking problem, known as Generalized Labeled multi-Bernoulli (GLMB) filter. This filter can estimate not only the number of the targets but also their trajectories, simultaneously [12]. It has been applied to several problems as tracking with merged measurements [13], track-before-detect [14,15], extended targets [16], cell biology [17,18], sensor scheduling [19], spawning targets [20], distributed data fusion [21], field robotics [22,23] and computer vision [24]. The GLMB filter for multitarget tracking with two sensors has been developed in [25,26]. An efficient implementation of the GLMB filter based on Gibbs sampling whose complexity depends linearly on the total number of measurements and quadratically on the number of hypothesized targets has been presented in [27]. This method has been extended to the multi-scan GLMB filter [28] and the multi-sensor GLMB filter [9].

In the multitarget tracking problem, clutter and detection profile are notable sources of uncertainty [29]. Clutter is the set of false measurements that do not originate from any true target and detection profile models the ability of the sensor to detect targets. Knowledge of these parameters are essential in Bayesian multitarget estimation. Mismatches in parameters of clutter and detection models lead to poor performance of filtering algorithms. While these parameters are unknown and randomly time-varying, they are normally assumed to be known in advance. This assumption is unrealistic in most practical applications and these parameters need to be estimated from training data or manually tuned [29].

Since the adaptability of the tracker to these unknown parameters are important in practice, several RFS filters have been proposed in the literature to perform multitarget tracking with mismatches in clutter and detection probability. Some of the proposed methods that accommodate the unknown clutter rate are given in [30–33]. A filter which bootstraps the clutter estimator of [29] into the CPHD filter [6] has been proposed in [34]. Several approaches for dealing with unknown detection probability have been presented in the literature, such as [29,35,36]. However, none of these filters can output target tracks. While the GLMB filter can output tracks, and has been applied to several problems without prior knowledge of clutter rate, as in [37–39], it is still computationally expensive. A low computational cost bootstrapping method using GLMB filter has been given in [40] for multisensor multitarget tracking with unknown detection probability.

Multisensor multitarget tracking with jointly unknown clutter rate and detection profile is far more complicated than those with a single unknown parameter. The use of multiple sensors leads to multidimensional ranked assignment problem which is the main hurdle in the implementation of the GLMB filter [9]. Furthermore, exploiting background information from training data for the multitarget estimation at each time frame is insufficient due to the time-varying nature of the two mentioned unknown parameters.

This work is aimed to contribute an efficient method for multitarget tracking that not only produces target trajectories but also estimates the jointly unknown clutter rate and detection profile online with low computational cost. By using a simple combination of the two well-known filters, the CPHD and GLMB filters, this method is not only fast in estimating the unknown parameters but also producing trajectories of the targets. Specifically, these two mentioned unknown parameters would be estimated separately by using the $\lambda-$CPHD and p_D-CPHD filters before feeding to the GLMB filter for the purpose of tracking trajectories. The preliminary results of this work are reported in [40]. Particularly, in [40], the unknown detection probability is treated by the p_D-CPHD filter before boostraped into the GLMB filter with known clutter rate. The soundness and effectiveness of the proposed solution are demonstrated in Section 4 via a multiple marine ships tracking application.

The remainder of this work is presented as follows. The backgrounds on GLMB filtering will be given in Section 2. The proposed bootstrapping method will be introduced in Section 3 followed by numerical studies in Section 4. Some concluding marks in Section 5.

2. Background

Some fundamentals on multitarget state-space model, the CPHD filter, and GLMB filter will be summarized in this section. Following the convention in [11], single target states are denoted with lower-case letters (i.e., x) while upper-case letters denote multitarget states (i.e., X). The corresponding spaces are denoted by blackboard bold letters ($\mathbb{X}, \mathbb{L}, \mathbb{Z}, etc$). The sequence of variable $X_i, X_{i+1}, ..., X_j$ is abbreviated by $X_{i:j}$. In this work the inner product $\int f(x)g(x)dx$ is rewritten as $\langle f, g \rangle$. Given a set S, the finite subsets of S is written as $\mathcal{F}(S)$, and $1_S(\cdot)$ denotes the indicator function of S. For a finite set X, $|X|$ represents its the number of elements, and the product $\prod_{x \in X} f(x)$ for some real-valued function f is denoted by the multitarget exponential f^X, with $f^{\emptyset} = 1$. Further, the generalized Kronecker-delta function δ_Y whose arguments can be arbitrary sets, vectors, integers, etc., is defined as follows

$$\delta_Y[X] = \begin{cases} 1 & if \quad X = Y \\ 0 & otherwise. \end{cases} \tag{1}$$

2.1. Multitarget States

As mentioned in Section 1, algorithms using non-labeled RFS cannot produce trajectories without using heuristic techniques [10]. The Labeled RFS framework, introduced in [11,41], is a principled approach to produce target tracks. Moreover, it is the only method that can produce trajectories from the filtering density [10]. In the labeled RFS frame work, a labeled target at time k is represented by a kinematic target state vector x_k in state space $\mathbb{X}$ and its unique label ℓ_k in the (discrete) label space $\mathbb{L}$, and hence $\boldsymbol{x} = (x, \ell) \in \mathbb{X} \times \mathbb{L}$. This unique label is characterized by two parameters: time of target birth τ and the index of individual targets born at the same time ρ, i.e., $\ell_k = (\tau, \rho) \in \mathbb{L}$ [11]. Hence, formally, a trajectory of each target is a sequence of consecutive labeled states with the same label [11]. Note that the label space for all targets born up to time k is the disjoint union $\mathbb{L}_k = \mathbb{L}_{k-1} \uplus \mathbb{B}_k$ where $\mathbb{B}_k$ is the label space for targets born at time k, and $\mathbb{L}_{k-1}$ is the label space of the targets born prior to time k. To distinguish the unlabeled states from labeled ones, the normal and bold letters (e.g., $x, X, \boldsymbol{x}, \boldsymbol{X}$) are used, respectively. Suppose that at time k, there are N targets with corresponding states $\boldsymbol{x}_{k,1}, \ldots, \boldsymbol{x}_{k,N}$, then the multitarget state can be represented as follows:

$$\boldsymbol{X}_k = \{\boldsymbol{x}_{k,1}, \ldots, \boldsymbol{x}_{k,N}\} \in \mathcal{F}(\mathbb{X} \times \mathbb{L}_k) \tag{2}$$

Definition 1. *[11] Let $\mathcal{L} : \mathbb{X} \times \mathbb{L} \to \mathbb{L}$ be the projection $\mathcal{L}(x; \ell) = \ell$, and hence $\mathcal{L}(\boldsymbol{X}) = \{\mathcal{L}(\boldsymbol{x}) : \boldsymbol{x} \in \boldsymbol{X}\}$ is the set of labels of $\boldsymbol{X}$. A labeled RFS with space $\mathbb{X}$ and (discrete) label space $\mathbb{L}$ is an RFS on $\mathbb{X} \times \mathbb{L}$ such that each realization $\boldsymbol{X}$ has distinct labels, i.e., $|\mathcal{L}(\boldsymbol{X})| = |\boldsymbol{X}|$.*

Since each target in a multitarget state has a distict label, $\delta_{|\boldsymbol{X}|}(|\mathcal{L}(\boldsymbol{X})|) = 1$, the distinct label indicator can be defined as follows [11]

$$\Delta(\boldsymbol{X}) \triangleq \delta_{|\boldsymbol{X}|}(|\mathcal{L}(\boldsymbol{X})|). \tag{3}$$

2.2. Standard Multitarget Dynamic Model

Given a multitarget state $\boldsymbol{X}_k$ at time k, each state $(x_k, \ell_k) \in \boldsymbol{X}_k$ can either exist with probability $P_{S,k+1|k}(\boldsymbol{x}_k)$ and evolve to a new state $\boldsymbol{x}_{k+1}$ at next time step $k+1$ with probability density $f_{k+1|k}(x_{k+1}|x_k, \ell_k)\,\delta_{\ell_k}(\ell_{k+1})$ or disappear with probability $1 - P_{S,k+1|k}(\boldsymbol{x}_k)$. Let $\boldsymbol{S}_{k+1|k}(\boldsymbol{x})$ be the labeled Bernoulli RFS of the surviving target with state $\boldsymbol{x}$ from time k to time $k+1$ and $\boldsymbol{B}_{k+1}$ be the labeled multi-Bernoulli RFS of the new-born targets at time $k+1$, then the multitarget state $\boldsymbol{X}_{k+1}$ is the union of the surviving targets and the new-born ones,

$$\boldsymbol{X}_{k+1} = \bigcup_{\boldsymbol{x}_k \in \boldsymbol{X}_k} \boldsymbol{S}_{k+1|k}(\boldsymbol{x}_k) \cup \boldsymbol{B}_{k+1}, \tag{4}$$

Following the convention in [9], in this work, the set $\boldsymbol{B}_{k+1}$ is distributed according to the labeled multi-Bernoulli (LMB) density. Furthermore, for simplicity, the subscript k for the current time is omitted, and the next time step $k+1$ is indicated by the subscript $'+'$.

Assuming that the appearance, disappearance, and movement of each target are independent of the others, the multitarget transition density (The Mahler's Finite Set Statistics (FISST) notion of density is used in this paper for consistency with the probability density [42]) is [11,41]

$$f(\mathbf{X}_+|\mathbf{X}) = f_S(\mathbf{X}_+ \cap (\mathbb{X} \times \mathbb{L})|\mathbf{X}) f_{B,+}(\mathbf{X}_+ - (\mathbb{X} \times \mathbb{L})) \tag{5}$$

in which the distribution of new-born targets is given by

$$f_{B,+}(\boldsymbol{B}_+) = \Delta(\boldsymbol{B}_+)\left[1_{\mathbb{B}_+} r_{B,+}\right]^{\mathcal{L}(\boldsymbol{B}_+)} [1 - r_{B,+}]^{\mathbb{B}_+ - \mathcal{L}(\boldsymbol{B}_+)} p_{B,+}^{\boldsymbol{B}_+}, \tag{6}$$

where $r_{B,+}(\ell)$ is the birth probability of new target with new-born label ℓ, and $p_{B,+}(\cdot;\ell)$ is the distribution of its kinematic state [11]. The distribution of the survival targets is

$$\begin{aligned} f_{S,+}(\mathbf{S}|\mathbf{X}) =& \Delta(\mathbf{S})\,\Delta(\mathbf{X})\,1_{\mathcal{L}(\mathbf{X})}(\mathcal{L}(\mathbf{S}))\,[\Upsilon(\mathbf{S};\cdot)]^{\mathbf{X}} \\ \Upsilon(\mathbf{S};x,\ell) =& \sum_{(x_+,\ell_+)\in \mathbf{S}} \delta_\ell(\ell_+) P_S(x,\ell) f_+(x_+|x,\ell) + (1 - 1_{\mathcal{L}(\mathbf{S})}(\ell)(1 - P_S(x,\ell)). \end{aligned} \tag{7}$$

2.3. Standard Multitarget Observation Model

Assuming that there are M sensors, each state $(x,\ell) \in \mathbf{X}$ can be either detected by sensor $s, s = 1, \ldots M$ with probability of detection $P_D^{(s)}(x,\ell)$ and generate an observation $z^{(s)} \in Z^{(s)}$ with likelihood $g_D^{(s)}\left(z^{(s)}|x,\ell\right)$, or being miss detected with probability $1 - P_D^{(s)}(x,\ell)$. The set of multitarget observations collected by the sth-sensor at time k is $Z_k^{(s)} = \left\{z_1^{(s)}, \ldots, z_M^{(s)}\right\} \in \mathcal{F}(\mathbb{Z})$, with $\mathbb{Z}$ being the observation space. Note that, the $s^{th}-$sensor can also receive spurious measurements or false alarms at each time step. Let $D^{(s)}(x)$ be the set of measurements generated by target with state x at time k, the multitarget observation at the current time k is the superposition of all observations of detected targets modeled by multi-Bernoulli RFS, i.e., $D^{(s)}(\mathbf{X}) = \bigcup_{x\in \mathbf{X}} D^{(s)}(x)$ and the clutter modeled by either Poisson or i.i.d. clutter RFS $C^{(s)}$.

$$Z^{(s)} = D^{(s)}(\mathbf{X}) \cup C^{(s)} \tag{8}$$

The likelihood function of a multitarget state $\mathbf{X}$ for sensor s is given as follows [9],

$$g^{(s)}(Z^{(s)}|\mathbf{X}) \propto \sum_{\theta^{(s)}\in\Theta^{(s)}} 1_{\Theta^{(s)}(\mathcal{L}(\mathbf{X}))}\left(\theta^{(s)}\right)\left[\Upsilon_{Z^{(s)}}^{(s,\theta^{(s)}(\mathcal{L}(x)))}(x)\right]^{\mathbf{X}} \tag{9}$$

where $\Theta^{(s)}$ is the set of positive association map $\theta^{(s)}$ at time k, $\theta^{(s)}: \mathbb{L} \to \{0, 1, \ldots, |Z^{(s)}|\}$, such that $\left[\theta^{(s)}(i) = \theta^{(s)}(j)\right] \Rightarrow [i = j]$ (i.e., each observation in $Z^{(s)}$ is assigned to at most one target, then each target has a distinct label), $\Theta^{(s)}(J)$ is the subset of $\Theta^{(s)}$with domain J, and

$$\Upsilon_{Z^{(s)}}^{(s,j)}(\boldsymbol{x}) = \begin{cases} \dfrac{P_D^{(s)}(\boldsymbol{x}) g^{(s)}(z_j^{(s)}|\boldsymbol{x})}{\kappa^{(s)}(z_j^{(s)})}, & j = 1 : M^{(s)} \\ 1 - P_D^{(s)}(\boldsymbol{x}) & j = 0. \end{cases} \tag{10}$$

Using the assumption that the sensors are conditionally independent (More concisely, the sensors do not interfere or influence each other while taking measurements or detections. The measurement noise, missed detections and clutter from each sensor in a multitarget scenario are, therefore, independent from the others), and let us define the following abbreviations

$$Z \triangleq (Z^{(1)}, \ldots, Z^{(M)}), \tag{11}$$

$$\Theta \triangleq \Theta^{(1)} \times \cdots \times \Theta^{(M)}, \tag{12}$$

$$\Theta(J) \triangleq \Theta^{(1)}(J) \times \cdots \times \Theta^{(M)}(J), \tag{13}$$

$$\theta \triangleq (\theta^{(1)}, \ldots, \theta^{(M)}), \tag{14}$$

$$1_{\Theta(I)}(\theta) \triangleq \prod_{s=1}^{M} 1_{\Theta^{(s)}(I)}(\theta^{(s)}), \tag{15}$$

$$\Upsilon_Z^{(j^{(1)},\ldots,j^{(M)})}(x,\ell) \triangleq \prod_{s=1}^{M} \Upsilon_{Z^{(s)}}^{(s,j^{(s)})}(x,\ell), \tag{16}$$

then, the multi-sensor likelihood is written as

$$\begin{aligned} g(Z|\mathbf{X}) &= \prod_{s=1}^{M} g^{(s)}(Z^{(s)}|\mathbf{X}) \\ &\propto \sum_{\theta\in\Theta} 1_{\Theta(\mathcal{L}(\mathbf{X}))}(\theta) \left[\Upsilon_Z^{(\theta(\mathcal{L}(x)))}(x)\right]^{\mathbf{X}}. \end{aligned} \tag{17}$$

Obviously, the form of the multi-sensor likelihood $g(Z|\mathbf{X})$ in (17) and that of it its single-sensor counterpart in (9) are identical.

2.4. Multitarget Bayesian Recursion

Let $\pi_{k-1}(\cdot|Z_{1:k-1})$ denotes the multitarget density of the multitarget state at time $k-1$, where $Z_{1:k-1} = (Z_1, \ldots, Z_{k-1})$ is the set of all observation history up to time $k-1$. For simplicity, we obmit the dependence on past measurements, i.e, we use $\pi_{k-1}(\cdot|Z_{k-1})$ instead of $\pi_{k-1}(\cdot|Z_{1:k-1})$. The multitarget Bayes filter use the Chapman-Kolmogorov equation to predict the multitarget state to time k given posterior at time $k-1$ as follows [3]

$$\pi_{k|k-1}(\mathbf{X}_k|Z_{k-1}) = \int f_{k|k-1}(\mathbf{X}_k|\mathbf{X})\,\pi_{k-1}(\mathbf{X}|Z_{k-1})\,d\mathbf{X}, \tag{18}$$

where $f_{k|k-1}(\mathbf{X}_k|\mathbf{X})$ is defined as the multitarget transition kernel from time $k-1$ to time k, and the integral in Equation (18) is the set integral defined for any function $f : \mathcal{F}(\mathbb{X}\times\mathbb{L}) \to \mathbb{R}$,

$$\int f(\mathbf{X})\,\delta\mathbf{X} = \sum_{i=0}^{\infty} \frac{1}{i!} \int f(\{\mathbf{x}_1, \ldots, \mathbf{x}_i\})\,d(\mathbf{x}_1, \ldots, \mathbf{x}_i) \tag{19}$$

The multitarget state $\mathbf{X}_k$ is partially observed at time k, and the RFS Z_k is modeled by the multitarget likelihood function $g_k(Z_k|\mathbf{X}_k)$, thus the multitarget posterior at this time is given by Bayes rule:

$$\pi_k(\mathbf{X}_k|Z_k) = \frac{g_k(Z_k|\mathbf{X}_k)\,\pi_{k|k-1}(\mathbf{X}_k|Z_{k-1})}{\int g_k(Z_k|\mathbf{X})\,\pi_{k|k-1}(\mathbf{X}|Z_{k-1})\,d\mathbf{X}}. \tag{20}$$

3. GLMB Recursion with Bootstrapping Method

In this section, the generalized labeled multi-Bernoulli (GLMB) filter with its recursion is summarized. The proposed method for estimating unknown backgrounds before bootstrapping them into this filter is also introduced.

3.1. GLMB Filter

Definition 2. *A GLMB density is a labeled multitarget density given as follows [11]*

$$\pi\left(\mathbf{X}\right)=\Delta\left(\mathbf{X}\right)\sum_{\varrho\in\Xi,J\subseteq\mathbb{L}}\sum\omega^{\left(J,\varrho\right)}\delta_{J}\left[\mathcal{L}\left(\mathbf{X}\right)\right]\left[p^{\left(\varrho\right)}\right]^{\mathbf{X}},\tag{21}$$

where the discrete space Ξ *is the space of association map histories* $\Theta_{0:k}\triangleq\Theta_0\times\ldots\times\Theta_k$ *, each* $\varrho=(\theta_{1:k})\in\Xi$ *represents a history of the (multisensor) positive 1-1 map, the weight* $\omega^{(J,\varrho)}$ *and multitarget exponential* $\left[p^{(\varrho)}\right]^{\mathbf{X}}$ *satisfy*

$$\sum_{\varrho\in\Xi}\sum_{J\in\mathbb{L}}\omega^{\left(J,\varrho\right)}\delta_{J}\left[\mathcal{L}\left(\mathbf{X}\right)\right]=1,\tag{22}$$

$$\int p^{\left(\varrho\right)}\left(x,\ell\right)dx=1.\tag{23}$$

Noting that, in Equation (21), while $\omega^{(J,\varrho)}\left(\mathcal{L}\left(\mathbf{X}\right)\right)$ is a function of only the labels of the multitarget state $\mathbf{X}$, whereas $\left[p^{(\varrho)}\right]^{\mathbf{X}}$ depends on entire set $\mathbf{X}$.

The cardinality distribution $Pr\left(|\mathbf{X}|=n\right)$, existence probability $r\left(\ell\right)$ and probability density $p\left(x,\ell\right)$ of a track $\ell\in\mathbb{L}$ are given as follows [11]:

$$Pr\left(|\mathbf{X}|=n\right)=\sum_{\varrho\in\Xi}\sum_{J\in\mathbb{L}}\delta_{n}\left[|J|\right]\omega^{\left(J,\varrho\right)}\tag{24}$$

$$r\left(\ell\right)=\sum_{\varrho\in\Xi}\sum_{J\in\mathbb{L}}1_{J}\left(\ell\right)\omega^{\left(J,\varrho\right)}\tag{25}$$

$$p\left(x,\ell\right)=\frac{1}{r\left(\ell\right)}\sum_{\varrho\in\Xi}\sum_{J\in\mathbb{L}}1_{J}\left(\ell\right)\omega^{\left(J,\varrho\right)}p^{\left(\varrho\right)}\left(x,\ell\right)\tag{26}$$

3.1.1. The GLMB Recursion

Since the GLMB filter is an exact closed-form multitarget Bayes filter under the standard multitarget dynamic and observation models [12], and the form of the likelihood function in a single sensor and multisensor cases are identical, the GLMB filter can be implemented via two separate steps (update and prediction) or the combined step (joint-predict-update process). In this work, for the convenience of proposed method, the two step GLMB recursion will be presented.

a. GLMB update

Given the standard multitarget observation likelihood function (9), the posterior multitarget density is calculated as follows [11]

$$\pi^{(s)}(\mathbf{X}|Z^{(s)})=\Delta(\mathbf{X})\sum_{(J,\varrho)\in\mathcal{F}(\mathbb{L})\times\Xi}\sum_{\theta^{(s)}\in\Theta}\omega^{(s,J,\varrho,\theta^{(s)})}_{(Z^{(s)})}(\mathcal{L}(\mathbf{X}))\left[p^{(s,\varrho,\theta^{(s)})}(\cdot|Z^{(s)})\right]^{\mathbf{X}},\tag{27}$$

where

$$\omega^{(s,J,\varrho,\theta^{(s)})}_{Z^{(s)}}(L)=\frac{\Gamma^{(s,J,\varrho,\theta^{(s)})}_{Z^{(s)}}}{\sum_{(J,\varrho)\in\mathcal{F}(\mathbb{L})\times\Xi}\sum_{\theta^{(s)}\in\Theta}\Gamma^{(s,J,\varrho,\theta^{(s)})}_{Z^{(s)}}}\tag{28}$$

$$\Gamma_{Z^{(s)}}^{(s,J,\varrho,\theta^{(s)})} = \omega_{Z^{(s)}}^{(s,J,\varrho)}(L)\left[\bar{p}_{Z^{(s)}}^{(s,\varrho,\theta^{(s)})}\right]^{J} \tag{29}$$

$$p^{(s,\varrho,\theta^{(s)})}(x,\ell|Z^{(s)}) = \frac{p^{(s,\varrho)}(x,\ell)\Upsilon_{Z^{(s)}}^{(s)}(x,\ell;\theta^{(s)})}{\bar{p}_{Z^{(s)}}^{(s,\varrho,\theta^{(s)})}(\ell)} \tag{30}$$

$$\bar{p}_{Z^{(s)}}^{(s,\varrho,\theta^{(s)})}(\ell) = \langle p^{(s,\varrho)}(\cdot,\ell), \Upsilon_{Z^{(s)}}^{(s)}(\cdot,\ell;\theta^{(s)})\rangle \tag{31}$$

and $\Upsilon_{Z^{(s)}}^{(s)}(x,\ell;\theta^{(s)})$ is given in (10).

b. Prediction

Given the posterior multitarget density at current time is a GLMB filtering density with the form of (21), the predicted multitarget density at next time step is calculated under the standard multitarget dynamic model (4) as follows [11]:

$$\pi_{+}^{(s)}(\mathbf{X}_{+}) = \Delta(\mathbf{X}_{+}) \sum_{(J_{+},\varrho)\in\mathcal{F}(\mathbb{L}_{+})\times\Xi} \omega_{+}^{(s,J_{+},\varrho)}(\mathcal{L}(\mathbf{X}_{+}))\left[p_{+}^{(s,\varrho)}\right]^{\mathbf{X}_{+}} \tag{32}$$

where

$$\omega_{+}^{(s,J_{+},\varrho)}(L) = \omega_{B}^{(s)}(J_{+}\cap\mathbb{B})\omega_{S}^{(s,\varrho)}(J_{+}\cap\mathbb{L}), \tag{33}$$

$$p_{+}^{(s,\varrho)}(x,\ell) = 1_{\mathbb{L}}(\ell)p_{S}^{(s,\varrho)}(x,\ell) + (1-1_{\mathbb{L}}(\ell))\,p_{B}^{(s)}(x,\ell) \tag{34}$$

$$p_{S}^{(s,\varrho)}(x,\ell) = \frac{\langle P_{S}^{(s)}(\cdot,\ell)f(x|\cdot,\ell), p^{(s,\varrho)}(\cdot,\ell)\rangle}{\bar{p}_{S}^{(s,\varrho)}(\ell)}, \tag{35}$$

$$\bar{p}_{S}^{(s,\varrho)}(\ell) = \int\langle P_{S}^{(s)}(\cdot,\ell)f(x|\cdot,\ell), p^{(s,\varrho)}(\cdot,\ell)\rangle dx \tag{36}$$

$$\omega_{S}^{(s,\varrho)}(L) = [\bar{p}_{S}^{(s,\varrho)}]^{L}\sum_{J\subseteq\mathbb{L}} 1_{J}(L)\left[Q_{S}^{(s,\varrho)}(\ell)\right]^{J-L}\omega^{(s,J,\varrho)} \tag{37}$$

$$Q_{S}^{(s,\varrho)}(\ell) = \langle 1-P_{S}^{(s)}(\cdot,\ell), p^{(s,\varrho)}(\cdot,\ell)\rangle. \tag{38}$$

3.2. Adaptive to Unknown Backgrounds

In practice, the both the clutter rate and detection profile are unknown and unpredictably vary with time. Prior knowledge of background models, therefore, are typically unavailable. Mismatch in background models results in degradation of tracker performance [1]. In this section, based on the suite of methods for tackling the unknown clutter rate and detection probability introduced in [4], the bootstrapping method will be proposed.

A technique that accommodates the jointly unknown clutter rate λ and the unknown probability of detection p_D has been introduced in [29]. This technique considers clutter as an RFS of "generator targets" or "false targets", and incorporates the non-homogeneous and unknown detection probability into each target state. Each real target state $x \in \mathbb{X}$ is corresponded to an augmented state $x_a = (x,a)$, in which $a \in \mathbb{X}^d = [0,1]$ is the variable on the probability detecting x. The augmented multitarget state now can be described as follows

$$X_a = (x_{a,1},\ldots,x_{a,n}) = \{(x_1,a_1),\ldots,(x_n,a_n)\} \tag{39}$$

Similarly, the augmented generator target state is $x_c = [\bar{x}, a_c]$ with $\bar{x} \in \mathbb{X}^c$ be the generator target state, and $a_c \in \mathbb{X}^d = [0,1]$. The augmented generator multitarget state is

$$X_c = (x_{c,1},\ldots,x_{c,m}) = \{(\bar{x}_1,a_{c1}),\ldots,(\bar{x}_n,a_{cm})\} \tag{40}$$

Then the probability of detection is replaced by a and a_c, respectively.

$$p_{D,a}^{(s)}(x_a) = p_{D,a}^{(s)}(x, a) \triangleq a \tag{41}$$

$$p_{D,c}^{(s)}(x_c) = p_{D,c}^{(s)}(\bar{x}, a_c) \triangleq a_c \tag{42}$$

Assuming that the false and true targets are statistically independent, then each of the augmented generator targets can be modeled for their characteristics as appearances, disappearances, and transitions, together with likelihood, detection and missed detection. The multitarget state is then a combination of (augmented) actual targets and clutter generators. Meaning that, the augmented hybrid space $\mathbb{X}^h$ involving the multitarget state can be defined as follows [29]

$$\mathbb{X}^{(h)} = \left(\mathbb{X} \times \mathbb{X}^{(d)}\right) \uplus \left(\mathbb{X}^{(c)} \times \mathbb{X}^{(d)}\right) \triangleq \mathbb{X}_a^{(d)} \uplus \mathbb{X}_c^{(d)} \tag{43}$$

where " $\uplus$ " denotes the disjoint union, and " $\times$ " denotes the Cartesian product.

The multitarget state (4) and multitarget observation (8)) at time k now become the hybrid ones:

$$X_h = X_a \uplus X_c \tag{44}$$

$$Z_h = Z_a \uplus Z_c \tag{45}$$

with Z_a and Z_c be the augmented multitarget and augmented generator observations, respectively. The integral of a function $f^{(h)} : \mathbb{X}^{(h)} \to \mathbb{R}$ is given by [29]

$$\int_{\mathbb{X}^{(h)}} f^{(h)}(X_h)\, dx_h = \int_{\mathbb{X}_a^{(d)}} f_a^{(d)}(X_a)\, \delta X_a + \int_{\mathbb{X}_a^c} f_a^{(c)}(X_c)\, \delta X_c \tag{46}$$

Here, the set integral (19) has been applied to both augmented multitarget state and augmented generator multitarget state terms, i.e, [4]

$$\int f_a^{(d)}(X_a)\, \delta X_a = \sum_{n \geq 0} \frac{1}{n!} f_a(\{x_{a,1}, \ldots, x_{a,n}\})\, dx_{a,1}, \ldots, dx_{a,n} \tag{47}$$

$$\int f_c^{(d)}(X_c)\, \delta X_c = \sum_{m \geq 0} \frac{1}{m!} f_a^c(\{x_{c,1}, \ldots, x_{c,n}\})\, dx_{c,1}, \ldots, dx_{c,m} \tag{48}$$

Noting that the the measurement likelihood is kept unchanged

$$g(x_a) = g_a^{(s)}(x, a) \triangleq g^{(s)}(x) \tag{49}$$

$$g(x_c) = g_c^{(s)}(\bar{x}, a_c) \triangleq g^{(s)}(\bar{x}) \tag{50}$$

While the method proposed in [29] results in good estimates of targets, it do not produce the trajectories of the targets. Moreover, although this method is a closed-form solution of the CPHD recursion with jointly unknown clutter rate and detection profile, it is proposed for single-sensor multiple targets estimation solely. In this paper, we propose a method of using the technique introduced in [29] to estimate the mentioned unknown parameters then bootstrapping them into the GLMB filter for tracking on-the-fly. The structure of the proposed method is given in Figure 1.

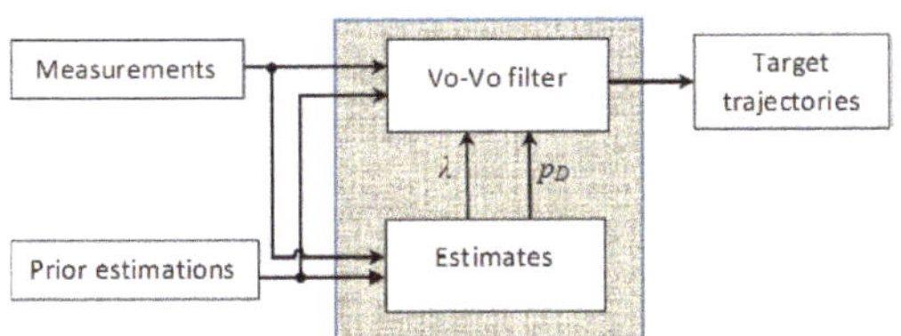

Figure 1. The proposed structure of the *B*-GLMB filter.

3.3. Implementation

Since after each filtering iteration, the number of components in the GLMB density grows at an exponential rate, the low weight terms should be truncated for tractability. In this work, we use Murty's ranked assignment algorithm to sample a given number of hypotheses of the multitarget density with the highest probability to be the correct ones. Then these components are propagated through the filtering recursion only. Although the use of Murty's algorithm leads to a cubic complexity in the product of the number of Doppler measurements, its implementation is reasonable because there are maximum 10 targets in this work.

4. Numerical Study

The advantages of multi-static Doppler radar such as lightweight, wide range of surveillance with high accuracy, and low power consumption lead to its broad applications in both civilian and military applications [43–45]. However,the number of the sensors in conjunction with the non-linear nature and and low observability of the Doppler type measurement leads to many numerical difficulties [44,45]. The use of multistatic Doppler-only measurements in a scenario of 10 receivers and one cooperative transmitter has been proposed in [46] and its extended version [47] for joint detection and tracking of one target.

This numerical study based on the model mentioned in [40] with 10 marine ships. Each ship at time k is represented by a $5-D$ state vector x_k in the surveillance of interest $x_k = \left[p_k^T, v_k^T, \alpha_k\right]^T$, where $p_k = [\mu_k, \lambda_k]^T$ and $v_k = \left[\dot{\mu}_k, \dot{\lambda}_k\right]^T$ denote the position and velocity in the longitude and latitude, respectively; α_k is the course of the target, and T denotes the transpose operation. The target dynamic model can be given as follows:

$$x_k = F_{k|k-1}(x_{k-1}) + Gn_k \tag{51}$$

where

$$F(x_{k-1}) = \begin{bmatrix} 1 & \frac{\sin(\alpha t)}{\alpha} & 0 & \frac{(\cos(\alpha t)-1)}{\alpha} & 0 \\ 0 & \cos(\alpha t) & 0 & -\sin(\alpha t) & 0 \\ 0 & -\frac{(\cos(\alpha t)-1)}{\alpha} & 1 & -\frac{\sin(\alpha t)}{\alpha} & 0 \\ 0 & \sin(\alpha t) & 0 & \cos(\alpha t) & 0 \\ 0 & 0 & 0 & 0 & 1 \end{bmatrix} x_{k-1}; \quad G = \begin{bmatrix} \frac{t^2}{2} & 0 & 0 \\ t & 0 & 0 \\ 0 & \frac{t^2}{2} & 0 \\ 0 & t & 0 \\ 0 & 0 & t \end{bmatrix}, \tag{52}$$

and t is sample period, n_k is a Gaussian noise vector of velocity and course noise components with zero-mean. Note that latitudinal and longitudinal measurements are in degrees $(^\circ)$, the distance, speed and time are given in nautical miles (M), knots (kn), and hours (h), respectively.

Remark 1. *Equation (52) is resulted from the assumption that the surveillance region is not very far from the Equator.*

The new births are assumed to be distributed with labeled multi-Bernoulli RFS distributions of parameters $f_B(x) = \left\{ r_B^{(i)}, p_B^{(i)} \right\}_{i=1}^{4}$ where $r_B^{(i)}$ is the ith common existence probability , and $p_B^{(i)}(x) = \mathcal{N}\left(x, \hat{x}_B^{(i)}, P_B\right)$ with

$$\begin{aligned}
\hat{x}_B^{(1)} &= [15.6^\circ N, 0, 113^\circ E, 0, 0]^T; \\
\hat{x}_B^{(2)} &= [13.2^\circ N, 0, 107.5^\circ E, 0, 0]^T \\
\hat{x}_B^{(3)} &= [18.2^\circ N, 0, 110.7^\circ E, 0, 0]^T; \\
\hat{x}_B^{(4)} &= [22.3^\circ N, 0, 118.8^\circ E, 0, 0]^T; \\
P_B &= diag\left(\left[2'N, 30\,(kn), 2'E, 30\,(kn), 6\pi/180\left(rads^{-1}\right)\right]\right)
\end{aligned}$$

Table 1 lists out the initial state of ten targets with random time of appearance and disappearance, and the average course is $\bar{\alpha} = 2\pi/180(rad/s)$.

The parameters of the dynamic model are given in Table 2.

Table 1. Target initial states.

Target	μ_k	λ_k	$\dot{\mu}_k$	$\dot{\lambda}_k$	α_k (rad/s)	Time of Birth (h)	Time of Beath (h)
1	18°12′15″	110°42′06″	32	−5	$-5\bar{\alpha}/8$	1	100
2	15°37′52″	113°57′14″	13	−9	$-\bar{\alpha}/2$	5	80
3	18°11′40″	110°41′43″	−18	0	$2\bar{\alpha}$	10	90
4	13°13′52″	107°29′31″	2	32	$-\bar{\alpha}/4$	20	100
5	22°17′11″	118°49′24″	6	−20	$-5\bar{\alpha}/6$	20	100
6	22°17′58″	118°48′05″	−22	6	$3\bar{\alpha}/4$	30	70
7	18°12′15″	110°42′06″	15	−30	$\bar{\alpha}/8$	30	70
8	15°35′57″	113°01′06″	−30	32	$3\bar{\alpha}/5$	45	85
9	13°11′44″	107°30′19″	28	−30	$5\bar{\alpha}/3$	55	100
10	15°36′04″	112°53′30″	30	5	$7\bar{\alpha}/4$	55	100

Table 2. Parameters of the Dynamic model.

Parameter	Symbol	Value
Sample period	t	0.15 (h)
Std. of speed noise	σ_V	2 (kn)
Std. of course noise	σ_α	$\pi/180$ (rads^{-1})
Common existence prob.	$r_B^{(1,2)}, r_B^{(3,4)}$	(0.04; 0.02)
Survival prob.	P_S	0.95
Number of targets	N	10

Consider the configuration of multiple Doppler sensors system including two spatially distributed receivers and one cooperative transmitter located as in Figure 2. Based on Doppler effect, this system can measure the speed of a target at a distance by calculating the altered frequency of the returned signals which originate from the emitting pulses of radio signals and being reflected to radar after reaching target [48].

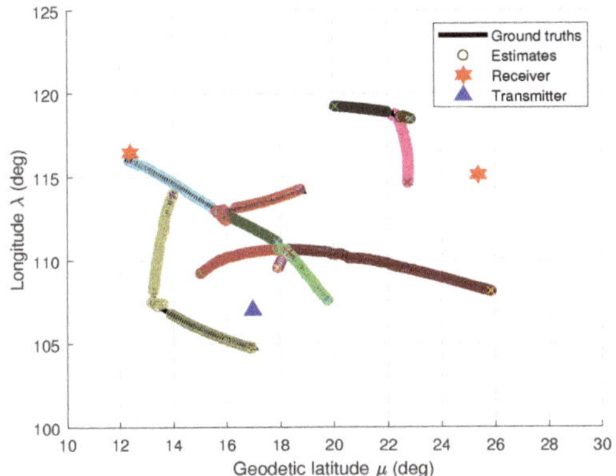

Figure 2. Configuration of multitarget tracking using MRS.

The observation of a target state x_k at the sth receiver using Doppler measurement is given by

$$z_k^{(s)} = -\nu_k^T \left(\frac{p_k - p_r^{(s)}}{\left\| p_k - p_r^{(s)} \right\|} + \frac{p_k - p_t}{\| p_k - p_t \|} \right) \frac{f_t}{c} + w_k, \tag{53}$$

in which p_k and ν_k have been defined above the Equation (51), $p_t = [\mu_t, \lambda_t]^T$ is the position of the transmitter, and $p_r^{(s)} = [\mu_r^{(s)}, \lambda_r^{(s)}]^T$ is the location of the $s^{th}-$receiver ; w_k is zero-mean Gaussian noise, $w_k \sim \mathcal{N}(0, \mathbf{Q}_k)$, with covariance $\mathbf{Q}_k = diag\left([1Hz^2]\right)$; and f_t is the signal frequency emitted from the transmitter, and c is the light speed.

Since the targets are dynamic in different directions, the value of observation $z_k^{(s)}$ in (53) can be negative or positive in the known interval $[-f_0, +f_0]$ of the Doppler sensor. In this work, the measurement space for two receivers have the same measurement space of $[-200Hz, 200Hz]$. The parameters of the observation model are given in Table 3. It can be seen that, not only the state equation but also the measurement one are highly nonlinear.

Table 3. Parameters of observation model.

Name	Symbol	Value
Transmitter	p_t	[16°58′16″N,107.02′48″E]
Receiver 1	$p_r^{(1)}$	[12°22′43″N, 116°28′25′E]
Receiver 2	$p_r^{(2)}$	[25°22′47″N, 115°07′19″E]
Transmit freq.	f_t	300 (Mhz)
Light speed	c	3×10^8 (m/s)
Detection prob.	p_D	[0.75; 0.98]
Clutter rate range	λ_c	[28; 60]
Surveil. area	S_r	[10° − 30°N, 100° − 125°E]

By using the proposed $B-$GLMB filter, the configuration of multiple marine ships tracking using multiple Doppler radars with ground truths and their tracking results are illustrated in Figure 2. For better visualization of multiple targets, each target is assigned to a distinct color. The results of longitudinal-latitudinal co-ordinate target trajectories are demonstrated in Figure 3.

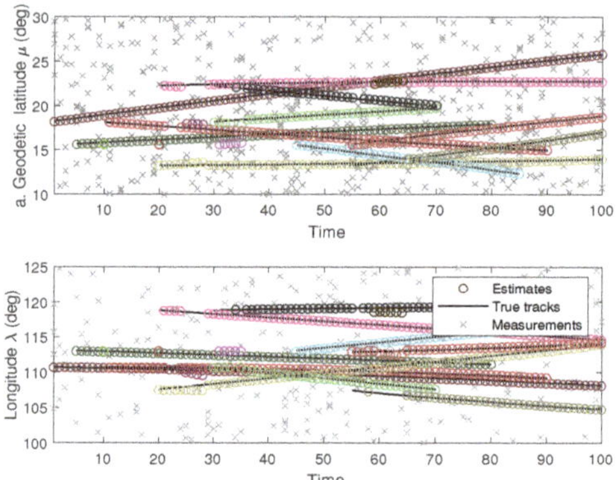

Figure 3. Tracking in longitudinal-latitudinal coordinates.

For evaluating the effectiveness of the proposed method comparing to the fixed-GLMB filter and the Joint-CPHD, 100 Monte - Carlo run has been used, and the distance, location and cardinality errors are calculated via Optimal Sub-Pattern Assignment, OSPA, [49] and shown in Figure 4a. By using this metric, the distance between the set of true multitarget states and that of estimated target states is calculated at each time step. For measuring the error between two set of tracks, the use of OSPA is insufficient, and OSPA$^{(2)}$ is needed. The OSPA$^{(2)}$errors [50] of the B-GLMB and fixed-GLMB filters are compared and plotted against time in Figure 4b, respectively.

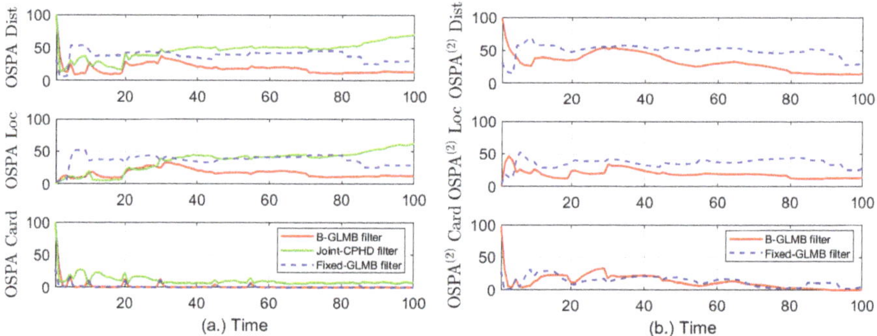

Figure 4. Evaluation of tracking errors using (**a**) OSPA, and (**b**) OSPA$^{(2)}$.

For the $B-$GLMB filter and joint-CPHD filter, the clutter rate fluctuates in the range of $\lambda_c = [28, 70]$, and the detection probability changes from 0.75 to 0.98, i.e., $p_D = [0.75, 0.98]$. The fixed-GLMB filter is used with fixed p_D of 0.75 and 0.98 and fixed λ of 28 and 70, respectively. The window length used in OSPA$^{(2)}$ to obtain the differences between the true and estimate sets of trajectories in Figure 4b is set at $w_l = 10$. Both the OSPA and OSPA$^{(2)}$ are used with cut-off parameter $c_0 = 0$ and $p = 1$.

Obviously from Figure 4a, the errors in distance and location between the set of true targets and the estimated ones using $B-$GLMB filter is the smallest values comparing to those of the fixed-GLMB and joint-CPHD filters. In addition, the errors in cardinality statistics using fixed-GLMB and $B-$GLMB are almost identical and better than error measured by joint-CPHD filter. The results of measuring errors between the set of true target tracks and that of the estimated tracks are given in Figure 4b. Once again, the effectiveness of the proposed method in reducing the errors in distances and locations of the target tracks is validated. The cardinality statistics for the $B-$GLMB filter, fixed-GLMB filter and joint-CPHD over 100 Monte Carlo run are given in Figure 5.

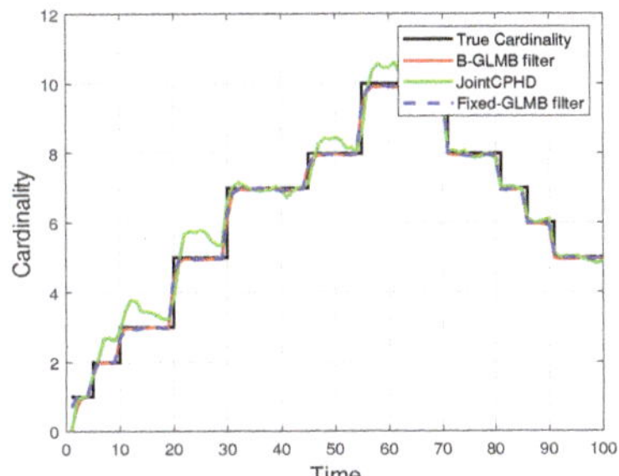

Figure 5. Cardinality tracking results.

5. Conclusions

This paper presented an efficient solution to the problem of tracking an unknown and time-varying number of marine ships from multiple sensors with unknown clutter rate and probability of detection. Particularly, these two unknown parameters are parallel estimated based on the $\lambda-$CPHD and the p_D-CPHD filters, then bootstrapped into the cutting-edge GLMB filter. By using the bootstrapping method, the proposed filter utilizes the advantages of the two former estimators in accommodating the unknown backgrounds and reduces the computational cost from tracking algorithm of the latter filter. The effectiveness and correctness of the proposed method are demonstrated in Section 4. From our best knowledge, this is the first principled online algorithm for tracking marine ships via multiple sensors with unknown backgrounds in Doppler measurements. For future work, one of the forcuses would be investigating the combination of multi-scan GLMB [28] and multisensor GLMB [9] filters for multisensor multitarget tracking.

Author Contributions: In this work, T.T.D.N. and W.L. contributed in software and discussion. C.-T.D. conceived and planned the conceptualization, data curation, formal analysis, investigation, and methodology. C.-T.D. also designed and directed the project, resources, supervision, validition and visualization. C.-T.D. wrote, revised and edited the original draft as well as the manuscript. All authors discussed the comments from reviewers before sending their responses.

Funding: This research was sponsored by the Joint CIPRS-MOET Scholarship.

Acknowledgments: The Authors acknowledge the administrative and technical support from their Supervisors and Universities.

Conflicts of Interest: The authors declare no conflict of interest in this study.

References

1. Bar-Shalom, Y.; Fortmann, T.E. *Tracking and Data Association*; Academic Press: San Diego, CA, USA, 1988; ISBN 978-012-079-760-8.
2. Blackman, S.; Popoli, R. *Design and Analysis of Modern Tracking Systems*; Artech House: Norwood, MA, USA, 1999; ISBN 978-158-053-006-4.
3. Mahler, R.S. *Statistical Multisource-Multitarget Information Fusion*; Artech House: Boston, MA, USA, 2007; ISBN 978-159-693-092-6.
4. Mahler, R.S. *Advances in Statistical Multisource-Multitarget Information Fusion*; Artech House: Boston, MA, USA, 2014; ISBN 978-16-0807-798-4.
5. Mahler, R.P. Multitarget Bayes filtering via first-order multitarget moments. *IEEE Trans. Aerosp. Electron. Syst.* **2003**, *39*, 1152–1178. [CrossRef]
6. Vo, B.-T.; Vo, B.-N.; Cantoni, A. Analytic implementations of the cardinalized probability hypothesis density filter. *IEEE Trans. Signal Process.* **2007**, *55*, 3553–3567. [CrossRef]
7. Mahler, R.P. PHD filters of higher order in target number. *IEEE Trans. Aerosp. Electron. Syst.* **2007**, *43*, 1523–1543. [CrossRef]

8. Vo, B.-N.; Vo, B.-T.; Pham, N.-T.; Suter, D. Joint Detection and Estimation of Multiple Objects from Image Observations. *IEEE Trans. Signal Process.* **2010**, *58*, 5129–5241. [CrossRef]
9. Vo, B.-N.; Vo, B.-T.; Beard, M. Multi-Sensor Multi-Object tracking with the Generalized Labeled Multi-Bernoulli Filter. *IEEE Trans. Signal Process.* **2019**, Accepted for publication. [CrossRef]
10. Mahler, R.S. "statistic 103" for multitarget tracking. *Sensors* **2019**, *19*, 202. [CrossRef] [PubMed]
11. Vo, B.-T.; Vo, B.-N. Labeled random finite sets and multi-object conjugate priors. *IEEE Trans. Signal Process.* **2013**, *61*, 3460–3475. [CrossRef]
12. Mahler, R.S. Exact Closed-Form Multitarget Bayes Filters. *Sensors* **2019**, *19*, 2818. [CrossRef]
13. Beard, M.; Vo, B.-T.; Vo, B.-N. Bayesian multi-target tracking with merged measurements using labelled random finite sets. *IEEE Trans. Signal Process.* **2015**, *63*, 1433–1447. [CrossRef]
14. Papi, F.; Kim, D.Y. A particle multi-target tracker for superpositional measurements using labeled random finite sets. *IEEE Trans. Signal Process.* **2015**, *63*, 4348–4358. [CrossRef]
15. Papi, F.; Vo, B.N.; Vo, B.T.; Fantacci, C.; Beard, M. Generalized Labeled Multi-Bernoulli Approximation of Multi-Object Densities. *IEEE Trans. Signal Process.* **2014**, *63*, 5487–5497. [CrossRef]
16. Beard, M.; Reuter, S.; Granström, K.; Vo, B.T.; Vo, B.N.; Scheel, A. Multiple extended target tracking with labeled random finite sets. *IEEE Trans. Signal Process.* **2016**, *64*,1638–1653. [CrossRef]
17. Nguyen, T.T.D.; Kim, D.Y. GLMB tracker with partial smoothing. *Sensors* **2019**, *19*, 4419. [CrossRef] [PubMed]
18. Hadden, W.J.; Young, J.L.; Holle, A.W.; McFetridge, M.L.; Kim, D.Y.; Wijesinghe, P.; Taylor-Weiner, H.; Wen, J.H.; Lee, A.R.; Bieback, K.; et al. Stem cell migration and mechanotransduction on linear stiffness gradient hydrogels. *Proc. Natl. Acad. Sci. USA* **2017**, *114*, 5647–5652. [CrossRef] [PubMed]
19. Beard, M.; Vo, B.-T.; Vo, B.-N.; Arulampalam, S. Void Probabilities and Cauchy-Schwarz Divergence for Generalized Labeled Multi-Bernoulli Models. *IEEE Trans. Signal Process.* **2017**, *65*, 5047–5061. [CrossRef]
20. Bryant, D.S.; Vo, B.-T.; Vo, B.-N.; Jones, B.A. A Generalized Labeled Multi-Bernoulli Filter with Object Spawning. *IEEE Trans. Signal Process.* **2018**, *66*, 6177–6189. [CrossRef]
21. Fantacci, C.; Vo, B.-N.; Vo, B.-T.; Battistelli, G.; Chisci, L. Robust fusion for multisensor multiobject tracking. *IEEE Signal Process. Lett.* **2018**, *25*, 640–644. [CrossRef]
22. Deusch, H.; Reuter, S.; Dietmayer, K. The labeled multi-Bernoulli SLAM filter. *IEEE Signal Process. Lett.* **2015**, *22*, 1561–1565. [CrossRef]
23. Moratuwage, D.; Adams, M.; Inostroza, F. δ-Generalised Labelled Multi-Bernoulli Simultaneous Localization and Mapping with an Optimal Kernel-Based Particle Filtering Approach. *Sensors* **2019**, *19*, 2290. [CrossRef]
24. Kim, D.Y.; Vo, B.-N.; Vo, B.-T.; Jeon, M. A labeled random finite set online multi-object tracker for video data. *Pattern Recognit.* **2019**, *90*, 377-389. [CrossRef]
25. Wei, B.; Nener, B.; Liu, W.; Ma, L. Centralized Multi-Sensor Multi-Target Tracking with Labeled Random Finite Sets. In Proceedings of the International Conference Control, Automation and Information Sciences (ICCAIS 2016), Ansan, Korea, 27–29 October 2016; pp. 82–87. [CrossRef]
26. Fantacci, C.; Papi, F. Scalable multisensor multitarget tracking using the marginalized $\delta-$GLMB density. *IEEE Signal Process. Lett.* **2016**, *23*, 863–867. [CrossRef]
27. Vo, B.-N.; Vo, B.-T.; Hoang. H.G. An efficient implementation of the Generalized Labeled Multi-Bernoulli filter. *IEEE Trans. Signal Process.* **2017**, *65*, 1975–1987. [CrossRef]
28. Vo, B.-N.; Vo, B.-T. A multi-scan labeled Random Finite Set model for multi-object state estimation. *IEEE Trans. Signal Process.* **2019**, *67*, 4948–4963. [CrossRef]
29. Mahler, R.S.; Vo, B.-T.; Vo, B.-N. CPHD filtering with unknown clutter rate and detection profile. *IEEE Trans. Signal Process.* **2011**, *59*, 3497–3513. [CrossRef]
30. Mahler, R.S.; El-Fallah, A. CPHD filtering with unknown probability of detection. In *Signal Processing, Sensor Fusion, and Target Recognition XIX*; International Society for Optics and Photonics: Orlando, FL, USA, 2010; p. 7697(76970F). [CrossRef]
31. Vo, B.-T.; Vo, B.-N.; Hoseinnezhad, R.; Mahler, R.P.S. Robust multi-Bernoulli filtering. *IEEE J. Sel. Top. Sign. Proces.* **2013**, *7*, 399–409. [CrossRef]
32. Correa, J.; Martin, A. Estimating detection statistics within a bayes-closed multi-object filter. In Proceedings of the 19th International Conference on Information Fusion (FUSION 2016), Heidelberg, Germany, 5–8 July 2016; pp. 811–819.
33. Si, W.; Zhu, H.; Qu, Z. Robust Poisson Multi-Bernoulli Filter With Unknown Clutter Rate. *IEEE Access* **2019**, *7*, 117871–117882. [CrossRef]

34. Beard, M.; Vo, B.-T.; Vo, B.-N. Multi-target filtering with unknown clutter density using bootstrap GMCPHD filter. *IEEE Signal Process. Lett.* **2013**, *20*, 323–326. [CrossRef]
35. Li, C.; Wang, W.; Kirubarajan, T.; Sun, J.; Lei, P. PHD and CPHD Filtering With Unknown Detection Probability. *IEEE Trans. Signal Process.* **2018**, *66*, 3784–3798. [CrossRef]
36. Wang, S.; Bao, Q.; Chen, Z. Refined PHD Filter for Multi-Target Tracking under Low Detection Probability. *Sensors* **2019**, *19*, 2842. [CrossRef]
37. Li, S.; Yi, W.; Hoseinnezhad, R.; Wang, B.; Kong, L. Multiobject tracking for generic observation model using labeled random finite sets. *IEEE Trans. Signal Process.* **2018**, *66*, 368–383. [CrossRef]
38. Kim, D.Y. Visual multiple-object tracking for unknown clutter rate. *IET Comput. Vision* **2018**, *12*, 728–734. [CrossRef]
39. Punchihewa, Y.G.; Vo, B.T.; Vo, B.N.; Kim, D.Y. with Labeled Random Finite Sets. *IEEE Trans. Signal Process.* **2018**, *66*, 3040–3055. [CrossRef]
40. Do, C.-T.; Nguyen, H.V. Tracking multiple targets from multistatic Doppler radar with unknown probability of detection . *Sensors* **2019**, *19*, 1672. [CrossRef] [PubMed]
41. Vo, B.-N.; Vo, B.-T.; Phung, D. Labeled random finite sets and the Bayes multitarget tracking filter. *IEEE Trans. Signal Process.* **2014**, *62*, 6554–6567. [CrossRef]
42. Vo, B.-N.; Singh, S.; Doucet, A. Sequential Monte Carlo methods for multi-target filtering with random finite sets. *IEEE Trans. Aerosp. Electron. Syst.* **2005**, *41*, 1224–1245. [CrossRef]
43. Chernyak, V.S. *Fundamentals of Multisite Radar Systems: Multistatic Radars and Multistatic Radar Systems*; Routledge: Boca Raton, FL, USA, 2018; ISBN 978-905-699-165-4.
44. Goodman, N.A.; Bruyere, D. Optimum and decentralized detection for multistatic airborne radar. *IEEE Trans. Aerosp. Electron. Syst.* **2007**, *43*. [CrossRef]
45. Smith, G.E.; Woodbridge, K.; Baker, C.J.; Griffiths, H. Multistatic micro-Doppler radar signatures of personnel targets. *IET Signal Process.* **2010**, *4*, 224–233 . [CrossRef]
46. Ristic, B.; Farina, A. Joint detection and tracking using multi-static Doppler-shift measurement. In Proceedings of the IEEE International Conference on Acoustics, Speech and Signal Processing (ICASSP 2012), Kyoto, Japan, 25–30 March 2012; pp. 3881–3884. [CrossRef]
47. Ristic, B.; Farina, A. Target tracking via multi-static Doppler shifts. *IET Radar Sonar Navig.* **2013**, *7*, 508–516. [CrossRef]
48. Guo, F.; Fan, Y.; Zhou, Y.; Xhou, C.; Li, Q. *Space Electronic Reconnaissance: Localization Theories and Methods*, 1st ed.; John Wiley & Sons: Solaris South Tower, Singapore, 2014; ISBN 978-111-854-219-4. .
49. Schuhmacher, D.; Vo, B.-T.; Vo, B.-N. A consistent metric for performance evaluation of multi-object filters. *IEEE Trans. Signal Process.* **2008**, *56*, 3447–3457. [CrossRef]
50. Beard, M.; Vo, B.-T.; Vo, B.-N. A solution for Large-scale Multi-Object Tracking. *arXiv* **2018**, arXiv:1804.06622.

Article

Optimal Target Assignment with Seamless Handovers for Networked Radars

Juhyung Kim [1], Doo-Hyun Cho [2], Woo-Cheol Lee [1], Soon-Seo Park [1] and Han-Lim Choi [3,*]

[1] Department of Aerospace Engineering, Korea Advanced Institute of Science and Technology, Daejeon 34141, Korea; jhkim@lics.kaist.ac.kr (J.K.); wclee@lics.kaist.ac.kr (W.-C.L.); sspark@lics.kaist.ac.kr (S.-S.P.)
[2] Mechatronics R&D Center, Samsung Electronics, Hwaseong 18448, Korea; dhcho@lics.kaist.ac.kr
[3] Department of Aerospace Engineering & KI for Robotics, Korea Advanced Institute of Science and Technology, Daejeon 34141, Korea
* Correspondence: hanlimc@kaist.ac.kr

Received: 8 August 2019; Accepted: 15 October 2019; Published: 19 October 2019

Abstract: This paper proposes a binary linear programming formulation for multiple target assignment of a radar network and demonstrates its applicability to obtain optimal solutions using an off-the-shelf mixed-integer linear programming solver. The goal of radar resource scheduling in this paper is to assign the maximum number of targets by handing over targets between networked radar systems to overcome physical limitations such as the detection range and simultaneous tracking capability of each radar. To achieve this, time windows are generated considering the relation between each radar and target considering incoming target information. Numerical experiments using a local-scale simulation were performed to verify the functionality of the formulation and a sensitivity analysis was conducted to identify the trend of the results with respect to several parameters. Additional experiments performed for a large-scale (battlefield) scenario confirmed that the proposed formulation is valid and applicable for hundreds of targets and corresponding radar network systems composed of five distributed radars. The performance of the scheduling solutions using the proposed formulation was better than that of the general greedy algorithm as a heuristic approach in terms of objective value as well as the number of handovers.

Keywords: target handover; seamless multi-target tracking; radar network systems; optimal scheduling; situational awareness

---

1. Introduction

The rapid development of computer and communications technologies since the 1980s led to a new doctrine in the military field of the United States under the name of Network-Centric Warfare (NCW) between the late 1990s and early 2000s [1,2]. The introduction of this concept enabled faster and better decision making on the battlefield, based on integrated situational awareness through the convergence and processing of information gathered by the networked sensors. Platforms in charge of attack or defense became able to respond quickly to enemy threats, using the integrated sensors and shooters, according to these decisions. The typical example of these networked systems-of-systems in the military field is the ballistic missile defense system [3]. The ballistic missile defense systems consist of precise surveillance radar networks with various types of platforms such as early warning radar and local air defense radar, and their combined intercept weapon systems [4].

One of the most dangerous enemy provocations that can be expected is simultaneous multiple ballistic missile attacks. To protect against such a situation in a timely manner, and to minimize damage, a very strictly constructed air defense system is necessary, one that can take into account precise information (missile type, trajectory, aim point, etc.) about the enemy missiles. In order for such an air defense system to perform properly, all sensing and intercept systems of the entire battlefield must systemically exchange information, and efficient decision-making should be performed based on that information. The typical processes for eliminating the ballistic missile threat are the target detection and identification, tracking and trajectory estimation, target evaluation, weapon–target assignment (WTA), and effective decision making, considering the flight phase of the ballistic missiles [5].

This paper proposes a novel sensor scheduling method to integrate heterogeneous sensor systems for a future battlefield where various type of sensors and intercept systems with diverse capabilities coexist. In particular, the main contribution of this paper is to provide the concept of seamless tracking that utilizes target handover between radars to have better situational awareness by using binary Mixed Integer Linear Programming (MILP) formulation. For simulations similar to real-world situations, it is assumed that early warning radar (EWR) catches and disseminates the entire battlefield situation, including target information. The time windows are generated considering the relative positions and velocities between radars and incoming targets. The time window thus generated represents the time period in which each radar can detect and track a target. Seamless tracking is a concept that allows continuous tracking by handing over targets to different radars that have not been assigned yet when encountering the limits of individual radars. In this study, a binary linear programming formulation was mainly used as a scheduling method to assign the tracking period to the appropriate time windows. To verify the effectiveness of the solution performed by off-the-shelf MILP solver, an additional heuristic approach was also implemented in the simulation experiment. For heuristic approach, we used and named First-In First-Out (FIFO) greedy algorithm that can implement a target handover situation. Many different MILP scheduling studies use the greedy algorithm together to compare performance [6,7]. Conversely, MILP formulations can sometimes be used to compare the performance of specially designed greedy algorithms [8–10].

The concept of how to operate multiple radar resources in a networked fashion is well documented in a paper by Green et al. [11]. Narykov et al. developed a sensor management algorithm for target tracking that uses multiple phased array radars to minimize the sensor system load [12]. Lian proposed a sensor selection optimization algorithm that can track multiple targets using a decentralized large-scale network within a labeled random finite set (RFS) framework [13]. Closer to the topic of this paper, Fu et al. proposed distributed sensor allocation for tracking multiple targets in wirelessly connected sensor networks; to improve the tracking performance, they solved the sensor fusion problem and the allocation optimization problem for the sensor and the whole target [14]. Yan introduced a method to optimize radar assignment for multiple targets, taking into account the limited time resources of each radar in the situation of detecting/tracking multiple targets with multiple networked multi-function phased array radars. This was a way to maintain the detection performance of the entire radar network even in overload situations that exceeded the tracking capability of individual radars [15]. Sherwani and Griffiths proposed a method to control the tracking parameters in order to construct an information sharing system that integrates multi-function radar networks, which are inherently limited in resource management, into one system [16]. Severson and Paley optimized radar resource management for ballistic missile reconnaissance and tracking through a decentralized consensus-based approach. Through this approach, each radar could determine their preferred radar–target allocation by balancing the radar load and minimizing the use of total radar [17]. They later solved the problem of optimal sensor coordination and tracking allocation so that multiple shipboard radars could integrate so to expand their search area and the number of tracking targets [18]. Regarding the radar scheduling using the concept of time windows, Chaolong et al. [19], Jang and Choi [20], Duan et al. [21], and Qiang et al. [22] introduced time window into the multi function phased array radar's task scheduling problems.

However, still the aforementioned studies do not contain a methodology for mathematical optimization considering the handover. The research most closely related to target handover for seamless tracking are studies on track to track correlation between the radar track and onboard IR track picture [23]. According to Lewis and Tabaczynski [24], handover technology was first achieved in 2003 between radars, and RF-to-RF and RF-to-IR handover was achieved in 2005. As the data association and sensor fusion technology developed, target handover technique is also evolving. On this basis, we are dealing with the long time-frame seamless tracking for multiple targets.

The rest of the paper is organized as follows. Section 2 introduces concepts for this study and describes the problem in detail. In Section 3, the problem is formally stated and explained in detail. Numerical simulation results are provided and discussed in Section 4. Finally, Section 5 discusses the conclusions of this study.

2. Preliminaries

2.1. Mission Overview

The problem of radar network resource management for ballistic missile defense is to deal with the schedule assignments of individual radars to precisely track the target. In this paper, we concentrate on the resource management problem of local radars that can perform precise target tracking, assuming that there is an EWR systems that can observe the whole battlefield situation. The objective function of radar resource assignment in a multi-target multi-radar situation should consider: (1) target priority for each radar; (2) continuity of target tracking; and (3) maximization of the number of tracked targets for the entire networked radar systems.

The decision maker in Figure 1 performs local radar resource management. Prior knowledge (predicted target trajectory) for resource management can be obtained through the EWRs in sensor systems, which can observe a relatively large area compared to local radar.

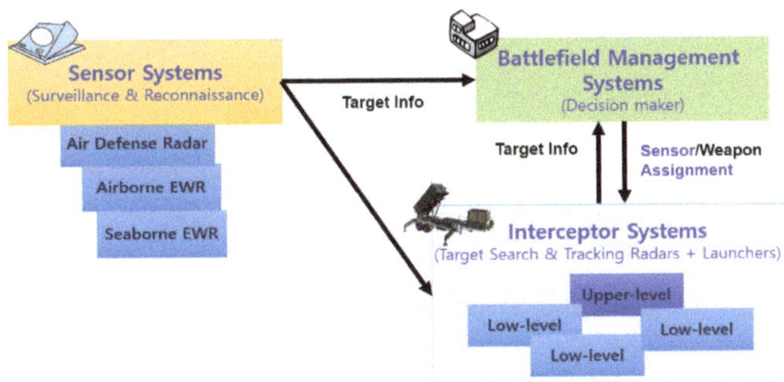

Figure 1. Schematic of system related to scheduling problem.

2.2. Key Notions in Scheduling

2.2.1. Radar

The radar parameters reflected in the resource management algorithm are the maximum number of targets that can be tracked simultaneously and the coverage of the radar. Radars are limited in the number of targets that can be tracked to the maximum according to the characteristics of each radar, and the maximum number of targets being tracked can be estimated according to the minimum tracking performance requirement. In this paper, to deal with large scale problems, the maximum number of targets per radar is arbitrarily assumed. If multiple radars with limited coverage are placed in different locations with different azimuth angles, they will have different time windows for the same target. For multiple target situations, the time window becomes more complex, and this makes the multiple target multiple radar resource management problem difficult. In this paper, the radar coverage is determined by radar position, tilt angle, azimuth direction range, altitude angular range, and distance direction range.

2.2.2. Target Priority

The target importance needs to be assessed using a priority-based metric that reflects the relative distance and remaining time between the radar and the target. Therefore, even if the same target is tracked by two or more radars, the target importance is different for each radar.

In this paper, since it is difficult to quantify the degree of threat according to the type of target, the target importance is calculated using the time remaining until the target hits the surface and the distance between the radar and the target. Here, the impact time of the target reflects the urgency to engage the target. Thus, it sets a higher priority when the remaining time becomes smaller. For a fast target, the remaining time will decrease very quickly, and thus the increasing rate of the tracking value over time will be higher than those of other targets. The distance between the target and the radar is a factor that reflects the Signal-to-Noise Ratio (SNR) and hence the expected tracking performance. Therefore, the target priority used in this study reflects the expected performance and urgency. The tracking value (v_t), determined by remaining time to impact (τ) and the distance from the radar ($dist$), is calculated as follows [25].

$$v_t = \left(1 - \frac{1}{1 + e^{-(\tau - \tau_0)/\alpha_\tau}}\right) + \left(1 - \frac{\beta_{dist}}{1 + e^{-(dist - dist_0)/\alpha_{dist}}}\right) \tag{1}$$

where τ_0, α_τ, $dist_0$, α_{dist}, and β_{dist} are parameters to determine the shape of the sigmoid function. $\tau_0 = 100$, $\alpha_\tau = 15$, $dist_0 = 500$, $\alpha_{dist} = 100$, and $\beta_{dist} = 0.8$ are used in this work.

Equation (1) decreases non-linearly (sigmoid) as the distance increases and reflects the change in average tracking performance according to SNR when the target is tracked in a single radar with a fixed resource.

2.2.3. Ballistic Target

The ballistic missile model is simulated including phases of boost, free-flight, and reentry as described in [26]. The acceleration acting on the ballistic target in each phase is expressed as follows.

$$\begin{aligned} \text{Boost phase} : \mathbf{a} &= \mathbf{a}_{thrust} + \mathbf{a}_{drag} + \mathbf{a}_{gravity} \\ \text{Free} - \text{flight phase} : \mathbf{a} &= \mathbf{a}_{gravity} \\ \text{reentry phase} : \mathbf{a} &= \mathbf{a}_{drag} + \mathbf{a}_{gravity} \end{aligned} \tag{2}$$

where

$$\begin{aligned} \mathbf{a}_{thrust} &= -\frac{\mathbf{T}}{m}\mathbf{u}_T \\ \mathbf{a}_{gravity} &= -\frac{\mu}{\| \mathbf{x} \|^3}\mathbf{x} \\ \mathbf{a}_{drag} &= -\frac{\rho(h) \| \mathbf{v} \|}{2\beta}\mathbf{v} \end{aligned} \tag{3}$$

Here, acceleration regarding Coriolis force can be added according to the coordinate system [26]. In Equation (3), $\mathbf{T}$ stands for the thrust magnitude, m stands for target mass, $\mathbf{u}_T$ stands for the unit vector which indicates thrust direction, μ stands for the Earth's gravitational constant, $\mathbf{x}$ stands for the vector from the Earth center to the target, ρ stands for the air density, h denotes target altitude, β stands for the ballistic coefficient, and $\mathbf{v}$ denotes the target velocity vector. Based on this, the trajectories of the ballistic targets in the Earth-Centered Earth-Fixed (ECEF) coordinate systems were generated, as can be seen in Figure 2.

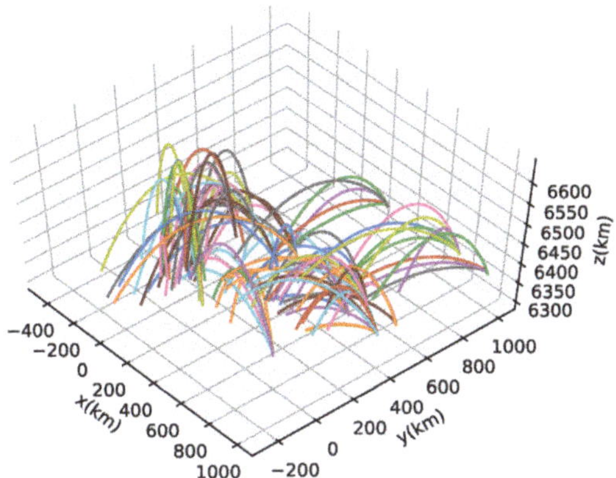

Figure 2. Randomly generated sample trajectories of ballistic targets.

2.2.4. Time Window

Each radar has a time window if the target trajectory for each target is within the coverage of the radar. The time window consists of the release (start) time and the due (end) time; the times at which the target enters and leaves the coverage of the radar are set as the release time and the due time of the time window, respectively. If Radar r has a time window for Target t and tracking is performed, the observation start time and the observation progress time are determined, and the constraint of the problem is specified so that tracking is performed only within the corresponding time window.

2.2.5. Handover

To update the state of the target for as long as possible, target needs to be tracked through multiple radars. Suppose Radar r_1 and r_2 are close to each other and lie in the same direction. In the scenario in which Target t enters coverage of r_1, enters coverage of r_2, leaves coverage of r_1 first, and then leaves coverage of r_2, the scheduling that r_1 observes first and that r_2 then observes can be thought proper. We define "Radar r_1 hand over target to Radar r_2" for the situation in which Radar r_1 tracks the target with Radar r_2 until stable measurement can be obtained after Radar r_2 starts target tracking.

2.3. *Toy Model Implementation*

A toy version of the scheduling problem with three targets and two radars is shown in Figure 3. Given knowledge about each target's trajectory and a set of radars, time windows for each target–radar pair can be calculated to ensure the maximum radar coverage. The time windows can be calculated simply by checking whether a target is inside the coverage of the radar or not; the coverage and assignments of each radar are colored differently depending on the radar. Suppose that each radar can track only a single target at a given time ($n^{capa} = 1$), and the quality of measurement from a single radar is sufficiently high that simultaneous tracking by multiple radars is not needed in the given instance. The assignment results for the situation in Figure 3a are obtained as in Figure 3b. For Target t_1, since Time Window $TW_{1,2}$ includes Time Window $TW_{1,1}$, only Radar r_2 tracks the target. For Target t_2, r_2 tracks it because the time window exists only for r_1. For t_3, r_1 tracks t_3 first because Time Window $TW_{3,1}$ starts before Time Window $TW_{3,2}$, and then r_1 hands over the target to r_2 at Interval 7. For stable tracking, both radars r_1 and r_2 simultaneously measure the target during the handover period in Interval 7. The planning horizon is divided into intervals of equal length for checking tracking status. Let us look at the results for Intervals 5 and 6 in Figure 3b. Because, from the assumption of the problem, each radar can track only a single target, the assignment in Time Window $TW_{3,1}$ starts from the release time of Interval 6, γ_6^{Int}, rather than the release time of $TW_{3,1}$, $\gamma_{3,1}^{TW}$, and the time interval corresponding to Interval 5 of Time Window $TW_{3,1}$ is excluded. The assignment to Time Window $TW_{3,1}$ continues until $\delta_{3,1}^{TW}$. After the handover from r_1 to r_2, r_2 tracks t_3 until Time Window $TW_{3,2}$ is finished.

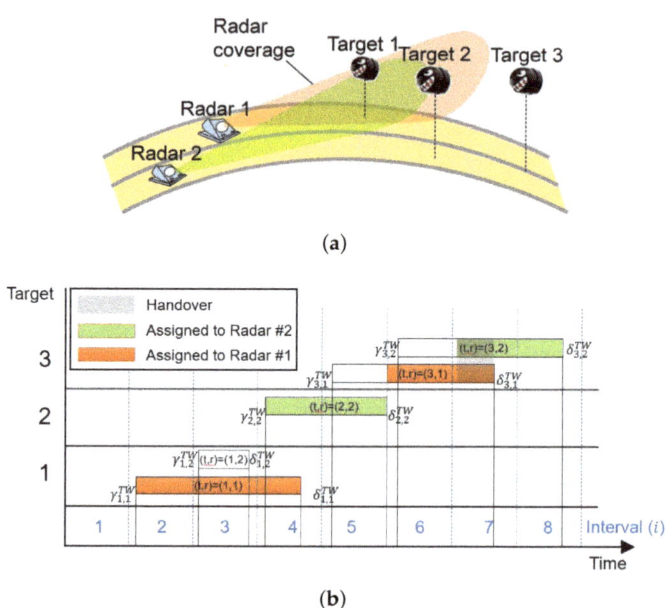

Figure 3. Conceptual diagram for sensor assignment considering target handover (refer to Table 1 for definitions of the symbols). (**a**) Physical circumstance description. (**b**) Description of time windows and handover procedure for seamless tracking.

2.4. Assumptions

Before embodying the problem, several assumptions must be made in order to implement seamless handover between multiple radars for multiple targets.

- First, communication between the radars is fast enough to ensure appropriate information sharing. Communication connections using satellites or terrestrial optical cables should be a prerequisite.
- Second, since numerous researches have been conducted on sensor fusion and data association techniques for the handover of ballistic target information [23,24,27–30], it is regarded that the targets are handed over smoothly, and filtering problems related to target processing and sensor fusion that occurs are not covered in this study. The methodological and technical problems that may arise in the process of handing over targets between radars are not discussed. Please note that there is an early warning radar (EWR) featuring handover capability has recently been introduced in the market [31].
- Third, it is assumed that ballistic missile information is provided by EWR so that the time window for each missile is within the entire mission planning horizon. In addition, the EWR is equipped with a target separation and data association capability in the ground-to-air-level clutter environment.

3. Problem Formulation

Scheduling for the general multi-target and multi-radar model is formulated as the following equations. The parameters and decision variables for the objective function and for the constraints are described in Tables 1 and 2.

Table 1. List of parameters.

Notation	Physical Meaning
$\gamma_{t,r}^{TW}$	Start time of time window
$\delta_{t,r}^{TW}$	End time of time window
γ_i^{Int}	Start time of Interval i
δ_i^{Int}	End time of Interval i
$\tau^{p,min}$	Minimum tracking assignment time
τ^{HO}	Target handover time
ω_t	Target priority(importance of target)
n^R	Number of radars
n^T	Number of targets
n^{capa}	Simultaneous tracking capability of each radar

Table 2. List of decision variables.

Notation	Value	Physical Meaning
$\tau_{t,r}^s$	$\in \mathbb{R}^+$	Start time of tracking
$\tau_{t,r}^p$	$\in \mathbb{R}^+$	Tracking duration time
x_t	$\in \{0,1\}$	Whether Target t is being allocated (tracked) or not
$x_{t,r}$	$\in \{0,1\}$	Whether Radar r tracks the target t or not
y_{t,r_1,r_2}	$\in \{0,1\}$	Whether Radar r_1 and r_2 handover the target t or not
y'_{t,r_1,r_2}	$\in \{0,1\}$	Support variable for y_{t,r_1,r_2}
$\theta_{t,r,i}$	$\in \{0,1\}$	Whether Radar r tracks the Target t in interval i or not

The objective function is the sum of target–radar–interval assignment $\theta_{t,r,i}$, tracking duration $\tau_{t,r}^p$, and target assignment x_t minus target–radar assignment $x_{t,r}$ with appropriate weight values for each term in the above formulation. The terms of the objective function have the following roles: the first term identifies the importance of the target–radar pair over time, the second term maximizes the tracking duration of the whole assignment, the third maximizes the number of targets to track, and the last one minimizes the number of handovers between different radars.

Maximize

$$c_1 \sum_{\theta \in \Theta} w_{t,r,i}\theta_{t,r,i} + c_2 \sum_{t \in T, r \in R} w_t \tau_{t,r}^p + c_3 \sum_{t \in T} w_t x_t - c_4 \sum_{t \in T, r \in R} w_t x_{t,r} \tag{4}$$

subject to

$$x_t = \max\{x_{t,1}, ..., x_{t,n^R}\} \quad \forall t \in T \tag{5}$$

$$\tau_{t,r}^p \leq M x_{t,r} \qquad \forall t \in T, r \in R \tag{6a}$$

$$\tau^{p,min} x_{t,r} \leq \tau_{t,r}^p \qquad \forall t \in T, r \in R \tag{6b}$$

$$\gamma_{t,r}^{TW} \leq \tau_{t,r}^s \qquad \forall t \in T, r \in R \tag{7a}$$

$$\tau_{t,r}^s + \tau_{t,r}^p \leq \delta_{t,r}^{TW} \qquad \forall t \in T, r \in R \tag{7b}$$

$$\delta_i^{Int} - \tau_{t,r}^s \leq M\theta_{t,r,i}^{start} \qquad \forall t \in T, r \in R, i \in I \tag{8a}$$

$$\tau_{t,r}^s + \tau_{t,r}^p - \gamma_i^{Int} \leq M\theta_{t,r,i}^{end} \qquad \forall t \in T, r \in R, i \in I \tag{8b}$$

$$\theta_{t,r,i} \leq x_{t,r} \qquad \forall t \in T, r \in R, i \in I \tag{8c}$$

$$\theta_{t,r,i} \leq \theta_{t,r,i}^{start} \qquad \forall t \in T, r \in R, i \in I \tag{8d}$$

$$\theta_{t,r,i} \leq \theta_{t,r,i}^{end} \qquad \forall t \in T, r \in R, i \in I \tag{8e}$$

$$\left(x_{t,r} + \theta_{t,r,i}^{start} + \theta_{t,r,i}^{end} - 2\right) \leq \theta_{t,r,i} \qquad \forall t \in T, r \in R, i \in I \tag{8f}$$

$$\sum_{r \in R} \theta_{t,r,i} \leq 2 \qquad \forall t \in T, i \in I \tag{9}$$

$$\sum_{t \in T} \theta_{t,r,i} \leq n^{capa} \qquad \forall r \in R, i \in I \tag{10}$$

$$(\tau_{t,r_1}^s + \tau_{t,r_1}^p) - (\tau_{t,r_2}^s + \tau^{HO}) \leq My'_{t,r_1,r_2} \tag{11a}$$

$$y_{t,r_1,r_2} \leq x_{t,r_1} \tag{11b}$$

$$y_{t,r_1,r_2} \leq x_{t,r_2} \tag{11c}$$

$$y_{t,r_1,r_2} \leq y'_{t,r_1,r_2} \tag{11d}$$

$$x_{t,r_1} + x_{t,r_2} + y'_{t,r_1,r_2} - 2 \leq y_{t,r_1,r_2} \tag{11e}$$

$$(\tau_{t,r_2}^s + \tau^{HO}) - (\tau_{t,r_1}^s + \tau_{t,r_1}^p) = M(1 - y_{t,r_1,r_2}) \tag{11f}$$

$$\forall t \in T, \quad r_1, r_2 \in R, \quad \delta_{t,r_1}^{TW} < \delta_{t,r_2}^{TW} \qquad \text{for (11a)-(11f)}$$

$$y_{t,r_1,r_2} + x_{t,r} \leq 1 \tag{11g}$$

$$\forall t \in T, \quad r_1, r, r_2 \in R, \quad \gamma_{t,r_1}^{TW} < \gamma_{t,r}^{TW} < \gamma_{t,r_2}^{TW}, \quad \delta_{t,r_1}^{TW} < \delta_{t,r}^{TW} < \delta_{t,r_2}^{TW} \qquad \text{for (11g)}$$

The constraints in Equation (5) bind target–radar assignment indicators to a single variable with an OR operator.

The constraints in Equation (6) represent the lower bound of the tracking duration for each target–radar pair if the corresponding binary indicator variable $x_{t,r}$ equals 1. M in the equations is a very large positive number and used to effectively activate the constraint only when the variables multiplied to this M take zero [32]. Briefly, $\tau^{p,min} x_{t,r} \leq \tau_{t,r}^p$ if $x_{t,r} = 1$, otherwise $\tau_{t,r}^p$ becomes 0.

The constraints in Equation (7) ensure the lower and upper bounds of the start and end time for each target–radar pair; the assignment should be inside the corresponding time window.

Let us call the constraints in Equation (8) the occupying constraints. For target–radar pair (t, r), $\theta_{t,r,i}$ indicates whether an assignment exists or not in Interval i. Briefly, $\theta_{t,r,i}$ equals 1 if the following conditions are met simultaneously: $y_{t,r} = 1$, $\tau_{t,r}^s \leq \delta_i^{Int}$, and $\gamma_i^{Int} \leq \tau_{t,r}^s + \tau_{t,r}^p$.

The constraints in Equation (9) limit the maximum number of radars used for tracking a single target to two; two radars are assigned when the handover occurs, otherwise a single radar tracks the target. Simply, the handover of a single target will only occur between two radars. The constraints in Equation (10) limit the capability of simultaneous tracking for each radar.

The constraints in Equation (11) are for the handover. To decide the handover indicator variable y_{t,r_1,r_2}, we define the support variable y'_{t,r_1,r_2} as in Equation (11a); y'_{t,r_1,r_2} is 1 if the end time of Radar r_1 tracking Target t is equal to or higher than the sum of start time and handover duration for Radar r_2. The reason for using the support variable is that both target–radar pairs (t, r_1) and (t, r_2) must be assigned the schedule simultaneously as well as satisfying the handover time constraint (Equation (11b)–(11f)). The constraints in Equation (11g) ensure that radars with duplicating and smaller time windows are excluded from the assignment.

4. Results and Discussion

In this section, computational results for test instances of the multiple radar resource scheduling problem described in Section 3 are reported. The optimization for the mixed-integer linear problem was solved by Gurobi 8.0.1 based on Python 3.6.2 and the optimization for the heuristic problem was solved with the same Python environment. The computation was conducted by a desktop with an Intel Core^{TM} i7-6700K, 4.00 GHz CPU and 16 GB of RAM.

Two experimental models were tested to verify the effectiveness of the algorithm, as shown in Figure 4. The first scenario verified the effectiveness of the exact algorithm using a local-scale model for easy parameter modification. The second scenario verified the practical applicability of the algorithm by introducing two approaches for the large-scale (battlefield) scenario, which considered far more virtual targets and radars than were used in the local-scale model.

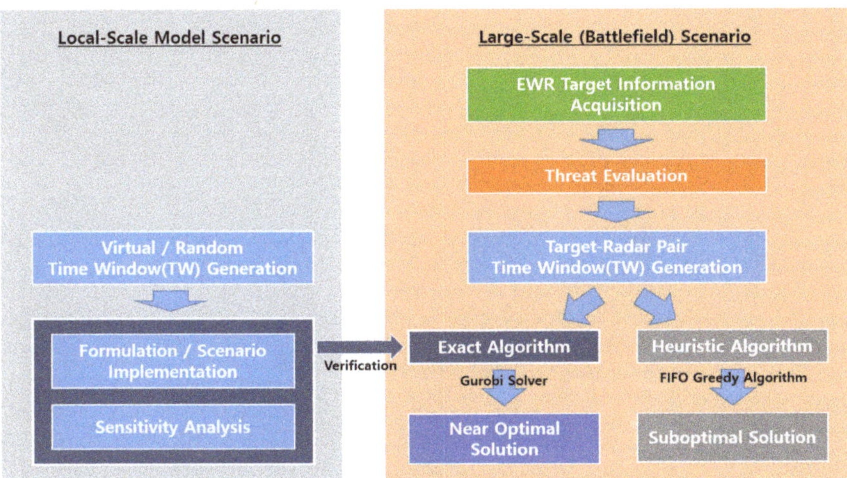

Figure 4. Overview of the numerical experiment.

4.1. Local Scale Scenario Experiment

4.1.1. Algorithm Verification

We verified the algorithm using a local-scale model to confirm that the objective function and constraints work well. The local-scale model allowed us to arbitrarily set the number of targets and radars. Among the four objective function terms, the first one "target importance" was assumed to be constant for this simple simulation. To check for changes according to the number of available radars, we first tested two different cases of 10 radars for a single target, and 20 radars for two targets. Figure 5a shows the tracking results for a single target using 10 radars with different time windows. The size of each time window was set to be randomly generated within a maximum of 60 s and a minimum of 30 s. Other parameters used here are shown in Table 3.

Table 3. Parameter set for local-scale model simulation.

Planning horizon	160 s
Minimum tracking assignment time	7 s
Target handover time	3 s
Simultaneous tracking capability of each radar (n^{capa})	2

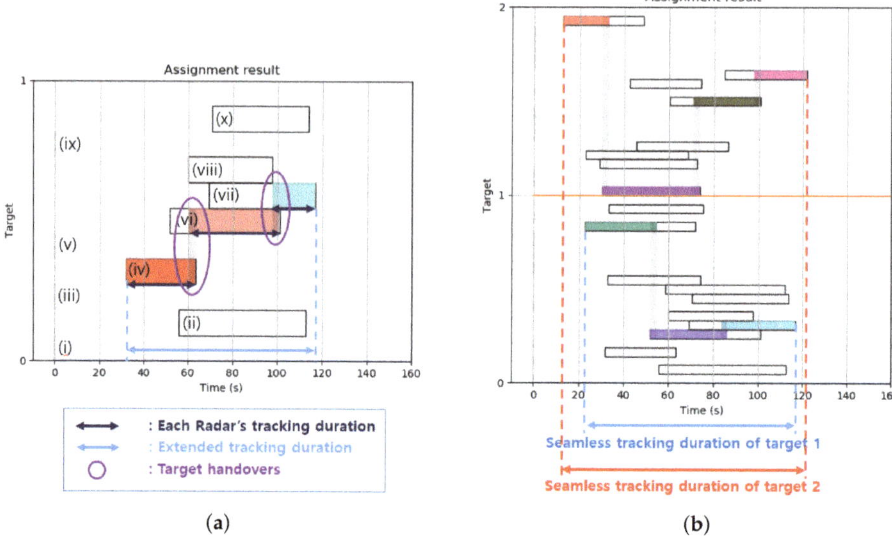

Figure 5. Optimal scheduling assignment results obtained using simple local-scale model. (**a**) One target and 10 radars. (**b**) Two targets and 20 radars.

As can be seen in Figure 5a, among the 10 radars, six radars were involved in target tracking, because the probability that each radar can participate in tracking the target was set as equal to or less than 60%. Two target handovers occurred in the relevant sections of (iv)–(vi) at 60 s and (vi)–(vii) at 97 s, and the time window was selected to keep track of the target for as long as possible while maintaining the minimum takeover time as designed in the objective function. What is unique here is that, although Time Window (ii) is longer than any of the others, the solver assigned targets to Time Windows (vi) and (vii) to track the target as long as possible and at the same time to meet the constraint of the minimum tracking assignment time. On top of the conditions given in the results shown in Figure 5a, Figure 5b shows the results of the tracking assignment of 20 radars for two targets, as well as results for adding one more target and 10 more radars. These results also show that radar resources were well assigned to reflect the designed objective function and the constraints, such that the first target required two handovers and the second target required three handovers to achieve maximum tracking of each target.

Thus far, we verified two of the four terms of the objective function in Equation (4), namely maximization of target tracking time and minimization of the number of target handovers, as well as the constraints, are working well. In the above test model, since the target number was set too small, the third term of the objective function, that is, the test result required to maximize the number of targets to track, could not be confirmed. Therefore, in the following experiment, to see how all the terms of the objective function can be demonstrated, we increased the number of targets and limited the number of radar. This involved one of the key parameters in Table 1, n^{capa}, the simultaneous tracking capability of each radar.

Using the parameters in Table 3, Figure 6a depicts the optimal scheduling results for tracking 10 targets with three radars. For each target, depending on the detection probability of 60% mentioned above, we can confirm that 1–3 radars were assigned to all targets except for the fourth target, which could not be detected and tracked in this simulation condition. Figure 6b depicts what happens when the radar's simultaneous tracking capability (n^{capa}) is adjusted to 3. The most noticeable thing is that the tracking durations of the fourth, sixth and seventh targets increased dramatically, as shown by the red colored arrows in Figure 6. Especially, it was possible to track the fourth target only in the time window of the first radar, as shown in Figure 6a; however, as n^{capa} increased, the target could be tracked in all available time window sections of the first and second radar, as shown in Figure 6b. In addition, considering the number of targets being tracked at 70 s in Figure 6, it can be seen in Figure 6b that seven targets could be tracked, while six targets could be tracked in Figure 6a. Thus, although there is a difference in degree, as the tracking ability improved, the tracking time for the entire target generally improved, as shown in Table 4.

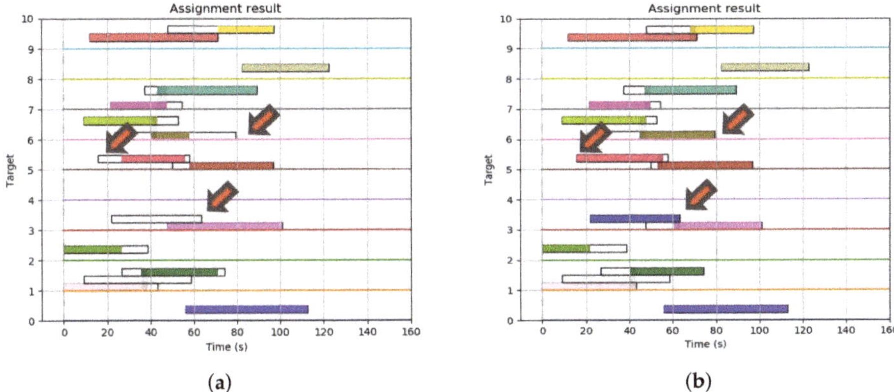

(**a**) (**b**)

Figure 6. Optimal scheduling assignment result for 10 targets and three radars with different simultaneous tracking capability. (**a**) When the radars can track two targets simultaneously ($n^{capa} = 2$). (**b**) When the radars can track three targets simultaneously ($n^{capa} = 3$).

Table 4. Tracking duration time according to simultaneous tracking capability.

Target Number	Total Tracking Duration (s)		
	$n^{capa} = 2$	$n^{capa} = 3$	increments
Target 1	56.8	56.8	0
Target 2	71.1	74.3	+3.2
Target 3	26.7	21.8	−4.9
Target 4	53.4	79.2	+25.8
Target 5	0	0	0
Target 6	70.2	81	+10.8
Target 7	48.8	70.1	+21.3
Target 8	67.9	67.9	0
Target 9	40.2	40.2	0
Target 10	85.4	85.4	0

This is the result, for certain targets, of slightly increasing or decreasing that tracking time according to the terms of the objective function in Equation (4) and the constraint "simultaneous tracking capability of radar (n^{capa})", written in Equation (10). The values in this table are the time taken from the moment the target was first detected by one radar to the moment it was lost after being handed over to another radar. One more noticeable point in Figure 6 is that, in the case of the 10th target, the minimum time required for the handover was not met because of the limitation of the simultaneous tracking capability, as shown in Figure 6a, while the target handover can be seen to have been smoothly accomplished in Figure 6b due to the increase in the tracking duration time of the third radar.

Figure 7 shows how the tracking of a target actually changed in the time window of each radar as the simultaneous tracking capability (n^{capa}) changed.

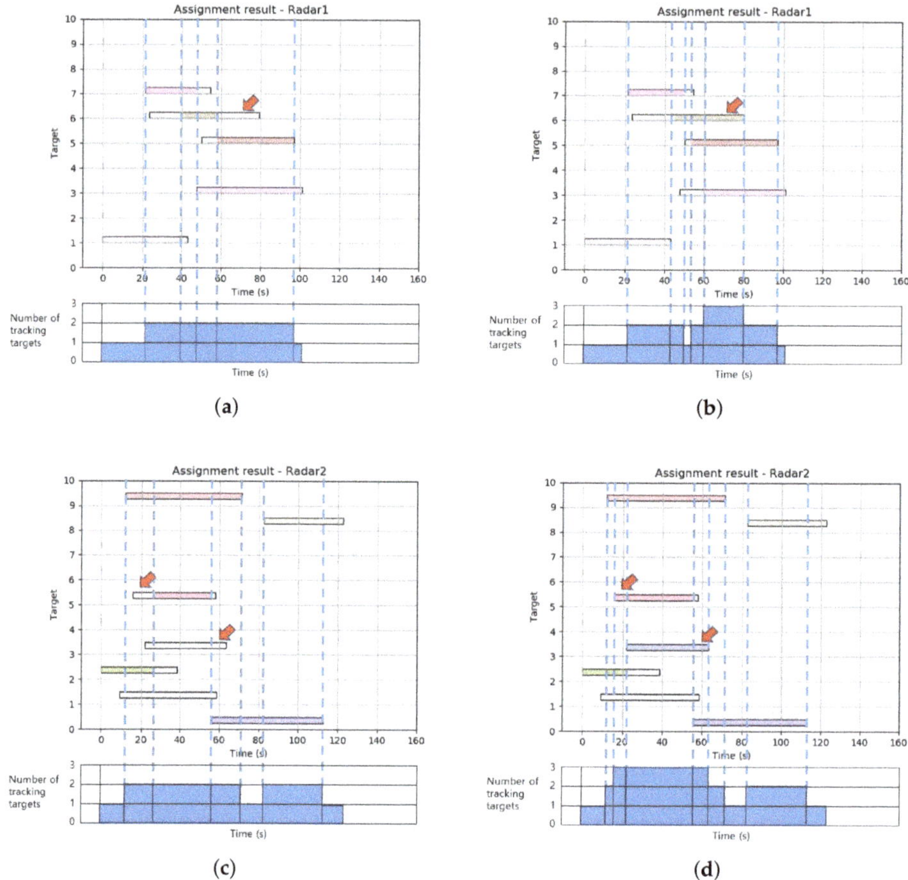

Figure 7. Optimal scheduling assignment result of each radar site according to the change of simultaneous tracking capability. (**a**) Assignment result of first radar when $n^{capa} = 2$. (**b**) Assignment result of first radar when $n^{capa} = 3$. (**c**) Assignment result of second radar when $n^{capa} = 2$. (**d**) Assignment result of second radar when $n^{capa} = 3$.

In Figure 7a,b, which are the assignment results for Radar 1, it is confirmed that the number of targets to be tracked throughout the whole planning horizon did not exceed a maximum of 2, in Figure 7a, and 3, in Figure 7b. Looking more closely at Figure 7b, we can observe that three targets were being tracked at the same time only between about 60 and 80 s as the tracking duration time of the seventh target expanded. Similarly, in Figure 7c,d, the tracking duration for the fourth and sixth targets expanded with increased simultaneous tracking capability, and thus three targets were being tracked simultaneously between about 15 and 60 s.

4.1.2. Parameter Sensitivity Analysis

As explained in the previous case, the objective value of the optimal scheduling problem depends on changes in the value of a particular parameter. Therefore, we performed a sensitivity analysis to determine how the parameters affect the outcome of the objective function. Three parameters were determined to affect the results. Sensitivity analysis was performed by fixing the remaining parameters while adjusting one target parameter, as shown in Table 5. The target parameter for the first sensitivity analysis was the simultaneous tracking capability (n^{capa}) of the radar, as shown in Table 5.

Table 5. Parameter setting for sensitivity analysis.

Parameters	Case 1	Case 2	Case 3
Simultaneous tracking capability	1 to 10	5	5
Minimum tracking assignment time	10	1 to 10	10
Handover time	10	10	1 to 10

Figure 8 shows how the results varied with the simultaneous tracking capability. As shown in Table 4, the result of the objective function initially increased when n^{capa} increased. However, it can be seen that, after a certain level, the result of the objective function was not significantly affected. This tendency was only the result of a given condition, and, when the condition changed, a point that was not affected by the change of n^{capa} could be changed. Specifically, if the simultaneous tracking capability of the radar covered the number of targets, the influence of n^{capa} would be insignificant. If the number of targets to be tracked were greater than the simultaneous tracking capability of the radar, when the value of n^{capa} is high, the objective value would also rise, as shown in Figure 8, until the radars can cover all the targets.

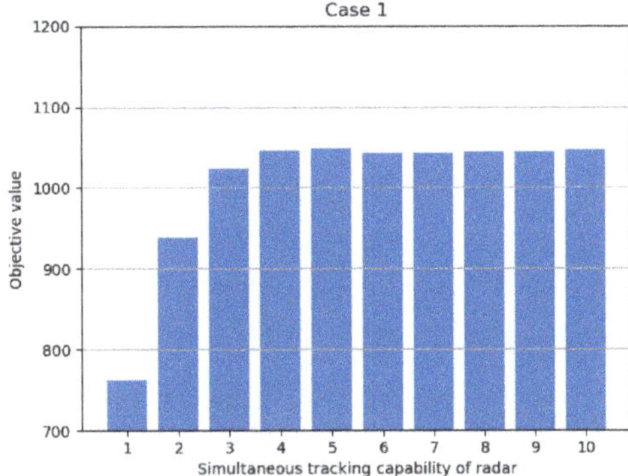

Figure 8. Objective value with respect to number of simultaneous tracking capability.

The second parameter that affects the value of the objective function is the minimum tracking assignment time. This parameter is the minimum time required for a radar to track a target, and physically refers to the time it takes for the radar filter system to stabilize the target tracking. Figure 9 shows the objective value according to the change of the minimum tracking assignment time. As can be seen in the figure, smaller minimum tracking assignment times led to higher levels of assignment, but the assignment was not affected after a certain level of minimum tracking assignment time.

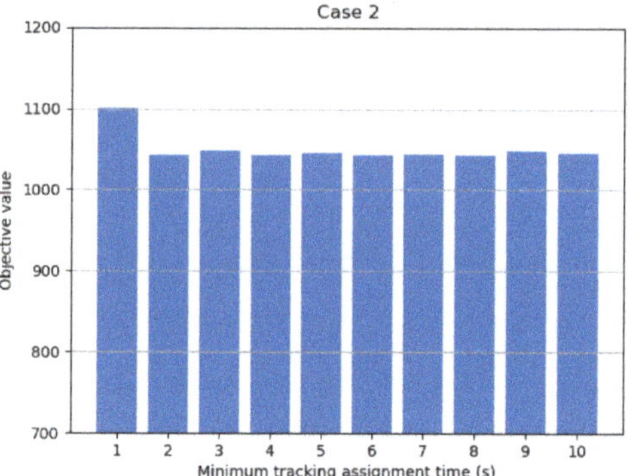

Figure 9. Objective value with respect to minimum tracking assignment time.

The third parameter for the sensitivity analysis was the time required to hand over the target between networked radars. In fact, this is a simulation parameter, and in a real environment it is very likely to be affected by the physical state of the network and filtering system. However, since it is a parameter that has an important influence on the simulation results, it was selected as a parameter in the sensitivity analysis to grasp its influence and overall tendency. As shown in Figure 10, the objective value decreased linearly as the handover time increased. This result shows that longer handover times led to less efficient overall target tracking. Therefore, a shorter handover time is better. In other words, the network system actually should be constructed so as to minimize the time required for target transmission, as well as the time required for convergence between a transmitted target and a self-detected target.

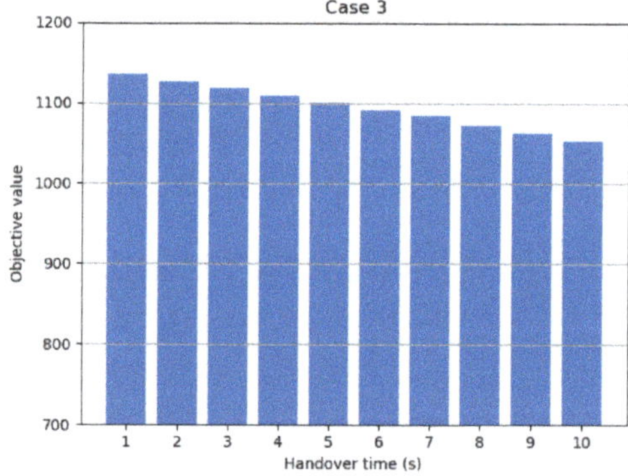

Figure 10. Objective value with respect to handover time.

4.2. Battlefield Scenario Experiment

In this experiment, scheduling optimization was performed assuming a situation in which 100 enemy ballistic missiles of four different types from four different launch sites flocked toward five friendly radar sites distributed appropriately. This is a much worse situation than that of local-scale model problem. Through this experiment, we verified the effectiveness and practical applicability of the optimal scheduling technique that employs the seamless handover method proposed in this study. Table 6 and Figure 11 show the parameters and conceptual diagram for this experiment, respectively. Compared to the local-scale model, the optimization planning horizon was increased to 1000 s and the number of targets and radars increased to 100 and 5, respectively. The most important parameter—the simultaneous tracking capability—was increased to 20 so that five radars could cover all the 100 targets.

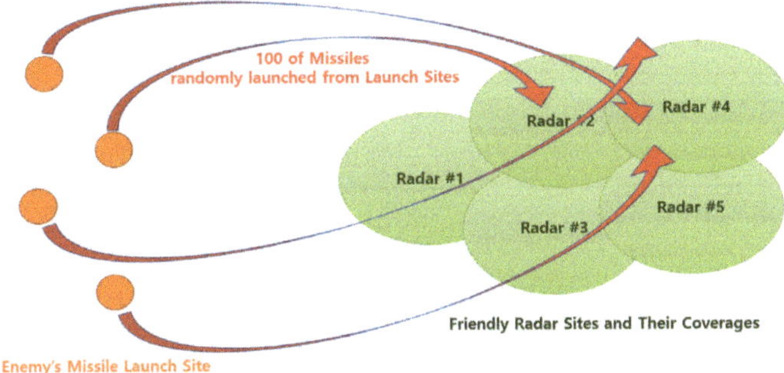

Figure 11. Conceptual diagram for the battlefield scenario.

Table 6. Parameter set for battlefield scenario experiment.

Number of target (n^T)	100
Number of radar (n^R)	5
Planning horizon	1000 s
Minimum tracking assignment time	7 s
Target handover time	3 s
Simultaneous tracking capability of each radar (n^{capa})	20

One of the most different aspects compared to the local-scale model is the importance of target, which is the first term of Equation (4). It is reflected in the objective function for this scenario unlike the previous experiment. The problem was solved with the assumption that the importance of target is uniform in the previous experiment. However, in this scenario, the target distance from radar and the response time available for the target were taken into consideration, as written in Equation (1), as in a real situation. Another difference related to the time window creation. In the local-scale model, a random function was used to generate a time window between arbitrary times selected by the user. However, in this experiment, it was assumed that the early warning radar provides the trajectory information of the ballistic missiles, so that the time windows could be created for radars located in various regions.

4.2.1. Weights of Objective Function Sensitivity Analysis

The objective function used in Equation (2) can be said to have some form of weighted sum. To check the dominance of each term of the objective function, the optimum value of individual objective was checked, as shown in Table 7.

Table 7. Solution ranges of each term of objective function and proposed coefficient setting.

Term	Opt. of Ind. Objective	Coeff. Value for Normalization
Target priority	18,610.8	$c_1 = 1$
Maximization of tracking time	7621.6	$c_2 = 2.44$
Maximization of the number of tracked target	98.0	$c_3 = 189.9$

As shown in Table 7, the first term, the priority of the target, was the dominant term that had the greatest influence on the objective function value. The third, the maximization number of tracked target, was found to have very little effect compared to the others. The fourth term, the minimization of the number of handover, is not included in this table because it acted as a penalty term. To analyze properly, the objective function needed to be normalized. In this study, considering the penalty terms, instead of dividing the objective by those optimum values, the most influential objective's coefficient (c_1) was set to 1 and the remaining coefficients were normalized accordingly.

Based on the coefficients determined above, a sensitivity analysis was conducted according to the penalty term, the minimization of the number of handover, as shown in Figure 12. It is trivial that the objective function value decreased with increasing c_4. One interesting point is that the number of handover decreased step-wise. These results indicate that, if c_4 is smaller than necessary, the overall objective function value can be high, but there are many unnecessary handovers. Therefore, choosing c_4 at which it starts to no longer decrease is the best decision to maximize the objective function value and reduce the number of handovers. Thus, c_4 of 35 was chosen for this case.

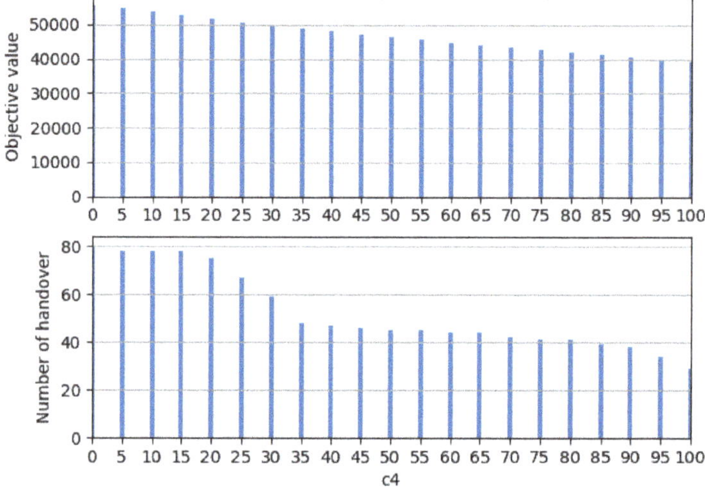

Figure 12. Sensitivity with respect to the penalty term.

4.2.2. Complexity Analysis

In general, the Mixed Integer Linear Program (MILP) problem is known as NP-hard or NP-complete problem. It is also known that NP-hard problem has exponential computational complexity [33]. Therefore, in this section, we look at how the complexity changes according to the parameters that affect the computational complexity, and to what extent we can use this algorithm using MILP formulation. To achieve that, we confirmed how the complexity appeared according to the number of targets, the number of radars, and the number of targets that can be simultaneously detected by radars (n^{capa}).

First, the most prominent in Figure 13 is the exponentially increasing computation time, as previously predicted. The most important parameter for analyzing here is the n^{capa}. The greater is the radar's ability to track simultaneously, the shorter id the calculation time due to the less load, in which case a gentle exponential curve is drawn. On the contrary, when the radar's simultaneous tracking capability is low, the calculation time explodes, and it is confirmed that the calculation is very slow in an overload situation exceeding a certain number of targets. On the other hand, the calculation time increase is more sensitive to the number of targets than to the number of radars. Based on this, it can be concluded that, when designing a radar network, it is very advantageous for the target and sensor assignment of multiple targets if we increase the simultaneous tracking capability of each radar. Based on these data, we can also establish the algorithm re-planning cycle that is envisioned in the future.

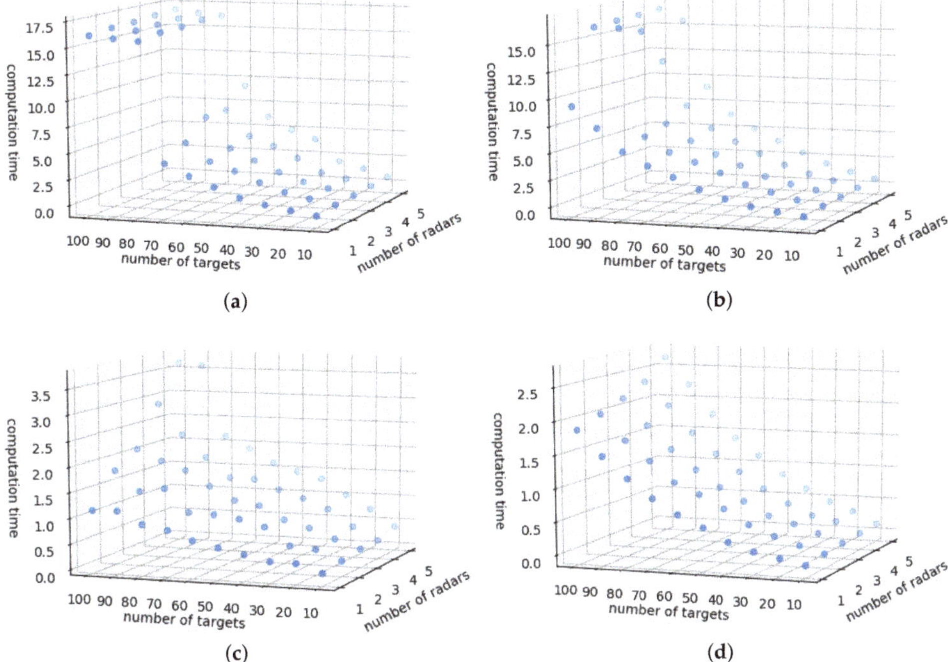

Figure 13. MILP formulation complexity as of computation time according to the change of simultaneous tracking capability. (**a**) Computation time when $n^{\text{capa}} = 10$. (**b**) Computation time when $n^{\text{capa}} = 15$. (**c**) Computation time when $n^{\text{capa}} = 20$. (**d**) Computation time when $n^{\text{capa}} = 25$.

4.2.3. Numerical Simulations for Comparison

In this scenario, the proposed MILP solution was compared with a heuristic technique, termed the First-in First-out (FIFO) greedy algorithm (Appendix A). This greedy heuristic is an extension of a sequential greedy algorithm for assignment [34] to take into account the handover requirement. Note that the greedy scheme can be a good reference algorithm as it works very well in many domains and also guarantees some optimality gap when the objective function satisfies certain conditions [34,35]. Detailed theoretical analysis of the greedy heuristic is omitted as it is out of the focus of this paper.

The overall procedure is well described in Figure 4.

Figure 14 shows the number of targets being simultaneously tracked by each radar over the planning horizon for the exact and heuristic algorithms. As can be seen in the figure, the simultaneous tracking load of each radar clearly increased between 400 and 650 s because targets were the most frequent and concentrated at that time. In terms of an objective value, the result of the proposed formulation solved by Gurobi commercial MILP solver returned 48861.9, while the heuristic algorithm gave a value of 45,332.4, an approximately 8% difference in performance. This was noticeably exhibited mainly in the simultaneous tracking loads of Radars 4 and 5, as shown in Figure 14. The reason for this is that the heuristic approach to solving this problem is to maximize the time that each radar tracks in a greedy manner. This phenomenon is explained by the local optima convergence, which is a typical disadvantage of the heuristic approach, and therefore shows an assignment result that is not properly distributed. Meanwhile, 48 handovers took place between the radars in the case of exact algorithm while 53 handovers occurred in the case of heuristic algorithm. When we compared performance with these results, we considered two main things: the number of handovers that act as the the penalty function in the objective function and the objective value obtained. The simulation results are more than simply comparing the objective values and having a low number of handovers. The absence of unnecessary handovers is much more advantageous in terms of radar operation. Although there are differences in the number of handovers depending on how to solve the problem, the target can be tracked for a much longer period of time than in the case without handovers between radars, and the resulting time margin would provide valuable time for the preparation of the next battle for each interceptor. The computation time of exact algorithm case was approximately 4.27 s longer because the solution using the Gurobi solver investigated as many cases as possible to find the optimal solution.

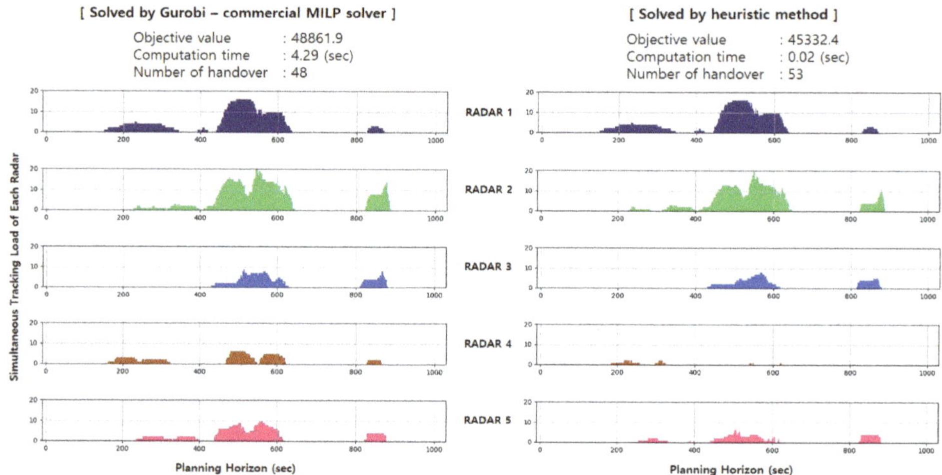

Figure 14. Simulation result comparison in the sense of the Number of simultaneously tracked target for each radar.

5. Conclusions

Using a local-scale model and real world example, we experimented with optimal scheduling of multiple radars for multiple targets and derived appropriate results. Simulation results show that the objective function of the proposed formulation is valid and effective for the real world situation by using target handover. The experiment results show that the proposed exact algorithm solved using Gurobi exhibited better results than that of the heuristic method in terms of performance, number of handovers and tracking load for entire systems. This paper opens the possibility of solving the problem of seamless multi-target tracking of multiple radar network against a large number of missile attacks. The results are especially helpful in preventing situations in which radars with limited detection area are unnecessarily tracking multiple targets at the same time. The resulting margin of tracking capability will increase the survivability in such situations. For future work, first, we will try to find adaptive parameters that may not be confined to a specific situation for each coefficient of objectives constituting the objective function. Second, we will expand and connect this sensor assignment problem into the weapon target assignment problem for an anti-air defense system which is composed of multiple radars and multiple interceptor systems. Third, we will continue to study the methodology to apply the Reinforcement Learning (RL) technique to this problem.

Author Contributions: Conceptualization, J.K., D.-H.C., W.-C.L., S.-S.P. and H.-L.C.; methodology, D.-H.C. and J.K.; software, D.-H.C., J.K., W.L. and S.-S.P.; validation, J.K. and D.-H.C.; formal analysis, J.K.; resources, D.-H.C.; data curation, J.K.; writing—original draft preparation, J.K. and D.-H.C.; writing—review and editing, J.K. and H.-L.C.; visualization, J.K. and D.-H.C.; and supervision, project administration, and funding acquisition, H.-L.C.

Funding: This work was conducted at the High-Speed Vehicle Research Center of KAIST with the support of the Defense Acquisition Program Administration and the Agency for Defense Development under contract UD170018CD. The first author is supported by the scholarship program of the Republic of Korea Navy.

Acknowledgments: The authors would like to thank the Editor and the anonymous reviewers for their valuable comments and suggestions.

Conflicts of Interest: The authors declare no conflict of interest.

Appendix A. Pseudo-Code of FIFO Greedy Algorithm

Algorithm A1 FIFO greedy algorithm

Input: Target information, Time windows information, Radar tracking capability
Output: Target tracking schedules for each radar
procedure SCHEDULE FIFO
  TW data $\leftarrow$ array with start and end time elements
  $\triangleright$ TW data : Set of time windows generated based on targets' movement and radar's location
  $t \leftarrow$ number of target
  $r \leftarrow$ number of radar
  $\tau \leftarrow \tau_0, \tau_f$
  $v \leftarrow 0{,}1$ $\triangleright$ v : Indicator that tells whether it is the release time(0) or the due time(1) of TW
  for t, r, τ, v in *TW data* **do**
    $t, r, \tau, v \leftarrow \phi$
    for each *TW data* in the scenario **do**
      $[t, r] \leftarrow$ target and radar pair of TW
      $\tau_o \leftarrow$ release time of TW
      $v \leftarrow 0$
      TW data $\leftarrow$ *TW data* $\bigcup(t, r, \tau_0, v)$
      $\tau_f \leftarrow$ due time of TW
      $v \leftarrow 1$
      TW data $\leftarrow$ *TW data* $\bigcup(t, r, \tau_f, v)$
    end for
  end for

  Sort *TW data* based on ascending time order

  for t, r, τ_0, v in *TW data* **do**
    if v is 0 **then** $\triangleright v = 0 : \tau$ is release time
      if target t is not on tracking **then**
        $\tau_o, \tau_f \leftarrow TW_o[t, r], TW_f[t, r]$
        if radar's resource available from τ_o to τ_f **then**
          Assign Radar r to Target t between τ_o and τ_f
        end if
      else $r_{last} \leftarrow$ last radar on target
        if Target t is on tracking but not reached the end of $TW_f[t, r]$ **then**
          Do not handover yet
          if the correlated TW of the two radars satisfies enough handover time **then**
            Perform handover
          else Do not handover
          end if
        end if
      end if
    else $\triangleright v = 1 : \tau$ is due time
      Terminate Radar r's tracking
    end if
  end for
end procedure

References

1. Owens, W.A. The Emerging U.S. System-of-Systems. *Naval Inst. Proc.* **1996**, *121*, 7.
2. Alberts, D.S.; Garstka, J.J.; Stein, F.P. *Network Centric Warfare: Developing and Leveraging Information Superiority*; Assistant Secretary of Defense (C3I/Command Control Research Program): Washington, DC, USA, 2000; pp. 87–114.
3. Dahmann, J.S.; Baldwin, K.J. Understanding the current state of US defense systems of systems and the implications for systems engineering. In Proceedings of the 2008 2nd Annual IEEE Systems Conference, Montreal, QC, Canada, 7–10 April 2008; pp. 1–7.
4. Barcio, B.T.; Ramaswamy, S.; Macfadzean, R.; Barber, K.S. Object-oriented analysis, modeling, and simulation of a notional air defense system. *Simulation* **1996**, *66*, 5–21. [CrossRef]
5. *Countering Air and Missile Threats*; Joint Publication 3-01; United States Joint Chief of Staff: Washington, DC, USA, 2018.
6. Cho, D.H.; Kim, J.H.; Choi, H.L.; Ahn, J. Optimization-Based Scheduling Method for Agile Earth-Observing Satellite Constellation. *J. Aerosp. Inf. Syst.* **2018**, *15*, 611–626. [CrossRef]
7. Khalafa, A.; Djouania, K.; Hamama, Y.; Alaylid, Y. Mixed-integer linear programming (MILP) for optimisation of medical equipment maintenance schedules. In Proceedings of the 2015 22nd Iranian Conference on Biomedical Engineering (ICBME), Tehran, Iran, 25–27 November 2015; pp. 205–209.
8. Zakaria, R.; Moalic, L.; Caminada, A.; Dib, M. Greedy Algorithm for Relocation Problem in One-Way Carsharing Systems. In Proceedings of the MOSIM 2014, 10ème Conférence Francophone de Modélisation, Optimisation et Simulation, Nancy, France, 5–7 November 2014.
9. Che, A.; Zeng, Y.; Lyu, K. An efficient greedy insertion heuristic for energy-conscious single machine scheduling problem under time-of-use electricity tariffs. *J. Clean. Prod.* **2016**, *129*, 565–577. [CrossRef]
10. Eles, A.; Cabezas, H.; Heckl, I. Heuristic algorithm utilizing mixed-integer linear programming to schedule mobile workforce. *Chem. Eng. Trans.* **2018**, *70*, 895–900.
11. Johnson, B.W.; Green, J.M. Naval network-centric sensor resource management. In *7th ICCRTS*; Naval Postgraduate School: Monterey, CA, USA, 2002.
12. Narykov, A.S.; Krasnov, O.A.; Yarovoy, A. Algorithm for resource management of multiple phased array radars for target tracking. In Proceedings of the IEEE 16th International Conference on Information Fusion, Istanbul, Turkey, 9–12 July 2013; pp. 1258–1264.
13. Lian, F.; Hou, L.; Wei, B.; Han, C. Sensor Selection for Decentralized Large-Scale Multi-Target Tracking Network. *Sensors* **2018**, *18*, 4115. [CrossRef] [PubMed]
14. Fu, Y.; Ling, Q.; Tian, Z. Distributed sensor allocation for multi-target tracking in wireless sensor networks. *IEEE Trans. Aerosp. Electron. Syst.* **2012**, *48*, 3538–3553. [CrossRef]
15. Yan, J.; Pu, W.; Liu, H.; Zhou, S.; Bao, Z. Cooperative target assignment and dwell allocation for multiple target tracking in phased array radar network. *Signal Proc.* **2017**, *141*, 74–83. [CrossRef]
16. Sherwani, H.; Griffiths, H.D. Tracking parameter control in multifunction radar network incorporating information sharing. In Proceedings of the IEEE 2016 19th International Conference on Information Fusion (FUSION), Heidelberg, Germany, 5–8 July 2016; pp. 319–326.
17. Severson, T.A.; Paley, D.A. Distributed multi-target search and track assignment using consensus-based coordination. In Proceedings of the IEEE SENSORS, Baltimore, MD, USA, 3–6 November 2013; pp. 1–4.
18. Severson, T.A.; Paley, D.A. Optimal sensor coordination for multitarget search and track assignment. *IEEE Trans. Aerosp. Electron. Syst.* **2014**, *50*, 2313–2320. [CrossRef]
19. Ying, C.; Wang, Y.; He, J. Study on time window of track tasks in multifunction phased array radar tasks scheduling. In Proceedings of the IET International Radar Conference, Guilin, China, 20–22 April 2009; pp. 20–22.
20. Jang, D.S.; Choi, H.L.; Roh, J.E. A time-window-based task scheduling approach for multi-function phased array radars. In Proceedings of the IEEE 2011 11th International Conference on Control, Automation and Systems, Gyeonggi-do, Korea, 26–29 October 2011; pp. 1250–1255.
21. Duan, Y.; Tan, X.S.; Qu, Z.G.; Wang, J.; Peng, X.Y. A scheduling algorithm for phased array radar based on adaptive time window. In Proceedings of the 2017 IEEE 3rd Information Technology and Mechatronics Engineering Conference (ITOEC), Chongqing, China, 3–5 October 2017; pp. 935–941.

22. Qiang, S.W.; Xing, H.; Jian, G. A task scheduling simulation for phased array radar with time windows. In Proceedings of the 2017 IEEE 6th International Conference on Computer Science and Network Technology (ICCSNT), Dalian, China, 21–22 October 2017; pp. 165–169.
23. Maurer, D.E. Information handover for track-to-track correlation. *Inf. Fus.* **2003**, *4*, 281–295. [CrossRef]
24. Lewis, B.M.; Tabaczynski, J.A. The Role of Serious Games in Ballistic Missile Defense. *Lincoln Lab. J.* **2019**, *23*, 11–24.
25. Jang, D.S.; Cho, D.H.; Choi, H.L. An approach to multiple radar resource assignment problem for managing highspeed multi target. In Proceedings of the KSAS 2016 Fall Conference, Jeju, Korea, 16–18 November 2016; pp. 910–911.
26. Li, X.R.; Jilkov, V.P. Survey of maneuvering target tracking: II. Ballistic target models. In Proceedings of the International Symposium on Optical Science and Technology, San Diego, CA, USA, 29 July–3 August 2001; Volume 4473, pp. 559–581.
27. Maurer, D.E.; Schirmer, R.W.; Kalandros, M.K.; Peri, J.S. Sensor fusion architectures for ballistic missile defense. *Johns Hopkins APL Tech. Digest* **2006**, *27*, 19–31.
28. Serakos, D. On BMD Target Tracking: Data Association and Data Fusion. *arXiv* **2014**, arXiv:1408.3596.
29. Song, T.L. Multi-target tracking filters and data association: A survey. *J. Inst. Control Robot. Syst.* **2014**, *20*, 313–322. [CrossRef]
30. Zhang, Z.; Fu, K.; Sun, X.; Ren, W. Multiple Target Tracking Based on Multiple Hypotheses Tracking and Modified Ensemble Kalman Filter in Multi-Sensor Fusion. *Sensors* **2019**, *19*, 3118. [CrossRef] [PubMed]
31. ELM-2090 TERRA Strategic Early Warning Dual Band Radar System. Available online: https://www.iai.co.il/p/elm-2090-terra (accessed on 29 September 2019).
32. Griva, I.; Nash, S.G.; Sofer, A. *Linear and Nonlinear Optimization*; Siam: Philadelphia, PA, USA, 2009; Volume 108.
33. Till, J.; Engell, S.; Panek, S.; Stursberg, O. Applied hybrid system optimization: An empirical investigation of complexity. *Control Eng. Pract.* **2004**, *12*, 1291–1303. [CrossRef]
34. Cho, D.H.; Choi, H.L. Greedy Maximization for Asset-Based Weapon—Target Assignment with Time-Dependent Rewards. In *Cooperative Control of Multi-Agent Systems*; John Wiley & Sons, Ltd.: New York, NY, USA, 2017; Chapter 5, pp. 115–138.
35. Vazirani, V.V. *Approximation Algorithms*; Springer GmbH: Berlin/Heidelberg, Germany, 2003.

Article

GLMB Tracker with Partial Smoothing

Tran Thien Dat Nguyen [1,*] and Du Yong Kim [2]

1 School of Electrical Engineering, Computing and Mathematical Sciences, Curtin University, Bentley 6102, Australia
2 School of Engineering, RMIT University, Melbourne 3000, Australia
* Correspondence: t.nguyen172@postgrad.curtin.edu.au

Received: 9 August 2019; Accepted: 10 October 2019; Published: 12 October 2019

Abstract: In this paper, we introduce a tracking algorithm based on labeled Random Finite Sets (RFS) and Rauch–Tung–Striebel (RTS) smoother via a Generalized Labeled Multi-Bernoulli (GLMB) multi-scan estimator to track multiple objects in a wide range of tracking scenarios. In the forward filtering stage, we use the GLMB filter to generate a set of labels and the association history between labels and the measurements. In the trajectory-estimating stage, we apply a track management strategy to eliminate tracks with short lifespan compared to a threshold value. Subsequently, we apply the information of trajectories captured from the forward GLMB filtering stage to carry out standard forward filtering and RTS backward smoothing on each estimated trajectory. For the experiment, we implement the tracker with standard GLMB filter, the hybrid track-before-detect (TBD) GLMB filter, and the GLMB filter with objects spawning. The results show improvements in tracking performance for all implemented trackers given negligible extra computational effort compared to standard GLMB filters.

Keywords: labeled RFS; RTS smoother; GLMB filter

---

1. Introduction

While single-object tracking algorithms have been studied extensively for more than half a century, multi-object tracking is currently a trending topic in signal processing society due to its extensive applications. The challenges of the multi-object tracking problem arise in the context of miss-detection, false alarms, object thinning, and appearing processes. To tackle these problems, several frameworks have been put forward in the literature such as the Joint Probabilistic Data Association (JPDA) [1], multiple hypotheses tracking [2], and recently, Random Finite Sets (RFS) [2]. In particular, RFS forms the mathematical basis of many modern multi-object filters such as Probability Hypothesis Density (PHD) filter [3–7], cardinalized PHD (CPHD) filter [8–10], multi-Bernoulli filter [11,12], the Generalized Labeled Multi-Bernoulli (GLMB) filter [13–19], and its approximation the Labeled Multi-Bernoulli (LMB) filter [20,21]. In many applications, tracking algorithms rely on the standard point measurements to update the object states; in contrast, TBD [22–25] is an alternative approach that bypasses the detection module to directly exploit the observed spatial data. This technique is introduced under the RFS framework in Reference [26] with the development of the so-called separable likelihood model and, recently, in a hybrid (combination of standard observation and separable observation models) approach in Reference [27]. In terms of system modelling, in many multi-object tracking scenarios, it is sufficient to consider object thinning and appearing processes via survivals, deaths, and instantaneous birth models. However, in many practical applications, new objects are also generated from a set or a subset of existing objects. In the context of RFS-based filtering techniques, such spawning models have been proposed for CPHD filter in References [28,29] and for GLMB filter in Reference [30]. Because these spawning models correctly reflect the physical state of the systems with spawning objects, the accuracy of the estimate is improved.

The early works on practical smoothing algorithms for single-object tracking were introduced by Bryson and Frazier [31]; by Rauch, Tung, and Striebel [32]; and subsequently by Fraser and Potter [33]. Later on, many alternative smoothing algorithms for a nonlinear dynamic model were proposed in References [34–39]. Recently, the closed-form solution for the Gaussian Mixture (GM) forward-backward smoother was derived in Reference [3]. Furthermore, the smoother for the multi-sensor tracking problem is addressed in Reference [40], while the smoothing solution for maneuvering-object tracking is presented in Reference [41]. In Reference [42], a method for joint tracking smoothing of object trajectory based on function fitting is also proposed. In a multi-object tracking context, several smoothing techniques have been put forward in the literature despite the challenge of the large smoothing state space. In particular, smoothing for PHD filter is introduced in References [43–45], while smoothing for CPHD filter and Multi-Bernoulli filter are given respectively in References [46,47]. Multiple objects can also be tracked with a fixed-lag smoother via Interacting Multiple Model (IMM) in Reference [48]. Closed-form solution for forward-backward smoothing based on GLMB RFS is introduced in Reference [49], and recently, multi-scan smoothing technique was proposed in Reference [50] with an efficient implementation based on Gibbs sampling, which can easily handle 100 scans. This is an unprecedented advance over traditional multi-scan solutions, which can only handle about 10 scans.

The labeled RFS approach has several theoretical and practical advantages over unlabeled approaches. The first is that labeled RFS filters can provide trajectory estimates naturally without heuristics, whereas this is not possible with unlabeled RFS filters; see Reference [51]. The second is that labeled RFS can provide ancestry information in a principled manner, whereas unlabeled RFS does not have the mechanism to do this (even with smoothing) [30]. The third is that labeled RFS admits analytical solutions such as the GLMB densities that are still valid RFS densities after any truncation, whereas unlabeled RFS cannot; see [51]. The fourth is that the truncation error (or error bound) for labeled RFS, such as GLMB, is available analytically, whereas this is not available for unlabeled RFS [13]. Consequently, truncation-based unlabeled RFS algorithms are heuristics [51]. Numerically, labeled RFS filters such as the GLMB have been demonstrated to be scalable in the number of objects [52], number of scans [50], and number of sensors [53]. Hence, the GLMB is a versatile class of models for multi-sensor multi-object problems.

In this paper, we introduce a tracker based on the GLMB filters and a modification of the multi-scan estimator proposed in Reference [50]. After the forward GLMB filtering stage, a pre-smooth stage is implemented to eliminate short-term tracks, which are usually initiated by false births or spawns. The threshold to prune these tracks varies depending on the tracking scenario. Subsequently, a multi-scan estimator which consists of a standard single-object filter and an Rauch–Tung–Striebel (RTS) smoother is applied on each estimated trajectory to produce smoothed estimates. As the proposed multi-scan estimator operates only on the estimated trajectories, the complexity is much lower than the full smoothing solution proposed in Reference [50]. Especially, this proposed tracker can completely eliminate track fragmentation as the multi-scan estimator estimates the entire trajectories but not single-scan multi-object states as in standard GLMB filters [13,15,16]. We demonstrate the application of the proposed tracker on both a standard measurement model and a TBD measurement model as well as tracking scenario with object spawning.

The structure of this paper is as follows. In Section 2, we provide background information on labeled RFSs, the multi-object transition kernel, the observation models, and the single-object RTS smoother for linear and nonlinear dynamic models. In Section 3, we propose the tracker based on the GLMB filters and the multi-scan estimator. In Section 4, we first show the experimental results for tracking with the standard observation model in linear and nonlinear tracking scenarios. We then show the tracking results of the proposed algorithm with a hybrid observation model. Finally, we demonstrate the performance of the algorithm on tracking biological cells in an image sequence where spawning process occurs.

2. Background

2.1. The Labeled RFS

Throughout this article, we adhere to the following notations. The set exponential is denoted as $[h(\cdot)]^X = \prod_{x \in X} h(x)$ while the inner product notation is denoted as $\langle f, g \rangle = \int f(x)g(x)dx$. The generalization of the Kronecker delta is denoted as follows:

$$\delta_Y(X) = \begin{cases} 1 & X = Y \\ 0 & X \neq Y \end{cases}$$

The set inclusion function is written as follows:

$$1_Y(X) = \begin{cases} 1 & X \subseteq Y \\ 0 & otherwise \end{cases}$$

$\mathbf{X}$ denotes the labeled set of objects, while $\mathbf{x} = (x, l)$ denotes a single labeled object, specifically, $x \in \mathbb{X}$ and $l \in \mathbb{L}$, where $\mathbb{X}$ and $\mathbb{L}$ are respectively the kinematic state space and the discrete labels space at the current time step. $\mathcal{L}$ is a label extraction function, i.e., $\mathcal{L}(\mathbf{x}) = l$ and $\mathcal{F}(\mathbf{X})$ denote sets of finite subsets of $\mathbf{X}$. The "+" sign is used to indicate the next time step when applicable.

The Finite-Set Statistics (FISST) integration is defined as follows [54]:

$$f(X)\delta X = \sum_{i=0}^{\infty} \frac{1}{i!} \int_{\mathbb{X}^i} f(\{x_1, ..., x_i\}) d(x_1, ..., x_i)$$

In multi-object tracking problem, the cardinality of object sets varies when objects enter or leave the surveillance region. As RFS is a random set of points in the sense that the number of points in the set is random and the points themselves are also random and unordered [54], a set of random objects can be naturally characterized as a RFS. Being introduced systematically for the first time in Reference [13], the labeled RFS incorporates the identities of elements into the unlabeled counterpart. Precisely, with the state space $\mathbb{X}$ and marks space $\mathbb{L}$, the labeled RFS is a marked simple point process whereas each realization has a distinct label [13,15]. The distinct label property is satisfied when $\mathbf{X}$ has the same cardinality as its labels $\mathcal{L}(\mathbf{X})$. Given this, the distinct label indicator can be written as follows [16]:

$$\Delta(\mathbf{X}) = \delta_{|\mathbf{X}|}(|\mathcal{L}(\mathbf{X})|) \tag{1}$$

The introduction of labeled RFS to the multi-object tracking problem allows direct estimation of trajectories which cannot be done previously with conventional RFS without a separate labeling scheme.

2.2. The Multi-Object Transition Kernel

In standard tracking scenario, an existing object can either survive or die in the next time step. The surviving objects are modeled as an LMB RFS with a survival probability of $p_S(x, l)$, a disappearance probability of $q_S(x, l) = 1 - p_S(x, l)$, and a spatial distribution of $f_{S+}(x_+|x, l)$. The model for such surviving objects is given as follows [13,15,16]:

$$\mathbf{f}_{S+}(\mathbf{X}_{S+}|\mathbf{X}) = \Delta(\mathbf{X})\Delta(\mathbf{X}_{S+})1_{\mathcal{L}(\mathbf{X})}(\mathcal{L}(\mathbf{X}_{S+}))[\Phi_{S+}(\mathbf{X}_{S+}|\cdot)]^{\mathbf{X}} \tag{2}$$

where

$$\Phi_{S+}(\mathbf{X}_{S+}|x, l) = \sum_{(x_+, l_+) \in \mathbf{X}_{S+}} \delta_l(l_+) p_S(x, l) f_{S+}(x_+|x, l) + [1 - 1_{\mathcal{L}(\mathbf{X}_{S+})}(l)] q_S(x, l)$$

In addition, the new birth objects can instantaneously appear at each time steps and they are modeled with LMB RFS as follows [13,15,16]:

$$\mathbf{f}_{B+}(\mathbf{X}_{B+}) = \Delta(\mathbf{X}_{B+})w_B(\mathcal{L}(\mathbf{X}_{B+}))[p_{B+}]^{\mathbf{X}_{B+}} \tag{3}$$

$$w_B(\mathcal{L}(\mathbf{X}_{B+})) = 1_{\mathbb{B}+}(\mathcal{L}(\mathbf{X}_{B+}))[1 - r_{B+}]^{\mathbb{B}_+ - \mathcal{L}(\mathbf{X}_{B+})}[r_{B+}]^{\mathcal{L}(\mathbf{X}_{B+})}$$

Furthermore, in certain scenarios, new objects can also be generated from existing objects, which leads to the need of a spawning model in order to correctly predict the state of the system at the next time step. Recently, a spawning model for GLMB filter has been proposed in Reference [30]; we introduce this model again here as follows for the sake of completeness.

For spawned objects, the naming convention is given as follows: if at time step k the label of an object is l, then the spawned labels from l at the next time step is $l_{spawn} = (l, k+1, i)$, where i is the index to distinguish between different spawned objects from the same parent. Following this convention, the set of all spawned labels in the next time step is $\mathbb{S}_+ = \mathbb{L} \times \{k+1\} \times \mathbb{N}$, where $\mathbb{N}$ is the set of positive natural numbers [30].

For each spawned object with the label $l_{spawn} \in \mathbb{S}_+(\mathcal{L}(\mathbf{x}))$, it will either exist with the probability $p_T(\mathbf{x}; l_{spawn})$ and a spatial distribution $f_{T+}(x_+|\mathbf{x}; l_{spawn})$ or not with the probability $q_T(\mathbf{x}; l_{spawn}) = 1 - p_T(\mathbf{x}; l_{spawn})$.

The density of the set $\mathbf{P}$ of new spawned objects from $\mathbf{x}$ is formulated as follows [30]:

$$\mathbf{f}_{T+}(\mathbf{P}|x, l_{spawn}) = \Delta(\mathbf{P})1_{\mathbb{S}_+(\mathcal{L}(\mathbf{x}))}(\mathcal{L}(\mathbf{P}))[\Phi_{T+}(\mathbf{P}|x; \cdot)]^{\mathbb{S}_+(\mathcal{L}(\mathbf{x}))} \tag{4}$$

where

$$\Phi_{T+}(\mathbf{P}|x; l_{spawn}) = \sum_{(x_+, l_+)} \delta_{l_{spawn}}(l_+)p_T(x, l_{spawn})f_{T+}(x_+|x, l_{spawn}) + [1 - 1_{\mathcal{L}(\mathbf{P})}(l_{spawn})]q_T(x, l_{spawn})$$

Let $\mathbf{Q}$ be a labeled set of objects spawned from $\mathbf{X}$ with $\mathcal{L}(\mathbf{Q}) \subseteq \mathbb{S}_+(\mathcal{L}(\mathbf{X}))$. As all labels sets are disjoint, the FISST convolution theorem [54] can be applied.

$$\mathbf{f}_{T+}(\mathbf{Q}|\mathbf{X}) = \Delta(\mathbf{Q})1_{\mathbb{S}_+(\mathcal{L}(\mathbf{X}))}(\mathcal{L}(\mathbf{Q}))[\Phi_{T+}(\mathbf{Q}|\cdot)]^{\mathbf{X}} \tag{5}$$

where

$$\Phi_{T+}(\mathbf{Q}|\mathbf{x}) = [\Phi_{T+}(\mathbf{Q} \cap (\mathbb{X} \times \mathbb{S}_+(\mathcal{L}(\mathbf{x}))|x; \cdot)]^{\mathbb{S}_+(\mathcal{L}(\mathbf{x}))}$$

As new birth objects (given in Equation (3)) are independent of the previous time step objects, the overall transition model is given as follows:

$$\mathbf{f}(\mathbf{X}_+|\mathbf{X}) = \mathbf{f}_{S+}(\mathbf{X}_{S+}|\mathbf{X})\mathbf{f}_{T+}(\mathbf{Q}|\mathbf{X})\mathbf{f}_{B+}(\mathbf{X}_{B+}) \tag{6}$$

As the spawned objects depend upon the objects from previous time steps, the prediction step of the filtering stage needs to be done in a joint manner to capture the objects' dependency. As a result, approximation is needed to convert the joint object distribution to a standard GLMB density for each time step in order to keep the algorithm tractable.

In the scenario where the spawning process is not present, the multi-object transition kernel is reduced to the following:

$$\mathbf{f}(\mathbf{X}_+|\mathbf{X}) = \mathbf{f}_{S+}(\mathbf{X}_{S+}|\mathbf{X})\mathbf{f}_{B+}(\mathbf{X}_{B+}) \tag{7}$$

2.3. The Multi-Object Observation Models

In the RFS multi-object tracking framework, given a set of measurements $Z = \{z_{1:|Z|}\}$, we have a standard observation model of the following form: [54]

$$g(Z|\mathbf{X}) \propto \sum_{\theta \in \Theta(\mathcal{L}(\mathbf{X}))} \prod_{(x,l) \in \mathbf{X}} \psi_Z^{(\theta(l))}(x,l) \tag{8}$$

where

$$\psi_Z^{(\theta(l))}(x,l) = \delta_0(\theta(l)) q_D(x,l) + (1 - \delta_0(\theta(l))) \frac{p_D(x,l) g(z_{\theta(l)}|x,l)}{\kappa(z_{\theta(l)})}$$

$\kappa(\cdot)$ is the clutter intensity, $p_D(\cdot)$ and $q_D(\cdot)$ are respectively the detection and miss-detection probabilities, $g(z|x,l)$ is the likelihood that (x,l) generates measurement z, $\theta : \mathbb{L} \to \{0 : |Z|\}$ is a positive 1-1 map, and Θ is the entire set of such mappings.

For image observation, with the assumption that object template $T(\cdot)$ is not overlapped, i.e., $T(x_1) \neq T(x_2)$ given $x_1 \neq x_2$, the separable likelihood is given by the following [26]:

$$g(y|\mathbf{X}) = f_B(y) \prod_{\mathbf{x} \in \mathbf{X}} g_y(\mathbf{x}) \tag{9}$$

where y denotes the observed image, f_B denotes the likelihood of the entire set of $\mathbf{X}$, and $g_y(x)$ denotes the likelihood of a single object in the observed image. The designs of f_B and g_y vary according to the applications, characteristics of observed image, and object appearances.

First introduced in Reference [27], the concept of a hybrid TBD observation model takes advantage of both standard and separable likelihood models. Intuitively, while detected objects can be updated by the associated point measurements, the miss-detected objects can be updated directly from the image observation. This intuition can be described mathematically by defining the following:

$$\sigma_T(T(y)|x,l) \triangleq \frac{g_T(T(y)|x,l)}{g_T(T(y)|\emptyset)} \tag{10}$$

The hybrid likelihood can then be written as follows [27]:

$$g(y|\mathbf{X}) \propto \sum_{\theta \in \Theta(\mathcal{L}(\mathbf{X}))} \prod_{(x,l) \in \mathbf{X}} \varphi_y^{(\theta(l))}(x,l) \tag{11}$$

where

$$\varphi_y^{(\theta(l))}(x,l) - \psi^{(\theta(l))}(x,l|Z)[\sigma_T(T(y)|x,l)]^{\delta_0 \theta(l)}$$

2.4. The Single Object RTS Smoother

Given a set single object observation $\{z_{1:N}\}$, where $N \leq K$ with K is the total number of tracking time steps, the smoothed density of an object state at time $k \leq N$, $p(x_k|z_{1:N})$, is obtained as follows [36].

Initially, let the joint distribution of x_k and x_{k+1} be rewritten as follows:

$$p(x_k, x_{k+1}|z_{1:k}) = p(x_{k+1}|x_k) p(x_k|z_{1:k}) \tag{12}$$

Then, the distribution of x_k given x_{k+1} and $z_{1:k}$ is given as follows:

$$p(x_k|x_{k+1}, z_{1:k}) = \frac{p(x_k, x_{k+1}|z_{1:k})}{p(x_{k+1}|z_{1:k})} \tag{13}$$

where $p(x_{k+1}|z_{1:k}) = \int p(x_{k+1}|x_k) p(x_k|z_{1:k}) dx_k$

From the Markov state-space model, we have the following property: $p(x_k|x_{k+1}, z_{1:N}) = p(x_k|x_{k+1}, z_{1:k})$. Hence, we have the following:

$$p(x_k|x_{k+1}, z_{1:N}) = \frac{p(x_k, x_{k+1}|z_{1:k})}{p(x_{k+1}|z_{1:k})} \tag{14}$$

Then, the joint distribution of x_k and x_{k+1} given the measurements set $z_{1:N}$ is given as follows:

$$p(x_k, x_{k+1}|z_{1:N}) = p(x_k|x_{k+1}, z_{1:N})p(x_{k+1}|z_{1:N}) \tag{15}$$

Finally, the smoothed density of state x_k can then be obtained via the marginalization step as follows:

$$p(x_k|z_{1:N}) = \int p(x_k|x_{k+1}, z_{1:N})p(x_{k+1}|z_{1:N})dx_{k+1} \tag{16}$$

3. The Proposed Tracker

3.1. The Filtering Stage

For this tracker, we assume Gaussian distribution for the dynamic state of each object. At this first stage, the tracker carries out a standard multi-object filtering process to obtain the forward estimated labels and the measurements to label association history. In this subsection, we provide the forward filtering steps for both the GLMB filter (with standard measurements and hybrid measurements observations) and GLMB filter with object spawning.

3.1.1. GLMB Filter without Objects Spawning

The procedure to estimate the state of a set of objects with the standard GLMB filter without including the spawning model in the transition kernel is given as follows.

Given a GLMB prior [16]

$$\pi(\mathbf{X}) = \Delta(\mathbf{X}) \sum_{(I,\xi)\in\mathcal{F}(\mathbb{L})\times\Xi} \omega^{(I,\xi)}\delta_I(\mathcal{L}(\mathbf{X}))[p^{(\xi)}]^{\mathbf{X}} \tag{17}$$

and the standard observation model as in Equation (8), the filtering density in the next time step is given by the following [16]:

$$\pi_{Z_+}(\mathbf{X}) \propto \Delta(\mathbf{X}) \sum_{I,\xi,I_+,\theta_+} \omega^{(I,\xi)}\omega_{Z_+}^{(I,\xi,I_+,\theta_+)}\delta_{I_+}(\mathcal{L}(\mathbf{X}))[p_{Z_+}^{(\xi,\theta_+)}]^{\mathbf{X}} \tag{18}$$

where $I \in \mathcal{F}(\mathbb{L})$, $\xi \in \Xi$, $I_+ \in \mathcal{F}(\mathbb{L}_+)$, $\theta_+ \in \Theta_+$ where ξ is the tracks to measurement association history and Ξ is the entire space of ξ.

$$\omega_{Z_+}^{(I,\xi,I_+,\theta_+)} = 1_{\Theta_+(I_+)}(\theta_+)[1-\bar{P}_S^{(\xi)}]^{I-I_+}[\bar{P}_S^{(\xi)}]^{I\cap I_+}[1-r_{B+}]^{\mathbb{B}_+-I_+}[r_{B+}]^{\mathbb{B}_+\cap I_+}[\bar{\psi}_{Z_+}^{(\xi,\theta_+)}]^{I_+}$$

$$\bar{P}_S^{(\xi)}(l) = \langle p^{(\xi)}(\cdot,l), p_S(\cdot,l)\rangle$$

$$\bar{\psi}_{Z_+}^{(\xi,\theta_+)}(l_+) = \langle \bar{p}_+^{(\xi)}(\cdot,l_+), \psi_{Z_+}^{(\theta_+(l_+))}(\cdot,l_+)\rangle$$

$$p_{Z_+}^{(\xi,\theta_+)}(x_+,l_+) = \frac{\bar{p}_+^{(\xi)}(x_+,l_+)\psi_{Z_+}^{(\theta_+(l_+))}(x_+,l_+)}{\bar{\psi}_{Z_+}^{(\xi,\theta_+)}(l_+)}$$

$$\bar{p}_{+}^{(\xi)}(x_+,l_+) = 1_{\mathbb{L}}(\{l_+\})\frac{\langle p_S(\cdot,l_+)f_{S+}(x_+|\cdot,l_+),p^{(\xi)}(\cdot,l_+)\rangle}{\bar{P}_S^{(\xi)}(l_+)} + 1_{\mathbb{B}_+}(\{l_+\})p_{B+}(x_+,l_+)$$

In tracking scenarios where raw spatial detection are also available, the hybrid model in Equation (11) can be used to replace the standard observation model with the probability of miss-detection being scaled by the spatial observation likelihood, i.e., given the GLMB prior as in Equation (17). The filtering density is then given as follows [27]:

$$\pi_{y_+}(\mathbf{X}) \propto \Delta(\mathbf{X}) \sum_{I,\xi,I_+,\theta_+} \omega^{(I,\xi)}\omega_{y_+}^{(I,\xi,I_+,\theta_+)}\delta_{I_+}(\mathcal{L}(\mathbf{X}))[p_{y_+}^{(\xi,\theta_+)}]^{\mathbf{X}} \tag{19}$$

where

$$\omega_{y_+}^{(I,\xi,I_+,\theta_+)} = 1_{\Theta_+(I_+)}(\theta_+)[1-\bar{P}_S^{(\xi)}]^{I-I_+}[\bar{P}_S^{(\xi)}]^{I\cap I_+}[1-r_{B+}]^{\mathbb{B}_+-I_+}[r_{B+}]^{\mathbb{B}_+\cap I_+}[\bar{\varphi}_{y_+}^{(\xi,\theta_+)}]^{I_+}$$

$$\bar{\varphi}_{y_+}^{(\xi,\theta_+)}(l_+) = \langle \bar{p}_+^{(\xi)}(\cdot,l_+),\varphi_{y_+}^{(\theta_+(l_+))}(\cdot,l_+)\rangle$$

$$p_{y_+}^{(\xi,\theta_+)}(x_+,l_+) = \frac{\bar{p}_+^{(\xi)}(x_+,l_+)\varphi_{y_+}^{(\theta_+(l_+))}(x_+,l_+)}{\bar{\varphi}_{y_+}^{(\xi,\theta_+)}(l_+)}$$

3.1.2. GLMB Filter with Objects Spawning

For the prior density which is a GLMB density as in Equation (17) and the transition kernel defined in Equation (6), by applying the joint predict–update approach, a proposal density can be written as follows [30]:

$$\tilde{\pi}_+(\mathbf{X}_+|Z_+) \propto \Delta(\mathbf{X}_+) \sum_{I,\xi,I_+,\theta_+} \omega^{(I,\xi)}\tilde{\omega}_{Z_+}^{(I,\xi,I_+,\theta_+)}\delta_{I_+}(\mathcal{L}(\mathbf{X}_+))[\tilde{p}_{Z_+}^{(\xi,\theta_+)}]^{\mathbf{X}_+} \tag{20}$$

$$\tilde{\omega}_{Z_+}^{(I,\xi,I_+,\theta_+)} = [r_{B+}]^{\mathbb{B}_+\cap I_+}[1-r_{B+}]^{\mathbb{B}_+-I_+}[\bar{p}_S]^{I\cap I_+}[1-\bar{p}_S]^{I-I_+}[\bar{p}_T]^{\mathbb{S}_+\cap I_+}[1-\bar{p}_T]^{\mathbb{S}_+-I_+},$$

$$\tilde{p}_{Z_+}^{(\xi,\theta_+)}(x_+,l_+) = \frac{\tilde{p}_+^{(\xi)}(x_+,l_+)\psi_{Z_+}^{(\theta_+(l_+))}(x_+,l_+)}{\tilde{\psi}_{Z_+}^{(\xi,\theta_+)}(l_+)},$$

$$\tilde{p}_+^{(\xi)}(x_+,l_+) = 1_{\mathbb{B}_+}(\{l_+\})p_{B+}(x_+,l_+) + 1_{\mathbb{L}}(\{l_+\})\tilde{p}_S^{(\xi)}(x_+,l_+) + 1_{\mathbb{S}}(\{l_+\})\tilde{p}_T^{(\xi)}(x_+,l_+),$$

$$\tilde{p}_S^{(\xi)} = \frac{\langle p_S(\cdot,l_+)f_{S+}(x_+|\cdot,l),p^{(\xi)}(\cdot,l_+)\rangle}{\bar{p}_S^{(\xi)}(l_+)},$$

$$\tilde{p}_T^{(\xi)} = \frac{\langle p_T(l_+)f_{T+}(x_+|\cdot,l),p^{(\xi)}(\cdot,l)\rangle}{\bar{p}_T^{(\xi)}(l_+)}$$

$$\bar{p}_S^{(\xi)} = \langle p^{(\xi)}(\cdot,l),p_S(l_+)\rangle,$$

$$\bar{p}_T^{(\xi)} = \langle p^{(\xi)}(\cdot,l), p_T(l_+)\rangle,$$

$$\tilde{\psi}_{Z_+}^{(\xi,\theta_+)}(l_+) = \langle \tilde{p}_+^{(\xi)}(\cdot,l_+), \psi_{Z_+}^{(\theta_+(l_+))}(\cdot,l_+)\rangle.$$

From this proposal density, Gibbs' sampler is applied to select high weight hypotheses. These hypotheses are subsequently used to form a standard GLMB density [30]:

$$\hat{\pi}(\mathbf{X}_+|Z_+) = \Delta(\mathbf{X}_+)\sum_{I,\xi,I_+,\theta_+} \delta_{I_+}(\mathcal{L}(\mathbf{X}_+))\hat{\omega}_+^{(I,\xi,I_+,\theta_+)}(Z_+)[p_{B+}\psi_+^{(\theta_+)}(\cdot|Z_+)]^{\mathbf{X}_{B+}} \frac{[\hat{p}_+^{(I,\xi,I_+,\theta_+)}(\cdot|Z_+)]^{\mathbf{X}_{S+}\cup\mathbf{X}_{T+}}}{[\bar{p}_+^{(I,\xi,I_+,\theta_+)}(\cdot|Z_+)]^{I_+}} \tag{21}$$

$$\hat{\omega}_+^{(I,\xi,I_+,\theta_+)}(Z_+) = \frac{\omega_+^{(I,\xi)}(I_+)[\bar{p}_+^{(I,\xi,I_+,\theta_+)}(\cdot|Z_+)]^{I_+}}{\sum_{I,\xi,I_+,\theta_+} \omega_+^{(I,\xi)}(I_+)[\bar{p}_+^{(I,\xi,I_+,\theta_+)}(\cdot|Z_+)]^{I_+}}$$

$$\hat{p}^{(I,\xi,I_+,\theta_+)}(x_+,l_+|Z_+) \triangleq 1_{I_+}(\{l_+\})\int p_+^{(I,\xi,\theta_+)}(\{(x_+,l_+),(x_{1,+},l_{1,+}),...,(x_{n,+},l_{n,+})\}|Z_+)d(x_{1,+},...,x_{n,+})$$

$$\bar{p}^{(I,\xi,I_+,\theta_+)}(x_+,l_+|Z_+) \triangleq 1_{\mathbb{B}_+}(\{l_+\})\langle p_{B+}(\cdot,l_+), \psi_+^{(\theta_+)}(\cdot|Z_+)\rangle + (1-1_{\mathbb{B}_+}(\{l_+\}))\langle \hat{p}_{Z_+}^{(I,\xi,I_+,\theta_+)}(x_+,l_+),1\rangle$$

3.2. GLMB Multi-Scan Estimator

The concept of a multi-scan estimator is introduced in Reference [50]. Given a multi-scan GLMB from time step j to k, the cardinality distribution of the number of trajectories is given as follows:

$$\Pr(|\mathcal{L}(\mathbf{X}_{j:k})| = n) = \sum_{\xi,I_{j:k}} \delta_n[|I_{j:k}|]w_{j:k}^{(\xi)}(I_{j:k}) \tag{22}$$

One possible form of a multi-scan estimator is to determine the component with the highest weight $w_{j:k}^{(\xi)}(I_{j:k})$ given that it has the most probable cardinality by maximizing Equation (22). The expected trajectory estimate can then be computed from $p_{j:k}^{(\xi)}(\cdot,l)$ for each $l \in I_{j:k}$.

In this work, we proposed modifications to the multi-scan estimator in Reference [50], which can eliminate track fragmentation and improve localization performance. The set of all estimated trajectories is updated at each time step via the most significant hypothesis with the most probable cardinality in the GLMB density. At the time step when state estimation is required, the information of estimated trajectories is passed into the estimator. At this stage, trajectories pruning is applied to eliminate short-term tracks. Subsequently, standard filtering and RTS smoothing techniques are applied on each trajectory to produce smoothed state estimates. The significance of this estimator is that it allows the application of smoothing techniques to improve the tracking accuracy while completely eliminates track fragmentation as the entire trajectory is estimated as a whole. In addition, as the complexity of the estimator depends only on the number of estimated tracks, the additional computational effort of the estimator is negligible compared to GLMB filtering. The detailed implementation of the estimator is given as in following subsections.

3.2.1. Estimating the Trajectories

Given the GLMB density at the end of each filtering cycle, the GLMB filter estimate is the result of the maximum posteriori estimate of the cardinality with the means of the object states being conditioned on the estimated cardinality [15]. Given that the possible highest number of tracked objects is N_{max}, the cardinality distribution of the the objects set over a finite set of hypotheses $\{(I,\xi)_h\}_{h=1:H}$ is written as follows:

$$\rho(n)|_{n=0:N_{max}} = \sum_{\mathfrak{H}\in\{(I,\xi)_h\}_{h=1:H}} \omega^{(\mathfrak{H})}\delta_n(|I^{(\mathfrak{H})}|) \tag{23}$$

The estimated cardinality is given as follows:

$$\hat{N} = argmax(\rho) \tag{24}$$

The estimated hypothesis is as follows:

$$\hat{\mathfrak{H}} = argmax_{(\mathfrak{H})}\omega^{(\mathfrak{H})}\delta_{\hat{N}}(|I^{(\mathfrak{H})}|) \tag{25}$$

The information from the filtering stage needs to be captured to facilitate the multi-scan estimator. At this stage, we represent a set of estimated trajectories at time k with a set of tuples defined as $\hat{\mathfrak{T}}_k \triangleq \{(\hat{l}_1^k, \hat{b}_{\hat{l}_1^k}, \hat{\xi}_{\hat{l}_1^k}), ..., (\hat{l}_{\hat{N}}^k, \hat{b}_{\hat{l}_{\hat{N}}^k}, \hat{\xi}_{\hat{l}_{\hat{N}}^k})\}$, where $\hat{l}_{\hat{n}}^k$ is the label of estimated trajectory n at time k, $\hat{b}_{\hat{l}_{\hat{n}}^k}$ is its corresponding initial birth state (including the time of birth and initial kinematic state), and $\hat{\xi}_{\hat{l}_{\hat{n}}^k}$ is the corresponding association history. In addition, we also have set of tuples for all estimated trajectories $\hat{\mathfrak{T}}$ from time step 1 to current time step k. This set of tuples is updated at the end of each filtering time step via updating the association history and initial birth state of existing trajectories and adding new tuples to the set if the trajectories are new. The procedure to update the tuples set is given in Algorithm 1.

Algorithm 1 Updating trajectories tuples

Input: $\hat{\mathfrak{T}}_k = \{(\hat{l}_1^k, \hat{b}_{\hat{l}_1^k}, \hat{\xi}_{\hat{l}_1^k}), ..., (\hat{l}_{\hat{N}}^k, \hat{b}_{\hat{l}_{\hat{N}}^k}, \hat{\xi}_{\hat{l}_{\hat{N}}^k})\}$, $\hat{\mathfrak{T}} = \{(\hat{l}_1, \hat{b}_{\hat{l}_1}, \hat{\xi}_{\hat{l}_1}), ..., (\hat{l}_N, \hat{b}_{\hat{l}_N}, \hat{\xi}_{\hat{l}_N})\}$
Output: The updated trajectories tuples set $\hat{\mathfrak{T}}$

for $n = 1$ to $\hat{N}$
    if $\hat{l}_n^k \in \{\hat{l}_1, ..., \hat{l}_N\}$
        Replace the tuple of label $\hat{l}_n^k$ in $\hat{\mathfrak{T}}$ with $(\hat{l}_n^k, \hat{b}_{\hat{l}_n^k}, \hat{\xi}_{\hat{l}_n^k})$
    else
        $\hat{\mathfrak{T}} = \hat{\mathfrak{T}} \cup (\hat{l}_n^k, \hat{b}_{\hat{l}_n^k}, \hat{\xi}_{\hat{l}_n^k})$
    end
end

3.2.2. Trajectories Pruning

For a set of estimated trajectories tuples from filtering stage $\hat{\mathfrak{T}} = \{(\hat{l}_1, \hat{b}_{\hat{l}_1}, \hat{\xi}_{\hat{l}_1}), ..., (\hat{l}_N, \hat{b}_{\hat{l}_N}, \hat{\xi}_{\hat{l}_N})\}$ the lifetime of a trajectory with label $\hat{l}_n$ is the length of the corresponding association history, which is given as follows:

$$\tau(l_n) = f_{length}(\hat{\xi}_{\hat{l}_n}) \tag{26}$$

where $f_{length}(\cdot)$ is the function that determines the length of the vector in its argument. If the length of a track is shorter than the threshold value τ_t, i.e., $\tau(l_n) < \tau_t$, this trajectory will be removed from the set of estimated trajectories.

3.2.3. Numerical Implementation of Single-Object Smoother

For completeness, we outline here the detailed numerical implementation of the single-object RTS smoother for both linear and nonlinear dynamic models with Gaussian assumption on the distribution of the states.

Given a linear dynamic model of the form

$$x_+ = Fx + q, z = Hx + r$$

where x is the system state, F is the linear transformation matrix, H is the linear observation matrix, q and r are respectively the process and observation Gaussian noise, and z is the current time step measurement, the backward smoothing step over an interval $N \leq K$ (where K is the total number of tracking time steps) can be implemented with the standard RTS Smoother [32]. The details of the RTS smoother is given in Algorithm 2, where the superscript s denotes the smoothed results.

Algorithm 2 Single-object Rauch–Tung–Striebel (RTS) smoother

Input: The filtered mean and covariance $\{x_k, P_k\}_{k=1:N}, F, Q$
Output: The smoothed mean and covariance $\{x_k^s, P_k^s\}_{k=1:N}$

Initialization: $x_N^s = x_N$ and $P_N^s = P_N$
for $k = N - 1$ down to 1
$\bar{x}_{k+1} = Fx_k$
$\bar{P}_{k+1} = FP_kF^T + Q$
$D = P_{k+1}F(\bar{P}_{k+1})^{-1}$
$x_k^s = x_k + D(x_{k+1}^s - \bar{x}_{k+1})$
$P_k^s = P_k + D(P_{k+1}^s - \bar{P}_{k+1})D^T$
end

For a nonlinear dynamic model, the RTS smoother can also be applied via the unscented transformation [39]. Given the dynamic model

$$x_+ = f(x, q), y = h(x, r)$$

where f is the nonlinear state transition function and h is the nonlinear observation function, other variables are interpreted the same as in the linear model; the smoothed results can be inferred via the Unscented RTS (URTS) smoother [36]. The smoothing procedure is presented in Algorithm 3, and the readers are referred to Reference [39] for the detailed implementation of the unscented transform. Compared to the Sequential Monte Carlo method, unscented transform is less computationally expensive as the number of sigma points to approximate a Gaussian distribution is much lower than the number of particles to represent the entire density.

3.2.4. Forward Filtering-Backward Smoothing of Trajectories

In this step, by using the measurement association history, the initial birth information (the state and the time at birth) in the estimated trajectories tuples set, and the measurements set, we apply standard single-object filtering and backward RTS smoothing techniques to produce a set of smoothed distributions of the trajectories. In this work, spatial distributions of tracks are assumed to be Gaussian distributed; hence, the estimated spatial distribution of track labeled l at time k is represented by the mean m_k^l and the covariance P_k^l. The details of the procedure to produce the tracks distributions are given in Algorithm 4. The SingleObjectPrediction and SingleObjectUpdate functions are chosen according to the dynamic model, which can be Kalman prediction and Kalman update or their nonlinear variances. The linearity of the system also determines the SingleObjectSmoothing function, which takes the form of either Algorithm 2 or Algorithm 3 to smooth each individual trajectory. The output of the algorithm is the smoothed spatial distributions of all estimated trajectories, which is $\{m_k^{\hat{l}_n}, P_k^{\hat{l}_n}\}_{k_i^{\hat{l}_n}:k_e^{\hat{l}_n}}$. From this set of distributions, the mean values can be extracted to be used as the estimated states of the trajectories.

Algorithm 3 Single-object Unscented RTS (URTS) smoother

Input: The filtered mean and covariance $\{x_k, P_k\}_{k=1:N}$, $f(x_+|x)$, Q
Output: The smoothed mean and covariance $\{x_k^s, P_k^s\}_{k=1:N}$

Initialization: $x_N^s = x_N$ and $P_N^s = P_N$
for $k = N-1$ down to 1
  $\{W_{i-1}^{(m)}, W_{i-1}^{(c)}, [\tilde{X}_i^x; \tilde{X}_i^q]\} = \text{UnscentedTransform}\ (x_k, P_k, Q)$
  $\tilde{X}_{i+} = f(\tilde{X}_i^x, \tilde{X}_i^q)$
  $\bar{x}_{k+1} = \sum_i W_{i-1}^{(m)} \tilde{X}_{i+}$
  $\bar{P}_{k+1} = \sum_i W_{i-1}^{(c)} (\tilde{X}_{i+} - \bar{x}_{k+1})(\tilde{X}_{i+} - \bar{x}_{k+1})^T$
  $\bar{C}_{k+1} = \sum_i W_{i-1}^{(c)} (\tilde{X}_i^x - x_k)(\tilde{X}_{i+} - \bar{x}_{k+1})^T$
  $D = \bar{C}_{k+1} \left(\bar{P}_{k+1}\right)^{-1}$
  $x_k^s = x_k + D\left(x_{k+1}^s - \bar{x}_{k+1}\right)$
  $P_k^s = P_k + D\left(P_{k+1}^s - \bar{P}_{k+1}\right) D^T$
end

While the advantages are mentioned previously, this estimator is also subjected to certain drawbacks in challenging tracking scenarios. First, depending on the nature of the problem, the user needs to set an appropriate pruning threshold τ_l to prevent the estimator from deleting correct trajectories, especially when track identity switching is severe. Second, as the estimator relies on the latest hypothesis to produce estimates, the more this hypothesis deviates from the truth, the more inaccurate the entire estimation. In addition, in the case that wrong new tracks keep appearing in the set of trajectory estimates, overestimating of the number of tracks is also possible. However, the benefit from track fragmentation reduction and improvement of tracking accuracy given negligible computational effort is much more than the risk of incorrectly estimating the number of tracks, and the following simulation results are a strong demonstration of the benefits of our proposed tracker.

Algorithm 4 Trajectory forward filtering-backward smoothing

Input: $\hat{\mathfrak{T}} = \{(\hat{l}_1, \hat{b}_{\hat{l}_1}, \hat{\xi}_{\hat{l}_1}), ..., (\hat{l}_N, \hat{b}_{\hat{l}_N}, \hat{\xi}_{\hat{l}_N})\}, \{Z_1, ..., Z_K\}$
Output: The estimated trajectories $\{\{m_k^{\hat{l}_1}, P_k^{\hat{l}_1}\}_{k_i^{\hat{l}_1}:k_e^{\hat{l}_1}}, ..., \{m_k^{\hat{l}_N}, P_k^{\hat{l}_N}\}_{k_i^{\hat{l}_N}:k_e^{\hat{l}_N}}\}$

for $n = 1$ to N
  Initialize $\{\bar{m}^{l_n}_{k_i^{\hat{l}_n}}, \bar{P}^{l_n}_{k_i^{\hat{l}_n}}\}$ from the initial birth $\hat{b}_{\hat{l}_n}$
  $\{\tilde{m}^{l_n}_{k_i^{\hat{l}_n}}, \tilde{P}^{l_n}_{k_i^{\hat{l}_n}}\} = \text{SingleObjectUpdate}\ (\bar{m}^{l_n}_{k_i^{\hat{l}_n}}, \bar{P}^{l_n}_{k_i^{\hat{l}_n}}, z^{k_i^{\hat{l}_n}}_{\xi^{k_i^{\hat{l}_n}}_{\hat{l}_n}})$
  for k from $k_i^{\hat{l}_n} + 1$ to $k_e^{\hat{l}_n}$
    $\{\bar{m}_k^{l_n}, \bar{P}_k^{l_n}\} = \text{SingleObjectPrediction}\ (\tilde{m}_{k-1}^{l_n}, \tilde{P}_{k-1}^{l_n})$
    if $\hat{\xi}_{\hat{l}_n}^k = 0$
      $\{\tilde{m}_k^{l_n}, \tilde{P}_k^{l_n}\} = \{\bar{m}_k^{l_n}, \bar{P}_k^{l_n}\}$
    else
      $\{\tilde{m}_k^{l_n}, \tilde{P}_k^{l_n}\} = \text{SingleObjectUpdate}\ (\bar{m}_k^{l_n}, \bar{P}_k^{l_n}, z^k_{\hat{\xi}^k_{\hat{l}_n}})$
    end
  end
  $\{m_k^{\hat{l}_n}, P_k^{\hat{l}_n}\}_{k_i^{\hat{l}_n}:k_e^{\hat{l}_n}} = \text{SingleObjectSmoothing}\ (\{\tilde{m}_k^{\hat{l}_n}, \tilde{P}_k^{\hat{l}_n}\}_{k_i^{\hat{l}_n}:k_e^{\hat{l}_n}})$
end

4. Experimental Results

4.1. Simulation Results

4.1.1. Linear Dynamic Model

In this experiment, we use a constant velocity model for the dynamic of the system. The state vector consists of information regarding the planar position and the velocity of the objects, which is $x_k = \left[p_x, p_y, \dot{p}_x, \dot{p}_y\right]^T$; while the measurement vector contains the position of the object, which is $z_k = [z_x, z_y]^T$. The transition and observation models are given respectively as follows:

$$f_+(x_+|x) = \mathcal{N}(x_+; Fx, Q)$$

$$h(z|x) = \mathcal{N}(z; Hx, R)$$

where $F = \begin{bmatrix} I_2 & \Delta I_2 \\ 0_2 & I_2 \end{bmatrix}$, $Q = \sigma_v^2 \begin{bmatrix} \frac{\Delta^4}{4} I_2 & \frac{\Delta^3}{2} I_2 \\ \frac{\Delta^3}{2} I_2 & \Delta^2 I_2 \end{bmatrix}$, $H = \begin{bmatrix} I_2 & 0_2 \end{bmatrix}$, $R = \sigma_\epsilon^2 I_2$. Particularly, in this experiment, we set $\sigma_v = 5\,\mathrm{m/s}$ and $\sigma_\epsilon = 15$ m.

The surveillance region is the $[-1000, 1000]$ m $\times$ $[-1000, 1000]$ m area, the total time step is $K = 100$, and $\Delta = 1$. The ground truth plot for this experiment is given in Figure 1. The surviving probability is set to $p_S = 0.99$, and the detection probability is $p_D = 0.95$. Clutter rate is set to 66 false alarms per scan. The birth probability is set to $r_B = 0.03$. The states of expected births are $m_B^{(1)} = [0.1, 0, 0.1, 0]^T$, $m_B^{(2)} = [400, 0, -600, 0]^T$, $m_B^{(3)} = [-800, 0, -200, 0]^T$, and $m_B^{(4)} = [-200, 0, 800, 0]^T$. The covariance matrix at birth is $P_B = diag([10, 10, 10, 10])$. The number of hypotheses for GLMB filter is capped at 20,000 components. In this experiment, we smooth the entire tracking interval from $k = 1$ to $k = K$. The threshold for the smoother to prune the track is set to $\tau_t = 3$ time steps.

We conduct the experiment over 100 Monte Carlo runs. The means of the estimated Optimal Subpattern Assignment (OSPA) error [55] and OSPA^2 error [52,56] are given respectively in Figures 2 and 3. Figure 4 shows the GLMB filter and proposed tracker-estimated cardinality of objects set for each time step along with the true values.

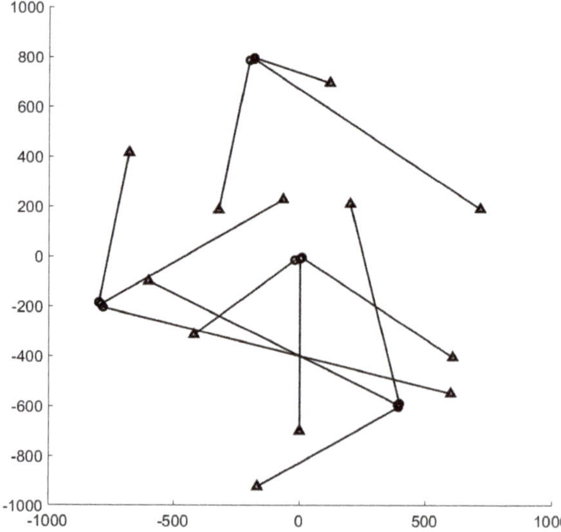

Figure 1. Ground truth for linear dynamic scenario (circle: track start position, triangle: track end position).

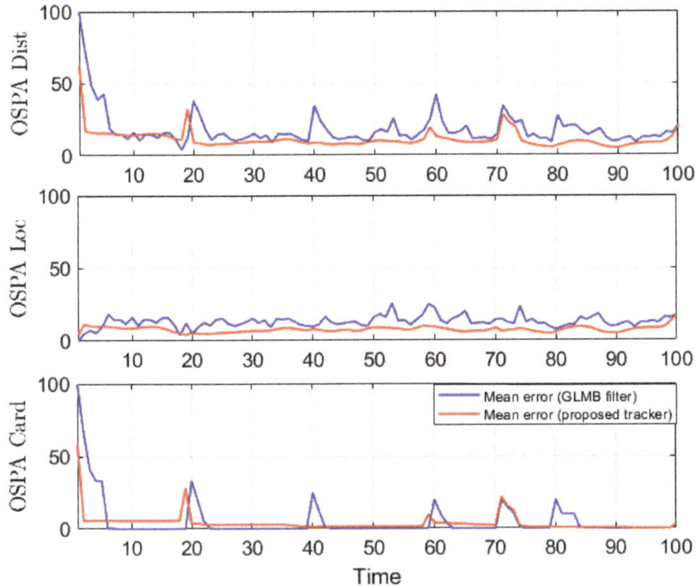

Figure 2. OSPA error for linear dynamic scenario.

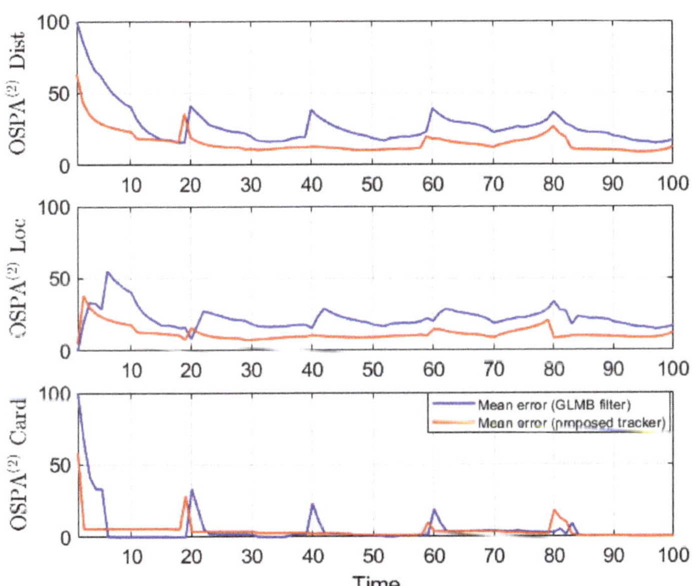

Figure 3. OSPA^2 error for linear dynamic scenario.

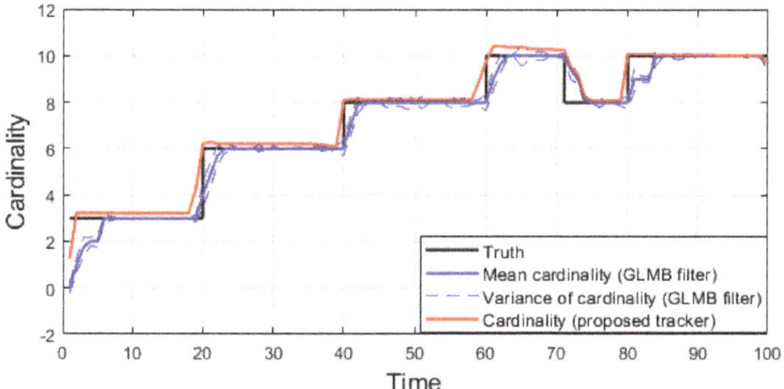

Figure 4. Estimated cardinality for linear dynamic scenario.

4.1.2. Nonlinear Dynamic Model

For the demonstration of the nonlinear tracking scenario, we use a constant turn model with 5-D state vector $x_k = \left[p_x, p_y, \dot{p}_x, \dot{p}_y, \omega\right]^T$, where ω is the object's turn rate. The transition density is given as follows:

$$f_+(x_+|x) = \mathcal{N}(x_+; F(\omega)x, Q)$$

where $F\left(\begin{bmatrix} p_x \\ \dot{p_x} \\ p_y \\ \dot{p_y} \\ \omega \end{bmatrix}\right) = \begin{bmatrix} 1 & \frac{\sin(\omega\Delta)}{\omega} & 0 & -\frac{1-\cos(\omega\Delta)}{\omega} & 0 \\ 0 & \cos(\omega\Delta) & 0 & -\sin(\omega\Delta) & 0 \\ 0 & \frac{1-\cos(\omega\Delta)}{\omega} & 1 & \frac{\sin(\omega\Delta)}{\omega} & 0 \\ 0 & \sin(\omega\Delta) & 0 & \cos(\omega\Delta) & 0 \\ 0 & 0 & 0 & 0 & 1 \end{bmatrix}$, $Q_\zeta = \begin{bmatrix} \sigma_\omega^2 GG^T & 0 \\ 0 & \sigma_v^2 \end{bmatrix}$ and $G = \begin{bmatrix} \Delta^2/2 & 0 \\ \Delta & 0 \\ 0 & \Delta^2/2 \\ 0 & \Delta \end{bmatrix}$.

In this experiment, we set $\sigma_\omega = \pi/180$ rad/s and $\sigma_v = 5$ m/s. The observation model is given as the bearing and range detection of the 2D vector $z_k = [\theta, r]^T$ with $\sigma_\theta = \pi/90$ rad and $\sigma_r = 5$ m.

The surveillance region is the half disc of the radius 2000 m with $K = 100$ time steps and $\Delta = 1$. The ground truth for this experiment is given in Figure 5. The surviving probability is set to $p_S = 0.99$ and the detection probability is $p_D = 0.95$. Clutter rate is set to 66 false alarms per scan. The expected birth states are $m_B^{(1)} = [-1500, 0, 250, 0, \pi/180]^T$, $m_B^{(2)} = [-250, 0, 1000, 0, \pi/180]^T$, $m_B^{(3)} = [250, 0, 750, 0, \pi/180]^T$, and $m_B^{(4)} = [1000, 0, 1500, 0, \pi/180]^T$ with $r_B = 0.02$, and the birth covariance is $P_B = diag([50, 50, 50, 50, \pi/30])$. The number of hypotheses is also capped at 20,000 components. The smoothing interval is the entire tracking sequence from $k = 1$ to $k = K$. We also set the track pruning threshold for the smoother to 3 time steps in this experiment.

For this scenario, we also test the performance of the tracker over 100 Monte Carlo runs. The means of OSPA error and OSPA^2 error of the estimates are plotted in Figures 6 and 7, respectively, while the set cardinality is shown in Figure 8.

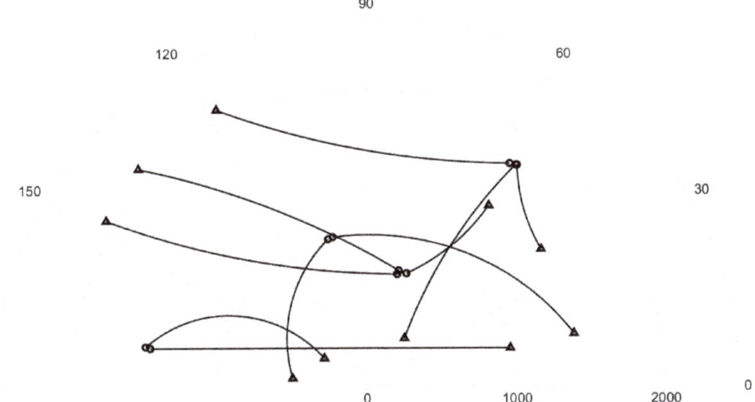

Figure 5. Ground truth for nonlinear dynamic scenario (circle: track start position, triangle: track end position).

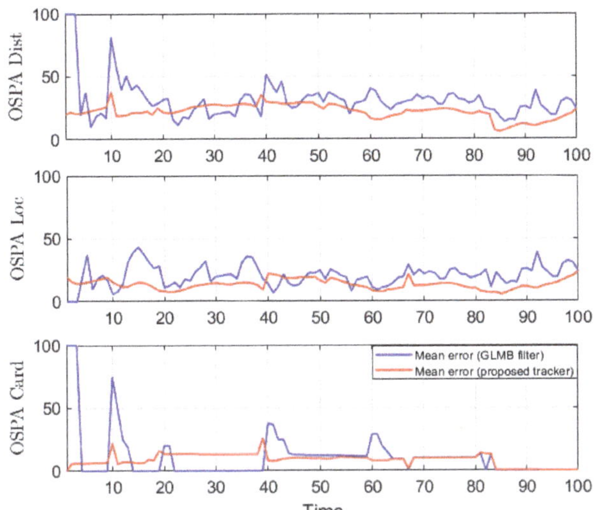

Figure 6. OSPA error for nonlinear dynamic scenario.

4.1.3. Hybrid TBD Observation Model

In this simulation, we use the hybrid measurement model to track objects following a linear dynamic motion model. The surveillance region is 100 × 100 pixels with image cell size of 1, total time step of $K = 100$, and $\Delta = 1$. The observation are the raw images, which are arrays of pixels. In particular, for a pixel i at the image coordinate $(a^{(i)}, b^{(i)})$, the array value is given as follows [26,27]:

$$y^{(i)} = \left[\sum_{x \in \mathbf{X}} \frac{I_k}{2\pi\sigma_h} \exp\left(-\frac{(a^{(i)} - p_x)^2 + (b^{(i)} - p_y)^2}{2\sigma_h^2} \right) \right] + w^{(i)} \tag{27}$$

where $w^{(i)} \backsim \mathcal{N}(0, \sigma_y)$ is Gaussian noise. In this experiment, we set $\sigma_h = 4$ and $\sigma_y = 1$. We choose the value of I_k such that the signal to noise ratio (SNR) varies over the range 7 to 10 dB. For the observation model from the perspective of the filter, we fix its SNR value to 10 dB. From the raw images, we then use hard-shareholding to extract the points measurements at each frame.

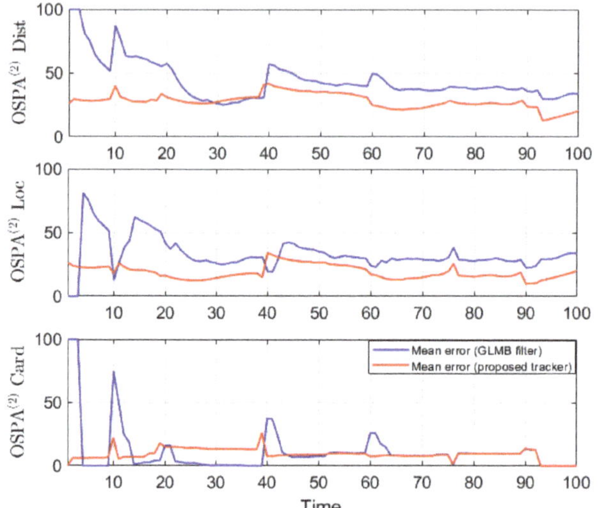

Figure 7. OSPA^2 error for nonlinear dynamic scenario.

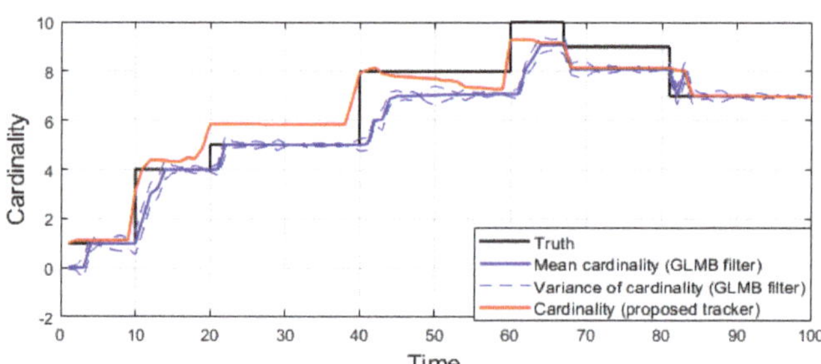

Figure 8. Estimated cardinality for nonlinear dynamic scenario.

The dynamic model and standard observation model are similar to the ones in Section 4.1.1 with $\sigma_v = 1$ pixel/s, $p_S = 0.98$, and $\sigma_\epsilon = 4$ pixels with a clutter rate of 10. The expected new births states are $m_B^{(1)} = [5, 0, 25, 0]^T$, $m_B^{(2)} = [5, 0, 90, 0]^T$, $m_B^{(3)} = [80, 0, 90, 0]^T$, $m_B^{(4)} = [5, 0, 5, 0]^T$, and $m_B^{(5)} = [90, 0, 30, 0]^T$ with the covariance of $P_B = diag([3, 2, 3, 2])$ and the probability r_B of 0.03. The ground truth location of objects is shown in Figure 9 while Figure 10 shows samples of raw image observation along with points detection. The implementation of the filtering phase is as the same as in Reference [27]. The smoothing interval is set to the entire tracking time with the track pruning threshold of the smoother set to 3 time steps.

This experiment is run over 100 Monte Carlo trials. The means of OSPA error and OSPA^2 error are shown respectively in Figures 11 and 12. The estimated cardinality is plotted in Figure 13.

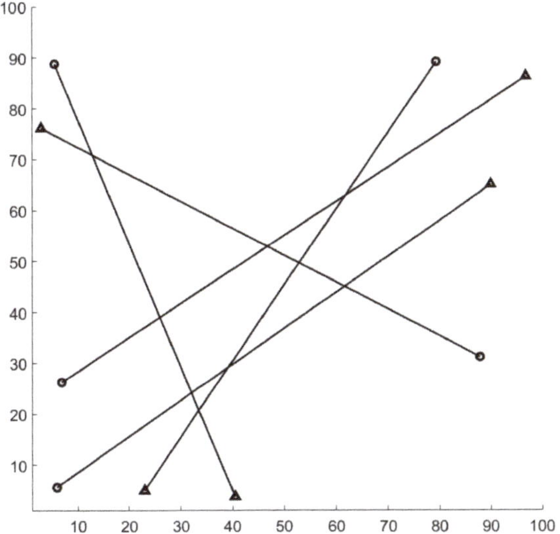

Figure 9. Ground truth for a hybrid track-before-detect (TBD) scenario (circle: track start position, triangle: track end position).

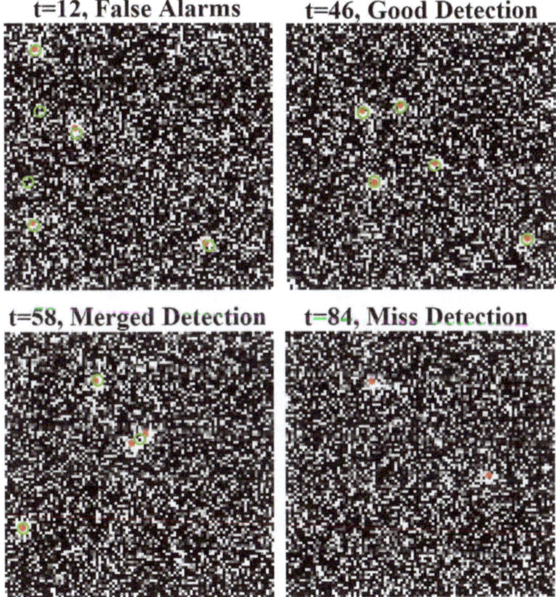

Figure 10. Samples of raw images and point observations for a hybrid TBD scenario (red asterisk: ground truth position, green circle: point detection).

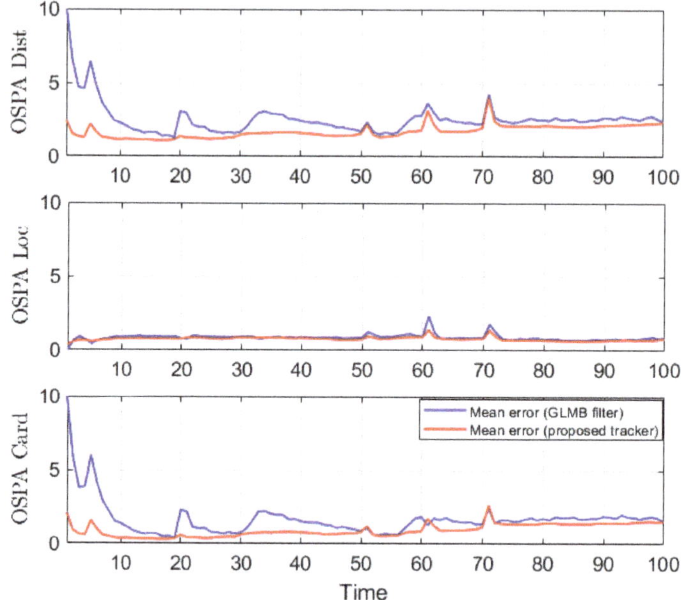

Figure 11. OSPA error for a hybrid TBD scenario.

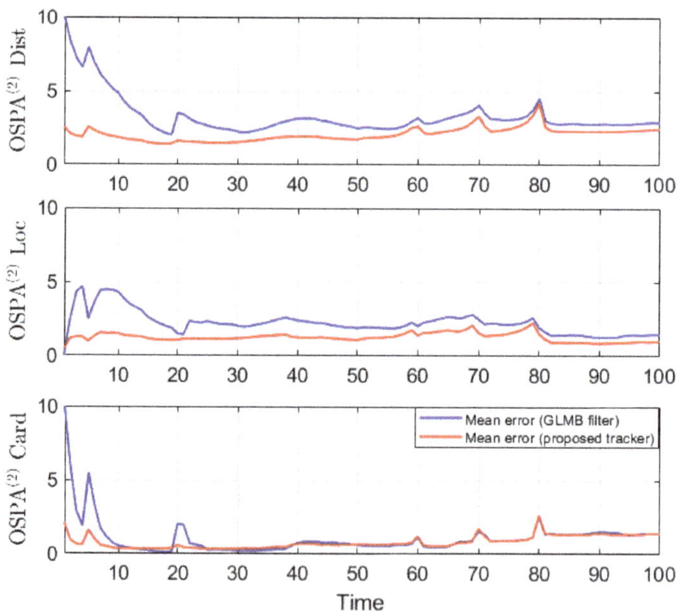

Figure 12. $OSPA^2$ error for a hybrid TBD scenario.

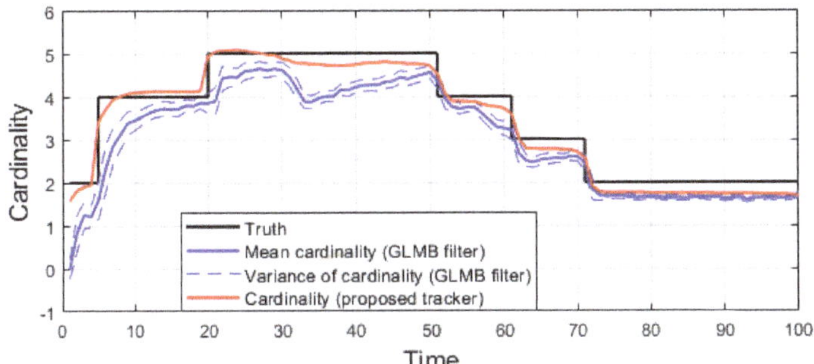

Figure 13. Estimated cardinality for a hybrid TBD scenario.

4.1.4. Discussion on the Simulation Results

For all simulated experiments, we observe lower OSPA and OSPA^2 errors for the proposed tracker compared to the GLMB filter results. In the first two experiments with the standard observation model, as the clutter rate is high, the filtered-only trajectories jiggle around the true paths due to false measurements. In Figures 2 and 6 as well as in Figures 3 and 7, the overall errors of the GLMB filter estimates are higher than of the proposed tracker estimates. The reduction of localization error contribute mainly to the improvement of the tracking performance. From the cardinality plots in Figures 4 and 8, on average, the proposed tracker slightly improves estimate cardinality performance as it is able to eliminate track fragmentation while eliminating incorrect tracks at some time steps.

In the hybrid TBD tracking experiment, as tracks are miss-detected due to low SNR, the proposed tracker improves tracking performance by eliminating track fragmentation. Not much localization error is reduced by the smoother step as the GLMB filter produces relatively good tracking results. The OSPA and OSPA^2 results presented in Figures 11 and 12 show slight improvement of the proposed tracker results compared to GLMB filter tracking results. However, the cardinality plot in Figure 13 clearly indicates that the proposed tracker is able to improve the estimated cardinality between time step 30 and 40.

The run time for all simulated scenario is given in Figure 14 in terms of the percentage of extra computational time of the proposed tracker over the computational time of the filtering step only. It is shown that the extra computational time is negligible in all three tracking scenarios with the extra computational time of the proposed tracker less than 0.5% of the filtering computational time. However, the main disadvantage is that the tracker needs to wait until the end of the smoothing interval to be able to produce tracking results.

4.2. Application to Cell Microscopy

In this experiment, we attempt to track biological cells from a sequence of images containing 90 frames by using the proposed tracker. A snapshot of the sequence is shown in Figure 15. In this application, we use the constant turn rate for the dynamic model as in Section 4.1.2 and the standard observation model as in Section 4.1.1. We also implement the measurement driven model as described in Reference [20]. For the first time step, the birth rate is set to a very high value ($\approx$1) to initialize objects. Subsequently, the birth rate is capped at 10^{-7}. The standard deviation of the turn rate noise is $\pi/90$ rad/s, and the standard deviation of the velocity noise is 5 pixels/frame. The number of hypotheses is capped at 10,000. The detection rate is set to 0.88, and the surviving rate and the spawning rate are 0.999 and 0.035, respectively. The clutter rate is set to 0.05. The cell spawning model is the same as described in Reference [57] with the covariance of the spawning model given as $Q_T = \begin{bmatrix} 40 & 0 & 0 & 0 & 0 \\ 0 & 5 & 0 & 0 & 0 \\ 0 & 0 & 40 & 0 & 0 \\ 0 & 0 & 0 & 5 & 0 \\ 0 & 0 & 0 & 0 & \pi/90 \end{bmatrix}$ and the smoothing interval set to the entire image sequence. In this application, we set the track pruning threshold of the estimator to 3 time steps.

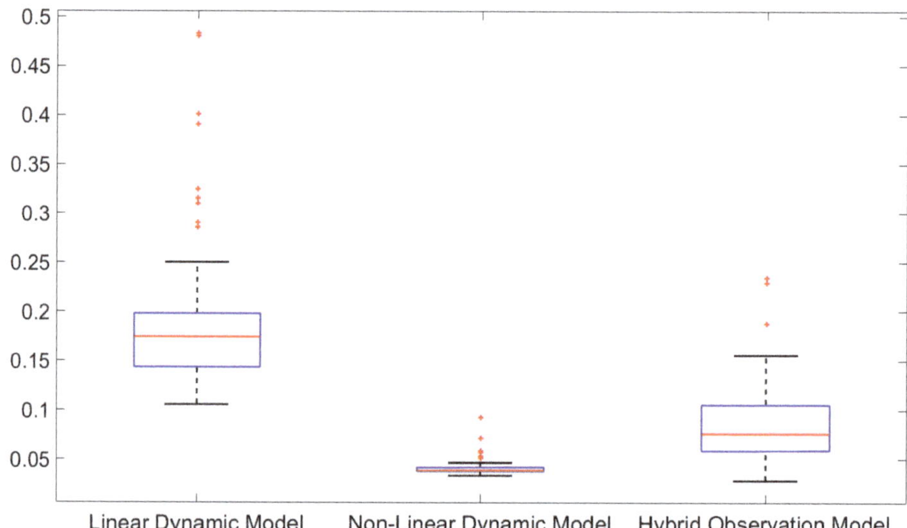

Figure 14. Percentage of smoothing time over filtering time.

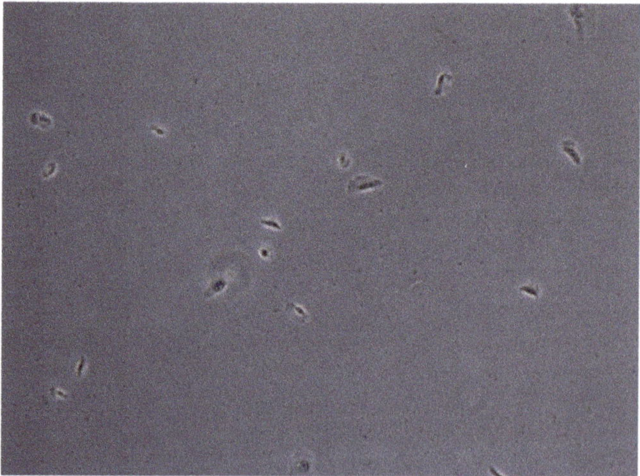

Figure 15. Snapshot of biological cell sequence.

From the tracking results, significant improvement is observed as the proposed tracker is able to eliminate incorrect spawned tracks. While the OSPA error in Figure 16 shows similar performance for the GLMB filter and the proposed tracker, the improvement is clearly reflected in the OSPA^2 cardinality error plots in Figure 17. From the cardinality plot in Figure 18, the estimated cardinality from our tracker is much closer to the true values as fewer incorrect spawned tracks are estimated. In this experiment, there is not much difference between the GLMB filter and the proposed tracker estimates localization error due to the mismatch between the dynamic model and actual motion of the cells. Finally, in Figure 19, we illustrate the improved tracking results in terms of tracking sequence for several time steps at a selected region where the cell splitting process occurs.

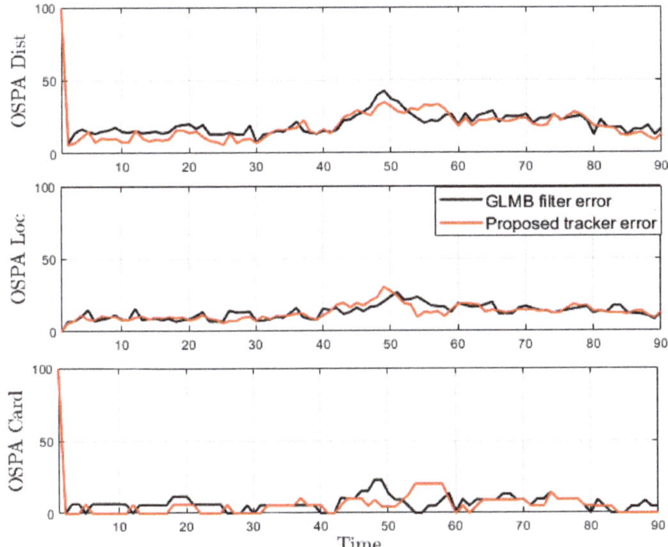

Figure 16. OSPA error for tracking biological cells.

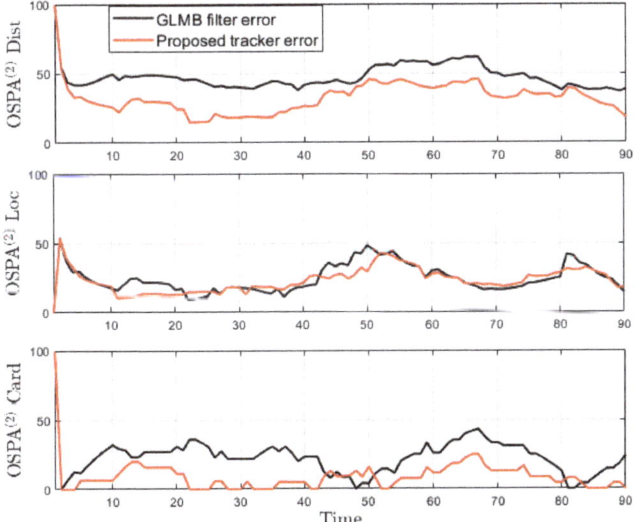

Figure 17. OSPA^2 error for tracking biological cells.

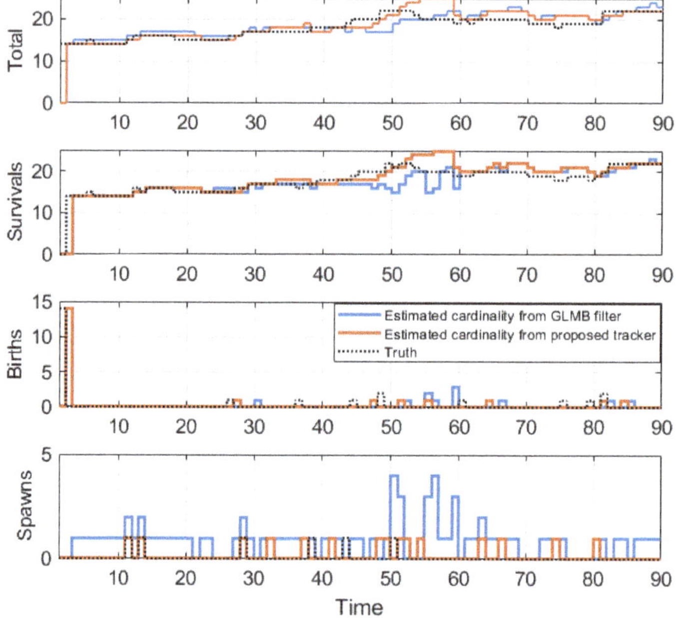

Figure 18. Estimated cardinality for tracking biological cells.

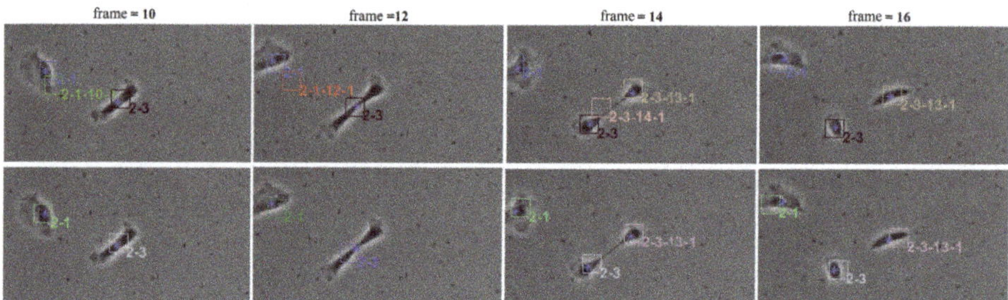

Figure 19. The tracked image sequences of biological cells with blue asterisks denoting points detection. Top row: Generalized Labeled Multi-Bernoulli (GLMB) filter tracking results. Bottom: Proposed tracker tracking results.

5. Conclusions

In this paper, we detailed the implementation of a new tracker based on GLMB filter and a modified multi-scan estimator. In addition to lowering the localization error by performing RTS smoother on each individual estimated trajectory, the proposed tracker can also reduce cardinality errors by deleting the short-term tracks via track management and by completely eliminating track fragmentation. The computation time is shown to contribute to less than 0.5% of the total tracking time, although a fixed delay time is needed before the tracker can produce the estimate. Therefore, in applications when real-time updates are not required, the proposed tracker can be used to improve the tracking results given negligible extra computation time. However, as the smoothing results strongly depend on the quality of the estimates obtained from the forward filtering step, if the filtered estimate experiences strong distortion, the performance of the proposed tracker degrades significantly.

Author Contributions: Methodology, T.T.D.N. and D.Y.K.; visualization, T.T.D.N.; writing—original draft, T.T.D.N.; writing—review and editing, T.T.D.N. and D.Y.K.

Funding: This work was funded by the Australian Research Council grant number DP160104662.

Acknowledgments: The authors acknowledge the administrative and technical support from Curtin University and RMIT University.

Conflicts of Interest: The authors declare no conflict of interest.

References

1. Fortmann, T.; Bar-Shalom, Y.; Scheffe, M. Sonar tracking of multiple targets using joint probabilistic data association. *IEEE J. Ocean. Eng.* **1983**, *8*, 173–184. [CrossRef]
2. Reid, D. An algorithm for tracking multiple targets. *IEEE Trans. Autom. Control* **1979**, *24*, 843–854. [CrossRef]
3. Vo, B.N.; Vo, B.T.; Mahler, R.P.S. Closed form solutions to forward-backward smoothing. *IEEE Trans. Signal Process.* **2012**, *60*, 2–17. [CrossRef]
4. Vo, B.N.; Ma, W.K. The Gaussian Mixture Probability Hypothesis Density Filter. *IEEE Trans. Signal Process.* **2006**, *54*, 4091–4104. [CrossRef]
5. Gao, Y.; Jiang, D.; Liu, M.Particle-Gating SMC-PHD filter. *Signal Process.* **2017**, *130*, 64–73. [CrossRef]
6. Wang, S.; Bao, Q.; Chen, Z. Refined PHD Filter for Multi-Target Tracking under Low Detection Probability. *Sensors* **2019**, *19*, 2842. [CrossRef] [PubMed]
7. Zheng, J.; Gao, M. Tracking Ground Targets with a Road Constraint Using a GMPHD Filter. *Sensors* **2018**, *18*, 2723. [CrossRef]
8. Vo, B.T.; Vo, B.N.; Cantoni, A. Analytic Implementations of the Cardinalized Probability Hypothesis Density Filter. *IEEE Trans. Signal Process.* **2007**, *55*, 3553–3567. [CrossRef]
9. Zhang, H.; Jing, Z.; Hu, S. Gaussian mixture CPHD filter with gating technique. *Signal Process.* **2009**, *89*, 1521–1530. [CrossRef]
10. Si, W.; Wang, L.; Qu, Z. Multi-Target Tracking Using an Improved Gaussian Mixture CPHD Filter. *Sensors* **2016**, *16*, 1964. [CrossRef]
11. Vo, B.T.; Vo, B.N.; Cantoni, A. The Cardinality Balanced Multi-Target Multi-Bernoulli Filter and Its Implementations. *IEEE Trans. Signal Process.* **2009**, *57*, 409–423. [CrossRef]
12. Hu, X.; Ji, H.; Liu, L. Adaptive Target Birth Intensity Multi-Bernoulli Filter with Noise-Based Threshold. *Sensors* **2019**, *19*, 1120. [CrossRef] [PubMed]
13. Vo, B.T.; Vo, B.N. Labeled Random Finite Sets and Multi-Object Conjugate Priors. *IEEE Trans. Signal Process.* **2013**, *61*, 3460–3475. [CrossRef]
14. Yang, B.; Wang, J.; Wang, W. An efficient approximate implementation for labeled random finite set filtering. *Signal Process.* **2018**, *150*, 215–227. [CrossRef]
15. Vo, B.N.; Vo, B.T.; Phung, D. Labeled Random Finite Sets and the Bayes Multi-Target Tracking Filter. *IEEE Trans. Signal Process.* **2014**, *62*, 6554–6567. [CrossRef]
16. Vo, B.N.; Vo, B.T.; Hoang, H.G. An Efficient Implementation of the Generalized Labeled Multi-Bernoulli Filter. *IEEE Trans. Signal Process.* **2017**, *65*, 1975–1987. [CrossRef]
17. Moratuwage, D.; Adams, M.; Inostroza, F. δ-Generalized Labeled Multi-Bernoulli Simultaneous Localization and Mapping with an Optimal Kernel-Based Particle Filtering Approach. *Sensors* **2019**, *19*, 2290. [CrossRef] [PubMed]
18. Liu, C.; Sun, J.; Lei, P.; Qi, Y. δ-Generalized Labeled Multi-Bernoulli Filter Using Amplitude Information of Neighboring Cells. *Sensors* **2018**, *18*, 1153. [CrossRef]
19. Do, C.T.; Van Nguyen, H. Tracking Multiple Targets from Multistatic Doppler Radar with Unknown Probability of Detection. *Sensors* **2019**, *19*, 1672. [CrossRef]
20. Reuter, S.; Vo, B.T.; Vo, B.N.; Dietmayer, K. The Labeled Multi-Bernoulli Filter. *IEEE Trans. Signal Process.* **2014**, *62*, 3246–3260. [CrossRef]
21. Xie, Y.; Song, T.L. Bearings-only multi-target tracking using an improved labeled multi-Bernoulli filter. *Signal Process.* **2018**, *151*, 32–44. [CrossRef]
22. Amrouche, N.; Khenchaf, A.; Berkani, D. Multiple target tracking using track before detect algorithm. In Proceedings of the 2017 International Conference on Electromagnetics in Advanced Applications (ICEAA), Verona, Italy, 11–15 September 2017; pp. 692–695. [CrossRef]
23. Grossi, E.; Lops, M.; Venturino, L. A Track-Before-Detect Algorithm with Thresholded Observations and Closely-Spaced Targets. *IEEE Signal Process. Lett.* **2013**, *20*, 1171–1174. [CrossRef]

24. Yi, W.; Morelande, M.R.; Kong, L.; Yang, J. An Efficient Multi-Frame Track-before-Detect Algorithm for Multi-Target Tracking. *IEEE J. Sel. Top. Signal Process.* **2013**, *7*, 421–434. [CrossRef]
25. Rathnayake, T.; Tennakoon, R.; Khodadadian Gostar, A.; Bab-Hadiashar, A.; Hoseinnezhad, R. Information Fusion for Industrial Mobile Platform Safety via Track-before-Detect Labeled Multi-Bernoulli Filter. *Sensors* **2019**, *19*, 2016. [CrossRef] [PubMed]
26. Vo, B.N.; Vo, B.T.; Pham, N.T.; Suter, D. Joint Detection and Estimation of Multiple Objects from Image Observations. *IEEE Trans. Signal Process.* **2010**, *58*, 5129–5141. [CrossRef]
27. Kim, D.Y.; Vo, B.N.; Vo, B.T.; Jeon, M. A labeled random finite set online multi-object tracker for video data. *Pattern Recognit.* **2019**, *90*, 377–389. [CrossRef]
28. Bryant, D.S.; Delande, E.D.; Gehly, S.; Houssineau, J.; Clark, D.E.; Jones, B.A. The CPHD Filter with Target Spawning. *IEEE Trans. Signal Process.* **2017**, *65*, 13124–13138. [CrossRef]
29. Lundgren, M.; Svensson, L.; Hammarstrand, L. A CPHD Filter for Tracking with Spawning Models. *IEEE J. Sel. Top. Signal Process.* **2013**, *7*, 496–507. [CrossRef]
30. Bryant, D.S.; Vo, B.; Vo, B.; Jones, B.A. A Generalized Labeled Multi-Bernoulli Filter with Object Spawning. *IEEE Trans. Signal Process.* **2018**, *66*, 6177–6189. [CrossRef]
31. Bryson, A.; Frazier, M. *Smoothing for Linear and Nonlinear Dynamic Systems*; Tech. Rep. TDR-63-119; Aeronautical Systems Division Wright Patterson AFB: Dayton, OH, USA, 1962.
32. Rauch, H.E.; Striebel, C.T.; Tung, F. Maximum likelihood estimates of linear dynamic systems. *AIAA J.* **1965**, *3*, 1445–1450. [CrossRef]
33. Fraser, D.; Potter, J. The optimum linear smoother as a combination of two optimum linear filters. *IEEE Trans. Autom. Control* **1969**, *14*, 387–390. [CrossRef]
34. Briers, M.; Doucet, A.; Maskell, S. Smoothing algorithms for state-space models. *Ann. Inst. Stat. Math.* **2009**, *62*, 61–89. [CrossRef]
35. Fearnhead, P.; Wyncoll, D.; Tawn, J. A sequential smoothing algorithm with linear computational cost. *Biometrika* **2010**, *97*, 447–464. [CrossRef]
36. Särkkä, S. Unscented Rauch–Tung–Striebel Smoother. *IEEE Trans. Autom. Control* **2008**, *53*, 845–849. [CrossRef]
37. Särkkä, S. Continuous-time and continuous–discrete-time unscented Rauch–Tung–Striebel smoothers. *Signal Process.* **2010**, *90*, 225–235. [CrossRef]
38. Kitagawa, G. Monte Carlo Filter and Smoother for Non-Gaussian Nonlinear State Space Models. *J. Comput. Graph. Stat.* **1996**, *5*, 1–25. [CrossRef]
39. Simon, J.; Julier, J.K.U. New extension of the Kalman filter to nonlinear systems. In Proceedings of the Signal Processing, Sensor Fusion, and Target Recognition VI, Orlando, FL, USA, 21–24 April 1997; Volume 3068. [CrossRef]
40. Chen, B.; Tugnait, J. Multisensor tracking of a maneuvering target in clutter using IMMPDA fixed-lag smoothing. *IEEE Trans. Aerosp. Electron. Syst.* **2000**, *36*, 983–991. [CrossRef]
41. Helmick, R.; Blair, W.; Hoffman, S. One-step fixed-lag smoothers for Markovian switching systems. *IEEE Trans. Autom. Control* **1996**, *41*, 1051–1056. [CrossRef]
42. Li, T.; Chen, H.; Sun, S.; Corchado, J.M. Joint Smoothing and Tracking Based on Continuous-Time Target Trajectory Function Fitting. *IEEE Trans. Autom. Sci. Eng.* **2019**, *16*, 1476–1483. [CrossRef]
43. Mahler, R.P.S.; Vo, B.T.; Vo, B.N. Forward-Backward Probability Hypothesis Density Smoothing. *IEEE Trans. Aerosp. Electron. Syst.* **2012**, *48*, 707–728. [CrossRef]
44. He, X.; Liu, G. Improved Gaussian mixture probability hypothesis density smoother. *Signal Process.* **2016**, *120*, 56–63. [CrossRef]
45. Nadarajah, N.; Kirubarajan, T.; Lang, T.; Mcdonald, M.; Punithakumar, K. Multitarget Tracking Using Probability Hypothesis Density Smoothing. *IEEE Trans. Aerosp. Electron. Syst.* **2011**, *47*, 2344–2360. [CrossRef]
46. Nagappa, S.; Delande, E.D.; Clark, D.E.; Houssineau, J. A Tractable Forward-Backward CPHD Smoother. *IEEE Trans. Aerosp. Electron. Syst.* **2017**, *53*, 201–217. [CrossRef]
47. Li, D.; Hou, C.; Yi, D. Multi-Bernoulli smoother for multi-target tracking. *Aerosp. Sci. Technol.* **2016**, *48*, 234–245. [CrossRef]
48. Koch, W. Fixed-interval retrodiction approach to Bayesian IMM-MHT for maneuvering multiple targets. *IEEE Trans. Aerosp. Electron. Syst.* **2000**, *36*, 2–14. [CrossRef]

49. Beard, M.; Vo, B.T.; Vo, B.N. Generalised labeled multi-Bernoulli forward-backward smoothing. In Proceedings of the 2016 19th International Conference on Information Fusion (FUSION), Heidelberg, Germany, 5–8 July 2016; pp. 688–694.
50. Vo, B.N.; Vo, B.T. A Multi-Scan Labeled Random Finite Set Model for Multi-Object State Estimation. *IEEE Trans. Signal Process.* **2019**, *67*, 4948–4963. [CrossRef]
51. Mahler, R. Exact Closed-Form Multitarget Bayes Filters. *Sensors* **2019**, *19*, 818. [CrossRef]
52. Beard, M.; Vo, B.T.; Vo, B.N. A Solution for Large-Scale Multi-Object Tracking. *arXiv* **2018**, arXiv:1804.06622.
53. Vo, B.N.; Vo, B.T.; Beard, M. Multi-Sensor Multi-Object Tracking with the Generalized Labeled Multi-Bernoulli Filter. *IEEE Trans. Signal Process.* **2019**. [CrossRef]
54. Mahler, R.P.S. *Statistical Multisource-Multitarget Information Fusion*; Artech House: Norwood, MA, USA, 2007.
55. Schuhmacher, D.; Vo, B.T.; Vo, B.N. A Consistent Metric for Performance Evaluation of Multi-Object Filters. *IEEE Trans. Signal Process.* **2008**, *56*, 3447–3457. [CrossRef]
56. Beard, M.; Vo, B.T.; Vo, B.N. OSPA(2): Using the OSPA metric to evaluate multi-target tracking performance. In Proceedings of the 2017 International Conference on Control, Automation and Information Sciences (ICCAIS), Chiang Mai, Thailand, 31 October–1 November 2017; pp. 86–91. [CrossRef]
57. Nguyen, T.T.D.; Kim, D.Y. On-line Tracking of Cells and Their Lineage from Time Lapse Video Data. In Proceedings of the 2018 International Conference on Control, Automation and Information Sciences (ICCAIS), Hangzhou, China, 24–27 October 2018; pp. 291–296. [CrossRef]

Article

DOA Tracking Based on Unscented Transform Multi-Bernoulli Filter in Impulse Noise Environment

Sun-yong Wu [1,2], Jun Zhao [1], Xu-dong Dong [1], Qiu-tiao Xue [1,*] and Ru-hua Cai [1]

[1] School of Mathematics and Computing Science, Guilin University of Electronic Technology, Guilin 541004, China; wsy121991@guet.edu.cn (S.-y.W.); jun325709@163.com (J.Z.); 18871347936@163.com (X.-d.D.); ruhuac@guet.edu.cn (R.-h.C.)

[2] Guangxi Information Science Experiment Center, Guilin 541004, China

* Correspondence: xqt121991@guet.edu.cn; Tel.: +86-158-7835-4306

Received: 15 August 2019; Accepted: 16 September 2019; Published: 18 September 2019

Abstract: Aiming at the problem of multiple-source direction of arrival (DOA) tracking in impulse noise, this paper models the impulse noise by using the symmetric α stable (SαS) distribution, and proposes a DOA tracking algorithm based on the Unscented Transform Multi-target Multi-Bernoulli (UT-MeMBer) filter framework. In order to overcome the problem of particle decay in particle filtering, UT is adopted to select a group of sigma points with different weights to make them close to the posterior probability density of the state. Since the α stable distribution does not have finite covariance, the Fractional Lower Order Moment (FLOM) matrix of the received array data is employed to replace the covariance matrix to formulate a MUSIC spatial spectra in the MeMBer filter. Further exponential weighting is used to enhance the weight of particles at high likelihood area and obtain a better resampling. Compared with the PASTD algorithm and the MeMBer DOA filter algorithm, the simulation results show that the proposed algorithm can more effectively solve the issue that the DOA and number of target are time-varying. In addition, we present the Sequential Monte Carlo (SMC) implementation of the UT-MeMBer algorithm.

Keywords: direction-of-arrival (DOA) tracking; impulse noise; Multi-Bernoulli filter; particle filtering

---

1. Introduction

Multi-target Direction of Arrival (DOA) estimation is an essential issue in array processing and has a wide range of applications in source location, radar, sonar, and wireless communications [1,2]. Sparse representation and compressive sensing methods are used for DOA estimation of coprime array [3–6], while these methods are only applied in the case where the sources are stationary. In addition, difficulties also arise from the uncertainties of the source dynamics: the source may be moving or static. Thus, it is significant to extend the static DOA estimation algorithm to the dynamic DOA tracking algorithm.

The representative dynamic DOA tracking algorithms include the subspace tracking algorithm and the particle filter (PF) algorithm. The subspace tracking algorithm includes Projection Approximation Subspace Tracking (PAST) [7] and the Projection Approximation Subspace Tracking with Deflation (PASTD) [8]. In essence, these algorithms transform the determination of the eigensubspace into solving an unconstrained optimization problem, and combine the recursive least squares (RLS) theory to achieve effective tracking of the eigensubspace of time-varying sources. However, the RLS method is very sensitive to impulse noise, and the PAST algorithm's subspace tracking performance will degrade sharply in the impulse noise environment [9–11]. In an army of acoustic applications, such as underwater and room acoustic signal processing, the noise environment is non-Gaussian and is impulsive in nature [12,13]. Under investigation, it was found that α stable distribution ($0 < \alpha \leq 2$) is a suitable noise model to describe this type of noise [14]. In recent years, DOA estimation technology in impulse noise environment has developed rapidly [15–17]. The PF algorithm based on Bayesian recursive estimation can solve the target tracking problem by utilizing a priori DOA and current measurement information [18]. In [19], the author considers the particle filtering method to estimate the single target DOA by using the spatial spectral function based on FLOM matrix as the likelihood function in the impulse noise environment. However, those algorithms needs to know the number of targets in advance and cannot deal with the estimation problem of the time-varying sources DOA.

In practical applications, such as submarine tracking and sonar positioning, the number of the sources are dynamic. Mahler introduced the concept of random finite set (RFS) in [20]. A tutorial on Bernoulli filters is introduced in [21]. A track-before-detect (TBD) Bernoulli filter based on RFS is proposed for DOA tracking in single dynamic system in [22], but it cannot solve the DOA tracking in multiple target dynamic system. The Multi-target Multi-Bernoulli (MeMBer) filtering [23] is a filter developed under the RFS framework. The advantage is that it operates on the dimensions of a single target space, thus avoiding the computational complexity and data association problems of the joint filter. Choppala P B et al. studied the Bayesian multi-target tracking problem based on phased array sensor, and proposed the MUSIC spatial spectral as a pseudo-likelihood in the Multi-Bernoulli filter in [24]. However, the shortcoming of this algorithm is that impulse noise is not considered, and Gaussian noise model is not appropriate in practical applications.

Based on the above analysis, a particle filter algorithm of DOA tracking for Unscented Transform MeMBer (UT-MeMBer) in an impulse noise environment is proposed. UT is used to construct a new important density function, which makes the estimation accuracy higher when the particle degenerates. Since particles close to the real state are more likely to output a larger spatial spectral response, the magnitude of the spatial spectral response is used as a feature of pseudo-likelihood. Based on the FLOM matrix, this paper uses FLOM matrix to substitute the covariance matrix to obtain the corresponding MUSIC spatial spectrum as the particle likelihood function. Further exponential weighting can increase the weight of the particles, making resampling more efficient. The main advantage of the tracking algorithm is that the number and state of the target can be accurately tracked when the number and state of the sources are unknown in impulse noise environment.

The rest of the paper is organized as follows. In Section 2, the problem of the DOA tracking in impulsive noise environment is described. In Section 3, we outline the Multi-Bernoulli's Bayesian theory of DOA tracking. An improved algorithm for likelihood functions is introduced in Section 4. The UT-MeMBer DOA particle filter tracking algorithm is given in Section 5. We then show our simulation results in Section 6 and conclusion in Section 7.

2. Problem Formulation

2.1. Array Signal Model

Consider the case of P narrow farfield signals $s_p(t), p = 1, 2, \cdots P$ with different DOA $\theta_1, \theta_2, \ldots, \theta_P$ arriving at a uniform linear array (ULA) with M sensors at discrete time t. The DOA of the pth source can be written as θ_p. The received signal of the arrays can be expressed as

$$\boldsymbol{Z}(t) = \boldsymbol{A}(\theta)\boldsymbol{S}(t) + \boldsymbol{N}(t) \tag{1}$$

where $\boldsymbol{N}_{M\times 1}(t) = [\mathbf{n}_1(t), \mathbf{n}_2(t), \ldots, \mathbf{n}_M(t)]^T$ represents the impulsive noise vector which is not correlated with signals. $\boldsymbol{Z}_{M\times 1}(t) = [z_1(t), z_2(t), \ldots, z_M(t)]^T$ is the measurement at time t, $\boldsymbol{A}_{M\times P}(t) = [\boldsymbol{a}(\theta_1), \boldsymbol{a}(\theta_2), \ldots, \boldsymbol{a}(\theta_P)]^T$ is array manifold, $\boldsymbol{S}_{P\times 1}(t) = [s_1(t), s_2(t), \ldots, s_P(t)]^T$ denotes the acoustic sources matrix, and

$$\boldsymbol{a}\left(\theta_p\right) = \left[1, e^{-j\frac{2\pi}{\lambda}d\sin\theta_p}, \ldots, e^{-j\frac{2\pi}{\lambda}(M-1)d\sin\theta_p}\right] \tag{2}$$

is the steering vector with λ denoting the wavelength of the carrier, and d is the array space.

2.2. α Stable Distribution

Most of the traditional research methods estimating the DOA are based on Gaussian noise models. In practical situations, such as radar echo and low-frequency atmospheric noise, they consist of impulse noise with a short duration and large amplitude. The performance of the algorithm will drop significantly when the Gaussian noise model is still modeled in an impulse noise environment. The α stable distribution is a good example of such a type with significant spike noise and a Gaussian distribution. The α stable distribution's probability function does not have the closed form, which can be conveniently described by its characteristic function as

$$\phi(t) = e^{\{jat-\gamma|t|^{\alpha}[1+j\beta \text{sgn}(t)\varpi(t,\alpha)]\}} \tag{3}$$

where

$$\varpi = \begin{cases} \tan\frac{\alpha\pi}{2}, \alpha \neq 1 \\ \frac{2}{\pi}\log|t|, \alpha = 1 \end{cases} \tag{4}$$

$$\text{sgn}(t) = \begin{cases} 1, t > 0 \\ 0, t = 0 \\ -1, t < 0 \end{cases} \tag{5}$$

α is the characteristic exponent, whose size can affect the degree of impulse and the range is $0 < \alpha \leq 2$. γ is a dispersion parameter whose mean is consistent with the variance of the Gaussian distribution. β is a symmetric parameter, and the distribution at $\beta = 0$ is a symmetric α stable (SαS) distribution. a is the positional parameter. When $\alpha = 2, \beta = 0$, it is a Gaussian distribution model. When $\alpha = 1, \beta = 0$, it is the Cauchy distribution model. When $\alpha = 1/2, \beta = -1$, it is the Pearson distribution model. A crucial difference between the Gaussian distribution and the α stable distribution is that the latter does not have second-order statistics so that its covariance is inaccurate.

3. MeMBer Bayesian Theory of DOA Tracking

3.1. Multi-Target Bayesian Theory

Assume that the state of the sources at time k is $x_k = \left[\theta_k, \dot{\theta}_k\right]^T$, where θ_k is the DOA and moves at a speed of $\dot{\theta}_k$ rad/s. The state and number of sources are changing at time $k+1$, which can be described by RFS. From [20], the sources state set in multiple sources tracking can be regarded as an RFS, namely

$$\mathbf{X}_k = \left\{x_{k,1}, \cdots, x_{k,P(k)}\right\} \tag{6}$$

where $\mathbf{X}_k$ represents a set of sources at time k, and the element of the set may be one or more or null. $\mathbf{Z}_k$ denotes the measurement set generated by all sources received time k, and the element is only one.

Single-target Bayesian filtering can be extended to multi-target tracking by modeling the above source states and measured values. The single target posterior probability density function (pdf) $p_{k|k}(x_k|\mathbf{Z}_{1:k})$ is replaced by the joint multi-target posterior $p_{k|k}(\mathbf{X}_k|\mathbf{Z}_{1:k})$. The Bayes joint filter recursion includes two stages: prediction and update. The prediction and update at time k in [24] are

$$p_{k|k-1}(\mathbf{X}_k|\mathbf{Z}_{1:k-1}) = \int f_{k|k-1}(\mathbf{X}_k|\mathbf{X}_{k-1})p_{k-1|k-1}(\mathbf{X}_{k-1}|\mathbf{Z}_{1:k-1})\delta\mathbf{X}_{k-1} \tag{7}$$

and

$$p_{k|k}(\mathbf{X}_k|\mathbf{Z}_{1:k}) = \frac{g(\mathbf{Z}_k|\mathbf{X}_k)p_{k|k-1}(\mathbf{X}_k|\mathbf{Z}_{1:k-1})}{\int g(\mathbf{Z}_k|\mathbf{X}_k)p_{k|k-1}(\mathbf{X}_k|\mathbf{Z}_{1:k-1})\delta\mathbf{X}_k} \tag{8}$$

where δ is the set integral and $\mathbf{Z}_{1:k-1}$ represents all the measurement sets up to time $k-1$. $g(\mathbf{Z}_k|\mathbf{X}_k)$ is a multi-target joint likelihood function and $f_{k|k-1}(\mathbf{X}_k|\mathbf{X}_{k-1})$ is a multi-target state transition probability density function. $p_{k|k-1}(\mathbf{X}_k|\mathbf{Z}_{1:k-1})$ represents the multi-target joint prediction probability density and $p_{k|k}(\mathbf{X}_k|\mathbf{Z}_{1:k})$ is the multi-target joint posterior probability density function.

3.2. Multi-Target Multi-Bernoulli Filter

A Bernoulli set X has a probability $1-r$ of being a null set, and has a probability r of containing a single element x that is distributed via a pdf $s(\cdot)$. The probability of a Bernoulli RFS can be expressed in [21] as

$$\pi(X) = \begin{cases} 1-r, \ X = \varnothing \\ rs(X), \ X = \{x\} \\ 0, \ other \end{cases} \tag{9}$$

A Multi-Bernoulli RFS X can be considered as union of a fixed number of independent Bernoulli sets that have existence probability $r^{(j)} \in (0,1), j = 1, \ldots J$ and the pdf $s^{(j)}$, such that

$$\mathbf{X} = \bigcup_{j=1}^{J} X^{(j)} \tag{10}$$

where the j th Bernoulli set is described by its two parameters: the existence probability $r^{(j)}$ and the pdf $s\left(X^{(j)}\right)$. So a Multi-Bernoulli RFS can be characterized by a posterior parameter set $\left\{\left(r_{k|k}^{(j)}, s_{k|k}\left(\mathbf{X}_k^{(j)}\right)\right)\right\}_{j=1}^{J_k}$, where $J_{k|k}$ indicates the number of sources. $\mathbf{Z}_k = \left[z_{1,k}, z_{2,k}, \ldots, z_{M,k}\right]^T$ denotes the sensor measurement data and $\mathbf{Z}_k \in \mathcal{Z}$, in which $\mathcal{Z}$ is the measurement space of the sensor. Target birth and survival are determined by birth probabilities $p_{b,k}(\mathbf{X}_k)$ and survival probabilities $p_{s,k}(\mathbf{X}_k)$, respectively. The source motion model is represented by the transition probability density $f_{k|k-1}(\mathbf{X}_k|\mathbf{X}_{k-1})$, and the prior probability of Multi-Bernoulli is described as

$$p(\mathbf{X}_{k-1}|\mathbf{Z}_{1:k-1}) \approx \left\{r_{k-1|k-1}^{(j)}, s_{k-1|k-1}\left(\mathbf{X}_{k-1}^{(j)}\right)\right\}_{j=1}^{J_{k-1}} \tag{11}$$

According to Equation (7), the prediction part can be described as

$$\begin{aligned} p(\mathbf{X}_k|\mathbf{Z}_{1:k-1}) &\approx \left\{\hat{r}_{k|k-1}^{(j)}, \hat{s}_{k|k-1}\left(\mathbf{X}_k^{(j)}\right)\right\}_{j=1}^{J_{k|k-1}} \\ &= \left\{r_{P,k|k-1}^{(j)}, s_{P,k|k-1}\left(\mathbf{X}_{k|k-1}^{(j)}\right)\right\}_{j=1}^{J_{P,k|k-1}} \cup \left\{r_{B,k}^{(j)}, s_{B,k}\left(\mathbf{X}_k^{(j)}\right)\right\}_{j=1}^{J_{B,k}} \end{aligned} \tag{12}$$

where

$$\begin{aligned} \hat{r}_{k|k-1}^{(j)} &= \left(1 - r_{k-1|k-1}^{(j)}\right) \cdot \int p_{b,k}\left(\mathbf{X}_k^{(j)}\right) s_{k-1|k-1}\left(\mathbf{X}_{k-1}^{(j)}\right) \mathrm{d}\mathbf{X}_{k-1}^{(j)} \\ &+ r_{k-1|k-1}^{(j)} \cdot \int p_{s,k}\left(\mathbf{X}_{k-1}^{(j)}\right) s_{k-1|k-1}\left(\mathbf{X}_{k-1}^{(j)}\right) \mathrm{d}\mathbf{X}_{k-1}^{(j)} \end{aligned} \tag{13}$$

$$\begin{aligned} \hat{s}_{k|k-1}\left(\mathbf{X}_{k|k-1}^{(j)}\right) &= \frac{p_{s,k}\left(\mathbf{X}_{k-1}^{(j)}\right) r_{k-1|k-1}^{(j)} \cdot \int f_{k|k-1}\left(\mathbf{X}_k^{(j)}|\mathbf{X}_{k-1}^{(j)}\right) s_{k-1|k-1}\left(\mathbf{X}_{k-1}^{(j)}\right) \mathrm{d}\mathbf{X}_{k-1}^{(j)}}{r_{k|k-1}^{(j)}} \\ &+ \frac{p_{b,k}\left(\mathbf{X}_k^{(j)}\right) \cdot \left(1 - r_{k-1|k-1}^{(j)}\right) \cdot b_{k|k-1}\left(\mathbf{X}_k^{(j)}\right)}{r_{k|k-1}^{(j)}} \end{aligned} \tag{14}$$

where $J_{k|k-1} = J_{P,k|k-1} + J_{B,k}, J_{P,k|k-1} = J_{k-1}$. The number of Multi-Bernoulli parameter sets for survival sources and newborn sources are represented by $J_{P,k|k-1}$ and $J_{B,k}$, respectively. According to Equation (8), if the predicted Multi-Bernoulli parameter set can be expressed as $\left\{\hat{r}_{k|k-1}^{(j)}, \hat{s}_{k|k-1}\left(\mathbf{X}_k^{(j)}\right)\right\}_{j=1}^{J_{k|k-1}}$, then the update process can be expressed as

$$p(\mathbf{X}_k|\mathbf{Z}_{1:k}) \approx \left\{r_{k|k}^{(j)}, s_{k|k}\left(\mathbf{X}_{k|k-1}^{(j)}\right)\right\}_{j=1}^{J_k} \tag{15}$$

where

$$r_{k|k}^{(j)} = \frac{\hat{r}_{k|k-1}^{(j)} \int g\left(\mathbf{Z}_k \middle| \mathbf{X}_k^{(j)}\right) \hat{s}_{k|k-1}\left(\mathbf{X}_k^{(j)}\right) \mathrm{d}\mathbf{X}_k^{(j)}}{1 - \hat{r}_{k|k-1}^{(j)} + \hat{r}_{k|k-1}^{(j)} \int g\left(\mathbf{Z}_k \middle| \mathbf{X}_k^{(j)}\right) \hat{s}_{k|k-1}\left(\mathbf{X}_k^{(j)}\right) \mathrm{d}\mathbf{X}_k^{(j)}} \tag{16}$$

$$s_{k|k}\left(\mathbf{X}_k^{(j)}\right) = \frac{g\left(\mathbf{Z}_k \middle| \mathbf{X}_k^{(j)}\right) \hat{s}_{k|k-1}\left(\mathbf{X}_k^{(j)}\right)}{\int g\left(\mathbf{Z}_k \middle| \mathbf{X}_k^{(j)}\right) \hat{s}_{k|k-1}\left(\mathbf{X}_k^{(j)}\right) \mathrm{d}\mathbf{X}_k^{(j)}} \tag{17}$$

where $g(\mathbf{Z}_k|\mathbf{X}_k)$ denotes the likelihood function. If the covariance of the general sensor array at time k in Gaussian noise environment is $\boldsymbol{R}_k$, the likelihood function can be expressed as

$$g(\mathbf{Z}_k|\mathbf{X}_k) = \frac{1}{\pi^M \det(\boldsymbol{R}_k)} \exp\left(-(\boldsymbol{Z}_k - \boldsymbol{A}(\mathbf{X}_k)\boldsymbol{S}_k)^H \boldsymbol{R}_k^{-1} (\boldsymbol{Z}_k - \boldsymbol{A}(\mathbf{X}_k)\boldsymbol{S}_k)\right) \tag{18}$$

The frame of Formula (18) is not held in impulse noise, so we propose to replace the likelihood function with a spatial spectrum method.

4. Improved Algorithm for Likelihood Function

In the practical engineering application, to guarantee the real-time and effectiveness of the estimation, the observation matrix of the array is obtained with a limited number of snapshots. Assuming L observations at time k, the array covariance matrix is calculated as $\hat{\boldsymbol{R}}_k = \boldsymbol{X}(t_k)\boldsymbol{X}(t_k)^H/L$. We assume that the noise vector $\boldsymbol{N}(t)$ is independent to the target signal and has a SαS distribution with a characteristic exponent of α. From [25], if the array observation matrix $\mathbf{Z}_k$ at time k is obtained, the FLOM matrix is defined as

$$\psi_{i,j} = \mathbb{E}\left\{\mathbf{Z}_{i,j}(k)\left|\mathbf{Z}_{j,i}(k)\right|^{p-2} \mathbf{Z}^*{}_{j,i}(k)\right\} \; 1 < p < \alpha \leq 2 \tag{19}$$

where $\psi_{i,j}$ represents the (i,j)th element of $\boldsymbol{\Psi}_k$, and $(\cdot)^*$ represents conjugate operation. The dimension of matrix $\boldsymbol{\Psi}_k$ is $M \times M$. In [25], the authors derived the form of the FLOM matrix as

$$\boldsymbol{\Psi}_k = \boldsymbol{a}(\theta_k)\boldsymbol{R}_s\boldsymbol{a}^H(\theta_k) + r\boldsymbol{I}_M \tag{20}$$

where $\boldsymbol{R}_s$ and r represent the source and additive noise of the FLOM matrix, respectively. As can be seen from Equation (20), the (i,j)th FLOM matrix element is defined as

$$\psi_{i,j} - \frac{\sum_{l=1}^{L} \mathbf{Z}_i(k)\left|\mathbf{Z}_j(k)\right|^{p-2} \mathbf{Z}_j^*(k)}{L} \tag{21}$$

Fractional moment p must satisfy $1 < p < \alpha \leq 2$. The FLOM is used to replace the covariance matrix of the signal in impulse noise, and then the eigendecomposition is performed on $\boldsymbol{\Psi}_k$ in the MUSIC algorithm to obtain the noise subspace $\boldsymbol{U}_n$. The form of the FLOM-MUSIC spatial spectrum estimation function is

$$g(\mathbf{Z}_k|\mathbf{X}_k) = P_{FLOM-MUSIC}(\mathbf{X}_k) = \left|\frac{1}{\boldsymbol{a}^H(\mathbf{C}\mathbf{X}_k)\boldsymbol{U}_n{\boldsymbol{U}_n}^H\boldsymbol{a}(\mathbf{C}\mathbf{X}_k)}\right|^{\zeta} \tag{22}$$

where $\mathbf{C} = [\mathbf{1}, \mathbf{0}]$, and the $\mathbf{C}\mathbf{X}_k$ represents source azimuth information. $\boldsymbol{a}(\cdot)$ is a space vector, and $\zeta \in \mathbf{R}^+$ represents an exponential weighting of the spatial spectrum. The response of the traditional MUSIC spatial spectral beamformer in an impulse noise environment is distorted, which can result in a significant degradation in the performance of the resampling step. After being weighted, the particles can be moved to the high likelihood region to the resampling performance.

5. UT-MeMBer DOA Particle Filter Tracking Algorithm

In this section, we describe the particle filter implementation of the UT-MeMBer algorithm. From [22], if the multi-target probability density parameter set at time $k-1$ is $\left\{\left(r^{j}_{k-1|k-1}, s^{j}_{k-1|k-1}\right)\right\}_{j=1}^{J_{k-1}}$, then the spatial posterior probability density at time $k-1$ and can be expressed as:

$$s^{(j)}_{k-1|k-1}(\boldsymbol{x}) = \sum_{i=1}^{N_{k-1}} \omega^{(i,j)}_{k-1} \boldsymbol{x}^{(i,j)}_{k-1}, j = 1, \ldots, J_{k-1} \tag{23}$$

where $s^{j}_{k-1|k-1}$ is the spatial posterior probability density, which can be approximated as the weighted particle set $\left\{\omega^{(i)}_{k-1}, \boldsymbol{x}^{(i)}_{k-1}\right\}_{i=1}^{N_{k-1}}$. N_{k-1} is the total number of particles, where $\boldsymbol{x}^{(i)}_{k-1}$ represents the state of the i th particle, including angle and speed, i.e., $\boldsymbol{x}^{(i)}_{k-1} = \left[\theta_{k-1}, \dot{\theta}_{k-1}\right]^{T}$. $\omega^{(i)}_{k-1}$ denotes the weight, usually satisfying $\sum_{i=1}^{N_{k-1}} \omega^{(i)}_{k-1} = 1$.

According to (12), the spatial posterior probability density of the prediction step consists of two items and can be written as

$$s^{(j)}_{k|k-1}(\boldsymbol{x}) = \sum_{i=1}^{N_{k|k-1}} \omega^{(i,j)}_{k|k-1} \boldsymbol{x}^{(i,j)}_{k|k-1}, j = 1, \ldots, J_{k|k-1} \tag{24}$$

where $N_{k|k-1} = N_{k-1} + N_{B,k}$ and $J_{k|k-1} = J_{p,k|k-1} + J_{B,k}$ represent the number of predicted particles and predicted MeMBer parameter sets, respectively. All particles can be sampled from two parts:

$$\boldsymbol{x}^{(i,j)}_{k|k-1} = \begin{cases} \boldsymbol{x}^{(i,j)}_{k-1,UT}, & i = 1, \ldots, N_{k-1} \\ \beta_k(\boldsymbol{x}_k|\boldsymbol{Z}_{k-1}), & i = N_{k-1}+1, \ldots, N_{k-1}+N_{B,k} \end{cases} \tag{25}$$

Among them, $N_{B,k}$ denotes the number of newborn particles at time k, $\boldsymbol{x}^{(i,j)}_{k-1,UT}$ is obtained by UT of $\boldsymbol{x}^{(i,j)}_{k|k-1}$ [13]. Particle filtering suffers from missing sample diversity, resulting in depletion of the sampled particles. In order to solve this problem, the surviving particles will be subjected to UT operations. A set of sigma points with different weights are selected by UT operation, and then the posterior probability density of the state is approximated by these sigma points. The weight is

$$\omega^{(i,j)}_{k|k-1} = \begin{cases} \dfrac{p_s r^{(j)}_{k-1|k-1}}{r^{(j)}_{k|k-1}} \cdot \dfrac{f_{k|k-1}\left(\boldsymbol{x}^{(i,j)}_{k|k-1} \middle| \boldsymbol{x}^{(i,j)}_{k-1|k-1}\right)}{\rho_k\left(\boldsymbol{x}^{(i,j)}_{k|k-1} \middle| \boldsymbol{x}^{(i,j)}_{k-1|k-1}, \boldsymbol{Z}_k\right)} \cdot \omega^{(i,j)}_{k-1}, & i = 1, \ldots N_{k-1} \\ \dfrac{p_b\left(1-r^{(j)}_{k-1|k-1}\right)}{r^{(j)}_{k|k-1}} \cdot \dfrac{b_{k|k-1}\left(\boldsymbol{x}^{(i,j)}_{k|k-1}\right)}{\beta_k\left(\boldsymbol{x}^{(i,j)}, \boldsymbol{Z}_{k-1}\right)} \cdot \frac{1}{B}, & i = N_{k-1}+1, \ldots, N_{k-1}+B \end{cases} \tag{26}$$

where p_s and p_b represent the survival probability of particles and the newborn probability of particles, respectively. N_{k-1} is the number of surviving particles sampled from the transition probability density $f_{k|k-1}$, and B is the number of newborn particles from the proposal probability density β_k. If the prediction MeMBer parameter sets can be expressed as $\left\{r^{j}_{k|k-1}, \left\{\omega^{(i,j)}_{k|k-1}, \boldsymbol{x}^{(i,j)}_{k|k-1}\right\}_{i=1}^{N_{k|k-1}}\right\}_{j=1}^{J_{k|k-1}}$ at time k, then the update MeMBer parameter sets can be written as $\left\{r^{j}_{k|k}, \left\{\omega^{(i,j)}_{k}, \boldsymbol{x}^{(i,j)}_{k}\right\}_{i=1}^{N_k}\right\}_{j=1}^{J_k}$. The weight is

$$\omega^{(i,j)}_{k} = p_{D,k}\left(\boldsymbol{x}^{(i,j)}_{k|k-1}\right) \cdot g\left(\boldsymbol{Z}_k \middle| \boldsymbol{x}^{(i,j)}_{k|k-1}\right) \cdot \omega^{(i,j)}_{k|k-1} \tag{27}$$

where $p_{D,k}$ is the detection probability, and the likelihood function $g\left(Z_k \middle| x_{k|k-1}^{(i,j)}\right)$ calculated by the MUSIC algorithm can be expressed as

$$g\left(\mathbf{Z}_k \middle| \boldsymbol{x}_{k|k-1}^{(i,j)}\right) = P_{FLOM-MUSIC}\left(\mathbf{C}\boldsymbol{x}_{k|k-1}^{(i,j)}\right) = \left| \frac{1}{\boldsymbol{a}\left(\mathbf{C}\boldsymbol{x}_{k|k-1}^{(i,j)}\right)^H \boldsymbol{U}_{\mathbf{n}} \boldsymbol{U}_{\mathbf{n}}^H \boldsymbol{a}\left(\mathbf{C}\boldsymbol{x}_{k|k-1}^{(i,j)}\right)} \right|^{\zeta} \tag{28}$$

where $\mathbf{C} = [\mathbf{1},\mathbf{0}]$, and $\mathbf{C}\boldsymbol{x}_{k|k-1}^{(i,j)}$ represents the azimuth angle information, ζ is the exponential weighting factor. $\boldsymbol{U}_{\mathbf{n}}$ represents the noise subspace obtained by the MUSIC algorithm. The steps of the UT-MeMBer DOA particle filter tracking algorithm are shown in Algorithm 1.

Algorithm 1 UT-MeMBer DOA particle filter tracking algorithm

Input: $\left[\left\{r_{k-1|k-1}^{(j)}, \left\{\omega_{k-1}^{(i,j)}, \boldsymbol{x}_{k-1}^{(i,j)}\right\}_{i=1}^{N_{k-1}}\right\}_{j=1}^{J_{k-1}}, \mathbf{Z}_k\right]$

Time Update

1. Predict the existence probability: $r_{k|k-1}^{j} = r_{P,k|k-1}^{j} + r_{B,k}^{j}$.
where $r_{P,k|k-1}^{j} = r_{k-1}^{j} \cdot \sum_{i=1}^{N_{k-1}} \omega_{k-1}^{(i,j)} \cdot p_{s,k}\left(\boldsymbol{x}_{k-1}^{(i,j)}\right)$ denotes the existence probability of survival model,
$r_{B,k}^{j} = \left(1 - r_{k-1}^{j}\right) \cdot \sum_{i=1}^{N_{B,k}} \omega_{k-1}^{(i,j)} \cdot p_{b,k}\left(\boldsymbol{x}_{k-1}^{(i,j)}\right)$ represents the existence probability of newborn model.
2. Calculate the predicted state of surviving particles: $\left[\left\{\boldsymbol{x}_{k|k-1}^{(i,j)}\right\}_{i=1}^{N_{k-1}}\right] = UT\left[\left\{\boldsymbol{x}_{k-1}^{(i,j)}\right\}_{i=1}^{N_{k-1}}\right]$.
-Calculate the array flow matrix $A\left(\mathbf{C}\boldsymbol{x}_{k-1}^{(i,j)}\right)$;
-Calculate the amplitude of the signal $\mathbf{S} = \left[\boldsymbol{A}(\theta)^H \boldsymbol{A}(\theta)\right]^{-1} \boldsymbol{A}(\theta)^H \mathbf{Z}_k$;
-Calculate the noise variance $\sigma^2 = \frac{1}{P} \sum_{p=1}^{P} \left|\mathbf{Z}_k - \boldsymbol{A}(\theta)\mathbf{S}\right|^2$;
-Select a weighted sample point of $2n_x + 1$ for each particle $\boldsymbol{x}_{k-1}^{(i,j)}$, where
$\chi_0 = \boldsymbol{x}_{k-1}^{(i,j)}, \quad W_0 = \kappa/(n_x + \kappa) s = 0$
$\chi_s = \boldsymbol{x}_{k-1}^{(i,j)} + \left(\sqrt{(n_x + \kappa)\sigma^2}\right), \quad W_s = \kappa/2(n_x + \kappa) s = 1, \ldots, n_x$,
$\chi_s = \boldsymbol{x}_{k-1}^{(i,j)} - \left(\sqrt{(n_x + \kappa)\sigma^2}\right), \quad W_s = \kappa/2(n_x + \kappa) s = n_x + 1, \ldots, 2n_x$
$\kappa = 5$ is a secondary scaling parameter, $n_x = 2$.
-Each sigma point propagates through a nonlinear function: $\gamma_s = f_{k|k-1}(\chi_s), s = 1, \ldots, 2n_x$;
-Compute the mean and covariance of γ_s: $\overline{\psi} = \sum_{s=0}^{2n_x} W_s \gamma_s, P = \sum_{s=0}^{2n_x} W_s\left(\gamma_s - \overline{\psi}\right)\left(\gamma_s - \overline{\psi}\right)^T$;
-Obtain: $\boldsymbol{x}_{k|k-1}^{(i,j)} \sim N\left(\overline{\psi}, P\right)$;
3. Construct a newborn target weighted particle: $\boldsymbol{x}_{k|k-1}^{(i,j)} \sim \beta_k(\boldsymbol{x}_k|\mathbf{Z}_{k-1}), i = N_{k-1} + 1, \ldots, N_{k-1} + N_{B,k}$.
4. Calculate the prediction weight $\omega_{k|k-1}^{(i,j)}, i = 1, \ldots, N_{k|k-1}$ according to (26).
5. Unite weighted particle set:
$\left\{\left(\boldsymbol{x}_{k|k-1}^{(i,j)}, \omega_{k|k-1}^{(i,j)}\right)_{i=1}^{N_{k|k-1}}\right\}_{j=1}^{J_{k|k-1}} = \left\{\left(\boldsymbol{x}_{p,k-1}^{(i,j)}, \omega_{p,k-1}^{(i,j)}\right)_{i=1}^{N_{k-1}}\right\}_{j=1}^{J_{k-1}} \cup \left\{\left(\boldsymbol{x}_{B,k}^{(i,j)}, \omega_{B,k}^{(i,j)}\right)_{i=1}^{N_{B,k}}\right\}_{j=1}^{J_{B,k}}$
where $J_{k|k-1} = J_{k-1} + J_{B,k}$, $N_{k|k-1} = N_{k-1} + N_{B,k}$.

Measurements Update

6. For each particle $\boldsymbol{x}_{k|k-1}^{(i,j)}$, Calculate the likelihood function $g\left(\mathbf{Z}_k \middle| \boldsymbol{x}_{k|k-1}^{(i,j)}\right)$ according to (28).
7. Update existence probability:
$r_{k|k}^{j} = \dfrac{r_{k|k-1}^{j} \cdot \sum_{i=1}^{N_{k|k-1}} g\left(\mathbf{Z}_k \middle| \boldsymbol{x}_{k|k-1}^{(i,j)}\right) \omega_{k|k-1}^{(i,j)} p_{D,k}\left(\boldsymbol{x}_{k|k-1}^{(i,j)}\right)}{1 - r_{k|k-1}^{j} + r_{k|k-1}^{j} \cdot \sum_{i=1}^{N_{k|k-1}} g\left(\mathbf{Z}_k \middle| \boldsymbol{x}_{k|k-1}^{(i,j)}\right) \omega_{k|k-1}^{(i,j)} p_{D,k}\left(\boldsymbol{x}_{k|k-1}^{(i,j)}\right)}$. where $j = 1, \cdots, J_{k|k-1}$.
8. The updated weight is calculated by (27) and normalized $\omega_k^{(i,j)} = \widetilde{\omega}_k^{(i,j)} / \left(\sum_{j=1}^{J_{k|k-1}} \sum_{i=1}^{N_{k|k-1}} \widetilde{\omega}_k^{(i,j)}\right)$.

Resample Step

9. $\left\{\left(\boldsymbol{x}_{k|k-1}^{(i,j)}, \omega_{k|k-1}^{(i,j)}\right)_{i=1}^{N_{k|k-1}}\right\}_{j=1}^{J_{k|k-1}} \rightarrow \left\{\left(\boldsymbol{x}_{k}^{(i,j)}, \omega_{k}^{(i,j)}\right)_{i=1}^{N_k}\right\}_{j=1}^{J_k}$.

Output: $\left\{r_k^j, \left(\boldsymbol{x}_k^{(i,j)}, \omega_k^{(i,j)}\right)_{i=1}^{N_k}\right\}_{j=1}^{J_k}$.

Algorithm 1 gives the pseudo-code of UT-MeMBer DOA particle filter tracking algorithm. The prediction is made in steps 1–5. Step 6 calculates each predicted particle likelihood function which is replaced by the MUSIC spatial spectral function. The update existence probability is calculated in step 7. Step 8 calculates the normalized weight. Particle resampling is performed in step 9. The particle set $\left\{\left\{\omega_k^{(i,j)}, x_k^{(i,j)}\right\}_{i=1}^{N_k}\right\}_{j=1}^{J_k}$ approximates the spatial probability density function $s_{k|k}^j$, and the estimation of updated source can be expressed as $\overline{x}_k = \sum_{i=1}^{N} \omega_k^{(i,j)} \cdot x_k^{(i,j)}$.

6. Simulation Results

Since the traditional MUSIC algorithm cannot solve the multi-source tracking problem when target number is varying, this paper uses FLOM matrix to substitute the covariance matrix to obtain the corresponding MUSIC spatial spectrum, which can be as the particle likelihood function. We proposed a UT-MeMBer DOA tracking algorithm under RFS framework, which can be named as UT-MB-FLOM-MUSIC algorithm. The Generalized Signal to Noise Ratio (GSNR) is defined as

$$\mathrm{GSNR} = 10\log\left(\mathbb{E}\left\{\left|s(k)\right|^2\right\}/\gamma\right) \tag{29}$$

where γ represents the noise dispersion parameter, and GSNR represents the ratio of signal intensity and noise dispersion. In the simulation, different characteristic indices α describe the degree of impact of different noises.

In the following simulation experiments, the estimated performance is evaluated by the root mean square error (RMSE), which is defined as

$$\mathrm{RMSE} = \frac{1}{P}\sum_{p=1}^{P}\frac{1}{\mathrm{MC}}\sum_{j=1}^{\mathrm{MC}}\left(\sqrt{\frac{1}{K}\sum_{i=1}^{K}\left(x_{ij} - \overline{x}_{ij}\right)^2}\right) \tag{30}$$

where x_{ij} and $\overline{x}_{ij}$ represent the estimated values and real values of the azimuth angle in the jth Monte Carlo (MC) simulation experiment at time i, respectively, and P indicates the number of sources at time i.

Assuming that the sources $x_k = \left[\theta_k(t), \dot{\theta}_k(t)\right]^T$ move with a constant velocity $\dot{\theta}_k(t)$ rad/s, the constant velocity (CV) model is given as

$$x_k = F_k x_{k-1} + G v_k \tag{31}$$

where the transfer matrix F_k and G are defined by

$$F_k = \begin{bmatrix} 1 & \Delta T \\ 0 & 1 \end{bmatrix}; \; G = \begin{bmatrix} \Delta T^2/2 \\ \Delta T \end{bmatrix} \tag{32}$$

respectively, where $\Delta T = 1s$ denotes the time step, and v_k is a zero-mean real Gaussian process used to model the disturbance on the source velocity, i.e., $v_k \sim \mathcal{N}(\mathbf{0}, \boldsymbol{\Sigma}_k)$ with $\boldsymbol{\Sigma}_k = 1$.

Experimental conditions are as follows: The number of array elements is $M = 10$, $d = \lambda/2$, the observation time is $K = 50$ s , $L = 100$, GSNR = 10 dB, MC = 100, and $\xi = 5$. The source survival probability $p_{s,k}(x_k) = 0.99$, and the source detection probability $p_{D,k}(x_k) = 0.98$. In the UT-MB-FLOM-MUSIC algorithm prediction step, we assume that there are six new sources at each time, i.e., $J_{B,k} = 6$, all obeying a uniform distribution on $[-\pi/2\,, \pi/2]$ and each new source produces 300 particles, i.e., $N_{B,k} = 300$. In the update step, the MUSIC spatial spectral function is used to replace the likelihood function and is exponentially weighted, which improves the feasibility of the algorithm. In the impulse noise model, the noise is Gaussian noise when $\alpha = 2$. The DOA estimation method based on the MeMBer can be named as MB-MUSIC algorithm, and the DOA estimation method based on the MeMBer of FLOM vector can be named as MB-FLOM-MUSIC algorithm.

6.1. Scenario 1: The Number of Targets Is Not Time-Varying

Consider a linear multi-source scenario with two sources. Since the PASTD algorithm cannot track the time-varying target, all the target survival time are 1–50 s. The initial source state are $x_1 = [-30; -0.5]$, and $x_2 = [5; 0.5]$.

Figure 1a shows the RMSE of angles for four algorithms when running 100 MC at α = 2, GSNR = 10 dB, and Figure 1b shows two source trajectories for a single MC. It can be seen from Figure 1a that the UT-MB-FLOM-MUSIC algorithm proposed in this paper is obviously better than the traditional PASTD and has the highest accuracy when the number of targets is constant. It can be seen in Figure 1b that the algorithm can effectively track the target trajectory, while the PASTD algorithm deviates from the real trajectory at several times.

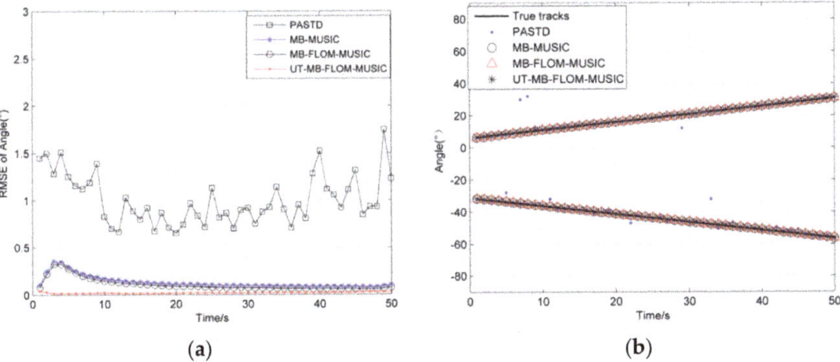

Figure 1. Root mean square error (RMSE) of angle under α = 2, L = 100 and Generalized Signal to Noise Ratio (GSNR) = 10 dB: (**a**) The RMSE of 100 MC; (**b**) source trajectory of Single MC.

We show the RMSE for tracking the multi-source motion when α = 1.3, GSNR = 10 dB, MC = 100, and L = 100 in Figure 2a. It can be seen from Figure 2a that the RMSE of the UT-MB-FLOM-MUSIC algorithm is smaller than that of the other three algorithms. The accuracy of the MB-MUSIC algorithm is significantly reduced in impulse noise, and the PAST algorithm is more accurate than MB-MUSIC. It can be seen from Figure 2b that the MB-MUSIC algorithm cannot effectively track the target trajectory in impulse noise, and the PASTD algorithm also has the problem of inaccurate target tracking. Based on the fact that the above target numbers are unchanged, we will analyze the target time-varying DOA tracking.

6.2. Scenario 2: The Number of Targets Is Time-Varying

Consider a linear multi-source scenario with three sources. The number of sources is time-varying due to births and deaths, the survival time of the four sources is 1–50 s, 10–50 s, 20–45 s, and the initial source states are $x_1 = [-30; -0.5]$, $x_2 = [5; 1.0]$, and $x_3 = [60; -2.0]$.

Figure 3a shows the RMSE of angles for three algorithms for running 100 MC at α = 2, L = 100 and GSNR = 10 dB, and Figure 3b shows three sources trajectory for a single MC. It can be seen from Figure 3 that the likelihood function of the MUSIC spatial spectrum instead of the Multi-Bernoulli particle filter update stage can effectively estimate the target number and motion state, and also verify the feasibility of the literature [14] in the Gaussian noise environment. Although the error is large at time 35, the overall error is below 2 degrees. It can also be seen from Figure 3a that the RMSE of the UT-MB-FLOM-MUSIC algorithm is also smaller than other algorithms even in the Gaussian noise environment.

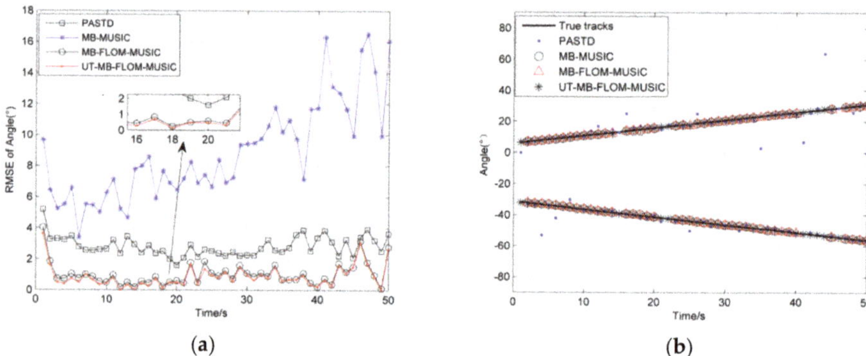

Figure 2. RMSE of angle under $\alpha = 1.3$, $L = 100$ and GSNR = 10 dB: (**a**) The RMSE of 100 MC; (**b**) source trajectory of Single MC.

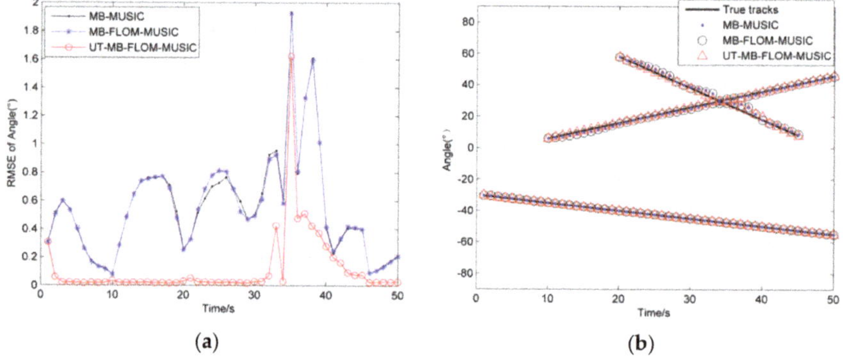

Figure 3. RMSE of angle under $\alpha = 2$, $L = 100$ and GSNR = 10 dB: (**a**) The RMSE of 100 MC; (**b**) source trajectory of Single MC.

Since Gaussian noise does not reflect true signal interference, the α stable distribution can reflect the impact of impulse noise. Figure 4 shows the RMSE and cardinality estimation error plots for three algorithms running 100 MC when the characteristic index α is different and the GSNR = 10 dB, $L = 100$. It can be seen from Figure 4a that, in $\alpha = 1.1 \sim 1.9$, the RMSE error of the three estimation algorithms first decreases, and finally tends to be flat. It also can be seen that the RMSE of the UT-MB-FLOM-MUSIC algorithm is significantly smaller than the MB-FLOM-MUSIC and MB-MUSIC algorithms when $\alpha = 1.1$ or $\alpha = 1.2$, so that the UT-MB-FLOM-MUSIC algorithm has a better effect when handling the impulse noise environment. Since the characteristic index is close to 2 when $\alpha = 1.8$ or $\alpha = 1.9$, Figure 4b shows that the cardinality estimation error of the three algorithms approaches 0. It also shows that it is feasible to use the MUSIC spatial spectrum as a substitute for the likelihood function when the noise environment is close to Gaussian noise while the MUSIC algorithm cannot effectively estimate the number of targets in an impulse noise environment.

In Figure 5, we show the RMSE and cardinality estimation for tracking the multi-source motion when $\alpha = 1.3$ and GSNR = 10 dB, MC = 100. It can be seen from Figure 5 that the RMSE of the UT-MB-FLOM-MUSIC algorithm is smaller than that of the other two algorithms. Although the RMSE will increase when the new target appears or disappears, it will decrease rapidly at the next time step. This phenomenon shows that the Multi-Bernoulli filter has a large recognition performance for the target and can quickly track the state of the target. Table 1 shows the RMSE and computing performance of the MB-MUSIC algorithm, MB-FLOM-MUSIC algorithm and the UT-MB-FLOM-MUSIC algorithm at one MC.

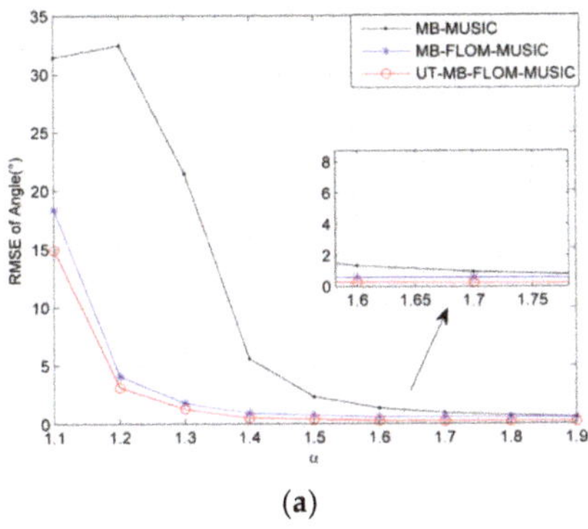

(a)

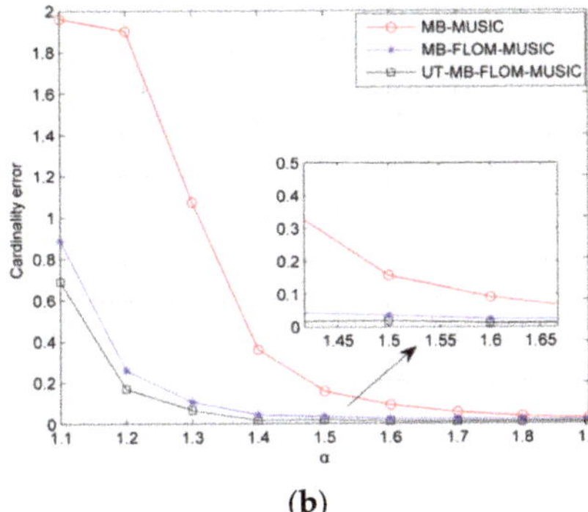

(b)

Figure 4. RMSE and cardinality error of angle under different α, $L = 100$, MC = 100 and GSNR = 10 dB: (**a**) The RMSE under different α; (**b**) cardinality error under different α.

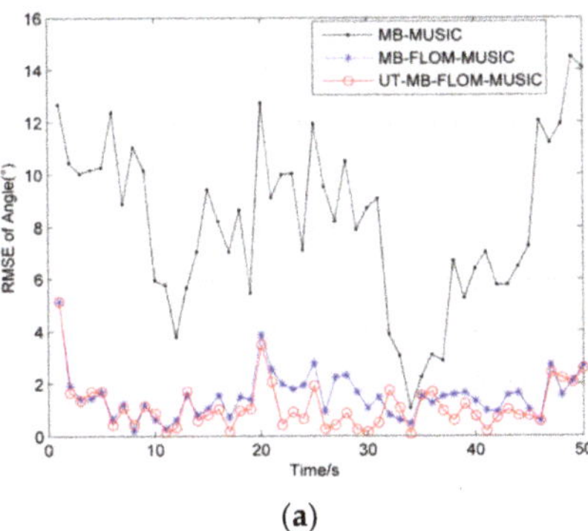

(a)

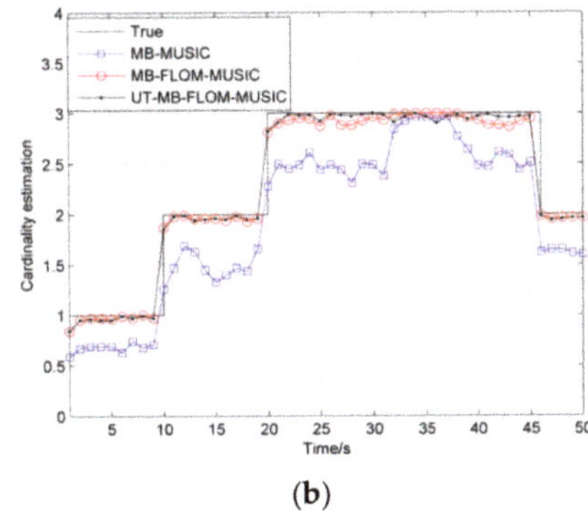

(b)

Figure 5. RMSE and Cardinality estimation of angle under α = 1.3 and GSNR = 10 dB, MC = 100: (**a**) RMSE of angle; (**b**) Cardinality estimation of angle.

Table 1. Running Time (CV model).

Algorithm	RMSE	Running Time/s
MB-MUSIC	7.6012	2.94
MB-FLOM-MUSIC	1.1396	9.59
UT-MB-FLOM-MUSIC	0.2698	114.67

The operating environment includes an Intel (R) Core (TM) i5-8500 CPU @ 3.00 GHz processor and a 64-bit operating system MATLAB 2014. It can be seen from Table 1 that the UT-MB-FLOM-MUSIC algorithm RMSE is smaller than other algorithms when the running time is too long.

Figure 6 analyzes the RMSE and probability of convergence (PROC) for three algorithms running 100 MC when α = 1.3 and GSNR = 0–16 dB. where PROC $= \frac{1}{K}\sum_{i=1}^{K}\sum_{j=1}^{MC} 1_{ij}/MC \times 100\%$, and 1_{ij} is defined as $1_{ij} = \begin{cases} 1, \left|x_{ij} - \overline{x}_{ij}\right| < \varepsilon \\ 0, otherwise \end{cases}$. let $\varepsilon = 1$. It can be seen from Figure 6a that the MB-FLOM-MUSIC and UT-MB-FLOM-MUSIC algorithms have higher accuracy than the MB-MUSIC in an impulse noise environment, and the UT-MB-FLOM-MUSIC algorithm has higher accuracy under the high GSNR. It can be seen form Figure 6b that as the SNR increases, the PROC increases. And at the same GSNR, the performance of the MB-FLOM-MUSIC algorithm is more significant.

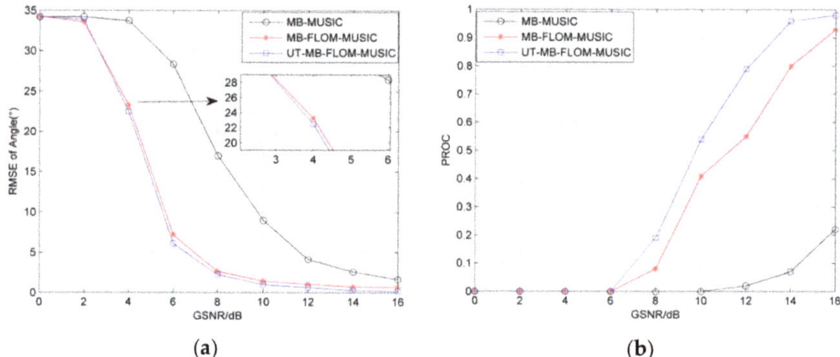

(a) (b)

Figure 6. RMSE and probability of convergence (PROC) of the angle under different GSNR, $\alpha = 1.3$ MC = 100 and $L = 100$: (**a**) RMSE of angle; (**b**) PROC at different GSNR.

Figure 7 shows the RMSE of three algorithms running 100 MC when $\alpha = 1.3$ and the snapshot number $L = 50, 100, 150$. It can be seen that the UT-MB-FLOM-MUSIC algorithm has the smallest RMSE and it works best when the snapshot number is $L = 150$.

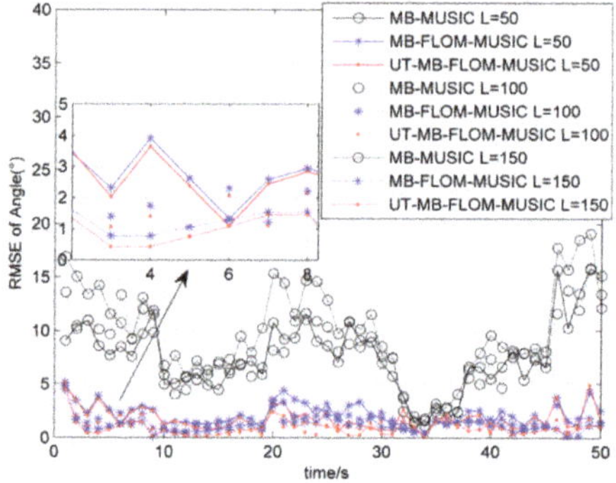

Figure 7. RMSE for source tracking under different snapshots, $\alpha = 1.3$ MC = 100 and GSNR = 10 dB.

6.3. Scenario 3: The Number of Targets Is Time-Varying and Maneuvering

Consider a nonlinear multi-source scenario with three sources. The number of sources is time-varying due to births and deaths, and the survival time of the three sources is 1–50 s, 10–50 s, 20–45 s, and the initial source state are $x_1 = [-30; -0.5]$, $x_2 = [5; 1.8]$, and $x_3 = [60; -2.0]$. The state transition matrix of the collaborative turning (CT) model is

$$F_k = \begin{bmatrix} 1 & \sin(T\omega)/\omega \\ 0 & \cos(T\omega) \end{bmatrix} \tag{33}$$

where $\omega = 0.25$ *rad* and other experimental conditions are the same as scenario 1.

Figure 8 shows the maneuvering target trajectory of three algorithms running one MC when $\alpha = 1.3$, $L = 100$, and GSNR = 10 dB. It can be clearly seen from Figure 8 that the three methods lose the target when the target crosses at time 33, but after time 36, the MB-FLOM-MUSIC algorithm and the UT-MB-FLOM-MUSIC algorithm can still capture the target state. Compared with the MB-FLOM-MUSIC algorithm, the target state estimation value of the UT-MB-FLOM-MUSIC algorithm is closer to the true value.

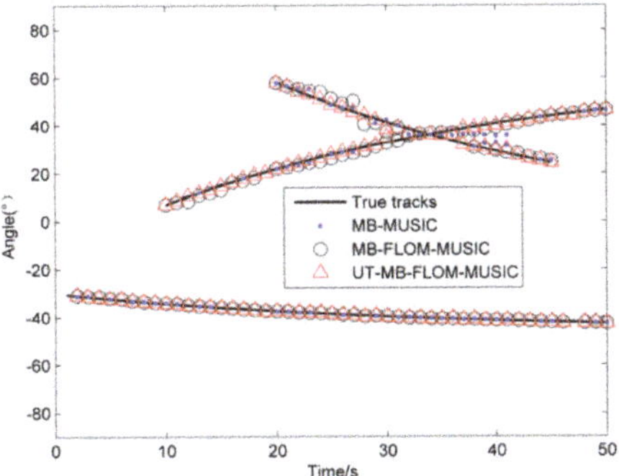

Figure 8. Target trajectory, $\alpha = 1.3$, L = 100, and GSNR = 10 dB.

In Figure 9, we show the RMSE and cardinality estimation for tracking the multi-source motion when $\alpha = 1.3$ and GSNR = 10 dB, MC = 100. It can be seen from Figure 9a that the RMSE of the UT-MB-FLOM-MUSIC algorithm is smaller than that of the other two algorithms. As can be seen from Figure 9b, when the target is maneuvering, the target is not captured by the three algorithms from time 33, but after time 36, the MB-FLOM-MUSIC algorithm and UT-MB-FLOM-MUSIC algorithm can still estimate the number of targets in time. Compared with the result of Figure 5b, the performance to capture targets of the UT-MB-FLOM-MUSIC algorithm is significantly weakened.

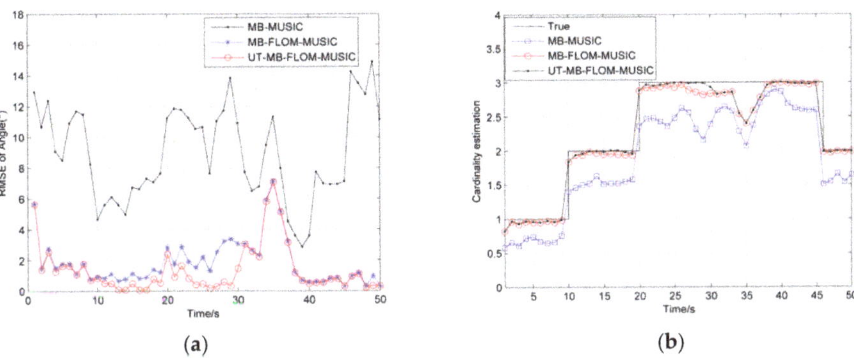

Figure 9. RMSE and cardinality estimation of angle under $\alpha = 1.3$ and GSNR = 10 dB, MC = 100: (**a**) RMSE of angle; (**b**) cardinality estimation of angle.

Table 2 shows the RMSE and computing performance of the MB-MUSIC algorithm, MB-FLOM-MUSIC algorithm and the UT-MB-FLOM-MUSIC algorithm. Compared with the results in Table 1, the RMSE and running time of the three algorithms are increased when the target is maneuvered. The RMSE of UT-MB-FLOM-MUSIC algorithm is smaller than other two algorithms when the running time is long.

Table 2. Running Time (CT model).

Algorithm	RMSE	Running Time/s
MB-MUSIC	8.7728	3.67
MB-FLOM-MUSIC	1.3198	10.73
UT-MB-FLOM-MUSIC	0.6102	135.30

7. Conclusions

A DOA tracking algorithm based on the UT-MeMBer particle filter in an impulse noise environment is proposed in this paper. Since the FLOM matrix is used instead of the covariance matrix, the spatial spectrum based on FLOM can well reflect the real DOA in impulse noise environment. For the persistent surviving particles, the sigma point is selected by UT to approximate the posterior density of the state to improve the performance of the particle. Then, the MUSIC spatial spectral function of the FLOM matrix is used to represent the likelihood function of the particle. And the weighting of the likelihood function can further increase the weight of the particles in the high likelihood region. The results show that the UT-MB-FLOM-MUSIC algorithm is more effective than the PASTD, MB-MUSIC, and MB-FLOM-MUSIC algorithms in an impulse noise environment. The advantage of this algorithm is that the MeMBer filter can operate the array data more directly, and can effectively track the target number of time-varying DOA. The shortcoming of this algorithm is that it takes a long time. Our future work will focus on how to improve the efficiency of the algorithm, maneuvering target tracking in other noisy environments, etc.

Author Contributions: The work presented here was carried out in collaboration between follow authors. S.-y.W., R.-h.C., and Q.-t.X. defined the research theme. J.Z. and X.-d.D. designed the methods and experiments, carried out the simulation experiments. J.Z. interpreted the results and wrote the paper.

Funding: This work is supported by National Natural Science Foundation of China (Grant 61561016, 61962012), Guangxi Natural Science Foundation (Grant 2016GXNSFAA380073), Guangxi Key Laboratory of Cryptography and Information Security (GCIS201611), Guangxi Colleges and Universities Key Laboratory of Data Analysis and Computation, and the GUET Excellent Gradute Thesis Program (Grant 17YJPYSS23).

Conflicts of Interest: The authors declare no conflict of interest.

References

1. Kim, H.; Viberg, M. Two decades of array signal processing research. *IEEE Signal Process. Mag.* **1996**, *13*, 67–94.
2. Van Trees, H.L. *Optimum Array Processing: Part IV of Detection, Estimation, and Modulation Theory*; John Wiley & Sons: Hoboken, NJ, USA, 2004.
3. Zhou, C.; Gu, Y.; Fan, X.; Shi, Z.; Mao, G.; Zhang, Y.D. Direction-of-Arrival Estimation for Coprime Array via Virtual Array Interpolation. *IEEE Trans. Signal Process.* **2018**, *22*, 5956–5971. [CrossRef]
4. Zhou, C.; Gu, Y.; Shi, Z.; Zhang, Y.D. Off-Grid Direction-of-Arrival Estimation Using Coprime Array Interpolation. *IEEE Signal Process. Lett.* **2018**, *25*, 1710–1714. [CrossRef]
5. Shi, Z.; Zhou, C.; Gu, Y.; Goodman, N.A.; Qu, F. Source estimation using coprime array: A sparse reconstruction perspective. *IEEE Sens. J.* **2017**, *17*, 755–765. [CrossRef]
6. Zhou, C.; Gu, Y.; Zhang, Y.D.; Shi, Z.; Jin, T.; Wu, X. Compressive Sensing based Coprime Array Direction-of-Arrival Estimation. *IET Commun.* **2017**, *11*, 1719–1724. [CrossRef]
7. Yang, B. Projection approximation subspace tracking. *IEEE Trans. Signal Process.* **1995**, *43*, 95–107. [CrossRef]
8. Yang, B. Convergence analysis of the subspace tracking algorithms PAST and PASTD. In Proceedings of the IEEE International Conference on Acoustics, Speech, and Signal Processing, Atlanta, GA, USA, 9 May 1996; pp. 1759–1762.
9. Chan, S.C.; Wen, Y.; Ho, K.L. A robust past algorithm for subspace tracking in impulsive noise. *IEEE Trans. Signal Process.* **2006**, *54*, 105–116. [CrossRef]
10. Liao, B.; Zhang, Z.G.; Chan, S.C. A New Robust Kalman Filter-Based Subspace Tracking Algorithm in an Impulsive Noise Environment. *IEEE Trans. Circuits Syst. II Express Briefs* **2010**, *57*, 740–744. [CrossRef]
11. Chan, S.C.; Zhang, Z.G.; Zhou, Y. A new adaptive Kalman filter-based subspace tracking algorithm and its application to DOA estimation. In Proceedings of the IEEE International Symposium on Circuits and Systems, Island of Kos, Greece, 21–24 May 2006; pp. 129–132.
12. Nikias, C.L.; Shao, M. *Signal Processing with Alpha-Stable Distributions and Applications*; John Wiley and Sons Inc.: Hoboken, NJ, USA, 1995; pp. 10–15.
13. Zha, D.; Qiu, T. Underwater sources location in non-Gaussian impulsive noise environments. *Digit. Signal Process.* **2006**, *16*, 149–163. [CrossRef]

14. Shao, M.; Nikias, C.L. Signal processing with fractional lower order moments: Stable processes and their applications. *IEEE Proc.* **1993**, *81*, 986–1010. [CrossRef]
15. Li, S.; He, R.; Lin, B.; Sun, F. DOA estimation based on sparse representation of the fractional lower order statistics in impulsive noise. *IEEE CAA J. Autom. Sin.* **2018**, *5*, 98–106. [CrossRef]
16. Shi, Y.; Mao, X.P.; Qian, C.; Liu, Y.T. Robust relaxation for coherent DOA estimation in impulsive noise. *IEEE Signal Process. Lett.* **2019**, *3*, 410–414. [CrossRef]
17. Zhang, J.; Qiu, T. A robust correntropy based subspace tracking algorithm in impulsive noise environments. *Digit. Signal Process.* **2017**, *62*, 168–175. [CrossRef]
18. Risfic, B.; Arulampalam, S.; Gordon, N. *Beyond the Kalman Filter: Particle Filters for Tacking Application*; Artech House: London, UK, 2004; pp. 35–62.
19. Zhong, X.; Premkumar, A.B.; Madhukumar, A.S. Particle filtering for acoustic source tracking in impulsive noise with alpha-stable process. *IEEE Sens. J.* **2012**, *13*, 589–600. [CrossRef]
20. Mahler, R. *Statistical Multi-Source Multi-Target Information Fusion*; Artech House, Inc.: London, UK, 2007; pp. 228–234.
21. Ristic, B.; Vo, B.T.; Vo, B.N.; Farina, A. A Tutorial on Bernoulli Filters: Theory, Implementation and Applications. *IEEE Trans. Signal Process.* **2013**, *61*, 3406–3430. [CrossRef]
22. Zhang, G.; Zheng, C.; Sun, S.; Liang, G.; Zhang, Y. Joint Detection and DOA Tracking with a Bernoulli Filter Based on Information Theoretic Criteria. *Sensors* **2018**, *18*, 3473. [CrossRef] [PubMed]
23. Vo, B.T.; Vo, B.N.; Cantoni, A. The cardinality balanced multi-target Multi-Bernoulli filter and its implementations. *IEEE Trans. Signal Process.* **2009**, *57*, 409–423.
24. Choppala, P.B.; Teal, P.D.; Frean, M.R. Adapting the Multi-Bernoulli filter to phased array observations using MUSIC as pseudo-likelihood. In Proceedings of the International Conference on Information Fusion, Salamanca, Spain, 7–10 July 2014; pp. 1–6.
25. Tsakalides, P.; Nikias, C.L. The robust covariation-based MUSIC (ROC-MUSIC) algorithm for bearing estimation in impulsive noise environments. *IEEE Trans. Signal Process.* **2002**, *44*, 1623–1633. [CrossRef]

Article

Refining Inaccurate Transmitter and Receiver Positions Using Calibration Targets for Target Localization in Multi-Static Passive Radar

Yongsheng Zhao, Dexiu Hu *, Yongjun Zhao, Zhixin Liu and Chuang Zhao

PLA Strategic Support Force Information Engineering University, Zhengzhou 450001, China
* Correspondence: paper_hdx@126.com

Received: 30 May 2019; Accepted: 24 July 2019; Published: 31 July 2019

Abstract: Transmitter and receiver position errors have been known to significantly deteriorate target localization accuracy in a multi-static passive radar (MPR) system. This paper explores the use of calibration targets, whose positions are known to the MPR system, to counter the loss in target localization accuracy arising from transmitter/receiver position errors. This paper firstly evaluates the Cramér–Rao lower bound (CRLB) for bistatic range (BR)-based target localization with calibration targets, which analytically indicates the potential of calibration targets in enhancing localization accuracy. After that, this paper proposes a novel closed-form solution, which includes two steps: calibration step and localization step. Firstly, the calibration step is devoted to refine the inaccurate transmitter and receiver locations using the BR measurements from the calibration targets, and then in the calibration step, the target localization can be accurately achieved by using the refined transmitter/receiver positions and the BR measurements from the unknown target. Theoretical analysis and simulation results indicate that the proposed method can attain the CRLB at moderate measurement noise level, and exhibits the superiority of localization accuracy over existing algorithms.

Keywords: multi-static passive radar; target localization; calibration target; bistatic range; transmitter and receiver position error; Cramér–Rao lower bound

---

1. Introduction

Passive radar technology, which allows operators to detect and localize potential targets using already existing transmitters such as commercial frequency modulation (FM) broadcast/digital audio broadcast (DAB)/terrestrial digital video broadcast (DVB-T) [1,2] and non-cooperative radar transmission [3,4], has been interesting to both civilian and military fields in the last few decades [5]. This sort of radar, compared to active radar technology, offers numerous advantages including lower cost, lower power usage and more covert surveillance capability, which suggests the possibility of employing passive radars on a wide range of concerned applications such as homeland security, costal surveillance and early warning system for vehicles detection, etc.

One of the remarkable characters of passive radar is the deployment of two receiving channels, with one for capturing the direct path signal from the transmitter and the other for collecting the potential target echoes [6]. By performing a cross-correlation operation between the direct path signal and the target echoes, the time delay (TD) could be measured, which holds information about the unknown target position. By multiplying with the signal propagation speed, the TD can be directly converted into the bistatic range (BR) [7]. Each BR measurement traces out an ellipsoid equation, with its foci located at the transmitter and the receiver positions. Theoretically, for multi-static passive radar (MPR), if enough BR measurements with respect to multiple transmitter-receiver pairs are available, the target position can be determined by solving the set of nonlinear ellipsoid equations.

However, due to the high nonlinearity implied in the BR measurement equations, determining the target position from the BR measurements obtained at a single time instant is not a trivial topic.

In recent years, numerous algorithms have been developed to address this challenging topic, mainly including iterative methods [8,9] and closed-form solution methods [10–14]. Iterative methods [8,9] rely on an initial position guess close to the true solution but such a good guess may not always be available in reality. By contrast, closed-form solution methods [10–14] have always been more attractive to researchers due to their computational efficiency, independence on initial guess and absence of divergence problem. Illuminated by Ho and Xu's two-step weighted least squares (2WLS) idea [15], these closed-form solution methods [10–14] generally follow the basic two-step framework below: in the first step, the non-linear BR equations are linearized into pseudo-linear ones by introducing some proper nuisance parameters and a coarse estimate of target position can be obtained from the pseudo-linear equation set via weighted least squares (WLS) minimization; then in the second step, the function relation between the target position and the introduced nuisance parameters is explored to refine the initial estimate.

Nevertheless, all the aforementioned methods are designed based on the assumption that the positions of the transmitters and receivers are exactly known, but such exact priori knowledge may not be available in reality. In fact, the positions of the transmitters and receivers are inevitably perturbed by errors to some extent, and these errors (also referred to as position uncertainties) are often non-negligible, especially when the antennas are mounted on moving platforms [16,17] or the exploited transmitters are highly non-cooperative (such as the hostile radar radiation whose position could usually only roughly determined by electronic reconnaissance technique [18,19]). On the other hand, Rui and Ho [20] quantitatively analyze the influence of the transmitter and receiver position error on the localization accuracy, indicating that the target localization accuracy can be very sensitive to the transmitter/receiver position error and a slight error in transmitter/receiver position could remarkably deteriorate the localization accuracy. More recently, some novel methods [21,22] that take the statistical distributions of transmitter/receiver position error into consideration are developed to reduce the target localization error, and they are shown to attain the Cramér–Rao lower bound (CRLB) under small measurement noise and transmitter/receiver position error assumption. Nevertheless, these methods [21,22] only present the solutions when the transmitter/receiver position errors exist, but cannot fundamentally compensate the localization performance loss arising from transmitter/receiver position error at the CRLB level.

The use of calibration sensors has been a common technique in wireless sensor network self-localization, where each sensor broadcasts signals and receives signals from other sensors so as to determine their positions collaboratively [23,24]. Syldatk [25,26] considers the calibration of ground sensor networks where an accurate calibration of sensor positions and orientations is required for target tracking. For source localization problem, Hasan first considered in [27] utilizing calibration sensors to improve the angle-based source localization performance. Ho [28], Yang [29] as well as Li [30] et al. further expanded and applied the calibration technique to range difference (RD)-based source localization problem, where additional RD measurements from the calibration sensors were incorporated to reduce the receivers' position error and thus improve the source localization accuracy. The successful use of calibration techniques in these fields inspires us the possibility of employing calibration technique in the target localization for multi-static passive radar. When it comes to BR-based target localization in multi-static passive radar, using a 'calibration target' with known position may also be able to mitigate the target localization performance loss arising from the transmitter/receiver position error. In theory, any target appearing in the radar coverage area and meanwhile broadcasting its position can be taken as a calibration target. Typically, for example, to avoid potential accidents and collisions, the commercial aircrafts will report their positions and other information to the ground stations and other aircraft by using the automatic dependent surveillance broadcast (ADS-B) system [31]. Hence, the commercial aircraft broadcasting ADS-B signal can be regarded as an off-the-shelf calibration target. If no such off-the-shelf calibration targets are available in the radar coverage area, we can manually launch some strong scatterers with known positions as calibration targets. However, despite that, up to now there exists no publication in the open literature that addresses refining inaccurate transmitter and receiver positions using calibration targets for target localization in multi-static passive radar.

Motivated by these facts, in this paper, taking the transmitter and receiver position error into consideration, we explore using calibration targets to counter the loss in BR-based target localization accuracy arising from the transmitter/receiver position error. We begin our work by deriving the CRLB for BR-based target position estimation when the BR measurements from the calibration targets are available. The interpretation on the CRLB demonstrates that the use of calibration targets can significantly mitigate the influence of the transmitter/receiver position error and dramatically enhance the localization accuracy, at least in the sense of CRLB. We then proceed to develop a novel localization method to alleviate the transmitter and receiver position error and enhance the localization accuracy using calibration targets. It mainly includes two processing stages, referred to as calibration stage and localization stage respectively. In the calibration stage, the BR measurements from the calibration targets are exploited to refine the inaccurate transmitter and receiver positions; in the localization stage, the refined transmitter and receiver positions and BR measurements corresponding to the unknown target are exploited to determine the target position. Both processing stages are closed-form, which brings the proposed method computational efficiency and high robustness. Furthermore, the accuracy of the proposed solution is shown analytically to reach the CRLB when the transmitters'/receivers'/calibration targets' position errors and the BR measurement noises are sufficiently small. Simulations will be conducted to verify the effectiveness and superiority of the proposed solution over existing methods.

Notations: There will be a lot of notations throughout this paper. Without exception, vectors (matrices) are denoted by bold lower (upper) case letters, respectively. Also, notations $(\cdot)^{\circ}$, $(\cdot)^{\mathrm{T}}$, $\|\cdot\|$, $(\cdot)^{-1}$, $\mathrm{E}(\cdot)$, $\mathbf{O}_{p\times q}$, $\mathbf{I}_{p\times p}$, $0_{p\times 1}$, $\mathrm{diag}(\cdot)$ and $\mathrm{tr}(\cdot)$, represent the true value of a noisy or an estimated variable, transpose operation, Euclidean norm, inverse of matrix, statistical expectation, a p-by-q zero matrix, an identity matrix of size p, a p-by-1 zero vector, diagonal matrix and the trace of a matrix, respectively.

The remainder of this paper is organized as follows. Section 2 is about the localization scenario in the presence of transmitter/receiver position error and calibration targets. In Section 3, the corresponding CRLB is evaluated, indicating the potential of calibration targets in improve localization accuracy. In Section 4, a closed-form solution is developed for target localization in the presence of transmitter/receiver position error and calibration targets, and theoretical performance analysis is also given. Section 5 describes the results of Monte Carlo simulations that compare the proposed solution with existing methods. Section 6 is the conclusion.

2. Problem Formulation

Address a typical multi-static passive radar localization scenario as presented in Figure 1, where M non-cooperative transmitters located at $\mathbf{s}^{o}_{t,m} = [x^{o}_{t,m}, y^{o}_{t,m}, z^{o}_{t,m}]^{\mathrm{T}}$ $(m = 1, 2, \ldots, M)$ are employed to illuminate the surveillance area, and N receivers located at $\mathbf{s}^{o}_{r,n} = [x^{o}_{r,n}, y^{o}_{r,n}, z^{o}_{r,n}]^{\mathrm{T}}$ $(n = 1, 2, \ldots, N)$ are deployed to determine a single target's position denoted by $\mathbf{u}^{o} = [x^{o}, y^{o}, z^{o}]^{\mathrm{T}}$. In fact, the exact positions of the transmitters and receivers might not be known, and only the inaccurate measured versions, i.e., $\mathbf{s}_{t,m} = [x_{t,m}, y_{t,m}, z_{t,m}]^{\mathrm{T}}$ and $\mathbf{s}_{r,n} = [x_{r,n}, y_{r,n}, z_{r,n}]^{\mathrm{T}}$, are available for processing. Formulaically, we arrive at

$$\mathbf{s}_{t,m} = \mathbf{s}^{o}_{t,m} + \Delta\mathbf{s}_{t,m} \tag{1}$$

$$\mathbf{s}_{r,n} = \mathbf{s}^{o}_{r,n} + \Delta\mathbf{s}_{r,n} \tag{2}$$

where $\Delta\mathbf{s}_{t,m}$ and $\Delta\mathbf{s}_{r,n}$ are the position error of the mth transmitter and the nth receiver respectively and also referred to as position uncertainty. Stacking (1) and (2) with respect to all the transmitters and receivers, yields a $3(M+N)$-by-1 transmitter and receiver position vector as

$$\mathbf{s} = \mathbf{s}^{o} + \Delta\mathbf{s} \tag{3}$$

where $\mathbf{s} = [\mathbf{s}_t^T, \mathbf{s}_r^T]^T$ with $\mathbf{s}_t = [\mathbf{s}_{t,1}^T, \mathbf{s}_{t,2}^T, \ldots, \mathbf{s}_{t,M}^T]^T$ and $\mathbf{s}_r = [\mathbf{s}_{r,1}^T, \mathbf{s}_{r,2}^T, \ldots, \mathbf{s}_{r,N}^T]^T$ is the noisy transmitter and receiver position vector, $\mathbf{s}^o = [(\mathbf{s}_t^o)^T, (\mathbf{s}_r^o)^T]^T$ with $\mathbf{s}_t^o = [(\mathbf{s}_{t,1}^o)^T, (\mathbf{s}_{t,2}^o)^T, \ldots, (\mathbf{s}_{t,M}^o)^T]^T$ and $\mathbf{s}_r^o = [(\mathbf{s}_{r,1}^o)^T, (\mathbf{s}_{r,2}^o)^T, \ldots, (\mathbf{s}_{r,N}^o)^T]^T$ is the true transmitter and receiver position vector, and $\Delta\mathbf{s} = [\Delta\mathbf{s}_t^T, \Delta\mathbf{s}_r^T]^T$ with $\Delta\mathbf{s}_t = [\Delta\mathbf{s}_{t,1}^T, \Delta\mathbf{s}_{t,2}^T, \ldots, \Delta\mathbf{s}_{t,M}^T]^T$ and $\Delta\mathbf{s}_r = [\Delta\mathbf{s}_{r,1}^T, \Delta\mathbf{s}_{r,2}^T, \ldots, \Delta\mathbf{s}_{r,N}^T]^T$ is the transmitter and receiver position error vector that can be assumed zero-mean Gaussian with covariance $\mathbf{Q}_s$ without loss of generality.

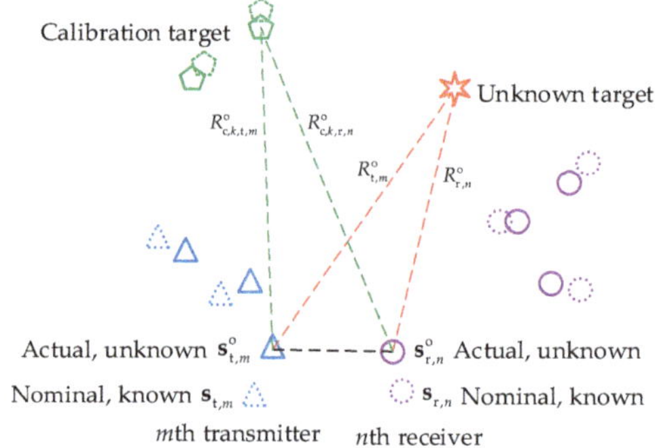

Figure 1. Practical scenario geometry of multi-static passive radar in the presence of transmitter/receiver position error and calibration targets.

Using the above notations, the distance from the mth transmitter to the target of interest is equal to

$$R^o_{t,m} = ||\mathbf{u}^o - \mathbf{s}^o_{t,m}|| \tag{4}$$

the distance from the target of interest to the nth receiver is equal to

$$R^o_{r,n} = ||\mathbf{u}^o - \mathbf{s}^o_{r,n}|| \tag{5}$$

and the baseline distance with respect to the mth transmitter and nth receiver is

$$R^o_{t,m,r,n} = ||\mathbf{s}^o_{t,m} - \mathbf{s}^o_{r,n}|| \tag{6}$$

According to this, the BR measurement with respect to the mth transmitter and nth receiver, i.e., the sum of the distances from the mth transmitter to the target and the target to the nth receiver, can be formulized as

$$\begin{aligned} r_{m,n} &= r^o_{m,n} + \Delta r_{m,n} \\ &= R^o_{t,m} + R^o_{r,n} - R^o_{t,m,r,n} + \Delta r_{m,n} \\ &= ||\mathbf{u}^o - \mathbf{s}^o_{t,m}|| + ||\mathbf{u}^o - \mathbf{s}^o_{r,n}|| - ||\mathbf{s}^o_{t,m} - \mathbf{s}^o_{r,n}|| + \Delta r_{m,n} \end{aligned} \tag{7}$$

where $r^o_{m,n} = R^o_{t,m} + R^o_{r,n} - R^o_{t,m,r,n}$ represents the true BR with respect to the mth transmitter and nth receiver, $\Delta r_{m,n}$ is the BR measurement noise. Herein, it is important to emphasize that the time delay measurement comes from the cross correlation operation between the target signal and the direct path reference signal [7], and its error characteristics are not affected by the transmitter/receiver position error since the transmitter/receiver position error is not involved in the estimation of time delay. Thus, the BR measurement noise is only related to the time delay measurement noise, and not related to the transmitter/receiver position error. Obviously, there will be MN BR measurements to be produced with respect to the M transmitters and N receivers, which can be recast into a MN-by-1 vector as

$$\mathbf{r} = \mathbf{r}^o + \Delta\mathbf{r} \tag{8}$$

where $\mathbf{r} = [\mathbf{r}_1^T, \mathbf{r}_2^T, \ldots, \mathbf{r}_M^T]^T$ with $\mathbf{r}_m = [r_{m,1}, r_{m,2}, \ldots, r_{m,N}]^T$ is the BR measurement vector, $\mathbf{r}^o = [(\mathbf{r}_1^o)^T, (\mathbf{r}_2^o)^T, \ldots, (\mathbf{r}_M^o)^T]^T$ with $\mathbf{r}_m^o = [r_{m,1}^o, r_{m,2}^o, \ldots, r_{m,N}^o]^T$ is the true BR vector, and $\Delta\mathbf{r} = [\Delta\mathbf{r}_1^T, \Delta\mathbf{r}_2^T, \ldots, \Delta\mathbf{r}_M^T]^T$ with $\Delta\mathbf{r}_m = [\Delta r_{m,1}, \Delta r_{m,2}, \ldots, \Delta r_{m,N}]^T$ is the BR measurement noise vector, which is usually assumed follow a Gaussian distribution with zero-mean and covariance $\mathbf{Q}_\mathbf{r}$.

As presented in Figure 1, to alleviate the transmitter/receiver position error and enhance localization accuracy, K calibration targets located at $\mathbf{c}_k^o = [x_{c,k}^o, y_{c,k}^o, z_{c,k}^o]^T$ $(k = 1, 2, \ldots, K)$ are employed, and the BRs among the calibration targets and the transmitter-receiver pairs are also measured. Similarly, the exact positions of the calibration targets are not known to us, and the nominal versions denoted by $\mathbf{c}_k = [x_{c,k}, y_{c,k}, z_{c,k}]^T$ $(k = 1, 2, \ldots, K)$ are given as

$$\mathbf{c}_k = \mathbf{c}_k^o + \Delta\mathbf{c}_k \tag{9}$$

where $\Delta\mathbf{c}_k$ is the position error of the kth calibration target. Collecting (9) for all the K calibration targets forms a $3K$-by-1 calibration target position vector as

$$\mathbf{c} = \mathbf{c}^o + \Delta\mathbf{c} \tag{10}$$

where $\mathbf{c} = [\mathbf{c}_1^T, \mathbf{c}_2^T, \ldots, \mathbf{c}_K^T]^T$ is the nominal calibration target position vector, $\mathbf{c}^o = [(\mathbf{c}_1^o)^T, (\mathbf{c}_2^o)^T, \ldots, (\mathbf{c}_K^o)^T]^T$ is the true calibration target position vector, and $\Delta\mathbf{c} = [\Delta\mathbf{c}_1^T, \Delta\mathbf{c}_2^T, \ldots, \Delta\mathbf{c}_K^T]^T$ is the calibration target position error vector that is usually supposed to obey Gaussian distribution with zero-mean and covariance $\mathbf{Q}_\mathbf{c}$. Herein, it should be pointed out that, the positions of calibration targets are generally considered to be more precise compared with those of the transmitters and receivers, although they are also contaminated by errors.

Then, the distance from the mth transmitter to the kth calibration target is given by

$$R_{c,k,t,m}^o = ||\mathbf{c}_k^o - \mathbf{s}_{t,m}^o|| \tag{11}$$

and the distance from the kth calibration target to the nth receiver is given by

$$R_{c,k,r,n}^o = ||\mathbf{c}_k^o - \mathbf{s}_{r,n}^o|| \tag{12}$$

Based on this, the BR measurement corresponding to the kth calibration target, mth transmitter and nth receiver can be modeled as

$$\begin{aligned} r_{c,k,m,n} &= r_{c,k,m,n}^o + \Delta r_{c,k,m,n} \\ &= R_{c,k,t,m}^o + R_{c,k,r,n}^o - R_{t,m,r,n}^o + \Delta r_{c,k,m,n} \\ &= ||\mathbf{c}_k^o - \mathbf{s}_{t,m}^o|| + ||\mathbf{c}_k^o - \mathbf{s}_{r,n}^o|| - ||\mathbf{s}_{t,m}^o - \mathbf{s}_{r,n}^o|| + \Delta r_{c,k,m,n} \end{aligned} \tag{13}$$

where $\Delta r_{c,k,m,n}$ represents measurement noise in $r_{c,k,m,n}$, $r_{c,k,m,n}^o = R_{c,k,t,m}^o + R_{c,k,r,n}^o - R_{t,m,r,n}^o$ represents the true BR with respect to the kth calibration target, mth transmitter and nth receiver. Collecting (13) for the set of K calibration targets, M transmitters and N receivers, results in a KMN-by-1 vector as

$$\mathbf{r}_c = \mathbf{r}_c^o + \Delta\mathbf{r}_c \tag{14}$$

where $\mathbf{r}_c = [\mathbf{r}_{c,1}^T, \mathbf{r}_{c,2}^T, \ldots, \mathbf{r}_{c,K}^T]^T$ with $\mathbf{r}_{c,k} = [\mathbf{r}_{c,k,1}^T, \mathbf{r}_{c,k,2}^T, \ldots, \mathbf{r}_{c,k,M}^T]^T$ and $\mathbf{r}_{c,k,m} = [r_{c,k,m,1}, r_{c,k,m,2}, \ldots, r_{c,k,m,N}]^T$ denotes the BR measurement vector from the calibration targets, $\mathbf{r}_c^o = [(\mathbf{r}_{c,1}^o)^T, (\mathbf{r}_{c,2}^o)^T, \ldots, (\mathbf{r}_{c,K}^o)^T]^T$ with $\mathbf{r}_{c,k}^o = [(\mathbf{r}_{c,k,1}^o)^T, (\mathbf{r}_{c,k,2}^o)^T, \ldots, (\mathbf{r}_{c,k,M}^o)^T]^T$ and $\mathbf{r}_{c,k,m}^o = [r_{c,k,m,1}^o, r_{c,k,m,2}^o, \ldots, r_{c,k,m,N}^o]^T$ denotes the corresponding true value vector, $\Delta\mathbf{r}_c = [\Delta\mathbf{r}_{c,1}^T, \Delta\mathbf{r}_{c,2}^T, \ldots, \Delta\mathbf{r}_{c,K}^T]^T$ with $\Delta\mathbf{r}_{c,k} = [\Delta\mathbf{r}_{c,k,1}^T, \Delta\mathbf{r}_{c,k,2}^T, \ldots, \Delta\mathbf{r}_{c,k,M}^T]^T$ and $\Delta\mathbf{r}_{c,k,m} = [\Delta r_{c,k,m,1}, \Delta r_{c,k,m,2}, \ldots, \Delta r_{c,k,m,N}]^T$ denotes the corresponding error vector, which is presumed to be a Gaussian random vector with zero mean and covariance $\mathbf{Q}_{rc}$.

Now, the purpose of this work is to determine the target position from the noisy BR measurements and the inaccurate transmitter/receiver positions. In particular, the calibration targets with known position and the corresponding BR measurements are also available to reduce the transmitter/receiver position error and improve localization accuracy.

3. Evaluation of the CRLB with Calibration Targets

The CRLB does not address the specific estimators employed, but simply reflects minimum possible variance that an unbiased estimator can achieve with existing observations. In this section, in order to justify the necessity of refining the inaccurate transmitter and receiver positions using calibration targets, we shall first set up the CRLB for the target localization problem described above. Besides the BR measurement noise, the position errors of transmitter, receivers and calibration targets are also included. From the localization scenario presented in Section 2, the deterministic but unknown parameters for the CRLB evaluation, collected into a $3(M + N + K + 1)$-by-1 vector $\boldsymbol{\varphi} = \left[(\mathbf{u}^o)^T, (\mathbf{s}^o)^T, (\mathbf{c}^o)^T\right]^T$, include the target position vector $\mathbf{u}^o$, the transmitter and receiver position vector $\mathbf{s}^o$, and the calibration target position vector $\mathbf{c}^o$; the observations, collected into a $(MN + KMN + 3M + 3N + 3K)$-by-1 vector $\mathbf{z} = \left[\mathbf{r}^T, \mathbf{r}_c^T, \mathbf{s}^T, \mathbf{c}^T\right]^T$, include the BR measurement vector $\mathbf{r}$ from the unknown target, the BR measurement vector $\mathbf{r}_c$ from the calibration targets, the inaccurate measured transmitter and receiver position vector $\mathbf{s}$, and the nominal calibration target position vector $\mathbf{c}$, which are Gaussian distributed and independent with one another. Based on this, the joint probability density function (pdf) of the observations parameterized by the unknown parameter vector is readily shown to be

$$\begin{aligned} p(\mathbf{z}|\boldsymbol{\varphi}) &= p(\mathbf{r}|\mathbf{u}^o,\mathbf{s}^o)\cdot p(\mathbf{r}_c|\mathbf{s}^o,\mathbf{c}^o)\cdot p(\mathbf{s}|\mathbf{s}^o)\cdot p(\mathbf{c}|\mathbf{c}^o) \\ &= \kappa\cdot\exp\Big[-\tfrac{1}{2}(\mathbf{r}-\mathbf{r}^o)^T\mathbf{Q}_r^{-1}(\mathbf{r}-\mathbf{r}^o)-\tfrac{1}{2}(\mathbf{r}_c-\mathbf{r}_c^o)^T\mathbf{Q}_{rc}^{-1}(\mathbf{r}_c-\mathbf{r}_c^o) \\ &\quad -\tfrac{1}{2}(\mathbf{s}-\mathbf{s}^o)^T\mathbf{Q}_s^{-1}(\mathbf{s}-\mathbf{s}^o)-\tfrac{1}{2}(\mathbf{c}-\mathbf{c}^o)^T\mathbf{Q}_c^{-1}(\mathbf{c}-\mathbf{c}^o)\Big] \end{aligned} \tag{15}$$

where κ is a constant with respect to the unknown parameters. By taking the logarithm of (15), partial derivatives with respect to the unknown parameters twice, and then expectation, the Fisher information matrix (FIM) can be calculated as

$$\begin{aligned} FIM(\boldsymbol{\varphi}) &= \mathrm{E}\left[\frac{\partial \ln p(\mathbf{z}|\boldsymbol{\varphi})}{\partial\boldsymbol{\varphi}}\left(\frac{\partial \ln p(\mathbf{z}|\boldsymbol{\varphi})}{\partial\boldsymbol{\varphi}}\right)\right] \\ &= \begin{bmatrix} \mathbf{X} & \mathbf{Y} & \mathbf{O}_{3\times3} \\ \mathbf{Y}^T & \mathbf{Z} & \mathbf{R}^T \\ \mathbf{O}_{3\times3} & \mathbf{R} & \mathbf{P} \end{bmatrix} \end{aligned} \tag{16}$$

where the blocks $\mathbf{X}$, $\mathbf{Y}$, $\mathbf{Z}$, $\mathbf{R}$ and $\mathbf{P}$ are respectively given by

$$\mathbf{X} = \left(\frac{\partial\mathbf{r}^o}{\partial\mathbf{u}^o}\right)^T\mathbf{Q}_r^{-1}\left(\frac{\partial\mathbf{r}^o}{\partial\mathbf{u}^o}\right) \tag{17}$$

$$\mathbf{Y} = \left(\frac{\partial\mathbf{r}^o}{\partial\mathbf{u}^o}\right)^T\mathbf{Q}_r^{-1}\left(\frac{\partial\mathbf{r}^o}{\partial\mathbf{s}^o}\right) \tag{18}$$

$$\mathbf{Z} = \mathbf{Q}_s^{-1} + \left(\frac{\partial\mathbf{r}^o}{\partial\mathbf{s}^o}\right)^T\mathbf{Q}_r^{-1}\left(\frac{\partial\mathbf{r}^o}{\partial\mathbf{s}^o}\right) + \left(\frac{\partial\mathbf{r}_c^o}{\partial\mathbf{s}^o}\right)^T\mathbf{Q}_{rc}^{-1}\left(\frac{\partial\mathbf{r}_c^o}{\partial\mathbf{s}^o}\right) \tag{19}$$

$$\mathbf{R} = \left(\frac{\partial\mathbf{r}_c^o}{\partial\mathbf{c}^o}\right)^T\mathbf{Q}_{rc}^{-1}\left(\frac{\partial\mathbf{r}_c^o}{\partial\mathbf{s}^o}\right) \tag{20}$$

$$\mathbf{P} = \mathbf{Q}_{\mathrm{c}}^{-1} + \left(\frac{\partial \mathbf{r}_{\mathrm{c}}^{\mathrm{o}}}{\partial \mathbf{c}^{\mathrm{o}}}\right)^{\mathrm{T}} \mathbf{Q}_{\mathrm{rc}}^{-1} \left(\frac{\partial \mathbf{r}_{\mathrm{c}}^{\mathrm{o}}}{\partial \mathbf{c}^{\mathrm{o}}}\right) \tag{21}$$

Denote $i_{m,n} = (m-1)N + n$, and $i_{k,m,n} = (k-1)MN + (m-1)N + n$. From the formulations of (7) and (13), the elements of the partial derivatives $\partial \mathbf{r}^{\mathrm{o}}/\partial \mathbf{u}^{\mathrm{o}}$, $\partial \mathbf{r}^{\mathrm{o}}/\partial \mathbf{s}^{\mathrm{o}}$, $\partial \mathbf{r}_{\mathrm{c}}^{\mathrm{o}}/\partial \mathbf{c}^{\mathrm{o}}$ and $\partial \mathbf{r}_{\mathrm{c}}^{\mathrm{o}}/\partial \mathbf{s}^{\mathrm{o}}$ in (17)–(21), can be determined as

$$\frac{\partial \mathbf{r}^{\mathrm{o}}}{\partial \mathbf{u}^{\mathrm{o}}}(i_{m,n}, 1:3) = \frac{(\mathbf{u}^{\mathrm{o}} - \mathbf{s}_{\mathrm{t},m}^{\mathrm{o}})^{\mathrm{T}}}{R_{\mathrm{t},m}^{\mathrm{o}}} + \frac{(\mathbf{u}^{\mathrm{o}} - \mathbf{s}_{\mathrm{r},n}^{\mathrm{o}})^{\mathrm{T}}}{R_{\mathrm{r},n}^{\mathrm{o}}} \tag{22}$$

$$\frac{\partial \mathbf{r}^{\mathrm{o}}}{\partial \mathbf{s}^{\mathrm{o}}} = \left[\begin{array}{cc} \frac{\partial \mathbf{r}^{\mathrm{o}}}{\partial \mathbf{s}_{\mathrm{t}}^{\mathrm{o}}} & \frac{\partial \mathbf{r}^{\mathrm{o}}}{\partial \mathbf{s}_{\mathrm{r}}^{\mathrm{o}}} \end{array}\right] \tag{23}$$

$$\frac{\partial \mathbf{r}^{\mathrm{o}}}{\partial \mathbf{s}_{\mathrm{t}}^{\mathrm{o}}}(i_{m,n}, 3m-2:3m) = \frac{(\mathbf{s}_{\mathrm{t},m}^{\mathrm{o}} - \mathbf{u}^{\mathrm{o}})^{\mathrm{T}}}{R_{\mathrm{t},m}^{\mathrm{o}}} - \frac{(\mathbf{s}_{\mathrm{t},m}^{\mathrm{o}} - \mathbf{s}_{\mathrm{r},n}^{\mathrm{o}})^{\mathrm{T}}}{R_{\mathrm{t},m,\mathrm{r},n}^{\mathrm{o}}} \tag{24}$$

$$\frac{\partial \mathbf{r}^{\mathrm{o}}}{\partial \mathbf{s}_{\mathrm{r}}^{\mathrm{o}}}(i_{m,n}, 3n-2:3n) = \frac{(\mathbf{s}_{\mathrm{r},n}^{\mathrm{o}} - \mathbf{u}^{\mathrm{o}})^{\mathrm{T}}}{R_{\mathrm{r},n}^{\mathrm{o}}} - \frac{(\mathbf{s}_{\mathrm{r},n}^{\mathrm{o}} - \mathbf{s}_{\mathrm{t},m}^{\mathrm{o}})^{\mathrm{T}}}{R_{\mathrm{t},m,\mathrm{r},n}^{\mathrm{o}}} \tag{25}$$

$$\frac{\partial \mathbf{r}_{\mathrm{c}}^{\mathrm{o}}}{\partial \mathbf{c}^{\mathrm{o}}}(i_{k,m,n}, 3k-2:3k) = \frac{(\mathbf{c}_{k}^{\mathrm{o}} - \mathbf{s}_{\mathrm{t},m}^{\mathrm{o}})^{\mathrm{T}}}{R_{\mathrm{c},k,\mathrm{t},m}^{\mathrm{o}}} + \frac{(\mathbf{c}_{k}^{\mathrm{o}} - \mathbf{s}_{\mathrm{r},n}^{\mathrm{o}})^{\mathrm{T}}}{R_{\mathrm{c},k,\mathrm{r},n}^{\mathrm{o}}} \tag{26}$$

$$\frac{\partial \mathbf{r}_{\mathrm{c}}^{\mathrm{o}}}{\partial \mathbf{s}^{\mathrm{o}}} = \left[\begin{array}{cc} \frac{\partial \mathbf{r}_{\mathrm{c}}^{\mathrm{o}}}{\partial \mathbf{s}_{\mathrm{t}}^{\mathrm{o}}} & \frac{\partial \mathbf{r}_{\mathrm{c}}^{\mathrm{o}}}{\partial \mathbf{s}_{\mathrm{r}}^{\mathrm{o}}} \end{array}\right] \tag{27}$$

$$\frac{\partial \mathbf{r}_{\mathrm{c}}^{\mathrm{o}}}{\partial \mathbf{s}_{\mathrm{t}}^{\mathrm{o}}}(i_{k,m,n}, 3m-2:3m) = \frac{(\mathbf{s}_{\mathrm{t},m}^{\mathrm{o}} - \mathbf{c}_{k}^{\mathrm{o}})^{\mathrm{T}}}{R_{\mathrm{c},k,\mathrm{t},m}^{\mathrm{o}}} - \frac{(\mathbf{s}_{\mathrm{t},m}^{\mathrm{o}} - \mathbf{s}_{\mathrm{r},n}^{\mathrm{o}})^{\mathrm{T}}}{R_{\mathrm{t},m,\mathrm{r},n}^{\mathrm{o}}} \tag{28}$$

$$\frac{\partial \mathbf{r}_{\mathrm{c}}^{\mathrm{o}}}{\partial \mathbf{s}_{\mathrm{r}}^{\mathrm{o}}}(i_{k,m,n}, 3n-2:3n) = \frac{(\mathbf{s}_{\mathrm{r},n}^{\mathrm{o}} - \mathbf{c}_{k}^{\mathrm{o}})^{\mathrm{T}}}{R_{\mathrm{c},k,\mathrm{r},n}^{\mathrm{o}}} - \frac{(\mathbf{s}_{\mathrm{r},n}^{\mathrm{o}} - \mathbf{s}_{\mathrm{t},m}^{\mathrm{o}})^{\mathrm{T}}}{R_{\mathrm{t},m,\mathrm{r},n}^{\mathrm{o}}} \tag{29}$$

for $k = 1, 2, \ldots, K$, $m = 1, 2, \ldots, M$ and $n = 1, 2, \ldots, N$, and zeros elsewhere.

By definition, the CRLB of $\boldsymbol{\varphi}$, denoted by $\mathbf{CRLB}_{\mathrm{c}}(\boldsymbol{\varphi})$, is given as $\mathrm{FIM}(\boldsymbol{\varphi})^{-1}$, where only the upper left 3-by-3 block is for the target position $\mathbf{u}^{\mathrm{o}}$. Invoking the partitioned matrix inversion formula as well as the matrix inversion lemma [32] twice on (16), leads to the CRLB of $\mathbf{u}^{\mathrm{o}}$ as

$$\mathbf{CRLB}_{\mathrm{c}}(\mathbf{u}^{\mathrm{o}}) = \mathbf{X}^{-1} + \mathbf{X}^{-1}\mathbf{Y}(\mathbf{Z} - \mathbf{Y}^{\mathrm{T}}\mathbf{X}^{-1}\mathbf{Y} - \mathbf{R}^{\mathrm{T}}\mathbf{P}^{-1}\mathbf{R})^{-1}\mathbf{Y}^{\mathrm{T}}\mathbf{X}^{-1} \tag{30}$$

For comparison purposes, the CRLB of $\mathbf{u}^{\mathrm{o}}$ with transmitter/receiver position error but without calibration derived in [22], denoted by $\mathbf{CRLB}_{\mathrm{s}}(\mathbf{u}^{\mathrm{o}})$, is also given below

$$\mathbf{CRLB}_{\mathrm{s}}(\mathbf{u}^{\mathrm{o}}) = \mathbf{X}^{-1} + \mathbf{X}^{-1}\mathbf{Y}(\widehat{\mathbf{Z}} - \mathbf{Y}^{\mathrm{T}}\mathbf{X}^{-1}\mathbf{Y})^{-1}\mathbf{Y}^{\mathrm{T}}\mathbf{X}^{-1} \tag{31}$$

where $\widehat{\mathbf{Z}} = \mathbf{Q}_{\mathrm{s}}^{-1} + (\partial \mathbf{r}^{\mathrm{o}}/\partial \mathbf{s}^{\mathrm{o}})^{\mathrm{T}}\mathbf{Q}_{\mathrm{r}}^{-1}(\partial \mathbf{r}^{\mathrm{o}}/\partial \mathbf{s}^{\mathrm{o}})$. For the sake of comparison, we proceed to construct an equivalent form of $\mathbf{CRLB}_{\mathrm{c}}(\mathbf{u}^{\mathrm{o}})$ by denoting $\breve{\mathbf{Z}}$ as $\breve{\mathbf{Z}} = \mathbf{Z} - \mathbf{R}^{\mathrm{T}}\mathbf{P}^{-1}\mathbf{R}$. After invoking the matrix inversion lemma [32] to $(\partial \mathbf{r}_{\mathrm{c}}^{\mathrm{o}}/\partial \mathbf{s}^{\mathrm{o}})^{\mathrm{T}}\mathbf{Q}_{\mathrm{rc}}^{-1}(\partial \mathbf{r}_{\mathrm{c}}^{\mathrm{o}}/\partial \mathbf{s}^{\mathrm{o}}) - \mathbf{R}^{\mathrm{T}}\mathbf{P}^{-1}\mathbf{R}$ and some algebraic manipulations, we further represent $\breve{\mathbf{Z}}$ as

$$\breve{\mathbf{Z}} = \mathbf{Q}_{\mathrm{s}}^{-1} + \left(\frac{\partial \mathbf{r}^{\mathrm{o}}}{\partial \mathbf{s}^{\mathrm{o}}}\right)^{\mathrm{T}} \mathbf{Q}_{\mathrm{r}}^{-1} \left(\frac{\partial \mathbf{r}^{\mathrm{o}}}{\partial \mathbf{s}^{\mathrm{o}}}\right) + \left(\frac{\partial \mathbf{r}_{\mathrm{c}}^{\mathrm{o}}}{\partial \mathbf{s}^{\mathrm{o}}}\right)^{\mathrm{T}} \left(\mathbf{Q}_{\mathrm{rc}} + \left(\frac{\partial \mathbf{r}_{\mathrm{c}}^{\mathrm{o}}}{\partial \mathbf{c}^{\mathrm{o}}}\right) \mathbf{Q}_{\mathrm{c}} \left(\frac{\partial \mathbf{r}_{\mathrm{c}}^{\mathrm{o}}}{\partial \mathbf{c}^{\mathrm{o}}}\right)^{\mathrm{T}}\right)^{-1} \left(\frac{\partial \mathbf{r}_{\mathrm{c}}^{\mathrm{o}}}{\partial \mathbf{s}^{\mathrm{o}}}\right) \tag{32}$$

Using (32), we obtain an equivalent expression of $\mathbf{CRLB}_{\mathrm{c}}(\mathbf{u}^{\mathrm{o}})$ as

$$\mathbf{CRLB}_{\mathrm{c}}(\mathbf{u}^{\mathrm{o}}) = \mathbf{X}^{-1} + \mathbf{X}^{-1}\mathbf{Y}(\breve{\mathbf{Z}} - \mathbf{Y}^{\mathrm{T}}\mathbf{X}^{-1}\mathbf{Y})^{-1}\mathbf{Y}^{\mathrm{T}}\mathbf{X}^{-1} \tag{33}$$

Through the comparison of (31) and (33), it is readily to observe that the two CRLBs are identical in structure, except that $\breve{\mathbf{Z}}$ is substituted by $\widehat{\mathbf{Z}}$. More specifically, the use of calibration targets introduces an additional component into the bracketed matrix expression to be inverted as

$$\begin{aligned}\tilde{\mathbf{Z}} &= \breve{\mathbf{Z}} - \widehat{\mathbf{Z}} \\ &= (\partial \mathbf{r}_c^o / \partial \mathbf{s}^o)^T \mathbf{Q}_{rc}^{-1} (\partial \mathbf{r}_c^o / \partial \mathbf{s}^o) - \mathbf{R}^T \mathbf{P}^{-1} \mathbf{R} \\ &= (\partial \mathbf{r}_c^o / \partial \mathbf{s}^o)^T \left(\mathbf{Q}_{rc} + (\partial \mathbf{r}_c^o / \partial \mathbf{c}^o) \mathbf{Q}_c (\partial \mathbf{r}_c^o / \partial \mathbf{c}^o)^T\right)^{-1} (\partial \mathbf{r}_c^o / \partial \mathbf{s}^o)\end{aligned} \tag{34}$$

Using (34), we can rewrite $(\breve{\mathbf{Z}} - \mathbf{Y}^T\mathbf{X}^{-1}\mathbf{Y})^{-1}$ in (33) as $((\widehat{\mathbf{Z}} - \mathbf{Y}^T\mathbf{X}^{-1}\mathbf{Y}) + \tilde{\mathbf{Z}})^{-1}$. Invoking the matrix inversion lemma [32] to the term $((\widehat{\mathbf{Z}} - \mathbf{Y}^T\mathbf{X}^{-1}\mathbf{Y}) + \tilde{\mathbf{Z}})^{-1}$ in (33), we obtain after some algebraic manipulations,

$$\mathbf{CRLB}_c(\mathbf{u}^o) - \mathbf{CRLB}_s(\mathbf{u}^o) = \mathbf{X}^{-1}\mathbf{Y}\boldsymbol{\Gamma}\mathbf{Y}^T\mathbf{X}^{-1} \tag{35}$$

where

$$\boldsymbol{\Gamma} = \mathbf{H}^{-1}\mathbf{Y}(\mathbf{I} + \mathbf{Y}^T\mathbf{H}^{-1}\mathbf{Y})^{-1}\mathbf{Y}^T\mathbf{H}^{-1} \tag{36}$$

$$\mathbf{H} = (\tilde{\mathbf{Z}} - \mathbf{Y}^T\mathbf{X}^{-1}\mathbf{Y}) \tag{37}$$

$$\mathbf{Y} = \left(\frac{\partial \mathbf{r}_c^o}{\partial \mathbf{s}^o}\right)^T \mathbf{L}_{rc} \tag{38}$$

and $\mathbf{L}_{rc}$ is the Cholesky decomposition of $(\mathbf{Q}_{rc} + (\partial \mathbf{r}_c^o/\partial \mathbf{c}^o)\mathbf{Q}_c(\partial \mathbf{r}_c^o/\partial \mathbf{c}^o)^T)^{-1}$, i.e., $\mathbf{L}_{rc}\mathbf{L}_{rc}^T = (\mathbf{Q}_{rc} + (\partial \mathbf{r}_c^o/\partial \mathbf{c}^o)\mathbf{Q}_c(\partial \mathbf{r}_c^o/\partial \mathbf{c}^o)^T)^{-1}$. In form, the right side of (35) is just the performance enhancement because of the use of calibration targets. It is positive semi-definite (PSD) since it has a symmetric structure and $\mathbf{Y}^T$ is not full column rank. Even if the nominal positions of calibration targets and the corresponding BR measurements are very noisy, (35) can still remain PSD. In theory, only in the edge case when $(\mathbf{Q}_{rc} + (\partial \mathbf{r}_c^o/\partial \mathbf{c}^o)\mathbf{Q}_c^{-1}(\partial \mathbf{r}_c^o/\partial \mathbf{c}^o)^T)^{-1}$ tends to zero and then $\mathbf{L}_{rc} \to \mathbf{O}$ and $\mathbf{Y} \to \mathbf{O}$, the performance enhancement in (34) would tend to zero. However, this edge case hardly exists in reality. Thus, mathematically, we can arrive at

$$\mathbf{CRLB}_s(\mathbf{u}^o) \geq \mathbf{CRLB}_c(\mathbf{u}^o) \tag{39}$$

The matrix inequality $\mathbf{A} \geq \mathbf{B}$ means that $\mathbf{A} - \mathbf{B}$ is PSD. It can be further deduced from (39) that $\mathrm{tr}(\mathbf{CRLB}_s(\mathbf{u}^o)) \geq \mathrm{tr}(\mathbf{CRLB}_c(\mathbf{u}^o))$. The trace of $\mathbf{CRLB}_c(\mathbf{u}^o)$ and $\mathbf{CRLB}_s(\mathbf{u}^o)$ respectively represents minimum possible variance of target position estimation with and without using calibration targets. Therefore, we can conclude that using calibration targets brings potential enhancement to the target localization accuracy, at least at the CRLB level.

Example 1. To substantiate the evaluation on the CRLB presented above, a numerical example using a typical multi-static passive radar localization scenario was conducted, as presented in Figure 2. There are $M = 3$ transmitters, $N = 4$ receivers and $K = 3$ calibration targets in the scenario, and their true positions are listed in Table 1. The noise covariance matrix of the BR measurements from the unknown target are given by $\mathbf{Q}_r = \sigma_r^2\mathbf{V}_r$, where σ_r reflects BR measurement noise level and $\mathbf{V}_r$ is set to 1 in the diagonal elements and 0.5 elsewhere. The covariance matrix of the transmitter/receiver position error is given as $\mathbf{Q}_s = \sigma_s^2\mathbf{V}_s$ where σ_s reflects the transmitter/receiver position error level and $\mathbf{V}_s = \mathrm{diag}(5\mathbf{I}_{3M\times 3M}, \mathbf{I}_{3N\times 3N})$. The covariance matrix of calibration target position error is $\mathbf{Q}_c = \sigma_c^2\mathbf{V}_c$ where σ_c reflects the calibration target position error level and $\mathbf{V}_c = \mathbf{I}_{3K\times 3K}$, and the covariance matrix of the corresponding calibration BR measurement noise is $\mathbf{Q}_{rc} = \sigma_{rc}^2\mathbf{V}_{rc}$ where $\sigma_{rc} = \sigma_r$ reflects calibration BR measurement noise level and $\mathbf{V}_{rc}$ is set to 1 in the diagonal elements and 0.5 elsewhere. The target of interest is located at position $\mathbf{u}^o = [50000, 15000, 5000]^T$m. The effect of the calibration targets on the target localization accuracy, in the sense of CRLB, is presented in Figure 3.

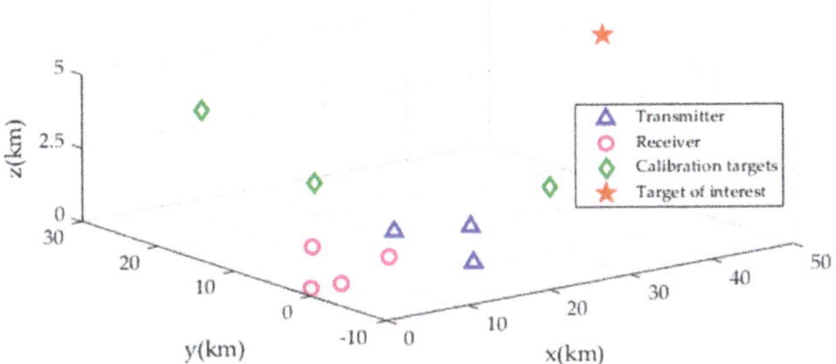

Figure 2. Localization scenario geometry for simulation.

Table 1. Positions (in meters) of the transmitters, receivers and calibration targets.

TX	$x^o_{t,m}$	$y^o_{t,m}$	$z^o_{t,m}$	RX	$x^o_{r,n}$	$y^o_{r,n}$	$z^o_{r,n}$	Calibration Targets	$x^o_{c,k}$	$y^o_{c,k}$	$z^o_{c,k}$
1	20,000	0	100	1	2000	2000	0	1	10,000	10,000	2500
2	15,000	5000	1000	2	2000	−2000	500	2	15,000	30,000	3000
3	15,000	−5000	2000	3	5000	5000	1000	3	20,000	−10,000	3500
–	–	–	–	4	5000	−5000	1500	–	–	–	–

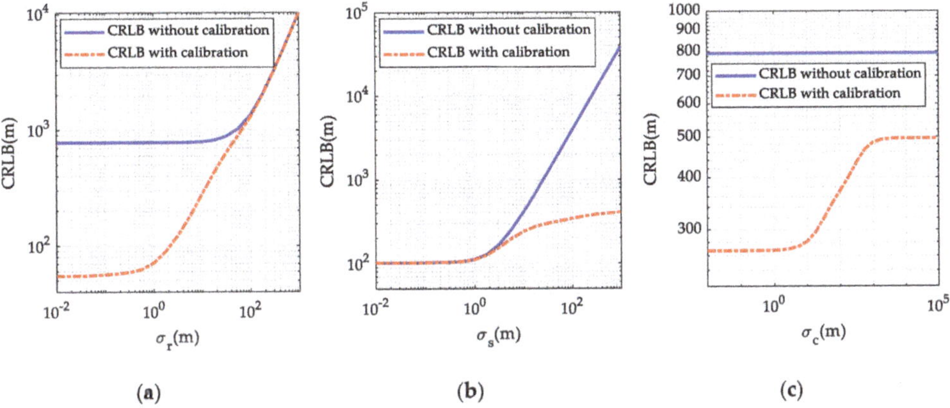

Figure 3. Comparison of the CRLBs with and without using calibration targets: (**a**) for different BR measurement noise level σ_r, (**b**) for different transmitter/receiver position error level σ_s; (**c**) for different calibration target position error level σ_c.

Figure 3a compares the CRLB curves with and without using calibration targets when the BR measurement noise level σ_r is varied from 10^{-2} m to 10^3 m while the transmitter/receiver position error level and calibration target position error level are fixed at $\sigma_s = 20$ m and $\sigma_c = 10$ m respectively. It can be observed from Figure 3a that the CRLB with calibration targets is generally below the one without, this coincides with the analytical conclusion given in (39). However, in the edge case where the BR measurement noise is very large, two CRLBs would tend to be the same. This is because in this case, the BR measurement noise dominates and effect of transmitter/receiver position error on the localization accuracy is relatively small. The CRLB curves versus the transmitter/receiver position error level σ_s are plotted in Figure 3b where the BR measurement noise level and calibration target position error level are fixed at $\sigma_r = 10$ m and $\sigma_c = 10$ m respectively. A similar trend, i.e., two CRLBs would tend to be the same, appears in Figure 3b, when the transmitter/receiver position error is sufficiently small. A reasonable explanation is that, in this case, the transmitter/receiver positions are known very accurately and their influence on the localization accuracy can be ignored compared to the BR measurement noise. The CRLB comparison versus calibration target position error level σ_c is provided in Figure 3c where $\sigma_r = 10$ m and $\sigma_s = 20$ m. Interestingly, the trend of CRLB curves implies that, even when the calibration target position error is extremely large, the CRLB with utilization of calibration targets are still remarkably below the one without. This justifies again the analysis under (35), and similar results have also been presented in previous studies [23–30] on source localization and sensor network localization issues. Generally, from Figure 3, the use of calibration targets brings a significant improvement in the localization accuracy in the normal case, at least at the CRLB level.

4. Proposed Localization Method

The evaluation of the CRLB in Section 3 has demonstrated the potential of calibration targets in improving localization accuracy. In what follows, we will proceed to develop a novel closed-form solution for the aforementioned practical localization scenario where the positions of transmitters and receivers are inaccurate but calibration targets are used to refine the transmitter/receiver position and enhance the localization accuracy. After that, a theoretical analysis will be performed to show that the proposed solution achieves the CRLB when satisfying some mild conditions.

4.1. Algorithm Development

The proposed solution mainly includes two processing stages, referred to as calibration stage and localization stage, respectively. The calibration stage is devoted to refining the inaccurate transmitter and receiver positions, and then the localization stage is devoted to determining the target position on the basis of the refined transmitter and receiver positions.

4.1.1. Calibration Stage

To make use of the BR measurements from the calibration targets, the calibration stage begins by reorganizing (13) as

$$r_{c,k,m,n} - ||\mathbf{c}_k^o - \mathbf{s}_{t,m}^o|| - ||\mathbf{c}_k^o - \mathbf{s}_{r,n}^o|| + ||\mathbf{s}_{t,m}^o - \mathbf{s}_{r,n}^o|| = \Delta r_{c,k,m,n} \tag{40}$$

Since only the erroneous versions of $\mathbf{c}_k^o$, $\mathbf{s}_{t,m}^o$ and $\mathbf{s}_{r,n}^o$ are available, we put $\mathbf{c}_k^o = \mathbf{c}_k - \Delta\mathbf{c}_k$, $\mathbf{s}_{t,m}^o = \mathbf{s}_{t,m} - \Delta\mathbf{s}_{t,m}$ and $\mathbf{s}_{r,n}^o = \mathbf{s}_{r,n} - \Delta\mathbf{s}_{r,n}$ into (40), and then expand it around erroneous values $\mathbf{c}_k$, $\mathbf{s}_{t,m}$ and $\mathbf{s}_{r,n}$ to the linear error terms as

$$\begin{aligned} &r_{c,k,m,n} - ||\mathbf{c}_k - \mathbf{s}_{t,m}|| - ||\mathbf{c}_k - \mathbf{s}_{r,n}|| + ||\mathbf{s}_{t,m} - \mathbf{s}_{r,n}|| - (\boldsymbol{\rho}_{c,k,t,m}^T + \boldsymbol{\rho}_{t,m,r,n}^T)\Delta\mathbf{s}_{t,m} - (\boldsymbol{\rho}_{c,k,r,n}^T - \boldsymbol{\rho}_{t,m,r,n}^T)\Delta\mathbf{s}_{r,n} \\ &= -(\boldsymbol{\rho}_{c,k,t,m} + \boldsymbol{\rho}_{c,k,r,n})^T\Delta\mathbf{c}_k + \Delta r_{c,k,m,n} \end{aligned} \tag{41}$$

where

$$\boldsymbol{\rho}_{c,k,t,m} = \frac{\mathbf{c}_k - \mathbf{s}_{t,m}}{||\mathbf{c}_k - \mathbf{s}_{t,m}||} \tag{42}$$

$$\boldsymbol{\rho}_{c,k,r,n} = \frac{\mathbf{c}_k - \mathbf{s}_{r,n}}{||\mathbf{c}_k - \mathbf{s}_{r,n}||} \tag{43}$$

$$\boldsymbol{\rho}_{t,m,r,n} = \frac{\mathbf{s}_{t,m} - \mathbf{s}_{r,n}}{||\mathbf{s}_{t,m} - \mathbf{s}_{r,n}||} \tag{44}$$

Stacking (41) for all the k, m and n, we can formulate them in matrix form as

$$\mathbf{h}_0 - \mathbf{G}_0\Delta\mathbf{s} = \Delta\mathbf{h}_0 \tag{45}$$

The elements of $\mathbf{h}_0$, $\mathbf{G}_0$ and $\Delta\mathbf{h}_0$ are given by

$$\mathbf{h}_0(i_{k,m,n}, 1) = r_{c,k,m,n} - ||\mathbf{c}_k - \mathbf{s}_{t,m}|| - ||\mathbf{c}_k - \mathbf{s}_{r,n}|| + ||\mathbf{s}_{t,m} - \mathbf{s}_{r,n}|| \tag{46}$$

$$\mathbf{G}_0 = \begin{bmatrix} \mathbf{G}_{0,\mathrm{t}} & \mathbf{G}_{0,\mathrm{r}} \end{bmatrix}, \mathbf{G}_{0,\mathrm{t}}(i_{k,m,n}, 3m-2:3m) = \boldsymbol{\rho}_{\mathrm{c},k,\mathrm{t},m}^{\mathrm{T}} + \boldsymbol{\rho}_{\mathrm{t},m,\mathrm{r},n}^{\mathrm{T}}, \mathbf{G}_{0,\mathrm{r}}(i_{k,m,n}, 3n-2:3n) = \boldsymbol{\rho}_{\mathrm{c},k,\mathrm{r},n}^{\mathrm{T}} - \boldsymbol{\rho}_{\mathrm{t},m,\mathrm{r},n}^{\mathrm{T}} \tag{47}$$

$$\mathbf{G}_{\mathrm{c}}(i_{k,m,n}, 3k-2:3k) = -(\boldsymbol{\rho}_{\mathrm{c},k,\mathrm{t},m} + \boldsymbol{\rho}_{\mathrm{c},k,\mathrm{r},n})^{\mathrm{T}} \tag{48}$$

$$\Delta\mathbf{h}_0(i_{k,m,n}, 1) = -(\boldsymbol{\rho}_{\mathrm{c},k,\mathrm{t},m} + \boldsymbol{\rho}_{\mathrm{c},k,\mathrm{r},n})^{\mathrm{T}}\Delta\mathbf{c}_k + \Delta r_{\mathrm{c},k,m,n} \tag{49}$$

for $i_{k,m,n} = (k-1)MN + (m-1)N + n$, $k = 0,1,\ldots,K-1$, $m = 0,1,\ldots,M-1$, $n = 0,1,\ldots,N-1$, and zeros elsewhere. Furthermore, the error vector $\Delta\mathbf{h}_0$ can be recast using a compact representation as follows

$$\Delta\mathbf{h}_0 = \mathbf{G}_{\mathrm{c}}\Delta\mathbf{c} + \Delta\mathbf{r}_{\mathrm{c}} \tag{50}$$

from which we have the mean $\mathrm{E}(\Delta\mathbf{h}_0) = 0_{KMN\times 1}$ and the covariance $\mathrm{cov}(\Delta\mathbf{h}_0) = \mathbf{G}_{\mathrm{c}}\mathbf{Q}_{\mathrm{c}}\mathbf{G}_{\mathrm{c}}^{\mathrm{T}} + \mathbf{Q}_{\mathrm{rc}}$. In (45), $\Delta\mathbf{s}$ represents the difference between the true and the nominal transmitter/receiver positions.

In order to refine the transmitter and receiver positions, $\Delta\mathbf{s}$ shall be estimated as accurately as possible. Recall that $\Delta\mathbf{s}$ is a Gaussian distributed random vector with mean $\mathrm{E}(\Delta\mathbf{s}) = 0_{3(M+N)\times 1}$ and covariance matrix $\mathrm{cov}(\Delta\mathbf{s}) = \mathbf{Q}_{\mathbf{s}}$, and it is independent of the error vector $\Delta\mathbf{h}_0$. Thus according to the Bayesian Gauss–Markov theorem [33], the linear minimum mean square error (LMMSE) estimate of $\Delta\mathbf{s}$ can be obtained from (45) as

$$\begin{aligned} \Delta\hat{\mathbf{s}} &= \mathrm{E}(\Delta\mathbf{s}) + \left(\mathrm{cov}(\Delta\mathbf{s})^{-1} + \mathbf{G}_0^{\mathrm{T}}\mathrm{cov}(\Delta\mathbf{h}_0)^{-1}\mathbf{G}_{\mathrm{s}}\right)^{-1}\mathbf{G}_0^{\mathrm{T}}\mathrm{cov}(\Delta\mathbf{h}_0)^{-1}(\mathbf{h}_0 - \mathbf{G}_0\mathrm{E}(\Delta\mathbf{s})) \\ &= \left(\mathbf{Q}_{\mathbf{s}}^{-1} + \mathbf{G}_0^{\mathrm{T}}(\mathbf{G}_{\mathrm{c}}\mathbf{Q}_{\mathrm{c}}\mathbf{G}_{\mathrm{c}}^{\mathrm{T}} + \mathbf{Q}_{\mathrm{rc}})^{-1}\mathbf{G}_0\right)^{-1}\mathbf{G}_0^{\mathrm{T}}(\mathbf{G}_{\mathrm{c}}\mathbf{Q}_{\mathrm{c}}\mathbf{G}_{\mathrm{c}}^{\mathrm{T}} + \mathbf{Q}_{\mathrm{rc}})^{-1}\mathbf{h}_0 \end{aligned} \tag{51}$$

Under the assumption that the noise in $\mathbf{G}_{\mathrm{c}}$ and $\mathbf{G}_0$ is sufficiently small to be ignored, the covariance matrix of $\Delta\hat{\mathbf{s}}$ can be given as

$$\mathrm{cov}(\Delta\mathbf{s} - \Delta\hat{\mathbf{s}}) = \left(\mathbf{Q}_{\mathbf{s}}^{-1} + \mathbf{G}_0^{\mathrm{T}}(\mathbf{G}_{\mathrm{c}}\mathbf{Q}_{\mathrm{c}}\mathbf{G}_{\mathrm{c}}^{\mathrm{T}} + \mathbf{Q}_{\mathrm{rc}})^{-1}\mathbf{G}_0\right)^{-1} \tag{52}$$

Using the estimate of transmitter and receiver position error in (51), we can refine the transmitter and receiver positions as

$$\hat{\mathbf{s}} = \mathbf{s} - \Delta\hat{\mathbf{s}} \tag{53}$$

Utilizing the fact $\mathbf{s} = \mathbf{s}^{\mathrm{o}} + \Delta\mathbf{s}$, we can rewrite $\hat{\mathbf{s}}$ in (53) as $\hat{\mathbf{s}} = \mathbf{s}^{\mathrm{o}} + \Delta\mathbf{s} - \Delta\hat{\mathbf{s}}$. Hence, the refined estimate of transmitter/receiver positions $\hat{\mathbf{s}}$ has a covariance matrix identical with (52). Forming the inverse of $\mathrm{cov}(\Delta\mathbf{s} - \Delta\hat{\mathbf{s}})$ and then comparing it to $\mathbf{Q}_{\mathbf{s}}^{-1}$ results in $\mathrm{cov}(\Delta\mathbf{s} - \Delta\hat{\mathbf{s}})^{-1} - \mathbf{Q}_{\mathbf{s}}^{-1} = \mathbf{G}_0^{\mathrm{T}}(\mathbf{G}_{\mathrm{c}}\mathbf{Q}_{\mathrm{c}}\mathbf{G}_{\mathrm{c}}^{\mathrm{T}} + \mathbf{Q}_{\mathrm{rc}})^{-1}\mathbf{G}_0$. It is natural to deduce that $\mathrm{cov}(\Delta\mathbf{s} - \Delta\hat{\mathbf{s}})^{-1} \geq \mathbf{Q}_{\mathbf{s}}^{-1}$ is PSD since $\mathbf{G}_0^{\mathrm{T}}(\mathbf{G}_{\mathrm{c}}\mathbf{Q}_{\mathrm{c}}\mathbf{G}_{\mathrm{c}}^{\mathrm{T}} + \mathbf{Q}_{\mathrm{rc}})^{-1}\mathbf{G}_0$ has a symmetric structure and $\mathbf{G}_{\mathrm{c}}$ is not full column rank. According to the PSD matrix property [34], $\mathrm{cov}(\Delta\mathbf{s} - \Delta\hat{\mathbf{s}})^{-1} \geq \mathbf{Q}_{\mathbf{s}}^{-1}$ is equivalent to $\mathbf{Q}_{\mathbf{s}} \geq \mathrm{cov}(\Delta\mathbf{s} - \Delta\hat{\mathbf{s}})$. That is to say, the refined positions of transmitters and receivers performs leastwise as well as, if not better than, the original ones, in terms of target localization accuracy.

4.1.2. Localization Stage

The localization stage starts by linearizing the BR equations from the unknown target. Firstly, reorganize (7) as

$$(r_{m,n} + R_{\mathrm{t},m,\mathrm{r},n}^{\mathrm{o}}) - R_{\mathrm{t},m}^{\mathrm{o}} = R_{\mathrm{r},n}^{\mathrm{o}} + \Delta r_{m,n} \tag{54}$$

Since we have obtained the refined estimate of $\mathbf{s}_{\mathrm{t},m}^{\mathrm{o}}$ and $\mathbf{s}_{\mathrm{r},n}^{\mathrm{o}}$ from calibration stage, we plug $\mathbf{s}_{\mathrm{t},m}^{\mathrm{o}} = \hat{\mathbf{s}}_{\mathrm{t},m} - (\Delta\mathbf{s}_{\mathrm{t},m} - \Delta\hat{\mathbf{s}}_{\mathrm{t},m})$ and $\mathbf{s}_{\mathrm{r},n}^{\mathrm{o}} = \hat{\mathbf{s}}_{\mathrm{r},n} - (\Delta\mathbf{s}_{\mathrm{r},n} - \Delta\hat{\mathbf{s}}_{\mathrm{r},n})$ into (54) and ignoring the second and higher error terms as

$$
\begin{aligned}
&2(\hat{\mathbf{s}}_{\mathrm{t},m}-\hat{\mathbf{s}}_{\mathrm{r},n})^{\mathrm{T}}\mathbf{u}^{\mathrm{o}}+2(r_{m,n}+R_{\mathrm{t},m,\mathrm{r},n})R^{\mathrm{o}}_{\mathrm{t},m}=(r_{m,n}+R_{\mathrm{t},m,\mathrm{r},n})^{2}+\hat{\mathbf{s}}^{\mathrm{T}}_{\mathrm{t},m}\hat{\mathbf{s}}_{\mathrm{t},m}-\hat{\mathbf{s}}^{\mathrm{T}}_{\mathrm{r},n}\hat{\mathbf{s}}_{\mathrm{r},n}-2R^{\mathrm{o}}_{\mathrm{r},n}\Delta r_{m,n}\\
&+2(\mathbf{u}^{\mathrm{o}}-\hat{\mathbf{s}}_{\mathrm{t},m}-R^{\mathrm{o}}_{\mathrm{r},n}\boldsymbol{\rho}_{\mathrm{t},m,\mathrm{r},n})^{\mathrm{T}}(\Delta\mathbf{s}_{\mathrm{t},m}-\Delta\hat{\mathbf{s}}_{\mathrm{t},m})-2(\mathbf{u}^{\mathrm{o}}-\hat{\mathbf{s}}_{\mathrm{r},n}-R^{\mathrm{o}}_{\mathrm{r},n}\boldsymbol{\rho}_{\mathrm{t},m,\mathrm{r},n})^{\mathrm{T}}(\Delta\mathbf{s}_{\mathrm{r},n}-\Delta\hat{\mathbf{s}}_{\mathrm{r},n})
\end{aligned}
\tag{55}
$$

By forming an auxiliary vector as $\boldsymbol{\theta}^{\mathrm{o}}=[(\mathbf{u}^{\mathrm{o}})^{\mathrm{T}},R^{\mathrm{o}}_{\mathrm{t},1},R^{\mathrm{o}}_{\mathrm{t},2},\ldots,R^{\mathrm{o}}_{\mathrm{t},M}]^{\mathrm{T}}$, we can collect (55) for all the m and n into a matrix form as

$$\mathbf{G}_1\boldsymbol{\theta}^{\mathrm{o}}=\mathbf{h}_1+\Delta\mathbf{h}_1 \tag{56}$$

where

$$\mathbf{G}_1=\begin{bmatrix}\mathbf{G}_{1,\mathrm{s}} & \mathbf{G}_{1,\mathrm{r}}\end{bmatrix} \tag{57}$$

$$\mathbf{G}_{1,\mathrm{s}}=2\begin{bmatrix}\hat{\mathbf{s}}_1\\ \hat{\mathbf{s}}_2\\ \vdots\\ \hat{\mathbf{s}}_M\end{bmatrix},\ \mathbf{s}_m=\begin{bmatrix}(\hat{\mathbf{s}}_{\mathrm{t},m}-\hat{\mathbf{s}}_{\mathrm{r},1})^{\mathrm{T}}\\ (\hat{\mathbf{s}}_{\mathrm{t},m}-\hat{\mathbf{s}}_{\mathrm{r},2})^{\mathrm{T}}\\ \vdots\\ (\hat{\mathbf{s}}_{\mathrm{t},m}-\hat{\mathbf{s}}_{\mathrm{r},N})^{\mathrm{T}}\end{bmatrix}, \tag{58}$$

$$\mathbf{G}_{1,\mathrm{r}}=2\begin{bmatrix}\mathbf{r}_1 & 0_{N\times1} & \cdots & 0_{N\times1}\\ 0_{N\times1} & \mathbf{r}_2 & \cdots & 0_{N\times1}\\ \vdots & \vdots & \ddots & \vdots\\ 0_{N\times1} & 0_{N\times1} & \cdots & \mathbf{r}_M\end{bmatrix},\ \mathbf{r}_m=\begin{bmatrix}r_{m,1}\\ r_{m,2}\\ \vdots\\ r_{m,N}\end{bmatrix} \tag{59}$$

$$\mathbf{h}_1=\begin{bmatrix}\mathbf{h}_{1,1}\\ \mathbf{h}_{1,2}\\ \vdots\\ \mathbf{h}_{1,M}\end{bmatrix},\ \mathbf{h}_{1,m}=\begin{bmatrix}(r_{m,1}+R_{\mathrm{t},m,\mathrm{r},1})^2+\hat{\mathbf{s}}^{\mathrm{T}}_{\mathrm{t},m}\hat{\mathbf{s}}_{\mathrm{t},m}-\hat{\mathbf{s}}^{\mathrm{T}}_{\mathrm{r},1}\hat{\mathbf{s}}_{\mathrm{r},1}\\ (r_{m,2}+R_{\mathrm{t},m,\mathrm{r},2})^2+\hat{\mathbf{s}}^{\mathrm{T}}_{\mathrm{t},m}\hat{\mathbf{s}}_{\mathrm{t},m}-\hat{\mathbf{s}}^{\mathrm{T}}_{\mathrm{r},2}\hat{\mathbf{s}}_{\mathrm{r},2}\\ \vdots\\ (r_{m,N}+R_{\mathrm{t},m,\mathrm{r},N})^2+\hat{\mathbf{s}}^{\mathrm{T}}_{\mathrm{t},m}\hat{\mathbf{s}}_{\mathrm{t},m}-\hat{\mathbf{s}}^{\mathrm{T}}_{\mathrm{r},N}\hat{\mathbf{s}}_{\mathrm{r},N}\end{bmatrix} \tag{60}$$

and the error vector $\Delta\mathbf{h}_1$ is related to the target position as

$$\Delta\mathbf{h}_1=\mathbf{B}_1\Delta\mathbf{r}+\mathbf{D}_1(\Delta\mathbf{s}-\Delta\hat{\mathbf{s}}) \tag{61}$$

where

$$\mathbf{B}_1=\begin{bmatrix}\mathbf{B}_{1,1} & \mathbf{O}_{N\times N} & \cdots & \mathbf{O}_{N\times N}\\ \mathbf{O}_{N\times N} & \mathbf{B}_{1,2} & \cdots & \mathbf{O}_{N\times N}\\ \vdots & \vdots & \ddots & \vdots\\ \mathbf{O}_{N\times N} & \mathbf{O}_{N\times N} & \cdots & \mathbf{B}_{1,M}\end{bmatrix} \tag{62}$$

$$\mathbf{D}_1=\begin{bmatrix}\mathbf{D}_{1,\mathrm{t},1} & \mathbf{O}_{N\times3} & \cdots & \mathbf{O}_{N\times3} & \mathbf{D}_{1,\mathrm{r},1}\\ \mathbf{O}_{N\times3} & \mathbf{D}_{1,\mathrm{t},2} & \cdots & \mathbf{O}_{N\times3} & \mathbf{D}_{1,\mathrm{r},2}\\ \vdots & \vdots & \ddots & \vdots & \vdots\\ \mathbf{O}_{N\times3} & \mathbf{O}_{N\times3} & \cdots & \mathbf{D}_{1,\mathrm{t},M} & \mathbf{D}_{1,\mathrm{r},M}\end{bmatrix} \tag{63}$$

with

$$\mathbf{B}_{1,m}=-2\mathrm{diag}\left(R^{\mathrm{o}}_{\mathrm{r},1},R^{\mathrm{o}}_{\mathrm{r},2},\ldots,R^{\mathrm{o}}_{\mathrm{r},N}\right) \tag{64}$$

$$\mathbf{D}_{1,\mathrm{t},m}=2\begin{bmatrix}(\mathbf{u}^{\mathrm{o}}-\hat{\mathbf{s}}_{\mathrm{t},m}-R^{\mathrm{o}}_{\mathrm{r},1}\boldsymbol{\rho}_{\mathrm{t},m,\mathrm{r},1})^{\mathrm{T}}\\ (\mathbf{u}^{\mathrm{o}}-\hat{\mathbf{s}}_{\mathrm{t},m}-R^{\mathrm{o}}_{\mathrm{r},2}\boldsymbol{\rho}_{\mathrm{t},m,\mathrm{r},2})^{\mathrm{T}}\\ \vdots\\ (\mathbf{u}^{\mathrm{o}}-\hat{\mathbf{s}}_{\mathrm{t},m}-R^{\mathrm{o}}_{\mathrm{r},N}\boldsymbol{\rho}_{\mathrm{t},m,\mathrm{r},N})^{\mathrm{T}}\end{bmatrix} \tag{65}$$

$$\mathbf{D}_{1,\mathrm{r},m} = -2\begin{bmatrix} (\mathbf{u}^{\mathrm{o}} - \hat{\mathbf{s}}_{\mathrm{r},1} - R^{\mathrm{o}}_{\mathrm{r},1}\boldsymbol{\rho}_{\mathrm{t},m,\mathrm{r},1})^{\mathrm{T}} & 0^{\mathrm{T}}_{3\times1} & \cdots & 0^{\mathrm{T}}_{3\times1} \\ 0^{\mathrm{T}}_{3\times1} & (\mathbf{u}^{\mathrm{o}} - \hat{\mathbf{s}}_{\mathrm{r},2} - R^{\mathrm{o}}_{\mathrm{r},2}\boldsymbol{\rho}_{\mathrm{t},m,\mathrm{r},2})^{\mathrm{T}} & \cdots & 0^{\mathrm{T}}_{3\times1} \\ \vdots & \vdots & \ddots & \vdots \\ 0^{\mathrm{T}}_{3\times1} & 0^{\mathrm{T}}_{3\times1} & \cdots & (\mathbf{u}^{\mathrm{o}} - \hat{\mathbf{s}}_{\mathrm{r},N} - R^{\mathrm{o}}_{\mathrm{r},N}\boldsymbol{\rho}_{\mathrm{t},m,\mathrm{r},N})^{\mathrm{T}} \end{bmatrix} \quad (66)$$

From the set of linear equations in (56), the WLS estimate of $\boldsymbol{\theta}^{\mathrm{o}}$, denoted by $\boldsymbol{\theta}$, which minimizes $\Delta\mathbf{h}_1^{\mathrm{T}}\mathbf{W}_1\Delta\mathbf{h}_1$ can be produced as

$$\boldsymbol{\theta} = (\mathbf{G}_1^{\mathrm{T}}\mathbf{W}_1\mathbf{G}_1)^{-1}\mathbf{G}_1^{\mathrm{T}}\mathbf{W}_1\mathbf{h}_1 \quad (67)$$

where $\mathbf{W}_1$ represents the weighting matrix and it can be computed by

$$\begin{aligned} \mathbf{W}_1 &= \left[\mathrm{E}(\Delta\mathbf{h}_1\Delta\mathbf{h}_1^{\mathrm{T}})\right]^{-1} \\ &= \left[\mathbf{B}_1\mathbf{Q}_\alpha\mathbf{B}_1^{\mathrm{T}} + \mathbf{D}_1\mathrm{cov}(\Delta\mathbf{s} - \Delta\hat{\mathbf{s}})\mathbf{D}_1^{\mathrm{T}}\right]^{-1} \end{aligned} \quad (68)$$

However, to compute $\mathbf{W}_1$, the unknown target position has to be acquired in advance. To resolve this contradiction, we preliminarily let $\mathbf{W}_1 = \mathbf{I}_{MN\times MN}$ and use (67) to compute a least squares estimate of $\boldsymbol{\theta}^{\mathrm{o}}$, and then use the estimated $\boldsymbol{\theta}^{\mathrm{o}}$ to update $\mathbf{W}_1$ for another repetition.

Based on the WLS theorem, it can be deduced that the estimate $\boldsymbol{\theta}$ is approximately unbiased and the corresponding covariance matrix can be obtained, given sufficiently small BR measurement noise and transmitter/receiver position error, as

$$\mathrm{cov}(\boldsymbol{\theta}) = (\mathbf{G}_1^{\mathrm{T}}\mathbf{W}_1\mathbf{G}_1)^{-1} \quad (69)$$

Next, the functional relation between the target position $\mathbf{u}^{\mathrm{o}}$ and the introduced nuisance parameters $R^{\mathrm{o}}_{\mathrm{t},1}, R^{\mathrm{o}}_{\mathrm{t},2}, \ldots, R^{\mathrm{o}}_{\mathrm{t},M}$, is explored to compute the final estimate of target position. To this end, reorganize the functional relation in (4) as

$$2(\mathbf{s}^{\mathrm{o}}_{\mathrm{t},m})^{\mathrm{T}}\mathbf{u}^{\mathrm{o}} = (\mathbf{u}^{\mathrm{o}})^{\mathrm{T}}\mathbf{u}^{\mathrm{o}} + (\mathbf{s}^{\mathrm{o}}_{\mathrm{t},m})^{\mathrm{T}}\mathbf{s}^{\mathrm{o}}_{\mathrm{t},m} - (R^{\mathrm{o}}_{\mathrm{t},m})^2 \quad (70)$$

Denoting the estimation error of $\boldsymbol{\theta}$ by $\Delta\boldsymbol{\theta}$, mathematically we arrive at

$$\mathbf{u}^{\mathrm{o}} = \boldsymbol{\theta}(1:3) - \Delta\boldsymbol{\theta}(1:3) \quad (71)$$

$$R^{\mathrm{o}}_{\mathrm{t},m} = \boldsymbol{\theta}(3+m) - \Delta\boldsymbol{\theta}(3+m) \quad (72)$$

Putting (71), (72) into the right side of (70) and $\mathbf{s}^{\mathrm{o}}_{\mathrm{t},m} = \hat{\mathbf{s}}_{\mathrm{t},m} - (\Delta\mathbf{s}_{\mathrm{t},m} - \Delta\hat{\mathbf{s}}_{\mathrm{t},m})$ into the both sides, we have after ignoring second-order error terms,

$$\begin{aligned} 2\hat{\mathbf{s}}_{\mathrm{t},m}^{\mathrm{T}}\mathbf{u}^{\mathrm{o}} &= \boldsymbol{\theta}(1:3)^{\mathrm{T}}\boldsymbol{\theta}(1:3) - \boldsymbol{\theta}(m+3)^2 + \hat{\mathbf{s}}_{\mathrm{t},m}^{\mathrm{T}}\hat{\mathbf{s}}_{\mathrm{t},m} \\ &- 2\boldsymbol{\theta}(1:3)\Delta\boldsymbol{\theta}(1:3) + 2\boldsymbol{\theta}(m+3)\Delta\boldsymbol{\theta}(m+3) + 2(\mathbf{u}^{\mathrm{o}} - \hat{\mathbf{s}}_{\mathrm{t},m})^{\mathrm{T}}(\Delta\mathbf{s}_{\mathrm{t},m} - \Delta\hat{\mathbf{s}}_{\mathrm{t},m}) \end{aligned} \quad (73)$$

The final estimate of target position should satisfy (73) and meanwhile retain as close as possible to the estimated values of target position in $\boldsymbol{\theta}$. In line with this principle, one has the following set of equations

$$\mathbf{G}_2\mathbf{u}^{\mathrm{o}} = \mathbf{h}_2 + \Delta\mathbf{h}_2 \quad (74)$$

where

$$\mathbf{G}_2 = \begin{bmatrix} \mathbf{I}_{3\times3} \\ 2\mathbf{s}_{t,1}^{\mathrm{T}} \\ 2\mathbf{s}_{t,2}^{\mathrm{T}} \\ \vdots \\ 2\mathbf{s}_{t,M}^{\mathrm{T}} \end{bmatrix} \tag{75}$$

$$\mathbf{h}_2 = \begin{bmatrix} \theta(1:3) \\ \theta(1:3)^{\mathrm{T}}\theta(1:3) - \theta(1+3)^2 + \hat{\mathbf{s}}_{t,1}^{\mathrm{T}}\hat{\mathbf{s}}_{t,1} \\ \theta(1:3)^{\mathrm{T}}\theta(1:3) - \theta(2+3)^2 + \hat{\mathbf{s}}_{t,2}^{\mathrm{T}}\hat{\mathbf{s}}_{t,2} \\ \vdots \\ \theta(1:3)^{\mathrm{T}}\theta(1:3) - \theta(M+3)^2 + \hat{\mathbf{s}}_{t,M}^{\mathrm{T}}\hat{\mathbf{s}}_{t,M} \end{bmatrix} \tag{76}$$

$$\Delta\mathbf{h}_2 = \mathbf{B}_2\Delta\theta + \mathbf{D}_2(\Delta\mathbf{s} - \Delta\hat{\mathbf{s}}) \tag{77}$$

$$\mathbf{B}_2 = \begin{bmatrix} -\mathbf{I}_{3\times3} & 0_{3\times1} & 0_{3\times1} & \cdots & 0_{3\times1} \\ -2\theta(1:3) & 2\theta(1+3) & 0 & \cdots & 0 \\ -2\theta(1:3) & 0 & 2\theta(2+3) & \cdots & 0 \\ \vdots & \vdots & \vdots & \ddots & \vdots \\ -2\theta(1:3) & 0 & 0 & \cdots & 2\theta(M+3) \end{bmatrix} \tag{78}$$

$$\mathbf{D}_2 = \begin{bmatrix} \mathbf{O}_{3\times3M} & \mathbf{O}_{3\times3N} \\ 2\mathrm{diag}\left\{(\mathbf{u}^o - \hat{\mathbf{s}}_{t,1})^{\mathrm{T}}, (\mathbf{u}^o - \hat{\mathbf{s}}_{t,2})^{\mathrm{T}}, \ldots, (\mathbf{u}^o - \hat{\mathbf{s}}_{t,M})^{\mathrm{T}}\right\} & \mathbf{O}_{3M\times3N} \end{bmatrix} \tag{79}$$

Invoking the WLS theorem again, one has the solution of target position, denoted by $\mathbf{u}$, from (74) as

$$\mathbf{u} = (\mathbf{G}_2^{\mathrm{T}}\mathbf{W}_2\mathbf{G}_2)^{-1}\mathbf{G}_2^{\mathrm{T}}\mathbf{W}_2\mathbf{h}_2 \tag{80}$$

where $\mathbf{W}_2$ is the weighting matrix and it is determined by

$$\begin{aligned} \mathbf{W}_2 &= \left[E(\Delta\mathbf{h}_2\Delta\mathbf{h}_2^{\mathrm{T}})\right]^{-1} \\ &= \Big[\mathbf{B}_2\mathrm{cov}(\theta)\mathbf{B}_2^{\mathrm{T}} + \mathbf{D}_2\mathrm{cov}(\Delta\mathbf{s} - \Delta\hat{\mathbf{s}})\mathbf{D}_2^{\mathrm{T}} + \mathbf{B}_2(\mathbf{G}_1^{\mathrm{T}}\mathbf{W}_1\mathbf{G}_1)^{-1}\mathbf{G}_1^{\mathrm{T}}\mathbf{W}_1\mathbf{D}_1\mathrm{cov}(\Delta\mathbf{s} - \Delta\hat{\mathbf{s}})\mathbf{D}_2^{\mathrm{T}} \\ &\quad + \mathbf{D}_2\mathrm{cov}(\Delta\mathbf{s} - \Delta\hat{\mathbf{s}})\mathbf{D}_1^{\mathrm{T}}\mathbf{W}_1\mathbf{G}_1(\mathbf{G}_1^{\mathrm{T}}\mathbf{W}_1\mathbf{G}_1)^{-1}\mathbf{B}_2^{\mathrm{T}}\Big]^{-1} \end{aligned} \tag{81}$$

But as presented in (81), the unknown target position is required in the computation of $\mathbf{W}_2$. Herein, to circumvent this dilemma, we preliminarily exploit the target position estimate contained in θ to form $\mathbf{W}_2$ and use (80) to estimate target position. After that we can utilize the estimated target position to update $\mathbf{W}_2$ for another repetition.

From the WLS theorem, the covariance matrix of $\mathbf{u}$ can be approximated, given sufficiently small BR measurement noise and transmitter/receiver position error, as

$$\mathrm{cov}(\mathbf{u}) = (\mathbf{G}_2^{\mathrm{T}}\mathbf{W}_2\mathbf{G}_2)^{-1} \tag{82}$$

4.2. Performance Analysis

As mentioned above, the CRLB traces out a lower bound for minimum possible variance that an unbiased estimator can achieve. Next, we will analyze the efficiency of the proposed solution by comparing its covariance matrix with the benchmark, i.e., CRLB. For derivation simplicity, we would compare their inverse, rather than directly compare the two separately. The CRLB has been presented in (33). By invoking the matrix inversion lemma [32] to (33) and using the definitions of $\mathbf{X}$ and $\mathbf{Y}$, we have after mathematical simplifications,

$$\mathbf{CRLB}_c(\mathbf{u}^o)^{-1} = \left(\frac{\partial \mathbf{r}^o}{\partial \mathbf{u}^o}\right)^T \mathbf{Q}_r^{-1}\left(\frac{\partial \mathbf{r}^o}{\partial \mathbf{u}^o}\right) - \left(\frac{\partial \mathbf{r}^o}{\partial \mathbf{u}^o}\right)^T \mathbf{Q}_r^{-1}\left(\frac{\partial \mathbf{r}^o}{\partial \mathbf{s}^o}\right)\breve{\mathbf{Z}}^{-1}\left(\frac{\partial \mathbf{r}^o}{\partial \mathbf{s}^o}\right)^T \mathbf{Q}_r^{-1}\left(\frac{\partial \mathbf{r}^o}{\partial \mathbf{u}^o}\right) \tag{83}$$

where the expression of $\breve{\mathbf{Z}}$ has been given in (32).

On the other hand, using (82), (81), (69), (68) and (52) successively, we can reformulate the inverse of $\text{cov}(\mathbf{u})$ as

$$\text{cov}(\mathbf{u})^{-1} = \mathbf{G}_3^T\mathbf{Q}_r^{-1}\mathbf{G}_3 - \mathbf{G}_3^T\mathbf{Q}_r^{-1}\mathbf{G}_4\bar{\mathbf{Z}}^{-1}\mathbf{G}_4^T\mathbf{Q}_r^{-1}\mathbf{G}_3 \tag{84}$$

where $\mathbf{G}_3 = \mathbf{B}_1^{-1}\mathbf{G}_1\mathbf{B}_2^{-1}\mathbf{G}_2$, $\mathbf{G}_4 = \mathbf{B}_1^{-1}\mathbf{D}_1$, and $\bar{\mathbf{Z}} = \mathbf{Q}_s^{-1} + \mathbf{G}_4^T\mathbf{Q}_r^{-1}\mathbf{G}_4 + \mathbf{G}_0^T(\mathbf{G}_c\mathbf{Q}_c\mathbf{G}_c^T + \mathbf{Q}_{rc})^{-1}\mathbf{G}_0$.

Comparing (83) with (84), we observe that $\mathbf{CRLB}_c(\mathbf{u}^o)^{-1}$ and $\text{cov}(\mathbf{u})^{-1}$ are identical in structure. Next, we proceed to prove their equivalency under the following conditions:

(C1) $||\Delta\mathbf{s}_{t,m}|| \ll ||\mathbf{c}_k^o - \mathbf{s}_{t,m}^o||$, $||\Delta\mathbf{s}_{t,m}|| \ll ||\mathbf{s}_{t,m}^o - \mathbf{s}_{r,n}^o||$, $||\Delta\mathbf{s}_{r,n}|| \ll ||\mathbf{c}_k^o - \mathbf{s}_{r,n}^o||$, $||\Delta\mathbf{s}_{r,n}|| \ll ||\mathbf{s}_{t,m}^o - \mathbf{s}_{r,n}^o||$, and $||\Delta\mathbf{c}_k|| \ll ||\mathbf{c}_k^o - \mathbf{s}_{t,m}^o||$, $||\Delta\mathbf{c}_k|| \ll ||\mathbf{c}_k^o - \mathbf{s}_{r,n}^o||$, for $k = 1, 2, \ldots, K$, $m = 1, 2, \ldots, M$ and $n = 1, 2, \ldots, N$;

(C2) $||\Delta r_{m,n}|| \ll ||\mathbf{u}^o - \mathbf{s}_{t,m}^o||$, $||\Delta r_{m,n}|| \ll ||\mathbf{u}^o - \mathbf{s}_{r,n}^o||$, $||\Delta r_{m,n}|| \ll ||\mathbf{s}_{t,m}^o - \mathbf{s}_{r,n}^o||$, and $||\Delta\mathbf{s}_{t,m} - \Delta\hat{\mathbf{s}}_{t,m}|| \ll ||\mathbf{u}^o - \mathbf{s}_{t,m}^o||$, $||\Delta\mathbf{s}_{r,n} - \Delta\hat{\mathbf{s}}_{r,n}|| \ll ||\mathbf{u}^o - \mathbf{s}_{r,n}^o||$ for $m = 1, 2, \ldots, M$ and $n = 1, 2, \ldots, N$;

The condition C1 implies the transmitter/receiver position error and the calibration target position error are negligibly small compared with the range between the calibration target and the transmitter/receiver. The condition C2 implies the BR measurement noise and the error in the refined transmitter/receiver position are negligibly small compared to the range between the calibration target and the transmitter/receiver. Using the conditions C1 and C2, we obtain, after some involved algebraic manipulations, that

$$\mathbf{G}_3 = \frac{\partial \mathbf{r}^o}{\partial \mathbf{u}^o}, \mathbf{G}_4 = -\frac{\partial \mathbf{r}^o}{\partial \mathbf{s}^o}, \mathbf{G}_0 = -\left(\frac{\partial \mathbf{r}_c^o}{\partial \mathbf{s}^o}\right), \mathbf{G}_c = -\left(\frac{\partial \mathbf{r}_c^o}{\partial \mathbf{c}^o}\right) \tag{85}$$

By this point, we can draw the conclusion that

$$\text{cov}(\mathbf{u})^{-1} = \mathbf{CRLB}_c(\mathbf{u}^o)^{-1} \tag{86}$$

That is, the proposed solution accomplishes the CRLB accuracy if the two conditions C1 and C2 are satisfied. In reality, localization scenarios, which satisfy the conditions C1 and C2, are not rare. These two conditions can be satisfied if the unknown target and the calibration targets are far from the transmitters and receivers, if not these conditions can still be satisfied if the BR measurement noise and the transmitter/receiver/calibration target position errors are sufficiently small.

5. Simulation Results

In this section, the efficiency and superiority of the proposed solution will be corroborated through Monte Carlo simulations. Amiri's method presented in [14], which does not consider the transmitter and receiver position error and Zhao's method proposed in [22], which considers the statistical distributions of transmitter/receiver position error but does not use any calibration targets, are chosen as references for comparison. The exact positions of transmitters/receivers/calibration targets are the same as those in Table 1. Localization accuracy is quantitatively evaluated using root mean squares error (RMSE), which comes from 1000 independent Monte Carlo runs. In each run, the zero-mean Gaussian random errors with covariance matrices $\mathbf{Q}_r = \sigma_r^2\mathbf{V}_r$, $\mathbf{Q}_{rc} = \sigma_{rc}^2\mathbf{V}_{rc}$, $\mathbf{Q}_s = \sigma_s^2\mathbf{V}_s$ and $\mathbf{Q}_c = \sigma_c^2\mathbf{V}_c$ are added to the BRs from unknown target, BRs from calibration targets, actual transmitter and receiver positions, and actual calibration target positions, respectively, in order to simulate a real localization scenario. The setting of $\mathbf{V}_r$, $\mathbf{V}_{rc}$, $\mathbf{V}_s$ and $\mathbf{V}_c$ are also the same as that in Example 1.

First of all, in order to intuitively show the difference between target localization with and without the use of calibration targets, we plot in Figure 4 the estimated target positions from each Monte Carlo run, which forms a scatter plot for target position estimation. For comparison, the scatterplots of Amiri's method and Zhao's method are also plotted. The transmitter/receiver position error level is set to $\sigma_s = 20$ m, the noise level of BR measurements from unknown target and calibration targets is set as $\sigma_r = \sigma_{rc} = 10$ m, and the calibration target position error level is set to be $\sigma_c = 10$ m. The true position of the unknown target is $\mathbf{u}^o = [50000, 15000, 5000]^T$m, which is marked with red pentagram in Figure 4 for comparison. By comparing the scatterplots of the methods, we find that with the use of calibration targets, the scattered dots of target position estimation are more closely around the target's true position, which intuitively illustrates the performance gain from the use of calibration targets. Without the use of calibration targets, considering the statistical distributions of transmitter/receiver position error can also reduce dispersion of estimated target position dots to some extent, but compared to using calibration targets, this degree of reduction in dispersion is not sufficiently impressive.

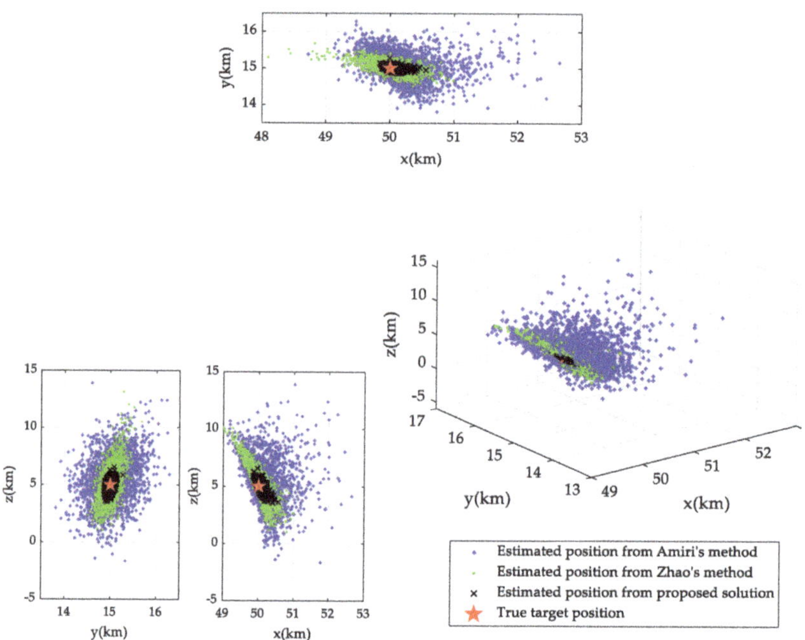

Figure 4. Scatter plots of estimated positions from different methods.

Now, in order to quantitatively evaluate the localization accuracy of the methods, we calculate the RMSE of the proposed solution under different error or noise conditions, and compare it with Amiri's method, Zhao's method, as well as the CRLB. As mentioned in Section 4.2, the localization accuracy of the proposed solution is related to the distance between the target and MPR system. Hence, in order to achieve a more comprehensive insight on the performance of the proposed solution, we consider two cases, i.e., the near-field case where the target is close to the MPR system, and the far-field case where the target is far away from the MPR system. The exact positions of transmitters/receivers/calibration targets remain the same as before. We first address the far-field target, whose position is set to $\mathbf{u}^o = [120000, 120000, 12000]^T$m. The results are presented in Figure 5.

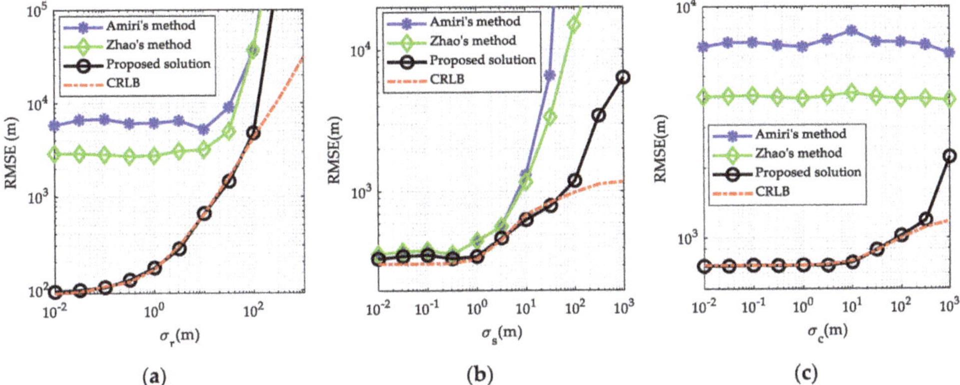

Figure 5. Comparison of the RMSEs among different localization methods in the far-field case: (**a**) with different BR measurement noise level σ_r and $\sigma_s = 20$ m, $\sigma_c = 10$ m; (**b**) with different transmitter/receiver position error level σ_s and $\sigma_r = 10$ m, $\sigma_c = 10$ m; (**c**) with different calibration target position error level σ_c and $\sigma_r = 10$ m, $\sigma_s = 20$ m.

Figure 5a plots the RMSE curves of the methods versus the BR measurement noise level. It shows that the localization RMSE of the proposed solution matches the CRLB very well and is about an order of magnitude lower than that of Amiri's method and Zhao's method at a low-to-moderate BR measurement noise level. Although it deviates from the CRLB when the BR measurement noise level is large, it is still much smaller than that of other two methods. The deviation from the CRLB, known as the thresholding phenomenon, is due to the ignored second order error terms in the design of the solution, which is invalid for large error levels. Owing to considering the statistical distributions of the transmitter/receiver position error, the RMSEs produced by Zhao's method is generally lower than that by Amiri's method. But compared with the use of the calibration targets in the proposed solution, the localization accuracy improvement brought by the consideration of transmitter/receiver position error in Zhao's method is not so significant. Figure 5b gives the RMSE curves of the methods versus the transmitter/receiver position error level. It can be seen that, the superiority of the proposed solution in localization accuracy is mainly reflected at moderate to high transmitter/receiver position error level. When the transmitter/receiver position error is small, the localization accuracy of the proposed solution and the other two methods is comparable. This again agrees very well with the theoretical performance in Section 3. Figure 5c compares the RMSEs from the methods with respect to different calibration target position error levels. As is illustrated in Figure 5c, the proposed solution always offers a remarkable advantage over the other two methods at different calibration target position error level, even when the calibration target position error is extremely large. This is in agreement with the previous simulation results for the CRLB in Section 3.

Next, the same set of simulations was repeated for a near-field target, whose position is set to be $\mathbf{u}^o = [12000, 1200, 1200]^T$m. The results are provided in Figure 6, from which we observe that the proposed solution still performs much better than the other methods. However, comparing with the corresponding results in Figure 5, we find the localization accuracy for near-field target is generally better than a far-field target, given the same noise and error levels. One reason may be that, when the target is close to the MPR system, the transmitters/receivers are far apart relative to the distance between the target and the MPR system. Thus, the localization geometry would become more regular and the corresponding geometric dilution of precision (GDOP) value would be smaller compared to the far-field case. However, on the other hand, comparing the thresholding values in Figures 5 and 6 indicates that the RMSE curves for the near-field target deviate from the CRLB at smaller values than those for the far-field target. This phenomenon is consistent with the analysis under (83) that the equivalency between the estimate variance and the CRLB is more affected by the BR measurement noises when the target is close to the MPR system.

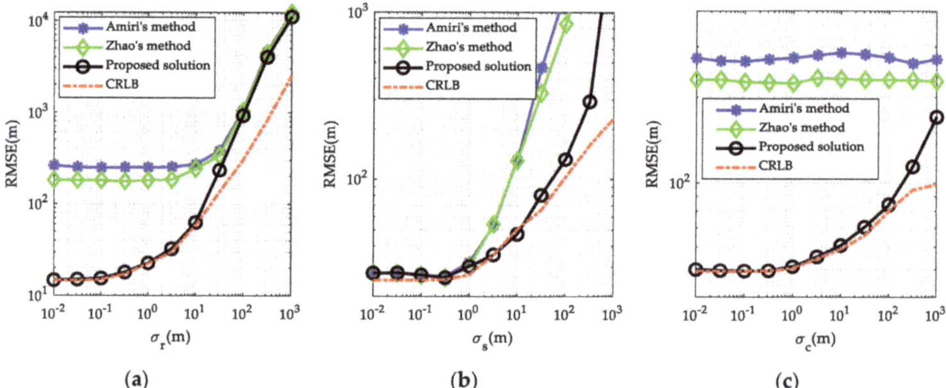

Figure 6. Comparison of the RMSEs among different localization methods in the near-field case: (**a**) with different BR measurement noise level σ_r and $\sigma_s = 20$ m, $\sigma_c = 10$ m; (**b**) with different transmitter/receiver position error level σ_s and $\sigma_r = 10$ m, $\sigma_c = 10$ m; (**c**) with different calibration target position error level σ_c and $\sigma_r = 10$ m, $\sigma_s = 20$ m.

At an intuitive level, the more calibration targets are used, the better the localization accuracy is. In what follows, we will quantitatively analyze the effect of number of calibration targets on the localization accuracy by varying the number of calibration targets from 1 to 10. The positions of the transmitters and receivers remain the same as before. The positions of calibration targets and unknown target are chosen randomly from the 50 km × 40 km × 5 km volume as presented in Figure 2. The simulation results are depicted in Figure 7.

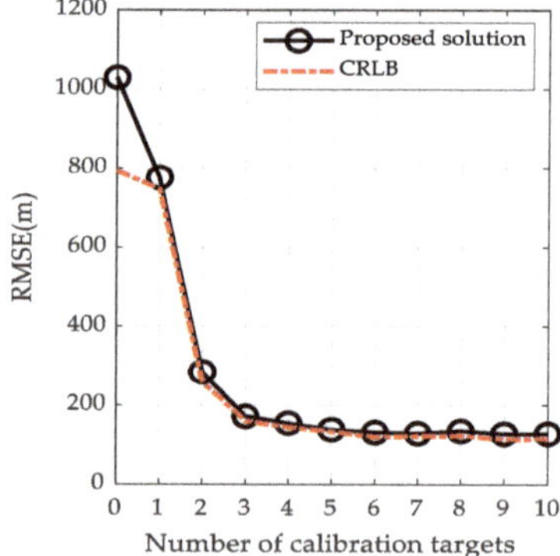

Figure 7. Localization accuracy versus the number of calibration targets.

Figure 7 shows the RMSE, as well as the CRLB, versus the number of calibration targets. As expected, when the number of calibration targets is small, the localization accuracy improves significantly as the number of calibration targets increases. However, it is seen that there is no obvious dependence on the number of calibration targets as soon as the number of calibration targets is larger than 3. This indicates that when the number of calibration targets reaches 3, the use of more calibration targets would only increase the computational expense and not remarkably enhance the localization accuracy. Therefore, in the absence of any other consideration, it is reasonable to set the number of calibration targets as 3.

6. Conclusions

This paper explores the use of calibration targets with known positions to refine the inaccurate transmitter/receiver positions and thus enhance target localization accuracy in MPR systems. We start our research by evaluating target localization CRLB in the presence of calibration targets, which justifies the potential of calibration targets in enhancing localization accuracy. Then, in order to fulfill this potential, a novel closed-form solution was designed for target localization using BR measurements from the unknown target as well as the calibration targets. The proposed solution was shown both analytically and numerically to attain the CRLB under some mild conditions, and verified to outperform existing methods in terms of localization accuracy. Furthermore, from the view of engineering practice, if the employed calibration targets are off-the-shelf, such as the commercial aircrafts broadcasting an ADS-B signal, the use of calibration targets would bring little added cost or complexity to the MPR system, but could bring a significant enhancement to target localization accuracy.

Author Contributions: Conceptualization, Y.Z. (Yongsheng Zhao); methodology, Y.Z. (Yongsheng Zhao) and D.H.; software, Z.L.; validation, Y.Z. (Yongjun Zhao); formal analysis, Y.Z. (Yongsheng Zhao); writing—original draft preparation, Y.Z. (Yongsheng Zhao); writing—review and editing, Y.Z. (Yongsheng Zhao); supervision, D.H.; project administration, C.Z.; funding acquisition, D.H. and Y.Z. (Yongjun Zhao).

Funding: This research was supported by National Natural Science Foundation of China, grant number 61703433.

Conflicts of Interest: The authors declare that they have no conflict of interest to this work.

References

1. Edrich, M.; Meyer, F.; Schroeder, A. Design and performance evaluation of a mature FM/DAB/DVB-T multi-illuminator passive radar system. *IET Radar Sonar Navig.* **2014**, *8*, 114–122. [CrossRef]
2. Tatiana, M.; Fabiola, C.; Enrico, T.; Annarita, D.L. Multi-Frequency Target Detection Techniques for DVB-T Based Passive Radar Sensors. *Sensors* **2016**, *16*, 1594–1618.
3. Wang, Y.; Bao, Q.; Wang, D.; Chen, Z. An experimental study of passive bistatic radar using uncooperative radar as a transmitter. *IEEE Geosci. Remote Sens. Lett.* **2015**, *12*, 1–5.
4. Piotr, S.; Michał, W.; Kulpa, K. Trial Results on Bistatic Passive Radar Using Non-Cooperative Pulse Radar as Illuminator of Opportunity. *Int. J. Electron. Telecommun.* **2012**, *58*, 171–176.
5. Olsen, K.E.; Asen, W. Bridging the gap between civilian and military passive radar. *IEEE Aerosp. Electron. Syst. Mag.* **2017**, *32*, 4–12. [CrossRef]
6. Palmer, J.; Palumbo, S.; Summers, A.; Merrett, D.; Searle, S.; Howard, S. An overview of an illuminator of opportunity passive radar research project and its signal processing research directions. *Digit. Signal Process.* **2011**, *21*, 593–599. [CrossRef]
7. Malanowski, M. Detection and parameter estimation of manoeuvring targets with passive bistatic radar. *IET Radar Sonar Navig.* **2012**, *6*, 739–745. [CrossRef]
8. Feng, Y.; Sun, H.Y.; Liu, D.L.; Ma, Y.H. Target location based on quasi-Newton algorithm in multistatic passive radar. In Proceedings of the IET International Radar Conference, Hangzhou, China, 14–16 October 2015.
9. Zhao, Y.S.; Zhao, Y.J.; Sun, D.; Zhao, C. Constrained total least squares localization algorithm for multistatic passive radar using bistatic range measurements, In Proceedings of the 19th International Radar Symposium (IRS), Bonn, Germany, 20–22 June 2018.
10. Dianat, M.; Taban, M.R.; Dianat, J.; Sedighi, V. Target localisation using least squares estimation for MIMO radars with widely separated antennas. *IEEE Trans. Aerosp. Electron. Syst.* **2013**, *49*, 2730–2741. [CrossRef]
11. Noroozi, A.; Sebt, M.A. Target localisation from bistatic range measurements in multi-transmitter multi-receiver passive radar. *IEEE Signal Process. Lett.* **2015**, *22*, 2445–2449. [CrossRef]
12. Einemo, M.; So, H.C. Weighted least squares algorithm for target localisation in multi-static passive radar. *Signal Process.* **2015**, *115*, 144–150. [CrossRef]
13. Park, C.H.; Chang, J.H. Closed-form localisation for multi-static passive radar systems using time delay measurements. *IEEE Trans. Wirel. Commun.* **2016**, *15*, 1480–1490. [CrossRef]

14. Amiri, R.; Behnia, F.; Zamani, H. Asymptotically efficient target localisation from bistatic range measurements in Multi-static passive radars. *IEEE Signal Process. Lett.* **2017**, *24*, 299–303. [CrossRef]
15. Chan, Y.T.; Ho, K.C. A simple and efficient estimator for hyperbolic location. *IEEE Trans. Signal Process.* **1994**, *42*, 1905–1915. [CrossRef]
16. Dawidowicz, B.; Kulpa, K.S.; Malanowski, M.; Misiurewicz, J.; Samczynski, P.; Smolarczyk, M. DPCA Detection of Moving Targets in Airborne Passive Radar. *IEEE Trans. Aerosp. Electron. Syst.* **2012**, *48*, 1347–1357. [CrossRef]
17. Brown, J.; Woodbridge, K.; Griffiths, H.; Stove, A. Passive bistatic radar experiments from an airborne platform. *IEEE Aerosp. Electron. Syst. Mag.* **2012**, *27*, 50–55. [CrossRef]
18. Thompson, E.C. Bistatic radar noncooperative illumination synchronization techniques. In Proceedings of the IEEE National Radar Conference, Dallas, TX, USA, 29–30 March 1989.
19. Matthes, D. Convergence of ESM sensors and passive covert radar. In Proceedings of the IEEE International Radar Conference, Arlington, TX, USA, 9–12 May 2005.
20. Rui, L.; Ho, K.C. Elliptic localisation: Performance study and optimum receiver placement. *IEEE Trans. Signal Process.* **2014**, *62*, 4673–4688. [CrossRef]
21. Chalise, B.K.; Zhang, Y.D.; Amin, M.G.; Himed, B. Target localisation in a multi-static passive radar system through convex optimization. *Signal Process.* **2014**, *102*, 207–215. [CrossRef]
22. Zhao, Y.S.; Hu, D.X.; Zhao, Y.J.; Liu, Z.X. Moving target localization in distributed MIMO radar with transmitter and receiver location uncertainties. *Chin. J. Aeronaut.* **2018**, in press. [CrossRef]
23. Moses, R.; Patterson, R. Self-calibration of sensor networks. In Proceedings of the SPIE—The International Society for Optical Engineering, Orlando, FL, USA, 7 August 2002; Volume 4743, pp. 108–119.
24. Ash, J.N.; Moses, R.L. On the relative and absolute positioning errors in self-localization systems. *IEEE Trans. Signal Process.* **2008**, *56*, 5668–5679. [CrossRef]
25. Syldatk, M.; Sviestins, E.; Gustafsson, F. Expectation maximization algorithm for calibration of ground sensor networks using a road constrained particle filter. In Proceedings of the 15th International Conference on Information Fusion, Singapore, 9–12 July 2012.
26. Syldatk, M.; Gustafsson, F. Simultaneous tracking and sparse calibration in ground sensor networks using evidence approximation. In Proceedings of the IEEE International Conference on Acoustics, Speech and Signal Processing, Vancouver, BC, Canada, 26–31 May 2013.
27. Hasan, S.M.; Sahr, J.D.; Hatke, G.F.; Keller, C.M. Passive source localization using an airborne sensor array in the presence of manifold perturbations. *IEEE Trans. Signal Process.* **2007**, *55*, 2486–2496.
28. Ho, K.C.; Yang, L. On the use of a calibration emitter for source localization in the presence of sensor position uncertainty. *IEEE Trans. Signal Process.* **2008**, *56*, 5758–5772. [CrossRef]
29. Yang, L.; Ho, K.C. Alleviating sensor position error in source localization using calibration emitters at inaccurate locations. *IEEE Trans. Signal Process.* **2009**, *58*, 67–83. [CrossRef]
30. Li, J.; Guo, F.; Yang, L.; Jiang, W.; Pang, H. On the use of calibration sensors in source localization using TDOA and FDOA measurements. *Digit. Signal Process.* **2014**, *27*, 33–43. [CrossRef]
31. Strohmeier, M.; Schafer, M.; Lenders, V.; Martinovic, I. Realities and challenges of nextgen air traffic management: The case of ADS-B. *IEEE Commun. Mag.* **2014**, *52*, 111–118. [CrossRef]
32. Petersen, K.B.; Pedersen, M.S. *The Matrix Cookbook*; Technical University of Denmark: Copenhagen, Denmark, 2012.
33. Kay, S.M. *Fundamentals of Statistical Signal Processing, Estimation Theory*; Prentice-Hall: Englewood Cliffs, NJ, USA, 1993.
34. Bellman, R. *Introduction to Matrix Analysis*; Mc Graw-Hill: New York, NY, USA, 1960.

Article

Refined PHD Filter for Multi-Target Tracking under Low Detection Probability

Sen Wang [1,*], Qinglong Bao [1] and Zengping Chen [2]

1 National Key Laboratory of Science and Technology on ATR, National University of Defense Technology, Changsha 410073, China

2 School of Electronics and Communication Engineering, SUN YAT-SEN University, Guangzhou 510275, China

* Correspondence: wangsen11@nudt.edu.cn; Tel.: +86-170-7746-1248

Received: 5 June 2019; Accepted: 24 June 2019; Published: 26 June 2019

Abstract: Radar target detection probability will decrease as the target echo signal-to-noise ratio (SNR) decreases, which has an adverse influence on the result of multi-target tracking. The performances of standard multi-target tracking algorithms degrade significantly under low detection probability in practice, especially when continuous miss detection occurs. Based on sequential Monte Carlo implementation of Probability Hypothesis Density (PHD) filter, this paper proposes a heuristic method called the Refined PHD (R-PHD) filter to improve multi-target tracking performance under low detection probability. In detail, this paper defines a survival probability which is dependent on target state, and labels individual extracted targets and corresponding particles. When miss detection occurs due to low detection probability, posterior particle weights will be revised according to the prediction step. Finally, we transform the target confirmation problem into a hypothesis test problem, and utilize sequential probability ratio test to distinguish real targets and false alarms in real time. Computer simulations with respect to different detection probabilities, average numbers of false alarms and continuous miss detection durations are provided to corroborate the superiority of the proposed method, compared with standard PHD filter, Cardinalized PHD (CPHD) filter and Cardinality Balanced Multi-target Multi-Bernoulli (CBMeMBer) filter.

Keywords: refined PHD filter; low detection probability; continuous miss detection; radar multi-target tracking; survival probability; target labels; posterior weight revision; sequential probability ratio test; hypothesis test

---

1. Introduction

The objective of Multi-Target Tracking (MTT) is to jointly estimate the number of targets and their individual states, and to provide target tracks or trajectories, from a sequence of measurements provided by sensing devices such as radar [1], sonar [2], or cameras [3]. Traditional MTT algorithms, including Joint Probabilistic Data Association Filter (JPDAF) [4] and Multiple Hypothesis Tracking (MHT) [5], always transform the multi-target tracking problem into multiple independent single-target tracking problems by data association processing according to a certain distance criterion. Association error resulting from complex scene deteriorates the tracking performance of JPDAF and MHT.

In recent years, multi-source multi-target information fusion theory based on Random Finite Sets (RFS) provides a unified and scientific mathematical basis for multi-sensor multi-target tracking problem [6,7]. Different from traditional heuristic methods, multi-target tracking methods based on RFS strictly describe target birth, death, spawning, miss detection and clutters in multi-target tracking process, directly estimate number and state of targets, and even provide target tracks or trajectories by modeling multi-target states and sensor measurements as RFS or labeled RFS, which has the best performance in Bayesian sense.

The multi-target Bayes filter is difficult to implement. Fortunately, some advanced approximations have been proposed, such as the Probability Hypothesis Density (PHD) filter [8–10], the Cardinalized PHD (CPHD) filter [11,12], the Multi-target Multi-Bernoulli (MeMBer) filter [13] and the Cardinality Balanced MeMBer (CBMeMBer) filter [14]. More recently, multi-target tracking algorithms based on labeled random finite sets have been proposed [15–23], and it can obtain track-valued estimates of individual targets without the need for post-processing, such as the Generalized Labeled Multi-Bernoulli (GLMB) filter [15] and Labeled Multi-Bernoulli (LMB) filter [16].

Since Sequential Monte Carlo (SMC) implementation and Gaussian Mixture (GM) implementation of PHD filter were proposed, the PHD filter has attracted significant attention in multi-target tracking research. To reduce the computational complexity of the PHD filter, several gating strategies were introduced to exclude clutter observation participating in filter updating [24,25]. To obtain target states from posterior PHD, several multi-target state extraction algorithms have been proposed, such as clustering [26,27] and data-driven methods [28–31]. To fuse information from multiple observation system, multi-sensor multi-target tracking filters based on PHD were proposed [32–34]. To track maneuvering targets, traditional multi-model method was introduced to PHD filter [35]. Faced with unknown backgrounds, such as unknown detection probability, unknown clutter parameter, several improved PHD filters can estimate background parameters while tracking [36,37]. In non-standard target observation model, several improved PHD filters were proposed to track extended target [38,39].

The standard PHD filter has considered the influence of the detection probability on multi-target tracking, but its performance degrades significantly under low detection probability in practice, especially when continuous miss detection occurs. For example, the posterior particle weights of a SMC-PHD filter will become small under continuous miss detection, and corresponding particles may be eliminated from the particle pool and then the undetected target will be lost.

Several recent works have made some attempts [40,41]. Based on the GM-PHD filter, the Refined GM-PHD (RGM-PHD) filter [40] was proposed to improve the performance of the GM-PHD filter under continuous miss detection. This method is effective in terms of various detection probabilities, false alarm rates and continuous miss detection rates. However, some key parameters of the RGM-PHD filter, including the penalty coefficient and the reward coefficient, are determined without explicit formula, which is difficult to be generalized to other applications. Also based on GM implementation of PHD filter, a novel target state estimate method was integrated into three improved GM-PHD filters [41], which results in better tracking performance in imperfect detection probability scenarios. However, lower bound of detection probability in simulations is set as 0.8, which can't sufficiently illustrate the effectiveness of the method under low detection probability.

In this paper, based on SMC implementation of PHD filter, we propose a heuristic method called Refined PHD (R-PHD) filter to improve multi-target tracking performance under low detection probability. First, survival probability dependent on target state is defined, which is based on the hypothesis that target enter and exit sensor Fields of View (FoV) usually occur at the boundary. Then, individual target and particle are assigned a unique label, which is utilized to confirm if miss detection occurs for each target and identify particles representing the undetected target. When miss detection occurs, posterior weights will be revised according to the prediction step. The key of the proposed method is to distinguish real targets and false alarms. This paper binarizes the likelihood function of individual extracted target, which is approximated as a random variable obeying two-point distribution. When extracted target is a real target, success probability of the two-point distribution is approximatively the detection probability. When extracted target is a false alarm, the success probability is approximatively a very small value. Then this paper transforms target confirmation problem into a hypothesis test problem, and utilizes Sequential Probability Ratio Test (SPRT) [42] to confirm real targets in real time. After target extraction at each time, we mark each extracted target as a real target or a false alarm, or make no decision according to test statistic.

The rest of the paper is organized as follows. Section 2 reviews probability hypothesis density filter and corresponding SMC implementation. Section 3 proposes the refined PHD filter in detail.

Computer simulations illustrating the effectiveness and the performance of the proposed method are provided in Section 4. Finally, Section 5 presents the conclusion.

2. Background

This section will introduce the probability hypothesis density filter. Furthermore, SMC implementation of the PHD filter will also be reviewed.

2.1. PHD Filter

The probability hypothesis density is defined as the first-order statistical moment of multiple-target posterior distribution. Similar to the constant-gain Kalman filter in single-target filtering, the PHD filter is the first-order moment approximation of the multi-target Bayes filter, which only recursively propagates first-order multi-target moments by time prediction and data-update steps. Suppose $D_{k-1|k-1}(x)$ is the PHD at time $k-1$, the predictor equation of the PHD filter can be expressed as

$$D_{k|k-1}(x) = b_{k|k-1}(x) + \int \left[p_S(x') \cdot f_{k|k-1}(x|x') + b_{k|k-1}(x|x')\right] \cdot D_{k-1|k-1}(x')dx', \tag{1}$$

where $f_{k|k-1}(x|x')$ is the single-target Markov transition density, $p_S(x')$ is the probability that a target with state x' at time $k-1$ will survive at time k, $b_{k|k-1}(x|x')$ is the PHD of targets at time k spawned by a single target x' at time $k-1$, and $b_{k|k-1}(x)$ is the PHD of new targets entering the scene at time k.

At time k the sensor collects a new multi-target measurement set $Z_k = \{z_1, \cdots, z_m\}$, if we assume that the predicted multi-target distribution is approximately Poisson, the closed-form formula of corrector equation of the PHD filter can be derived as

$$D_{k|k}(x) \approx \left[1 - p_D(x) + p_D(x) \sum_{z \in Z_k} \frac{L_z(x)}{\lambda \cdot c(z) + \int p_D(x) L_z(x) D_{k|k-1}(x)dx}\right] \cdot D_{k|k-1}(x), \tag{2}$$

where $L_z(x)$ is the single-target likelihood function, $p_D(x)$ is the probability that a target with state x at time k will be detected, λ is the average number of Poisson-distributed false alarms, the spatial distribution of which is governed by the probability density $c(z)$.

The expected number of targets can be estimated by rounding the integral of the PHD over the entire state space, and then the state-estimates of the targets can be obtained from the local maxima of the PHD.

2.2. SMC-PHD Filter

Up to now, PHD filters can be realized by SMC approximation or GM approximation. Compared with the GM-PHD filter, the SMC-PHD filter is suitable for problems involving non-linear non-Gaussian dynamics. Regardless of spawned targets, the following sequentially describes each of the SMC-PHD filter processing steps: initialization, prediction, correction, and state estimation.

Initialization: Suppose prior PHD at time 0 is

$$D_{0|0}(x) \approx \sum_{i=1}^{v_{0|0}} w_{0|0}^i \delta(x - x_{0|0}^i), \tag{3}$$

where $\delta(x)$ is Dirac delta function, $v_{0|0}$ is the number of particles, $x_{0|0}^i$ is the ith particle and $w_{0|0}^i$ is the corresponding weight.

Prediction: Suppose PHD at time $k-1$ can be approximated using a group of particles

$$D_{k-1|k-1}(x) \approx \sum_{i=1}^{v_{k-1|k-1}} w_{k-1|k-1}^i \delta(x - x_{k-1|k-1}^i). \tag{4}$$

The meaning of the variables in the above formula is similar to that of Equation (3). Then the predicted PHD at time k is

$$D_{k|k-1}(x) \approx \sum_{i=1}^{v_{k|k-1}} w^i_{k|k-1}\delta(x - x^i_{k|k-1}), \tag{5}$$

where $v_{k|k-1} = v_{k-1|k-1} + v^{birth}_{k|k-1}$ is the number of predicted particles, $v^{birth}_{k|k-1}$ is the number of appearing particles, $x^i_{k|k-1}, i = 1, \cdots, v_{k-1|k-1}$ is obtained by the single-target Markov transition density, $x^i_{k|k-1}, i = v_{k-1|k-1} + 1, \cdots, v_{k|k-1}$ is sampled from the probability density of the spontaneously appearing targets, $w^i_{k|k-1} = p_S(x^i_{k-1|k-1}) \cdot w^i_{k-1|k-1}, i = 1, \cdots, v_{k-1|k-1}$ is the weight corresponding to persisting particles, $w^i_{k|k-1} = 1/\rho, i = v_{k-1|k-1} + 1, \cdots, v_{k|k-1}$ is the weight corresponding to appearing particles, and the PHD filter requires ρ particles to adequately maintain track on any individual target.

Correction: After receiving the multi-target measurement set, the posterior PHD at time k can be approximated as

$$D_{k|k}(x) \approx \sum_{i=1}^{v_{k|k}} w^i_{k|k}\delta(x - x^i_{k|k}), \tag{6}$$

where $v_{k|k} = v_{k|k-1}$ and $x^i_{k|k} = x^i_{k|k-1}, i = 1, \cdots, v_{k|k}$ are the same as those of the predicted PHD, and ith particle weight can be calculated by

$$w^i_{k|k} = w^i_{k|k-1}p_D(x^i_{k|k-1})\sum_{z\in Z_k} \frac{L_z(x^i_{k|k-1})}{\lambda c(z) + \sum\limits_{e=1}^{v_{k|k-1}} w^e_{k|k-1}p_D(x^e_{k|k-1})L_z(x^e_{k|k-1})} + w^i_{k|k-1}\left[1 - p_D(x^i_{k|k-1})\right]. \tag{7}$$

The above particle weights are not equal, and the resampling technique can be utilized to replace them with new, equal weights.

State estimation: The expected number of targets at time k is $\hat{N}_{k|k} \approx round\left(\sum\limits_{i=1}^{v_{k|k}} w^i_{k|k}\right)$, and the state-estimates of the targets can be obtained by clustering [26,27], data-driven methods [28–31], and so on.

3. Refined PHD Filter

The standard SMC-PHD filter has considered the influence of the detection probability on multi-target tracking. It is indicated from Equation (7) that particle weight of the posterior PHD is a weighted sum of two terms [28–31]. The first term updates predicted particle weight according to the likelihood function, while the second term directly propagates predicted weight to the posterior PHD considering possible miss detection. Furthermore, the weights of these two terms are detection probability and probability of miss detection, respectively. However, the performance of the standard SMC-PHD filter degrades significantly under low detection probability in practice, especially when continuous miss detection occurs. This is because, continuous miss detection of a target makes the posterior weights small, which may eliminate corresponding particles from the particle pool and then lose the target. This paper proposes a heuristic method called Refined PHD (R-PHD) filter aiming to improve the performance of the SMC-PHD filter under low detection probability. In the proposed method, survival probability dependent on target state is defined, and each target is assigned a label. When miss detection occurs, posterior weights will be revised according to the prediction step. After state estimation of each step, target confirmation is conducted based on sequential probability ratio test.

3.1. Survival Probability Dependent on Target State

Measurement of a specific target is not collected due to either miss detection or death of the target. Multi-target tracking algorithms should take some compensation measures for the former reason, while do nothing for the latter. The way to distinguish between the two reasons is to consider the survival probability of the target. If this survival probability is larger than a threshold, algorithms can confirm the target is persisting and compensate miss detection. The earlier versions of the SMC-PHD filter consider the survival probability as a constant which is independent of target state and can't be used to judge whether the target survives. This paper defines a new survival probability dependent on target state, which is used as one of the conditions to revise posterior particle weights.

Intuitively, targets usually enter sensor FoV from the boundary and exit also from the boundary. The survival probability of a specific target can be very high when it is located in the middle of FoV. On the contrary, the survival probability of a specific target drops rapidly when it is located near the boundary of FoV and moves outwards. Without the loss of generality from an algorithmic viewpoint, this paper considers a rectangular FoV which possesses four boundaries, up and down, left and right, and then the survival probability of a target at time k is

$$p_S^k = \min\left\{p_{S,up}^k, p_{S,down}^k, p_{S,left}^k, p_{S,right}^k\right\}, \tag{8}$$

where $p_{S,up}^k, p_{S,down}^k, p_{S,left}^k, p_{S,right}^k$ are the survival probabilities of the target with respect to the four boundaries, respectively.

Suppose the particles representing the target at time k are $x_{k|k}^i, i = 1, \cdots, v_{k|k}$, where each particle is a four-dimensional vector $x_{k|k}^i = \left[p_x^i, v_x^i, p_y^i, v_y^i\right]^T$, representing the target position and velocity along the x-axis and y-axis, respectively, and then the target state and corresponding variance can be estimated as $mean\left(x_{k|k}^i\right)$ and $\mathrm{var}\left(x_{k|k}^i\right)$. If the particles follow Gaussian distribution and the four variables in particles are independent of each other, the state of this target follows Gaussian distribution $N\left(\left[p_x, v_x, p_y, v_y\right]^T, diag\left[\sigma_{px}^2, \sigma_{vx}^2, \sigma_{py}^2, \sigma_{vy}^2\right]\right)$, where $v_x = mean\left(v_x^i\right)$, $\sigma_{vx}^2 = \mathrm{var}\left(v_x^i\right)$ and so on. Based on the above discussion, the survival probability of the target with respect to the right boundary is

$$p_{S,right}^k = \Pr\left(u \le \frac{b_{right} - p_x - v_x T}{\sqrt{\sigma_{px}^2 + \sigma_{vx}^2 T^2}}\right), \tag{9}$$

where $u \sim N(0,1)$, b_{right} is the position of the right boundary, and T is the sampling period. Meanwhile, the survival probabilities of the target with respect to the other three boundaries have similar results.

It should be mentioned that a specific target and corresponding particles share the identical survival probability, which is used for the predictor equation and revising posterior particle weights.

3.2. Labeling Target and Particle

In order to confirm if miss detection occurs for each target and identify particles representing the undetected target, every target and particle has its own unique label. On the other hand, the standard SMC-PHD filter can only provide the point-valued estimates of the target states at each time, not track-valued estimates of individual targets due to no record of the target identities. Some principled solutions such as labeled RFS [15,16] were proposed, and produce track-valued estimate without post processing. This paper attaches a unique label to individual targets and particles, which can be used not only for trajectory extraction, but can also compensate miss detection. It should be pointed out that the particles representing a target can have several different labels, and particles with identical labels can also belong to different targets. Labels are assigned to individual targets and individual particles, considering the following principles:

Principles for labeling targets:

1. The label of one target is determined by the label with the largest number of particles belonging to this target.
2. If the label of one target is zero, a new positive number will be assigned to it as its label.

3. When there are multiple targets with the same label at time k, the optimal successor will be selected and keep its label unchanged while others will be assigned a new positive number sequentially.

Principles for labeling particles:

4. The label of appearing particles is initialized as zero.
5. Particles remain their labels unchanged when surviving.
6. The resampling technique doesn't change the labels of particles.
7. If the label of one target is zero, the corresponding particles with label zero will be also assigned a new label corresponding to the target's new label.
8. When there are multiple targets with the same label at time k, the label of the particles representing optimal successor will remain unchanged, while others will be changed with their targets.

It should be mentioned that principle 7 is consistent with principle 2, and principle 8 is consistent with principle 3. False alarm may have the same label as a real target. Consequently, the optimal successor should be selected from all the targets with the same label to inherit the label. Suppose the state of the only target with label l at time $k-1$ is $x_{l,k-1}$, the states of targets with the same label at time k are $x_{l,k}^{(n)}, n = 1, 2, \cdots$, then the optimal successor can be selected by comparing the single-target Markov transition density

$$\underset{n}{\operatorname{argmax}} f_{k|k-1}(x_{l,k}^{(n)} \big| x_{l,k-1}), \tag{10}$$

The detailed Algorithm 1 of labelling particles and targets at each time is provided as below:

Algorithm 1 Labelling Particles and Targets

Initialization: the initialization particles are labelled with zeros, and maximum label is set to $r = 0$.
Prediction: labels of the prediction particles are $l_{k|k-1}^i, i = 1, \cdots, v_{k|k-1}$, where $l_{k|k-1}^i = l_{k-1|k-1}^i, i = 1, \cdots, v_{k-1|k-1}$ and $l_{k|k-1}^i = 0, i = v_{k-1|k-1} + 1, \cdots, v_{k|k-1}$.
Correction: labels of the posterior particles are $l_{k|k}^i = l_{k|k-1}^i, i = 1, \cdots, v_{k|k}$, and the resampling technique doesn't change the labels of particles.
Trajectory extraction: $\hat{N}_{k|k}$ targets are extracted from the posterior PHD. The label of target $x_k^{(n)}$ can be determined by $\underset{l}{\operatorname{argmax}} \left| \left\{ l_{k|k}^i \middle| l_{k|k}^i = l, x_{k|k}^i \in x_k^{(n)}, i = 1, \cdots, v_{k|k} \right\} \right|, n = 1, \cdots, \hat{N}_{k|k}$, where $|X|$ represents the cardinality of set X.

For each target $x_k^{(n)}$, if its label is zero, then $r = r + 1$, set its label to r, and set $l_{k|k}^i \Big| l_{k|k}^i = 0, x_{k|k}^i \in x_k^{(n)}, i = 1, \cdots, v_{k|k}$ to r.

If there are multiple targets with the same label l, the optimal successor can be selected by Equation (10), and for other target similar operation will be performed like the scene that the label of target is zero.

3.3. Revision of Posterior Weights

The posterior particle weights of a specific target will become small if continuous miss detection occurs to it, which may eliminate corresponding particles from the particle pool and then lose the target. In order to maintain the target that is not detected due to low detection probability, this paper replaces the posterior weights with corresponding prediction weights. That is to say, Equation (7) is modified to $w_{k|k}^i = w_{k|k-1}^i$. However, a target can't be detected when it disappears from FoV. Therefore, this paper only considers the target whose survival probability is above a threshold p_S^{th}. Furthermore, only when the sum of posterior weights is less than half of the sum of corresponding prediction weights will this paper conduct revision operations. Revisions of posterior weights are performed after the correction step, and the detailed Algorithm 2 at each time is provided as below:

Algorithm 2 Revision of Posterior Weights

For each target $x_{k-1}^{(n)}, n = 1, \cdots, \hat{N}_{k-1|k-1}$ at previous time
  If the survival probability of $x_{k-1}^{(n)}$ is above the threshold: $p_S^{k-1,(n)} > p_S^{th}$, then
    Find the prediction weights and posterior weights corresponding to target $x_{k-1}^{(n)}$:
    $w_{k|k-1}^i, w_{k|k}^i, i \in I^{(n)}$, where $I^{(n)}$ is the set of the index representing target $x_{k-1}^{(n)}$.
    If $sum\left(w_{k|k}^i\right) < \frac{1}{2} sum\left(w_{k|k-1}^i\right), i \in I^{(n)}$, do
      $w_{k|k}^i = w_{k|k-1}^i, i \in I^{(n)}$
    End
  End
End

3.4. Target Confirmation Based on Sequential Probability Ratio Test

Revisions of posterior weights will bring a new problem: once Poisson-distributed false alarms are captured by the probability density of the spontaneously appearing targets, the proposed algorithm will regard them as targets and maintain corresponding particles and weights by prediction step although no measurement available afterwards. In order to distinguish real targets from false alarms captured by the probability density of newborn targets, the measurement of each target extracted from posterior PHD should be recorded. Suppose $x_{l,k}^{(n)}, n = 1, \cdots, \hat{N}_{k|k}$ is the target with label l at time k extracted from the posterior PHD, and the measurement set at time k collected by the sensor is $Z_k = \{z_1, \cdots, z_m\}$, then the likelihood function of Z_k with respect to $x_{l,k}^{(n)}$ is defined as

$$L_{l,k} = \max_{z \in Z_k} L_z(x_{l,k}^{(n)}). \tag{11}$$

Obviously, the parameter $L_{l,k}, k = 1, 2, \cdots$ can tell us whether the target with label l is a real target. For simplification, the proposed algorithm binarizes the above likelihood function as

$$L'_{l,k} = \begin{cases} 1, & L_{l,k} \geq L^{th} \\ 0, & L_{l,k} < L^{th} \end{cases}, \tag{12}$$

where L^{th} is the threshold judging if there is a measurement of one target. The probability that there exists corresponding measurement of target $x_{l,k}^{(n)}$ is

$$\begin{aligned} &\Pr\left(L'_{l,k} = 1\right) \\ &= \Pr\left(\max_{z \in Z_k} L_z(x_{l,k}^{(n)}) \geq L^{th}\right) \\ &= 1 - \Pr\left(L_z(x_{l,k}^{(n)}) < L^{th}, \forall z \in Z_k\right) \end{aligned} \tag{13}$$

Furthermore, the bigger the cumulative sum $\underset{k}{sum} L'_{l,k}$, the more we can confirm that the target with label l is a real target.

In order to confirm targets in real time, this paper proposes the method based on sequential probability ratio test. Without loss of generality, suppose the single-target likelihood function is Gaussian

$$L_z(x) = \sqrt{\frac{1}{(2\pi)^2 \det(C)}} \exp\left(-\frac{1}{2}(z - H(x))^T C^{-1}(z - H(x))\right), \tag{14}$$

where $z = [z_1, z_2]^T$, $C = diag\left[\sigma_1^2, \sigma_2^2\right]$ is the covariance matrix of observation noises, $H(x)$ is the deterministic state-to-measurement transform model, and target x is located at the coordinate origin, the probability that the likelihood function of single measurement is above the threshold is

$$\Pr\left(L_z(x) > L^{th}\right) = \Pr\left(\frac{z_1^2}{\sigma_1^2} + \frac{z_2^2}{\sigma_2^2} < -2\ln\left(2\pi L^{th}\sqrt{\sigma_1^2\sigma_2^2}\right)\right) = \iint\limits_{\frac{z_1^2}{\sigma_1^2}+\frac{z_2^2}{\sigma_2^2}<-2\ln\left(2\pi L^{th}\sqrt{\sigma_1^2\sigma_2^2}\right)} f(z_1, z_2)dz_1dz_2, \quad (15)$$

which indicates the measurement z lies inside the ellipse $z_1^2/\sigma_1^2 + z_2^2/\sigma_2^2 = -2\ln\left(2\pi L^{th}\sqrt{\sigma_1^2\sigma_2^2}\right)$ and where $f(z_1, z_2)$ is the spatial distribution of z. Suppose false alarms obey uniform distribution spatially, then the probability that the likelihood function of single clutter is above the threshold is $p_0 = S_e/S_{FOV}$, where S_e is the area of the above ellipse and S_{FOV} is the area of the whole FoV. If a real target is detected, the probability that the likelihood function of target measurement is above the threshold is $q_0 = \iint\limits_{\frac{z_1^2}{\sigma_1^2}+\frac{z_2^2}{\sigma_2^2}<-2\ln\left(2\pi L^{th}\sqrt{\sigma_1^2\sigma_2^2}\right)} L_z(x)dz_1dz_2$. Suppose $\sigma^2 = \sigma_1^2 = \sigma_2^2$, then the probability q_0 with respect to observation noise variance σ^2 under different thresholds is depicted in Figure 1, which indicates that q_0 is close to 1 under a suitable threshold.

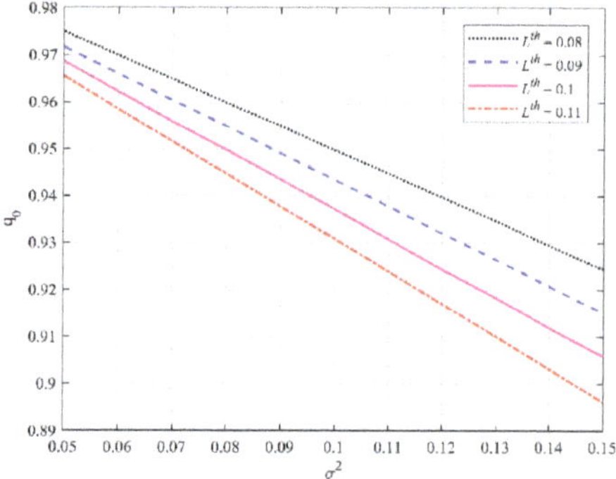

Figure 1. Probability that the likelihood function of target measurement is above the threshold.

Next, we consider the situation of multiple measurements. When the target $x_{l,k}^{(n)}$ is a false alarm, the measurement set Z_k can be organized as the union of measurements from targets and clutter. The likelihood function of Z_k with respect to $x_{l,k}^{(n)}$ is

$$L_{l,k} = \max\left\{\max_{z\in Z_k\backslash K_k} L_z(x_{l,k}^{(n)}), \max_{z\in K_k} L_z(x_{l,k}^{(n)})\right\}, \quad (16)$$

Therefore, Equation (13) is

$$\begin{aligned}
&\Pr\left(L'_{l,k} = 1\right) \\
&= 1 - \Pr\left(L_z(x_{l,k}^{(n)}) < L^{th}, \forall z \in Z_k\backslash K_k\right)\Pr\left(L_z(x_{l,k}^{(n)}) < L^{th}, \forall z \in K_k\right) \\
&\approx 1 - \Pr\left(L_z(x_{l,k}^{(n)}) < L^{th}, \forall z \in K_k\right) \\
&= 1 - \left(\Pr\left(L_z(x_{l,k}^{(n)}) < L^{th}, z \in K_k\right)\right)^D \\
&= 1 - (1 - p_0)^D
\end{aligned} \quad (17)$$

where D is the number of clutters, following $\Pr(D = d) = \lambda^d e^{-\lambda}/d!, d = 0, 1, \cdots$. The approximation is reasonable, because the false alarm always appears before or after corresponding real target, which results that it can neither be associated with the measurement of its corresponding real target nor that of other real targets. Considering D is a random variable, the expectation of Equation (17) is

$$E\left[1-(1-p_0)^D\right]=1-\sum_{d=0}^{\infty}(1-p_0)^d\frac{\lambda^d e^{-\lambda}}{d!}=1-e^{-p_0\lambda}. \tag{18}$$

On the other hand, when the target $x_{l,k}^{(n)}$ is a real target, the measurement set Z_k can be divided into three parts: the measurement generated from target $x_{l,k}^{(n)}$, measurements generated from other targets and clutters. The likelihood function of Z_k with respect to $x_{l,k}^{(n)}$ is

$$L_{l,k}=\max\left\{L_{\Theta_k(x_{l,k}^{(n)})}(x_{l,k}^{(n)}),\max_{z\in K_k}L_z(x_{l,k}^{(n)}),\max_{z\in Z_k\backslash K_k\backslash\Theta_k(x_{l,k}^{(n)})}L_z(x_{l,k}^{(n)})\right\}. \tag{19}$$

Therefore, Equation (13) is

$$\begin{aligned}
&\Pr\left(L'_{l,k}=1\right)\\
&=p_D(x_{l,k}^{(n)})\left[1-(1-q_0)(1-p_0)^D\Pr\left(L_z(x_{l,k}^{(n)})<L^{th},\forall z\in Z_k\backslash K_k\backslash\Theta_k(x_{l,k}^{(n)})\right)\right]\\
&+\left(1-p_D(x_{l,k}^{(n)})\right)\left[1-(1-p_0)^D\Pr\left(L_z(x_{l,k}^{(n)})<L^{th},\forall z\in Z_k\backslash K_k\right)\right]\\
&\approx p_D(x_{l,k}^{(n)})\left[1-(1-q_0)(1-p_0)^D\right]+\left(1-p_D(x_{l,k}^{(n)})\right)\left[1-(1-p_0)^D\right]\\
&\approx p_D(x_{l,k}^{(n)})+\left(1-p_D(x_{l,k}^{(n)})\right)\left[1-(1-p_0)^D\right]\\
&\approx p_D(x_{l,k}^{(n)})
\end{aligned} \tag{20}$$

in which we consider if the real target $x_{l,k}^{(n)}$ is detected. Three approximations are reasonable when all real targets are far from each other, q_0 is close to 1, and $p_D(x_{l,k}^{(n)})>>1-e^{-p_0\lambda}$, respectively.

In summary, random variable $L'_{l,k}$ obeys two-point distribution

$$\Pr\left(L'_{l,k}=1\right)=p,\Pr\left(L'_{l,k}=0\right)=1-p, \tag{21}$$

where success probability $p=p_1=1-e^{-p_0\lambda}$ when the target with label l is a false alarm, and $p=p_2=p_D(x_{l,k}^{(n)})$ when the target with label l is a real target. Then, target confirmation can be represented as a hypothesis test problem

$$H:p=p_1\leftrightarrow K:p=p_2. \tag{22}$$

SPRT tells us: when $\underset{k}{sum}L'_{l,k}\le A_n$ is true, to accept H, mark $x_{l,k}^{(n)}$ as a false alarm and eliminate corresponding particles; when $\underset{k}{sum}L'_{l,k}\ge B_n$ is true, to reject H and mark $x_{l,k}^{(n)}$ as a real target; otherwise, to make no decision and maintain particles of $x_{l,k}^{(n)}$. The parameters A_n and B_n are

$$\begin{aligned}
A_n&=\left(\tfrac{\beta}{1-\alpha}-n\ln\tfrac{1-p_2}{1-p_1}\right)/\ln\tfrac{p_2(1-p_1)}{p_1(1-p_2)}\\
B_n&=\left(\tfrac{1-\beta}{\alpha}-n\ln\tfrac{1-p_2}{1-p_1}\right)/\ln\tfrac{p_2(1-p_1)}{p_1(1-p_2)}
\end{aligned} \tag{23}$$

where α,β are Type I error rate and Type II error rate, respectively, and n is the cumulative time of the target with label l from emerging to current step.

It should be mentioned that confirmation of a real target always lags behind its emerging. Fortunately, we can make up point-valued estimates of the target at previous times.

3.5. Refined PHD Filter

The key modules of the refined PHD filter were explained in the previous subsections. Here, we summarize the overall steps of the proposed method in Algorithm 3.

Algorithm 3 Refined PHD Filter

Initialization: suppose prior PHD at time 0 is Equation (3), $l^i_{0|0} = 0, i = 1, \cdots, v_{0|0}$, and $r = 0$.
Prediction: the predicted PHD at time k is Equation (5), and labels of the prediction particles are $l^i_{k|k-1} = l^i_{k-1|k-1}, i = 1, \cdots, v_{k-1|k-1}$ and $l^i_{k|k-1} = 0, i = v_{k-1|k-1}+1, \cdots, v_{k|k-1}$.
Correction: the posterior particle weights at time k are calculated by Equation (7), and $l^i_{k|k} = l^i_{k|k-1}, i = 1, \cdots, v_{k|k}$.
Revision of Posterior Weights: execute revision of posterior weights introduced in Section 3.3, the revised weights are still represented as $w^i_{k|k}$, and the posterior PHD at time k is Equation (6).
State estimation: $\hat{N}_{k|k} \approx round\left(\sum_{i=1}^{v_{k|k}} w^i_{k|k}\right)$, resample with no change of particle labels, and $\hat{N}_{k|k}$ targets are extracted by k-means clustering: $x^{(n)}_k, n = 1, \cdots, \hat{N}_{k|k}$.
Trajectory extraction: determine the label of target $x^{(n)}_k$ and corresponding particles according to Section 3.2.
Survival Probability Calculation: calculate survival probability of individual target according to Section 3.1.
Target Confirmation: calculate test statistic $\underset{k}{sumL'}_{l,k}$, and for individual target, add it to the confirmation set, discard it, or make no decision according to Section 3.4.

4. Simulation

4.1. Simulation Scenery

In this section, we use computer simulations to demonstrate the effectiveness and performance of the proposed method. Suppose FoV is a two-dimensional region $[-50, 50] \times [0, 100]$ in which multiple targets appear or disappear at any time. The state equation and the measurement equation of single target can be represented as follows:

$$x_k = \begin{bmatrix} 1 & T & 0 & 0 \\ 0 & 1 & 0 & 0 \\ 0 & 0 & 1 & T \\ 0 & 0 & 0 & 1 \end{bmatrix} x_{k-1} + \begin{bmatrix} T^2/2 & 0 \\ T & 0 \\ 0 & T^2/2 \\ 0 & T \end{bmatrix} \begin{bmatrix} n_1 \\ n_2 \end{bmatrix}, \tag{24}$$

$$z_k = \begin{bmatrix} 1 & 0 & 0 & 0 \\ 0 & 0 & 1 & 0 \end{bmatrix} x_k + \begin{bmatrix} w_1 \\ w_2 \end{bmatrix}, \tag{25}$$

where target state $x_k = \left[p_{xk}, v_{xk}, p_{yk}, v_{yk}\right]^T$ consists of the target position and velocity along the x-axis and y-axis, only target position is measured represented as z_k, sampling period $T = 1$, and the process noise and the measurement noise are both zero mean Gaussian noises: $[n_1, n_2]^T \sim N\left([0,0]^T, diag[0.01, 0.01]\right)$, $[w_1, w_2]^T \sim N\left([0,0]^T, diag[0.09, 0.09]\right)$. This paper considers five targets with motion parameters showed in Table 1, and the total time of simulation is $T_{total} = 100$. Figure 2 depicts the simulation scenery in x-y coordinate system.

Table 1. Motion parameters of targets

Target	Initial State	Birth Time	Death Time
1	$[-50, 1.65, 100, -1.65]^T$	1	60
2	$[-50, 1.65, 0, 1.65]^T$	11	70
3	$[-50, 0.875, 30, 0.875]^T$	11	90
4	$[50, -1.16, 70, -1.16]^T$	31	90
5	$[50, -1.65, 50, 0]^T$	41	100

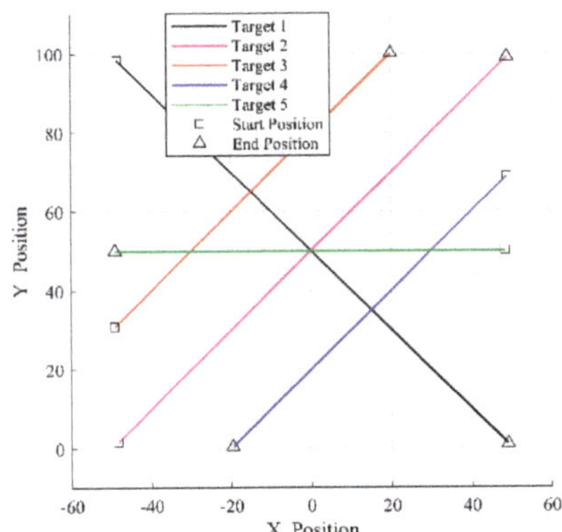

Figure 2. Simulation scenery in x-y coordinate system.

Cardinality and Optimal Sub-Pattern Assignment (OSPA) distance [43] between real set of target states and estimated set of target states are employed as performance evaluation criterions, and the cut-off factor and the order used in OSPA are $c = 10$, $p = 2$, respectively. The performance of the proposed Refined PHD (R-PHD) filter is evaluated in comparison with the standard PHD filter, CPHD filter, and CBMeMBer filter, and the filters here are all implemented with SMC implementations. Survival probability is set as 0.99 in PHD, CPHD and CBMeMBer. In R-PHD, the threshold L^{th} and p_S^{th} are set as 0.1 and 0.5 respectively, and Type I and II error rates are set as $\alpha = 0.1$ and $\beta = 0.1$, respectively. In all four filters, 1000 particles are used for per target, and the probability density of newborn targets is modeled as Gaussian mixture of target initial states with the covariance of $diag[1, 0.1, 1, 0.1]$.

4.2. Evaluation of Different Detection Probabilities

Figure 3 depicts the mean OSPA and cardinality versus time over $P = 200$ Monte-Carlo runs, where the detection probability is set as $p_D = 0.85$, independent of target state, and the average number of Poisson-distributed false alarms is set as $\lambda = 10$. Mean OSPA is depicted in Figure 3a, and each data point is calculated as

$$\frac{1}{P}\sum_{p=1}^{P} OSPA_{p,k}, \tag{26}$$

where $OSPA_{p,k}$ is OSPA distance at time k in pth Monte-Carlo trial. Mean cardinality is depicted in Figure 3b, and each data point is calculated as

$$\frac{1}{P}\sum_{p=1}^{P} \hat{N}_{p,k|k}, \tag{27}$$

where $\hat{N}_{p,k|k}$ is the estimated number of targets at time k in pth Monte-Carlo trial. Due to low detection probability, the measurements from targets are intermittent, and the PHD filter, CPHD filter and

CBMeMBer filter can't obtain excellent results. The mean OSPA of the R-PHD filter is usually smaller than that of the competing methods, and the mean cardinality of the R-PHD filter is closer to the ground truth. It is worth noting that OSPA distances at time 11 and 91 are apparently large in all filters, due to simultaneous birth or death of two targets. Figure 3 illustrates that the proposed method can effectively track multiple targets under low detection probability.

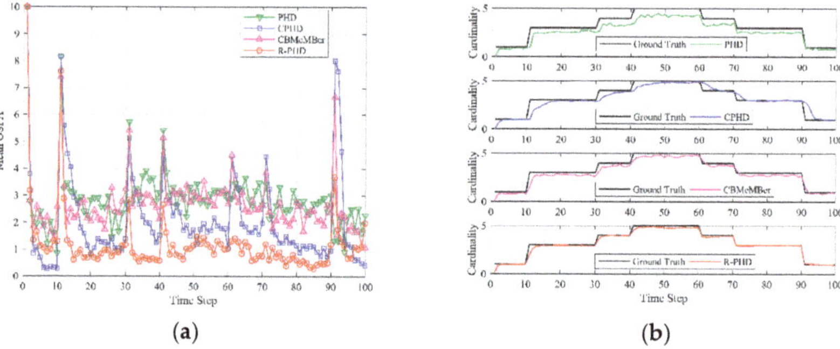

Figure 3. OSPA and cardinality performances of different methods versus time ($p_D = 0.85$, $\lambda = 10$): (**a**) mean OSPA; (**b**) mean cardinality.

Then, we compare multi-target tracking performances of different methods with respect to different detection probabilities from $p_D = 0.7$ to $p_D = 1$. Figure 4 illustrates the multi-target tracking results, where the average number of false alarms is set as $\lambda = 10$ for all simulations. Mean OSPA with respect to different detection probabilities is depicted in Figure 4a, and each data point is calculated as

$$\frac{1}{PT_{total}} \sum_{p=1}^{P} \sum_{k=1}^{T_{total}} OSPA_{p,k}, \tag{28}$$

where $OSPA_{p,k}$ is OSPA distance at time k in pth Monte-Carlo trial. Mean Root Mean Square Error (RMSE) of cardinality with respect to different detection probabilities is depicted in Figure 4b, and each data point is calculated as

$$\frac{1}{T_{total}} \sum_{k=1}^{T_{total}} \sqrt{\frac{1}{P} \sum_{p=1}^{P} \left(N_k - \hat{N}_{p,k|k}\right)^2}, \tag{29}$$

where $\hat{N}_{p,k|k}$ is the estimated number of targets at time k in pth Monte-Carlo trial, and N_k is real number of targets at time k. Figure 4 shows that mean OSPA and mean RMSE of cardinality both decrease monotonically as detection probability increases in the PHD filter, CPHD filter and CBMeMBer filter. In addition, the performance of the R-PHD filter is relatively stable, and the proposed method presents better performance than baselines when detection probability is no more than 0.95. It should be mentioned that when detection probability is 1, that is to say, there is no miss detection, the R-PHD filter has inferior performance than the other three methods, which can be explained by the Type I error rate in SPRT.

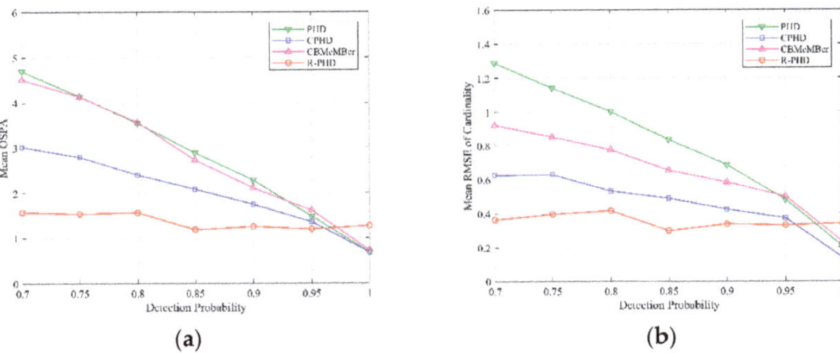

(a) (b)

Figure 4. OSPA and cardinality performances of different methods with respect to different detection probabilities from $p_D = 0.7$ to $p_D = 1$ ($\lambda = 10$): (**a**) mean OSPA; (**b**) mean RMSE of cardinality.

4.3. Evaluation of Different Average Numbers of False Alarms

Figure 5 depicts the mean OSPA and cardinality versus time over $P = 200$ Monte-Carlo runs, where the detection probability is set as $p_D = 0.95$, independent of target state, and the average number of Poisson-distributed false alarms is set as $\lambda = 20$. Mean OSPA is depicted in Figure 5a, and each data point is calculated using Equation (26). Mean cardinality is depicted in Figure 5b, and each data point is calculated using Equation (27). Showed in Figure 5a, the mean OSPA of the R-PHD filter is usually smaller than that of the PHD filter and CBMeMBer filter, while it is bigger than that of the CPHD filter at most steps. Figure 5b illustrates that the number of targets estimation of the CPHD filter always lags behind the ground truth when target birth or target death occurs.

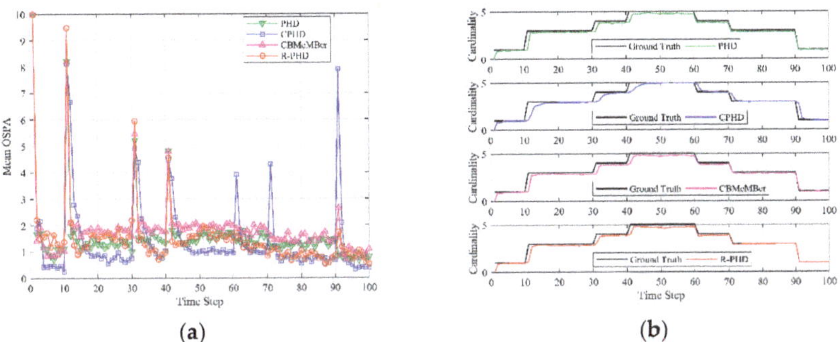

(a) (b)

Figure 5. OSPA and cardinality performances of different methods versus time ($p_D = 0.95$, $\lambda = 20$): (**a**) mean OSPA; (**b**) mean cardinality.

Figure 6 illustrates multi-target tracking performances of different methods with respect to different average numbers of false alarms from $\lambda = 10$ to $\lambda = 30$, where the detection probability is set as $p_D = 0.95$. Mean OSPA with respect to different average numbers of false alarms is depicted in Figure 6a, and each data point is calculated using Equation (28). Mean RMSE of cardinality with respect to different average numbers of false alarms is depicted in Figure 6b, and each data point is calculated using Equation (29). Figure 6 shows that multi-target tracking performances of different methods deteriorate slightly as average number of false alarms increases. Furthermore, the R-PHD filter outperforms the other three methods at $\lambda = 10$, while it has inferior OSPA performance compared to the PHD filter and CPHD filter and inferior cardinality performance compared to the CPHD filter at $\lambda = 30$. That is because the hypothesis $p_D >> 1 - e^{-p_0 \lambda}$ is no longer valid when the number of clutters is considerable. Generally, the proposed method can provide a satisfactory result under high average numbers of false alarms.

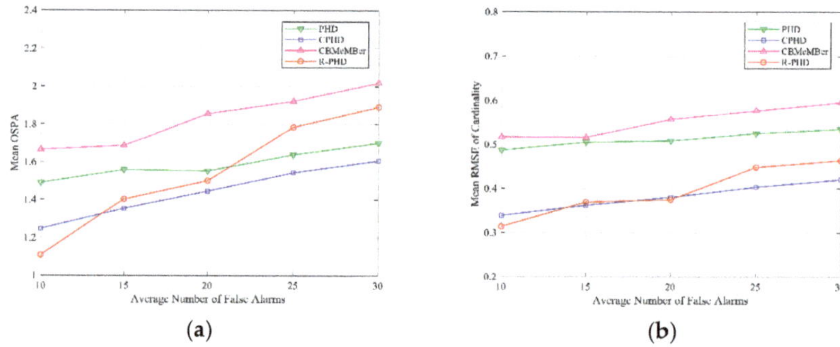

Figure 6. OSPA and cardinality performances of different methods with respect to different average numbers of false alarms from $\lambda = 10$ to $\lambda = 30$ ($p_D = 0.95$): (**a**) mean OSPA; (**b**) mean RMSE of cardinality.

4.4. Evaluation of Different Continuous Miss detection Durations

Next, we consider the scenario that targets are undetected for continuous steps. Figure 7 shows the multi-target tracking results of different methods under continuous miss detection during $45 \leq k \leq 51$, detection probability in other steps is set as $p_D = 0.9$ and average number of false alarms is set as $\lambda = 10$ in simulations. Mean OSPA in Figure 7a is obtained by averaging 200 trials of Monte-Carlo simulation using Equation (26), and mean cardinality in Figure 7b is obtained using Equation (27). Figure 7 demonstrates that the PHD filter and CBMeMBer filter lose all targets when continuous miss detection during $45 \leq k \leq 51$ occurs, which results that mean OSPA is up to the cut-off factor and mean cardinality is close to 0 from $k = 45$ to $k = 100$. The CPHD filter loses four targets when continuous miss detection occurs, while it can maintain one target after $k = 51$. Evidently, the proposed R-PHD filter can maintain all targets and its performance is almost immune to continuous miss detection.

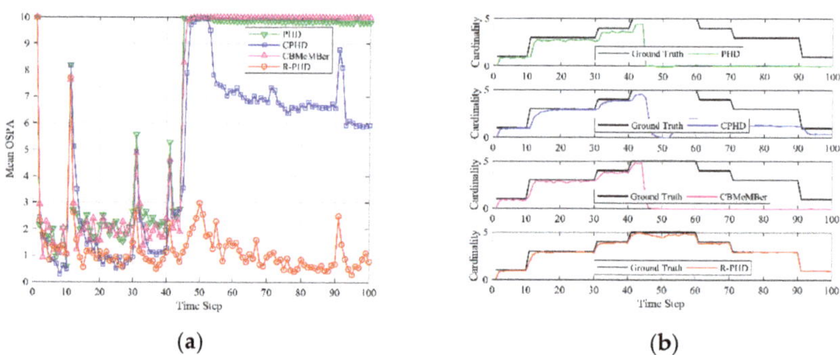

Figure 7. OSPA and cardinality performances of different methods versus time under continuous miss detection during $45 \leq k \leq 51$ ($p_D = 0.9$, $\lambda = 10$): (**a**) mean OSPA; (**b**) mean cardinality.

Figure 8 illustrates multi-target tracking performances of different methods with respect to different continuous miss detection durations from $s = 3$ to $s = 11$, where the detection probability is set as $p_D = 0.9$, average number of false alarms is set as $\lambda = 10$. Continuous miss detection duration is represented as s, and it always begins at time $k = 45$. That is to say, $s = 3$ indicates that targets are missed during $45 \leq k \leq 47$. The mean OSPA with respect to different continuous miss detection durations is depicted in Figure 8a, and each data point is calculated using Equation (28). Mean RMSE of cardinality with respect to different continuous miss detection durations is depicted in Figure 8b, and each data point is calculated using Equation (29). The performances of the PHD filter, CPHD filter and CBMeMBer filter deteriorate as continuous miss detection duration increases, while that of the proposed R-PHD filter is relatively stable and always superior than the other three methods. In conclusion, the proposed R-PHD filter can effectively track multiple targets when continuous miss detection occurs.

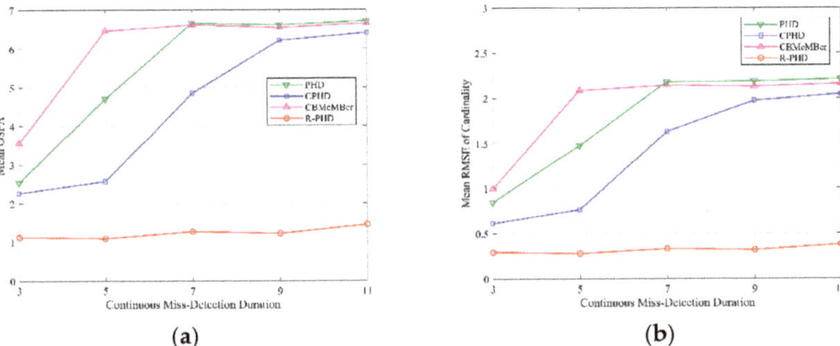

Figure 8. OSPA and cardinality performances of different methods with respect to different continuous miss detection durations from $s = 3$ to $s = 11$ ($p_D = 0.9$, $\lambda = 10$): (**a**) mean OSPA; (**b**) mean RMSE of cardinality.

5. Conclusions

In this paper, a heuristic method called the refined PHD filter is proposed to improve the multi-target tracking performance of the PHD filter under low detection probability in practice. First, survival probability dependent on target state is defined, which is one of the conditions of performing posterior weights revision. Then, we label individual targets and particles, which can be utilized to confirm if miss detection occurs for each target and identify particles representing the undetected target. In addition, it can provide track-valued estimates of individual targets. When miss detection occurs due to low detection probability, posterior particle weights will be revised according to the prediction step. In order to distinguish real targets and false alarms in real time, we regard the target confirmation problem as a hypothesis test problem and introduce sequential probability ratio test to judge the success probability of the two-point distribution. Simulation results with respect to various detection probabilities, average numbers of false alarms and continuous miss detection durations are provided, which indicates that the multi-target tracking performance of the R-PHD filter outperforms the competing methods.

Author Contributions: Conceptualization, S.W.; Data curation, S.W.; Methodology, Q.B.; Project administration, Z.C.; Software, Q.B.; Validation, S.W.; Visualization, S.W.; Writing—original draft, S.W.; Writing—review & editing, Q.B.

Funding: This research received no external funding and the APC was funded by National Key Laboratory of Science and Technology on ATR.

Acknowledgments: The authors would like to thank the Editor and the anonymous reviewers for their valuable comments and suggestions.

Conflicts of Interest: The authors declare no conflict of interest.

References

1. Jiang, D.; Liu, M.; Gao, Y.; Gao, Y.; Fu, W.; Han, Y. Time-matching random finite set-based filter for radar multi-target tracking. *Sensors* **2018**, *18*, 4416. [CrossRef] [PubMed]
2. Zhang, Q.; Xie, Y.; Song, T.L. Distributed multi-target tracking in clutter for passive linear array sonar systems. In Proceedings of the 2017 20th International Conference on Information Fusion (Fusion), Xi'an, China, 10–13 July 2017; pp. 1–8.
3. Scheidegger, S.; Benjaminsson, J.; Rosenberg, E.; Krishnan, A.; Granström, K. Mono-camera 3D multi-object tracking using deep learning detections and PMBM filtering. In Proceedings of the 2018 IEEE Intelligent Vehicles Symposium (IV), Changshu, China, 26–30 June 2018; pp. 433–440.

4. Bar-Shalom, Y. Extension of the probabilistic data association filter in multi-target tracking. In Proceedings of the 5th Symposium on Nonlinear Estimation, San Diego, CA, USA, 23–25 September 1974; pp. 16–21.
5. Reid, D.B. An algorithm for tracking multiple targets. In Proceedings of the 1978 IEEE Conference on Decision and Control including the 17th Symposium on Adaptive Processes, San Diego, CA, USA, 10–12 January 1979; pp. 1202–1211.
6. Mahler, R.P. Random-set approach to data fusion. In Proceedings of the Automatic Object Recognition IV, Orlando, FL, USA, 6–7 April 1994; pp. 287–296.
7. Goodman, I.R.; Mahler, R.P.; Nguyen, H.T. *Mathematics of Data Fusion*; Springer Science & Business Media: Heidelberg, Germany, 2013; Volume 37.
8. Mahler, R.P.S. Multitarget Bayes filtering via first-order multitarget moments. *IEEE Trans. Aerosp. Electron. Syst.* **2003**, *39*, 1152–1178. [CrossRef]
9. Vo, B.N.; Singh, S.; Doucet, A. Sequential Monte Carlo methods for multi-target filtering with random finite sets. *IEEE Trans. Aerosp. Electron. Syst.* **2005**, *41*, 1224–1245.
10. Vo, B.; Ma, W. The gaussian mixture probability hypothesis density filter. *IEEE Trans. Signal Process.* **2006**, *54*, 4091–4104. [CrossRef]
11. Mahler, R. PHD filters of higher order in target number. *IEEE Trans. Aerosp. Electron. Syst.* **2007**, *43*, 1523–1543. [CrossRef]
12. Vo, B.-T.; Vo, B.-N.; Cantoni, A. Analytic implementations of the cardinalized probability hypothesis density filter. *IEEE Trans. Signal Process.* **2007**, *55*, 3553–3567. [CrossRef]
13. Mahler, R.P.S. *Statistical Multisource-Multitarget Information Fusion*; Artech House: London, UK, 2007.
14. Vo, B.-T.; Vo, B.-N.; Cantoni, A. The cardinality balanced multi-target multi-bernoulli filter and its implementations. *IEEE Trans. Signal Process.* **2009**, *57*, 409–423. [CrossRef]
15. Papi, F.; Vo, B.N.; Vo, B.T.; Fantacci, C.; Beard, M. Generalized labeled multi-bernoulli approximation of multi-object densities. *IEEE Trans. Signal Process.* **2015**, *63*, 5487–5497. [CrossRef]
16. Reuter, S.; Vo, B.N.; Vo, B.T.; Dietmayer, K. The labeled multi-bernoulli filter. *IEEE Trans. Signal Process.* **2014**, *62*, 3246–3260. [CrossRef]
17. Vo, B.T.; Vo, B.N. Labeled random finite sets and multi-object conjugate priors. *IEEE Trans. Signal Process.* **2013**, *61*, 3460–3475. [CrossRef]
18. Vo, B.T.; Vo, B.N.; Phung, D. Labeled random finite sets and the bayes multi-target tracking filter. *IEEE Trans. Signal Process.* **2014**, *62*, 6554–6567. [CrossRef]
19. Vo, B.T.; Vo, B.N. Multi-scan generalized labeled multi-bernoulli filter. In Proceedings of the 2018 21st International Conference on Information Fusion (FUSION), Cambridge, UK, 10–13 July 2018; pp. 195–202.
20. Vo, B.T.; Vo, B.N. A multi-scan labeled random finite set model for multi-object state estimation. *arXiv* **2018**, arXiv:1805.10038.
21. Vo, B.T.; Vo, B.N.; Hoang, H.G. An efficient implementation of the generalized labeled multi-bernoulli filter. *IEEE Trans. Signal Process.* **2017**, *65*, 1975–1987. [CrossRef]
22. Beard, M.; Vo, B.T.; Vo, B.N. A solution for large-scale multi-object tracking. *arXiv* **2018**, arXiv:1804.06622.
23. Punchihewa, Y.G.; Vo, B.T.; Vo, B.N.; Kim, D.Y. Multiple object tracking in unknown backgrounds with labeled random finite sets. *IEEE Trans. Signal Process.* **2018**, *66*, 3040–3055. [CrossRef]
24. Zheng, Y.; Shi, Z.; Lu, R.; Hong, S.; Shen, X. An efficient data-driven particle PHD filter for multitarget tracking. *IEEE Trans. Ind. Inform.* **2013**, *9*, 2318–2326. [CrossRef]
25. Macagnano, D.; Abreu, G.T.F.D. Adaptive gating for multitarget tracking with gaussian mixture filters. *IEEE Trans. Signal Process.* **2012**, *60*, 1533–1538. [CrossRef]
26. Kanungo, T.; Mount, D.M.; Netanyahu, N.S.; Piatko, C.D.; Silverman, R.; Wu, A.Y.J.C.G. A local search approximation algorithm for k-means clustering. *Comput. Geom.* **2004**, *28*, 89–112. [CrossRef]
27. Everitt, B.S.; Dunn, G. *Applied Multivariate Data Analysis*; Arnold Wiley: London, UK, 2001; Volume 2.
28. Zhao, L.; Ma, P.; Su, X.; Zhang, H. A new multi-target state estimation algorithm for PHD particle filter. In Proceedings of the 2010 13th International Conference on Information Fusion, Edinburgh, UK, 26–29 July 2010; pp. 1–8.
29. Li, T.; Corchado, J.M.; Sun, S.; Fan, H.J.C.J.O.A. Multi-EAP: Extended EAP for multi-estimate extraction for SMC-PHD filter. *Chin. J. Aeronaut.* **2016**, *30*, 368–379. [CrossRef]
30. Schikora, M.; Koch, W.; Streit, R.; Cremers, D. *A Sequential Monte Carlo Method for Multi-target Tracking with the Intensity Filter*; Springer: Berlin, Germany, 2012.

31. Li, T.; Sun, S.; Bolić, M.; Corchado, J.M. Algorithm design for parallel implementation of the SMC-PHD filter. *Signal Process.* **2016**, *119*, 115–127. [CrossRef]
32. Li, T.; Prieto, J.; Fan, H.; Corchado, J.M. A robust multi-sensor PHD filter based on multi-sensor measurement clustering. *IEEE Commun. Lett.* **2018**, *22*, 2064–2067. [CrossRef]
33. Liu, L.; Ji, H.; Zhang, W.; Liao, G. Multi-sensor multi-target tracking using probability hypothesis density filter. *IEEE Access* **2019**, *7*, 67745–67760. [CrossRef]
34. Li, T.; Elvira, V.; Fan, H.; Corchado, J.M. Local-diffusion-based distributed SMC-PHD filtering using sensors with limited sensing range. *IEEE Sens. J.* **2019**, *19*, 1580–1589. [CrossRef]
35. Yuan, C.; Wang, J.; Lei, P.; Sun, Z.; Xiang, H. Maneuvering multi-target tracking based on PHD filter for phased array radar. In Proceedings of the IET International Radar Conference 2015, Hangzhou, China, 14–16 October 2015; pp. 1–4.
36. Li, C.; Wang, W.; Kirubarajan, T.; Sun, J.; Lei, P. PHD and CPHD filtering with unknown detection probability. *IEEE Trans. Signal Process.* **2018**, *66*, 3784–3798. [CrossRef]
37. Chen, X.; Tharmarasa, R.; Pelletier, M.; Kirubarajan, T. Integrated clutter estimation and target tracking using poisson point processes. *IEEE Trans. Aerosp. Electron. Syst.* **2012**, *48*, 1210–1235. [CrossRef]
38. Han, Y.; Han, C. Two measurement set partitioning algorithms for the extended target probability hypothesis density filter. *Sensors* **2019**, *19*, 2665. [CrossRef] [PubMed]
39. Shen, X.; Song, Z.; Fan, H.; Fu, Q. RFS-based extended target multipath tracking algorithm. *IET Radar Sonar Navig.* **2017**, *11*, 1031–1040. [CrossRef]
40. Yazdian-Dehkordi, M.; Azimifar, Z. Refined GM-PHD tracker for tracking targets in possible subsequent missed detections. *Signal Process.* **2015**, *116*, 112–126. [CrossRef]
41. Gao, L.; Liu, H.; Liu, H. Probability hypothesis density filter with imperfect detection probability for multi-target tracking. *Optik* **2016**, *127*, 10428–10436. [CrossRef]
42. Gao, Y.; Liu, Y.; Li, X.R. Tracking-aided classification of targets using multihypothesis sequential probability ratio test. *IEEE Trans. Aerosp. Electron. Syst.* **2018**, *54*, 233–245. [CrossRef]
43. Schuhmacher, D.; Vo, B.T.; Vo, B.N. A consistent metric for performance evaluation of multi-object filters. *IEEE Trans. Signal Process.* **2008**, *56*, 3447–3457. [CrossRef]

Article

Doppler Data Association Scheme for Multi-Target Tracking in an Active Sonar System

Yu Yao [1,*], Junhui Zhao [1] and Lenan Wu [2]

[1] School of Information Engineering, East China Jiaotong University, Nanchang 330031, China; junhuizhao@hotmail.com

[2] School of Information Science and Engineering, Southeast University, Nanjing 210096, China; wuln@seu.edu.cn

* Correspondence: shell8696@hotmail.com; Tel.: +86-1317-789-6959

Received: 18 March 2019; Accepted: 26 April 2019; Published: 29 April 2019

Abstract: In many wireless sensors, the target kinematic states include location and Doppler information that can be observed from a time series of range and velocity measurements. In this work, we present a tracking strategy for comprising target velocity components as part of the measurement supplement procedure and evaluate the advantages of the proposed scheme. Data association capability can be considered as the key performance for multi-target tracking in an active sonar system. Then, we proposed an enhanced Doppler data association (DDA) scheme which exploits target range and target velocity components for linear multi-target tracking. If the target velocity measurements are not incorporated into target kinematic state tracking, the linear filter bank for the combination of target velocity components can be implemented. Finally, a significant enhancement in the multi-target tracking capability provided by the proposed DDA scheme with the linear multi-target combined probabilistic data association method is demonstrated in a sonar underwater scenario.

Keywords: Doppler data association (DDA); Doppler measurement; kinematic state estimation; multi-target tracking; tracking performance

1. Introduction

Besides position measurements, Doppler measurement can offer supplementary statistics about target state, which would enhance tracking performance [1]. The problem of multi-radar tracking using both position and radial velocity measurements was discussed in References [2,3]. The authors presented the track-while-scan algorithm of maneuvering targets in a clutter environment. The filtering algorithm was nonlinear and adaptive. The measurement of two or more different radial velocity components allows the calculation of rectangular velocity components [4]. The main problem for multi-target tracking is distinguishing between measured values resulted from a specific target and measured values caused by other radar target echoes or interference [5,6].

The Doppler measurements are employed in a couple with the target range information as supplementary state information, which is used to overcome this problem of recognizing range-overlapped targets. Literature [7] put forward an advanced joint probabilistic data association scheme, which utilizes Doppler measurements along with range measurements via a nonlinear programming method. Compared with the traditional joint probabilistic data association method, the tracking performance of the modified joint probabilistic data association algorithm is obviously improved for multi-target tracking in noisy jamming environment. Reference [8] developed a method by using an interacting multiple model estimator including several extended Kalman filter elements to deal with Doppler measurements. Several kinds of particle filters were used to deal with the component of target velocity measurements [9]. Similar methods have been presented in [2,8] as well (and references therein). The authors proposed that they have to employ nonlinear filters at the receiver to process the additional Doppler measurements.

From References [10,11], the error cross-correlation between transformed range measurements and Doppler measurements has been addressed. The two-step optimal estimator is another kind of sequential filtering technique, which was employed for multi-target tracking which comprises a Doppler measurements procedure [12]. Reference [13] presented a sequential filtering algorithm to improve the performance of multi-target tracking. However, the approach was heavily dependent on the use of extended Kalman filters for calculating the target estimated and predicted state. If target velocity measurements are involved as part of the target state vectors, the interacting multiple model estimators must be employed to deal with the nonlinearity distortion between range and velocity measurement. The error of observation between location and velocity is correlated. The error is a key problem when the velocity component is to be integrated in the sensor system to enhance multi-target tracking capability. Reference [14] developed other types of nonlinear filters, for example the particle filters or unscented Kalman filters, which are employed to replace the extended Kalman filters for an improved performance. References [15,16] discussed the problem of joint detection and tracking of a target using multi-static Doppler-only measurements. The authors developed for this application of a multi-sensor Bernoulli particle filter with information gain-driven receiver selection. A consensus dual-stage nonlinear filter algorithm was presented to solve the Doppler-only target tracking problem in a distributed and scalable way [17]. Based on such a decomposition, a novel dual-stage filter for centralized multi-sensor Doppler-only tracking was proposed [18]. Reference [19] developed a consensus Gaussian mixture cardinalized probability hypothesis density filter to distributed multi-target tracking with range and Doppler sensors.

However, the supplementary computational load is a challenging problem. One of many thorny issues for multi-target tracking is distinguishing between observations achieved by a target of interest and observations derived from the jamming targets or interference noise. Reference [20] presented the data association problem for the multi-target tracking. Then, the target velocity measurements were employed in alliance with the range component as a supplementary discriminant of observation derived from overcoming the issue of discriminating dense objects [21]. Actually, among all the currently existing methods stated above, a remarkable attainment is the enhanced performance in measurement supplement profited from the combination of velocity observations.

Reference [22] developed hierarchical cognitive radar processing by applying the fully adaptive radar framework for cognition to a distributed radar network engaged in single target tracking. Two monostatic radar nodes are connected through a fusion center, and transmitted waveforms are adapted in real-time. However, the hardware requires the two radars to operate at the same pulse repetition frequency, limiting the degrees of freedom and affecting the velocity tracking accuracy. While previous research [23] has resulted in similar adaptive systems, this work presents a new approach to adaptive radar networks, treating each node as an independent instance of the framework for cognition. The authors in [24] formalized the work in [25,26] and presented a cognitive radar framework for a system engaged in target tracking. The model includes the higher-level tracking processor and specifies the feedback mechanism and optimization criterion used to obtain the next set of sensor data. The authors in [27] determined an optimal range for angle tracking radars based on evaluating the standard deviation of all kinds of errors in a tracking system. As the method increases in complexity, the usage of nonlinear filters might lead to system stability and capability problems.

In this paper, in order to improve the performance of multiple extended target tracking, we analyze the advantage of combining target velocity measurements into the target state vectors, then present an advanced Doppler data association (DDA) scheme which employs target range and velocity components for improving target tracking performance. The principal difference between the proposed DDA scheme and existing methods is that in the former, the target velocity component is utilized only for data association. A joint likelihood for target velocity and range observations is used for the measurement supplement while potential influences on the multi-target tracking from the target velocity component might not be overlooked. However, the proposed scheme offers a robust usage of velocity observations that would realize significant performance enhancements to those stated in the literature without the supplementary computational burden due to the use of nonlinear filters [28].

A linear framework is utilized for multi-target tracking, which results in the application of an advanced linear multi-target (LM) integrated probabilistic data association (IPDA) technique [29–33]. By comparing the LM tracking algorithm and the advanced LM integrated Doppler measurements association algorithm, the superiority of the Doppler measurements association method is verified in a sea environment and the multi-target tracking capability of approaches using and without using the proposed method are compared. Both theoretical analysis and simulation results demonstrate that the proposed system based on the Doppler measurements association method has a great performance enhancement and less operation processing burden compared with the traditional method.

The contributions of this paper can be summarized as follows:

(1) We present a tracking strategy for comprising target velocity components as part of the measurement supplement procedure and evaluate the advantages of the proposed method.
(2) We introduce a feasible scheme for multiple range-extended target tracking, which results in the development of an optimized LMIPDA algorithm.
(3) We analyze the tracking performance of the proposed algorithm in a sonar underwater scenario.

Our work is organized as follows. In Section 2, a system model for multi-target tracking in noisy jamming environment is discussed. In Section 3, we propose an advanced Doppler measurements association method and extend to the LMIPDA-Doppler measurements association scheme for multi-target tracking in the linear suboptimal framework. In Section 4, the tracking performance of the DDA method is analyzed. The simulation results demonstrating the proposed schemes are presented in Section 5, and conclusions are drawn in Section 6.

2. Multiple Target Models

2.1. Multiple Target Measurements

The trace of the k-th target can be expressed as

$$\mathbf{x}_{p+1}^{k} = \mathbf{F}_{p}\mathbf{x}_{p}^{k} + \mathbf{v}_{p}^{k} \tag{1}$$

where $\mathbf{x}_p^k$ is the k-th target kinematic state at time p, $\mathbf{F}_p$ denotes the transition matrix of target state and $\mathbf{v}_p^k$ describes the vector of additive white Gaussian noise (AWGN) with zero mean and covariance $\mathbf{Q}_p^k$. In the Cartesian coordinate system [1], the target kinematic state can be denoted by a vector of six components including position and velocity for each axis $\mathbf{x}_p^k = \begin{bmatrix} x_p & \dot{x}_p & y_p & \dot{y}_p & z_p & \dot{z}_p \end{bmatrix}'$. During the p-th scan, a set of m_p sonar observations $\mathbf{Z}_p = \left\{\mathbf{z}_{p,1}, \mathbf{z}_{p,2}, \ldots, \mathbf{z}_{p,m_p}\right\}$ are chosen from the system detections. Each observation's $\mathbf{z}_{p,i}, i = 1, 2, \ldots, m_p$ is a vector of four observed values which comprise both range and velocity components from the i-th target $\mathbf{z}_{p,i} = \left\{\mathbf{y}_{p,i}^{c}, y_{p,i}^{d}\right\}$. The accumulated measurement sequence up to the p-th scan can be denoted as

$$\mathbf{Z}^{p} = \left\{\mathbf{Z}_1, \mathbf{Z}_2, \ldots, \mathbf{Z}_p\right\} = \left\{\mathbf{Y}_c^{p}, \mathbf{Y}_d^{p}\right\}. \tag{2}$$

The target position observation has a linear correlation with the target kinematic state. The k-th target position observation during the p-th scan is denoted as:

$$\mathbf{y}_{p}^{c} = \mathbf{H}_{p}^{c}\mathbf{x}_{p}^{k} + \omega_{p}^{k}. \tag{3}$$

In Equation (3), $\mathbf{H}_p^c = \begin{bmatrix} 1 & 0 & 0 & 0 & 0 & 0 \\ 0 & 0 & 1 & 0 & 0 & 0 \\ 0 & 0 & 0 & 0 & 1 & 0 \end{bmatrix}$ denotes the system transition matrix. ω_p^k describes the vector of AWGN with zero mean and covariance $\mathbf{R}_p^{c,k}$. The target velocity measurement from the k-th target is denoted as

$$y_p^d = h\left(\mathbf{x}_p^k\right) + n_p^k. \tag{4}$$

The measurement error n_p^k denotes AWGN with zero mean and covariance $\mathrm{R}_p^{d,k}$. According to the Reference [34], the term $h\left(\mathbf{x}_p^k\right)$ in Equation (4) can be expressed by $h\left(\mathbf{x}_p^k\right) = \frac{\left(x_p - x_p^s\right)\left(\dot{x}_p - \dot{x}_p^s\right) + \left(y_p - y_p^s\right)\left(\dot{y}_p - \dot{y}_p^s\right) + \left(z_p - z_p^s\right)\left(\dot{z}_p - \dot{z}_p^s\right)}{\sqrt{\left(x_p - x_p^s\right)^2 + \left(y_p - y_p^s\right)^2 + \left(z_p - z_p^s\right)^2}}$, here $\mathbf{x}_p^s = \left[x_p^s, \dot{x}_p^s, y_p^s, \dot{y}_p^s, z_p^s, \dot{z}_p^s\right]'$ describes the given state vector including sonar position and sonar velocity during the p-th scan. We assume that the system process noise v_j^k and measurement noises w_k^k and n_i^k are independent of each other for all j, p, i and k. From position and Doppler measurement models Equations (3) and (4), the conditional probability density functions (PDFs) can be written as

$$\begin{aligned} & p\left(\mathbf{y}_p^c \middle| \mathbf{x}_p^k\right) \sim \mathrm{N}\left(\mathbf{y}_p^c; \mathbf{H}_p^c \mathbf{x}_p^k, \mathbf{R}_p^{c,k}\right) \\ & p\left(y_p^d \middle| \mathbf{x}_p^k\right) \sim \mathrm{N}\left(y_p^d; h\left(\mathbf{x}_p^k\right), \mathrm{R}_p^{d,k}\right) \\ & p\left(\mathbf{y}_p^c, y_p^d \middle| \mathbf{x}_p^k\right) = p\left(\mathbf{y}_p^c \middle| \mathbf{x}_p^k\right) p\left(y_p^d \middle| \mathbf{x}_p^k\right) \\ & \sim \mathrm{N}\left(\begin{bmatrix} \mathbf{y}_p^c \\ y_p^d \end{bmatrix}; \begin{bmatrix} \mathbf{H}_p^c \mathbf{x}_p^k \\ h\left(\mathbf{x}_p^k\right) \end{bmatrix}, \begin{bmatrix} \mathbf{R}_p^{c,k} & 0 \\ 0 & R_p^{d,k} \end{bmatrix}\right) \end{aligned} \tag{5}$$

2.2. Clutter Measurements

During the p-th scan, the number of clutter measurements can be considered as an inhomogeneous Poisson distribution. Each clutter component is related to a range measurement and a velocity measurement pair. It is assumed that the target range and velocity measurements of each clutter measurement are independent of each other. The density of each clutter component can be denoted as a product of the PDF of spatial clutter $\rho_{p,i}^c$ and the PDF of clutter Doppler $\rho_{p,i}^d$ [34]

$$\rho_{p,i} \triangleq \rho\left(\mathbf{z}_{p,i}\right) = \rho\left(y_{p,i}^c\right)\rho\left(y_{p,i}^d\right). \tag{6}$$

It is worth noting that we consider $\rho\left(y_{p,i}^c\right)$ and $\rho\left(y_{p,i}^d\right)$ as known a priori in this paper.

3. Doppler Measurement Association

Based on the LM procedure [31], we develop the DDA method in a joint model for LM tracking for IPDA. To solve the LM tracking problem, we need to estimate the joint posterior density of individual target state $\mathbf{x}_p^k$ conditioned on measurement sequences up to the p-th scan $\mathbf{Z}^p$ as follows

$$p\left(\mathbf{x}_p^k, \chi_p^k \middle| \mathbf{Z}^p\right), k = 1, \ldots, K. \tag{7}$$

In Equation (7), χ_p^k denotes the existence of the k-th target at the p-th scan. K describes the total number of tracks. We can express the joint posterior density of the track state as follows:

$$p\left(\mathbf{x}_p, \chi_p \middle| \mathbf{Z}^l\right) = p\left(\chi_p \middle| \mathbf{Z}^p\right) p\left(\mathbf{x}_p, \middle| \chi_p, \mathbf{Z}^l\right). \tag{8}$$

When $l = p$, the above formula can be considered as an estimation model. And $l = p - 1$, the above formula is a prediction model. Based on the Bayesian theory, $p\left(\mathbf{x}_p, \chi_p \middle| \mathbf{Z}^l\right)$ can be calculated recursively.

At time p, given prior density $P\left(\chi_{p-1}\middle|\mathbf{Z}^{p-1}\right)$ and prior density of the target state $p\left(\mathbf{x}_{p-1},\middle|\chi_{p-1},\mathbf{Z}^{p-1}\right)$, the iterative procedure can be expressed by the following stages.

(1) Calculate the predicted prior density $P\left(\chi_p\middle|\mathbf{Z}^{p-1}\right)$ and the predicted prior density of the target state $p\left(\mathbf{x}_p\middle|\chi_p,\mathbf{Z}^{p-1}\right)$.
(2) Measurement selection $\mathbf{Z}_p = \left\{\mathbf{z}_{p,1},\mathbf{z}_{p,2},\ldots,\mathbf{z}_{p,m_p}\right\}$ via gating for the underlying track.
(3) Calculate the predicted density of the i-th target state measurement $p\left(\mathbf{x}_{p,i},\middle|\chi_p,\mathbf{Z}^{p-1}\right)$.
(4) Update the posterior density $P\left(\chi_p\middle|\mathbf{Z}^p\right)$ and the posterior density $p\left(\mathbf{x}_p,\middle|\chi_p,\mathbf{Z}^p\right)$.

An illustrative diagram for an iteration process of the above four stages is presented in Figure 1.

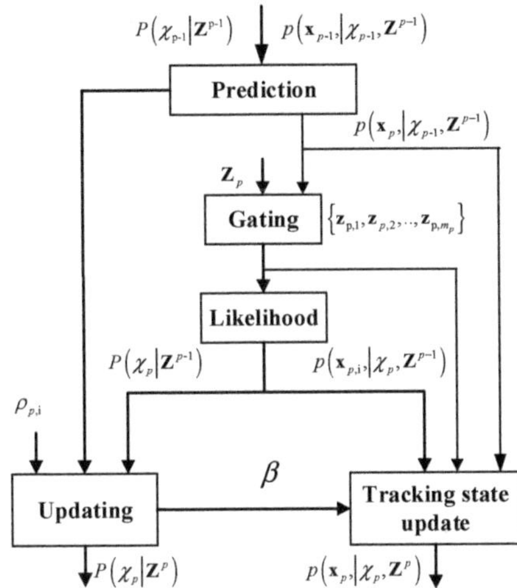

Figure 1. An illustrative diagram for an iteration process of synthetic tracking.

Step 1: Prediction

Based on the Gaussian hypothesis, every target measurement can be considered as a single Gaussian density function, which is expressed as:

$$p\left(\mathbf{x}_p\middle|\xi_p(c),\chi_p,\mathbf{Z}^{p-1}\right) \sim \mathrm{N}\left(\mathbf{x}_p;\hat{\mathbf{x}}_{p|p-1}\left(\xi_p(c)\right),P_{p|p-1}\left(\xi_p(c)\right)\right) \tag{9}$$

where $\xi_p(c)$ denotes event that the c-th out of C_p track components is true at the k-th sonar scan. $\hat{\mathbf{x}}_{p|p-1}\left(\xi_p(c)\right)$ and $P_{p|p-1}\left(\xi_p(c)\right)$ describes the mean and covariance of predictive prior density of the target track state, respectively.

Step 2: Measurement Selection

The $\mathbf{Z}_p = \left\{\mathbf{z}_{p,1},\mathbf{z}_{p,2},\ldots,\mathbf{z}_{p,m_p}\right\}$ measurement selection procedure is implemented at a component level. During the p-th scan, the c-th of C_p components chooses its measurements using a range measurement confirmation gate, which is concentrated at the expected range measurement $\hat{\mathbf{y}}_p^c\left(\xi_p(c)\right)$ as follows

$$\left(\mathbf{y}_{p,i}^c - \hat{\mathbf{y}}_p^c\left(\xi_p(c)\right)\right)'\left[\mathbf{S}_p^c\left(\xi_p(c)\right)\right]^{-1}\left(\mathbf{y}_{p,i}^c - \hat{\mathbf{y}}_p^c\left(\xi_p(c)\right)\right) \le \gamma, \tag{10}$$

where γ denotes a fixed threshold, $\mathbf{y}_{p,i}^c, i = 1,\ldots,m_p$ describes the i-th confirmed measurement and $\mathbf{S}_p^c\left(\xi_p(c)\right)^2$ describes the innovation covariance of the c-th of C_p components. In the gating process, only the range measurements can be utilized to produce a tracking gate for choosing a set of confirmed observations.

Step 3: Predictive Measurement PDF

$p\left(\mathbf{x}_p, \middle| \chi_p, \mathbf{Z}^{p-1}\right)$ is assumed to be a sum of C_p mutually exclusive components $\sum_{c=1}^{C_p} p\left(\xi_p(c)\middle|\chi_p, Z^{p-1}\right)p\left(\mathbf{x}_p\middle|\xi_p(c), \chi_p, \mathbf{Z}^{p-1}\right)$. Consequently, the predicted measurement PDF $\Lambda_{p,i}$ under the assumption of each measurement's $\mathbf{z}_{p,i}$ can be expressed as

$$\begin{aligned} \Lambda_{p,i} &\triangleq p\left(\mathbf{x}_{p,i}, \middle|\chi_p, \mathbf{Z}^{p-1}\right) \\ &= \sum_{c=1}^{C_p} p\left(\xi_p(c)\middle|\chi_p, Z^{p-1}\right)p\left(\mathbf{y}_{p,i}^c, y_{p,i}^d\middle|\xi_p(c), \chi_p, \mathbf{Z}^{p-1}\right) \end{aligned} . \tag{11}$$

In Equation (11), the term $p\left(\mathbf{y}_{p,i}^c, y_{p,i}^d\middle|\xi_p(c), \chi_p, \mathbf{Z}^{p-1}\right) \approx p\left(\mathbf{y}_{p,i}^c\middle|\xi_p(c), \chi_p, \mathbf{Z}^{p-1}\right)p\left(y_{p,i}^d\middle|\xi_p(c), \chi_p, \mathbf{Z}^{p-1}\right)$, which is the measurement likelihood function conditioned on the event $\xi_p(c)$. The joint likelihood can be expressed as a product of target range likelihood and target velocity likelihood.

Step 4: The Probability of Target Existence Update

From the Reference [30], the data association factor can be written as:

$$\delta_p = P_d P_g \left(1 - \sum_{i=1}^{m_p} \frac{\Lambda_{p,i}}{\rho_{p,i}}\right) \tag{12}$$

where P_d and P_g denote the probability of detection and the probability of the tracking gate, respectively. $\rho_{p,i}$ denotes the clutter density of a measurement in Equation (6). Then, we can write the probability of target existence as follows:

$$P\left\{\chi_p\middle|\mathbf{Z}^p\right\} = \frac{\left(1-\delta_p\right)P\left\{\chi_p\middle|\mathbf{Z}^{p-1}\right\}}{1-\delta_p P\left\{\chi_p\middle|\mathbf{Z}^{p-1}\right\}} \tag{13}$$

and the data association probabilities can be expressed as:

$$\beta_{p,i} \triangleq \frac{1}{1-\delta_p}\begin{cases} 1 - P_d P_g, & i = 0 \\ P_d P_g \frac{\Lambda_{p,i}}{\rho_{p,i}}, & i > 0 \end{cases} . \tag{14}$$

It needs to be emphasized that the contribution of target Doppler information is embodied in the factor of data association Equation (12), the probability of target existence Equation (13), and the probabilities of data association Equation (14).

Step 5: Tracking Update

From the Reference [30], the prior density of the target state is updated as follows:

$$\begin{aligned} & p\left(\mathbf{x}_p, \middle|\chi_p, \mathbf{Z}^p\right) \\ &= \sum_{i=0}^{m_k} \sum_{c=1}^{C_p} P\left\{\xi_p(c)\middle|\chi_p, Z^{p-1}\right\}\beta_{p,i}\frac{\Lambda_{p,i}\left(\xi_p(c)\right)}{\Lambda_{p,i}} p\left(\mathbf{x}_p, \middle|\xi_p(c), \chi_p, \mathbf{z}_{p,i}, \mathbf{Z}^p\right) \\ &= \sum_{c=1}^{C_{p+1}} p\left(\xi_{p+1}(c)\middle|\chi_p, Z^p\right)p\left(x_p, \middle|\xi_{p+1}(c), \chi_p, Z^p\right) \end{aligned} . \tag{15}$$

The term $p\left(\mathbf{x}_{p},\middle|\xi_p(c),\chi_p,\mathbf{z}_{p,i},\mathbf{Z}^p\right)$ in (15) is the conditional posterior density, which can be denoted as

$$p\left(\mathbf{x}_{p},\middle|\xi_p(c),\chi_p,\mathbf{z}_{p,i},\mathbf{Z}^p\right)=\frac{p\left(\mathbf{y}_{p,i}^{c}\middle|\mathbf{x}_p\right)p\left(y_{p,i}^{d}\middle|\mathbf{x}_p\right)p\left(\mathbf{x}_p\middle|\xi_p(c),\chi_p,\mathbf{Z}^{p-1}\right)}{p\left(\mathbf{y}_{p,i}^{c}\middle|\xi_p(c),\chi_p,\mathbf{Z}^{p-1}\right)p\left(y_{p,i}^{d}\middle|\xi_p(c),\chi_p,\mathbf{Z}^{p-1}\right)}. \quad (16)$$

Because of the nonlinear relationship between target state and target Doppler $h\left(\mathbf{x}_p^k\right)$, we have to employ a nonlinear filter for solving Equation (16). In our work, in order to simplify the discussion, we choose not to use the Doppler measurements track state updates. The term $p\left(y_{p,i}^{d}\middle|\mathbf{x}_p\right)$ can be approximated as $p\left(y_{p,i}^{d}\middle|\xi_p(c),\chi_p,\mathbf{Z}^{p-1}\right)$. Therefore, the conditional posterior density can be rewritten as

$$p\left(\mathbf{x}_{p},\middle|\xi_p(c),\chi_p,\mathbf{z}_{p,i},\mathbf{Z}^p\right)=\mathrm{N}\left(\mathbf{x}_p\left(\xi_p(c)\right);\mathbf{x}_{p|p}^{c,j},\mathbf{P}_{p|p}^{c,j}\right). \quad (17)$$

It is worth noting that the conditional posterior density becomes the standard Kalman filter form. That is to say, $p\left(\mathbf{x}_{p},\middle|\xi_p(c),\chi_p,\mathbf{z}_{p,i},\mathbf{Z}^p\right)$ represents the output of a standard Kalman filter with predicted target state $\hat{\mathbf{x}}_{p|p-1}\left(\xi_p(c)\right)$ and covariance $P_{p|p-1}\left(\xi_p(c)\right)$ as stated in step 1.

Considering the problem of multi-target tracking, the LM scheme views all observations from other targets as clutter. Therefore, the density of clutter can be modulated by contributions from other targets. During the p-th scan, the LM scheme estimates a revised density of clutter for all tracking gates, which is employed to compute the data association factor, the target existence probability and measurement association probabilities for all K targets. As a result, the probability P_i^k that the i-th measurement related to the k-th target can be expressed as

$$\begin{aligned}P_i^k&=P\left(\theta_{p,i}^k,\chi_p^k\middle|\mathbf{Z}^{p-1}\right)\\&=P_d^kP_g^kP\left(\chi_p^k\middle|\mathbf{Z}^{p-1}\right)\frac{\Lambda_{p,i}^k}{\sum_{i=1}^{m_p}\Lambda_{p,i}^k}\end{aligned}. \quad (18)$$

In Equation (18), $\theta_{p,i}^k$ denotes the event that the i-th measurement is resulted from the k-th track at time p. $\Lambda_{p,i}^k$ is presented in Equation (11). $P\left(\chi_p^k\middle|\mathbf{Z}^{p-1}\right)$ describes the predicted prior probability of the k-th target. Owing to the presence of multi-target, the modified clutter density $\Omega_{p,i}^k$ in the gate of the k-th track can be expressed as:

$$\Omega_{p,i}^k=\rho_{p,i}+\sum_{\delta=1,\delta\neq k}^{K}\Lambda_{p,i}^{\delta}\frac{P_i^{\delta}}{1-P_i^{\delta}}. \quad (19)$$

Hence, relying on a linear multi-target scheme, we can obtain the data association factor of the k-th target at time p as follows:

$$\delta_p^k=P_d^kP_g^k\left(1-\sum_{i=1}^{m_p^k}\frac{\Lambda_{p,i}^k}{\Omega_{p,i}^k}\right). \quad (20)$$

The measurement association probability of the k-th target at time p can be expressed as

$$\beta_{p,i}^k=\frac{1}{1-\delta_p^k}\begin{cases}1-P_d^kP_g^k, & i=0\\ P_d^kP_g^k\frac{\Lambda_{p,i}^k}{\Omega_{p,i}^k}, & i>0\end{cases}. \quad (21)$$

Therefore, LM-IPDA scheme with DDA can be acquired. The DDA method can be used in the joint IPDA algorithm for target tracking in clutter. The joint IPDA algorithm recursively updates both the probability of target existence and target state estimate. The probability of target existence is used as a track quality measure for false track discrimination. It is worth noting that the contribution of multi-target Doppler information is also embodied in the factor of data association Equation (20), the probability of target existence Equation (18), and the probabilities of data association Equation (21).

4. Performance Evaluation

During the p-th scan, the current tracks, which can be described by mean $\hat{\mathbf{x}}_{p|p}$ and covariance $\mathbf{P}_{p|p}$, are updated by using the current observations. All observations, which lie outside the validation gates $\left(\mathbf{y}_{p,i}^{c}-\hat{\mathbf{y}}_{p}^{c}\right)'\left[\mathbf{S}_{p}^{c}\right]^{-1}\left(\mathbf{y}_{p,i}^{c}-\hat{\mathbf{y}}_{p}^{c}\right)>\gamma$, are considered as irrelevant observations. Any irrelevant observation is processed by the track initiation module, where tentative tracks are formed based on the observations from two successive scans, that is, a window spanned by the maximum expected target velocity over a scan period selects all possible pairs of observations from two successive scans. Consequently, the observation pairs form tentative tracks by using the two-point initiation technique [35].

An initial target existence probability $P\left\{\chi_p\middle|\mathbf{Z}^p\right\}$ will be assigned to all new tracks. All tracks can be updated recursively by using a new measurement set, the corresponding probabilities of target existence are updated too. During the tracking procedure, the probabilities of target existence are evaluated against specific thresholds for track confirmation and termination. A track exists if its probability of target existence is greater than the predefined threshold of track confirmation. A track disappears if its probability of target existence is below the predefined threshold of track termination. Multiple tracks, which are close to each other, would be combined into a single track. Only the confirmed tracks are displayed to the observer. We evaluate multi-target tracking performance based on the three criteria as follows:

A. Number of confirmed true tracks (NCTT) and Number of confirmed false tracks (NCFT)

A track in correspondence of the true state of a target can be considered as a true track. The j-th track denoted by mean $\hat{\mathbf{x}}_{p|p}^{j}$ and covariance $\mathbf{P}_{p|p}^{j}$ has relation with the j-th target state $\mathbf{x}_{p}^{j}$, a constant threshold ψ is employed to satisfy $\left(\hat{\mathbf{x}}_{p|p}^{j}-\mathbf{x}_{p}^{i}\right)'\left(\mathbf{P}_{p|p}^{j}\right)^{-1}\left(\hat{\mathbf{x}}_{p|p}^{j}-\mathbf{x}_{p}^{i}\right)\leq\psi$. On the contrary, a track that does not associate with any true target state can be viewed as a false track. NCTT and NCFT are important measures for the multi-target tracking performance.

B. Target resolution

A possible issue with target state measure is that a track related to closely spaced targets would be viewed as one true track. Assuming the sampling frequency f_s, the carrier frequency f_c and the speed of electromagnetic wave c, the range resolution is $\Delta R\triangleq\frac{c}{2f_s}$ and the velocity resolution is $\Delta v\triangleq\frac{c\Delta f_d}{2f_c}$, where Δf_d denotes the Doppler shift resolution. We use mean square error (MSE) performance to define target resolution.

C. The ability to capture the target

The measure calculates the frequency that a confirmed track associates with an observable true target state. We use MSE performance to define the ability of capturing the target.

5. Simulation

The performance enhancement of multi-target tracking provided by the proposed DDA scheme with respect to different criteria is demonstrated in this section. We compare the tracking performance of the LMIPDA algorithm using the DDA scheme with nonlinear filtering method mentioned in [12] in a sonar underwater target tracking scenario.

5.1. Sonar Underwater Tracking

We propose the underwater scenario as follows: A warm water surrounding with depth (120 m) and acoustic velocity (1460 m/s) was employed for the simulated experiment. Sound transmission was represented by using multipath expansions with propagation properties for instance water refraction, attenuation and spreading as presented in [36]. The proposed system was denoted by an active sonar array, which was appropriate for transmitting and receiving short wave sound signals. The signal transferred 1/3 second pulses, each 5 s over an 8 min period. Transmit signals were composed of either continuous wave (CW) or phase modulated pulse waveforms that signified two diverse location and velocity resolutions. Generally, the continuous waveform had improved velocity resolution and worse location resolution than the phase modulated pulse waveforms. In theory, the CW waveforms have a location resolution (220 m) and velocity resolution (0.1 m/s) while the phase modulated pulses have location resolution (0.6 m) and velocity resolution (10 m/s). We considered that the target Doppler was limited to the interval [−20, +20] m/s to make allowances for the target speeds up to 30 knots. The proposed sonar systems were able to detect multiple targets and ranges in the interval [220, 4200] m.

There are ten targets in following scenario, including four range-spread targets and six point targets. The range-spread target is considered as a linear time-invariant filter with random impulse response, where the amplitude of each range cell is a zero-mean Gaussian variable. Target kinematic state and target position state are modeled by Equations (1) and (3), respectively. The transition matrix of target kinematic state $\mathbf{F}_p$ and the covariance of the noise $\mathbf{Q}_p$ can be described as follows

$$\mathbf{F}_p = diag(\mathbf{F}_1, \mathbf{F}_1, \mathbf{F}_1), \mathbf{F}_1 = \begin{bmatrix} 1 & T \\ 0 & 1 \end{bmatrix} \tag{22}$$

and

$$\mathbf{Q}_P = diag(\mathbf{Q}_1, \mathbf{Q}_1, \mathbf{Q}_1), \mathbf{Q}_1 = \begin{bmatrix} T^3/3 & T^2/2 \\ T^2/2 & T \end{bmatrix} \tag{23}$$

where T denotes data sampling interval for tracking. The sampling interval of transmit impulse was 5 s. The parameter T varied owing to a limited acoustic transmission period and time-variant ranges of multiple targets. We considered the noise parameter of the proposed system process was $q = 0.1$. The initial parameters of all targets are presented in Table 1.

Table 1. The initial parameters of all targets.

Target	Category	Location	Velocity	Acceleration	Azimuth
1	Submarine	[−400, 200, −30]	8.6 m/s	4.5 m/s^2	51.3 deg
2	Submarine	[−600, 300, −50]	11.6 m/s	5.3 m/s^2	200.3 deg
3	Surface ship	[−1200, 1300, 0]	7.6 m/s	3.2 m/s^2	150.2 deg
4	Small boat	[−650, 450, 0]	9.6 m/s	3.4 m/s^2	
5	Point buoy	[−150, 800, −20]	3.6 m/s	0	
6	Point buoy	[−890, 1400, −20]	3.6 m/s	0	
7	Point buoy	[−90, 30, −100]	3.6 m/s	0	
8	Point bottom	[−920, 890, −100]	6.1 m/s	0	
9	Point bottom	[−1920, 1890, −100]	6.1 m/s	0	
10	Point bottom	[−80, 90, −100]	6.1 m/s	0	

We considered the targets 5–10 had constant velocity and the targets 1–4 were in accelerated motion. The range-spread submarine was characterized by a cluster of ten scattering points organized in a given form of length 60 m that was corresponding to the heading of the object. Similarly, the range-spread larger ship was characterized by a cluster of thirteen scattering points organized in a given form of length 100 m. The three-dimensional (3D) track trajectory of all targets is shown in Figure 2.

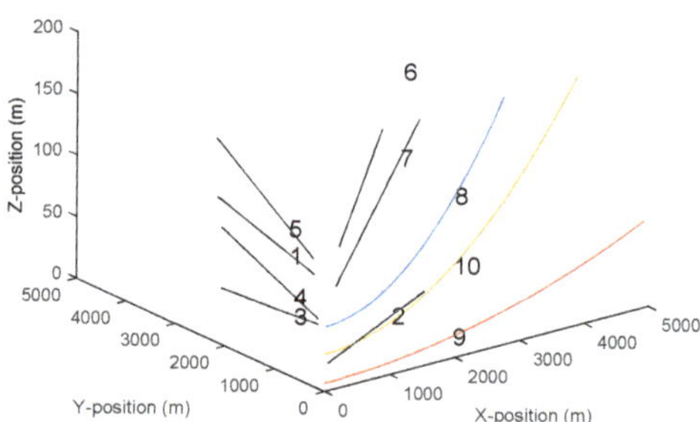

Figure 2. The 3D trajectory of ten targets diagram.

The transmit waveform illuminated sonar targets, which was reflected by certain targets. The backscattering signals were received by the proposed system. A cluster of transmission pathways comprising three bounces, which were calculated in every direction gave rise to 185 paths for every scattering point. We modeled all objects as a point scatter or a series of point-like targets.

A sonar signal processing method, which changes with waveform mode, can be utilized to calculate the response of the proposed transceiver to the whole set of sonar echoes. With the approach, a clustering procedure integrates all sonar echoes that fall within the similar sonar resolution cell into a particular cluster of sonar echoes. The approximation errors for the clustered sonar echoes can be calculated from the target scattering signal power, the interference noise characteristics and the measurement resolution, and are much smaller than the related measurement resolution. In our work, the approximation errors of the range and Doppler for both CW and phase modulated pulse mode are presented in Table 2.

Table 2. Measurement error.

Transmit Types	Range Error	Doppler Error
CW	95.3 m	0.025 m/s
Phase modulated pulse	0.85 m	8.5 m/s

The measurement errors of the bearing and elevation were about 2 deg for the two transmit types. It is worth noting that the clustering procedure increased the measuring errors on the basis of the number of the clustered sonar echoes and spread in measurement space. It influenced the range measurements the most because of signal echoes received at the sonar system along a multipath with a similar the frequency domain but a relative spread in the time domain.

The measured values of range, bearing and elevation can be mapped into Cartesian location values [37]. The clustering procedure integrated sets of multiple target detections at a distance of 12 m from each other into a clustered detection. The multipath from a range-spread target to the proposed system gave rise to false images looking like a cluster of reproductions below the water bottom or above the water surface. For the aims of multi-target tracking research, every clustered observation which size is larger than 15 m below water bottom can be excluded to eliminate the numerous images of the sonar target. We used the additional target-clustering procedures in combination with the filtering of detections to solve the problem of multiple target detections. However, multiple target images were still produced for a point target or the spatially spread scattering points corresponding to a range-spread target because of the multipath from the sonar system to the target.

The detection threshold can be deduced from the corresponding false alarm probability, which relates to the clutter quantitative value. For a given false alarm probability $p_{fa} = 0.0001$, the expected value of clutter measurements varied with the number of measuring resolution cells. There were 120 clutter detections per scan.

The target position measurements are shown in Figure 3. Figure 4 presents Doppler measurements versus range for CW mode. The multiple target detections in a sea clutter environment were observed by the sonar system over 100 scans for a single simulated experiment. There were two possibilities. The first was that the target lay outside the proposed system's field of view. The second was that the received signal strength was lesser than the given detection threshold. Not all targets were clearly observed during each sonar scan. Multiple target detections were not perceived for targets 1 and 2 during certain parts of the simulated experiment. The range-spread larger target 3 approached the dim point target 4 during the 38th–44th sonar scan and approached two stationary point targets 5 and 6 to offer a sonar sea multi-target tracking environment.

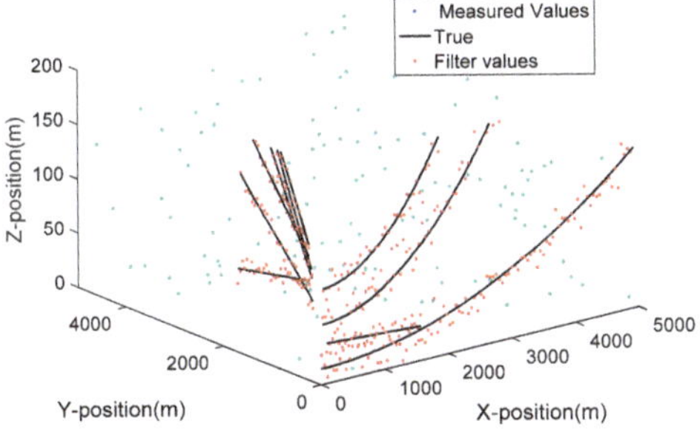

Figure 3. The target position measurements.

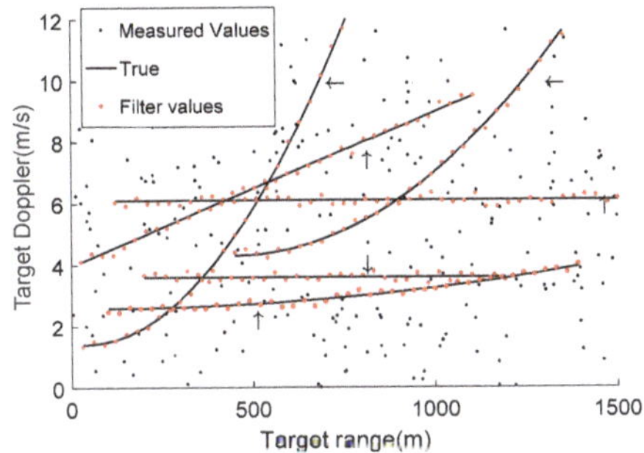

Figure 4. Doppler measurements versus range for continuous wave (CW) mode.

5.2. Performance Evaluation

The simulation parameters associated with the proposed schemes, comprising the initial target existence probability, track confirmation threshold and track termination threshold were assumed to obtain the greatest capability. These experiment parameters are shown in Table 3.

The results were obtained from 600 Monte Carlo simulations for both transmit modes. During each simulation run, three trackers (the extended Kalman filtering technique, LMIPDA, LMIPDA-DDA) were applied to the same measurement data employing three schemes: (1) Updating tracks via target range measurements based on extended Kalman filtering as presented in [12], (2) updating tracks via target range measurements based on the regular trackers LMIPDA, and (3) updating tracks via both target range measurements and the supplementary target velocity measurements based on the advanced trackers LMIPDA-DDA.

Table 3. Tracker parameter settings.

Simulation Parameters	Extended Kalman Filtering	LMIPDA	LMIPDA-DDA
Thresholds for track confirmation	0.95	0.95	0.95
Thresholds for track termination	0.005	0.005	0.005
Merge threshold	4	4	4
Initial probability of target existence	0.08	0.08	0.08
Detection probability	0.98	0.98	0.98
System process noise	0.1	0.1	0.1
Process noise for Doppler state	1 m/s	1 m/s	1 m/s
Target speed	800 m/s	800 m/s	800 m/s

The comparisons of NCTT across three methods are illustrated in Figure 5a,b. The average computation time is presented in Table 4.

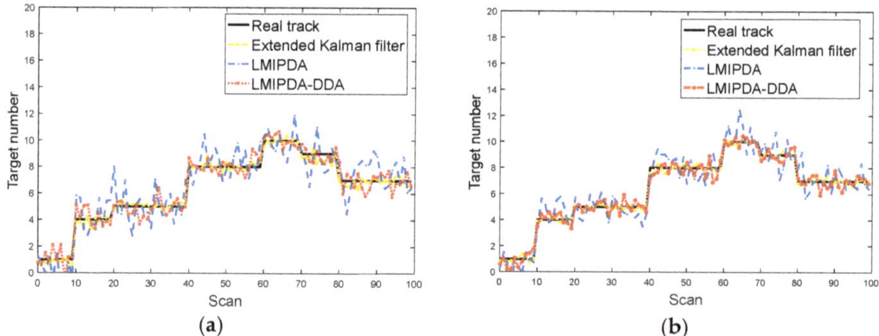

Figure 5. The performance of the number of confirmed true tracks (NCTT): (**a**) CW mode; (**b**) pulse mode.

Table 4. The average computation times in CW and pulse transmission methods.

Method	NCTT	CPU	Peak CPU
CW mode			
LMIPDA	1.245	1.545	5.142
LMIPDA-DDA	0.0041	1.301	3.978
Extended Kalman filtering	4.643	8.183	9.112
Pulse mode			
LMIPDA	1.345	1.312	4.265
LMIPDA-DDA	0.001	1.532	3.532
Extended Kalman filtering	3.053	6.423	9.023

As shown in Figure 5 and Table 4, the LMIPDA-DDA algorithm had an enhanced confirmation response to multi-target observations. Meanwhile, the accuracy of NCTT provided by the LMIPDA-DDA algorithm was obviously superior to the LMIPDA algorithm. The LMIPDA method played an important part on single scan tracking but was not robust enough for abrupt changes in target tracking, while the LMIPDA-DDA method played an important part on multiple scan tracking. The extended Kalman filtering method as stated in [12] was slightly better than the proposed DDA scheme. The nonlinear filtering technique was optimal in this case.

As shown in Table 4, a faster calculating speed can be obtained. With Doppler data association, the average operation times of the LMIPDA-DDA algorithm were lower. Furthermore, the top operation times of the LMIPDA-DDA algorithm were also lower than other methods due to data association. Faster termination of false tracks can be observed. As can be expected, the nonlinear filtering technique spent the most time to implement iterated operation. However, if the distribution of target velocity component is hard to distinguish from the distribution of false target velocity component, the tracking capability of data association method would reduce to that of the traditional method. It is worth noting that if the proposed system switched between CW and phase modulated pulse transmission modes, the measurement errors of target range and Doppler would vary as presented in Table 2. The errors would undoubtedly impact the multi-target tracking performance under consideration. To simplify the discussion, the problem has been ignored in this paper.

The comparison of NCFT across methods is shown in Figure 6. As can be seen from Figure 6, a remarkable reduction in NCFT was seen for both CW and pulse modes. The presence of confirmed false tracks can severely limit confidence level and usage of trackers; better discrimination of false tracks allows for reduction of track confirmation threshold, which leads to better response to target measurements. When the transmitter was switched to pulse mode, target Doppler measurements got larger errors as present in Table 2. As can be expected, a remarkable reduction in NCFT was still obtained from Figure 6b. That is because the distribution of target Doppler measurements could be discriminated from that of false Doppler measurements.

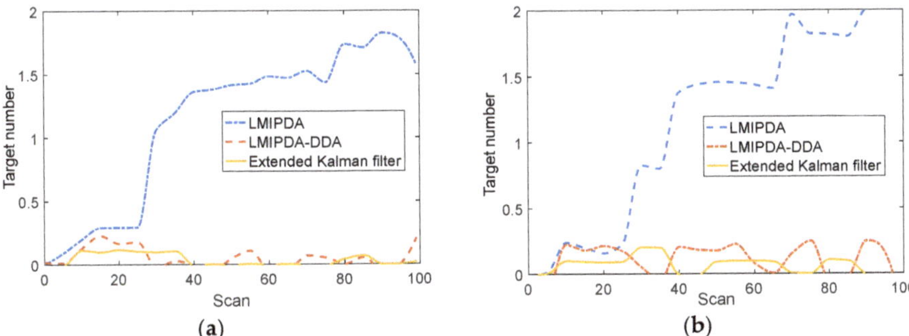

Figure 6. The performance of the number of confirmed false tracks (NCFT): (**a**) CW mode; (**b**) pulse mode.

The comparison of the ability of capturing the target across methods is presented in Figures 7 and 8. Performance difference for targets 1–3 is given in this section. The rest of targets show a similar trend. The proposed DDA method had a better confirmation response to target measurements in most cases.

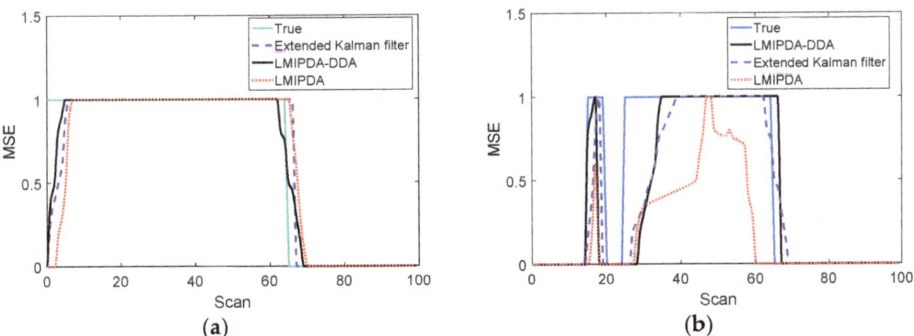

Figure 7. *Cont.*

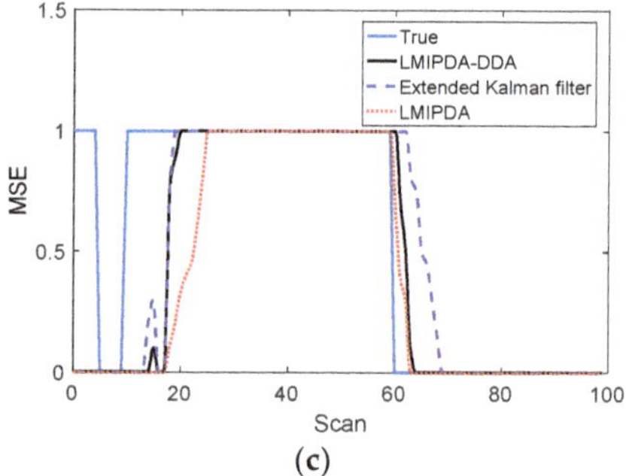

Figure 7. The capabilities of capturing targets 1–3 in CW mode; (**a**) target 1; (**b**) target 2; (**c**) target 3.

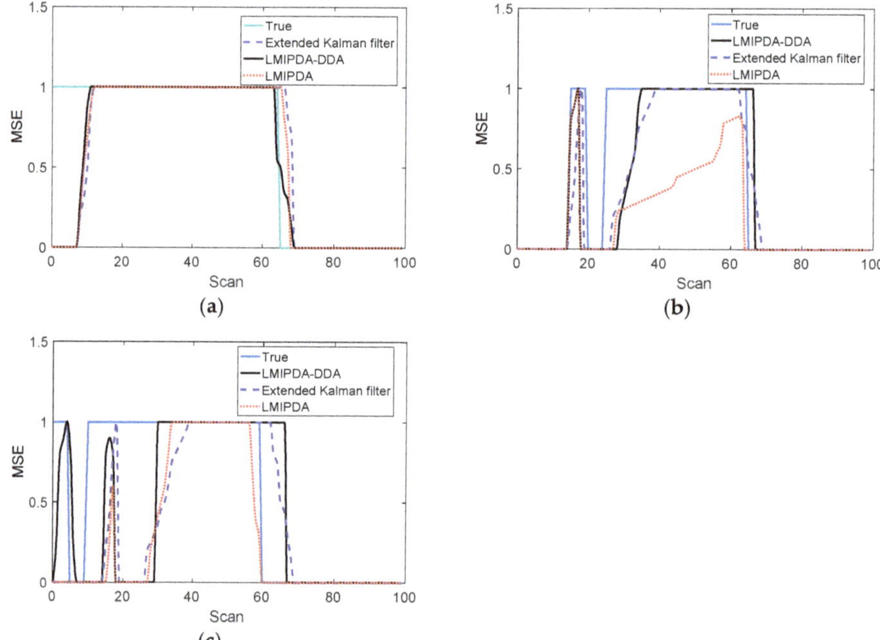

Figure 8. The capabilities of capturing targets 1–3 in pulse mode; (**a**) target 1; (**b**) target 2; (**c**) target 3.

As can be seen from Figure 7c, during the first 10 scans, only the proposed DDA method had captured the underlying target 3. We can explain the phenomenon by understanding that the LMIPDA-DDA algorithm works on single scan measurements, while the extended Kalman filtering technique stated in [12] works on multiple scan measurements. Hence, the former will be more adaptable to abrupt changes in target measurements.

Furthermore, as shown in Figure 8, the proposed DDA method and the extended Kalman filtering technique suffered from a longer true track confirmation delay compared with the CW mode case, as presented in Figure 7, because target information flow rate from the target measurements was lower. It is easy to see that, if the distribution of target Doppler measurements is indistinguishable from the distribution of false Doppler measurements, the performance of the trackers with DDA method would reduce to that of the trackers without DDA method.

The comparisons of target resolution capability across methods are given in Figure 9a,b.

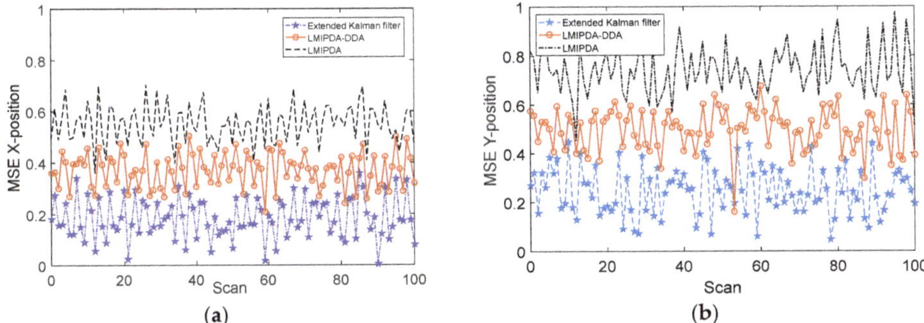

Figure 9. The comparisons of mean square error (MSE) performance across methods: (**a**) MSE in X-axial; (**b**) MSE in Y-axial.

As can be seen from Figure 9a,b, the target resolution capability offered by the LMIPDA-DDA algorithm was better than the LMIPDA algorithm at each scan. However, as can be expected, the performance provided by the nonlinear filtering technique was slightly better than the proposed DDA scheme. The simulation result demonstrated that, compared with the traditional tracking algorithm without DDA, the resolution capability of the sensor system provided by the LMIPDA-DDA algorithm was obviously improved. The key difference between the Doppler measurement association scheme and traditional methods without DDA was that in the Doppler measurement association scheme, the target velocity observed values were used for measurement association. The sonar sensor system improved by Doppler data association scheme had little influence on its operations in terms of system robustness and filter complexity but offered a significant decrease in the amount of false observations. The computational complexity was also greatly reduced. Therefore, if we make a trade-off between complication and performance gain, the proposed DDA scheme was superior to nonlinear filtering technique in this case.

6. Conclusions

In this paper, a sonar sensor system provided by a Doppler measurement association method was proposed for enhancing the performance of multi-target tracking in a noisy jamming environment. In the Doppler measurement association scheme, the target velocity measurements are utilized for calculating the observation likelihood, which are an important part for distinguishing true measurements from phony targets or clutter measurements. The sonar system improved by the proposed scheme has a tiny influence on system stability but offers a significant decrease in the amount of confirmed false targets. Meanwhile, the target resolution capability of the system provided by the LMIPDA-DDA algorithm is obviously improved, which can be realized without equipping a nonlinear filter bank. In the clutter environment, the traditional method that uses location-only component obtain a substantial NCTT, while the proposed method that combine target velocity components do not. The usefulness of the enhanced DDA scheme has been verified in a sea clutter environment.

Author Contributions: Conceptualization, Y.Y. and J.Z.; methodology, Y.Y.; software, J.Z.; validation, Y.Y., J.Z. and L.W.; formal analysis, Y.Y.; investigation, Y.Y.; resources, J.Z.; data curation, J.Z.; writing—original draft preparation, Y.Y.; writing—review and editing, J.Z.; visualization, Y.Y.; supervision, L.W.; project administration, L.W.; funding acquisition, Y.Y.

Funding: This work was supported by the National Natural Science Foundation of China (61761019, 61861017, 61861018, 61862024) and the Natural Science Foundation of Jiangxi Province (Jiangxi Province Natural Science Fund) (20181BAB211014, 20181BAB211013) and Foundation of Jiangxi Educational Committee of China (GJJ180352).

Conflicts of Interest: The authors declare no conflict of interest.

References

1. Mo, L.; Song, X.; Sun, Z.K.; Bar-Shalom, Y. Unbiased Converted Measurements for Tracking. *IEEE Trans. Aerosp. Electron. Syst.* **1998**, *34*, 1023–1027.
2. Farina, A.; Pardini, S. A Track-While-Scan Algorithm Using Radial Velocity in a Clutter Environment. *IEEE Trans. Aerosp. Electron. Syst.* **1978**, *14*, 769–779. [CrossRef]
3. Farina, A.; Pardini, S. A New Class of Track-While-Scan Algorithms Using the Radial Velocity Measurements. In Proceedings of the International Conference on Radar, Paris, France, 12 December 1978; pp. 118–128.
4. Farina, A.; Pardini, S. Multiradar Tracking Systems Using the Radial Velocity Measurements. *IEEE Trans. Aerosp. Electron. Syst.* **1979**, *15*, 555–563. [CrossRef]
5. Farina, A.; Studer, F.A. *Radar Data Processing: Introduction and Tracking Vol. 1.*; John Wiley: New York, NY, USA, 1985.
6. Farina, A.; Studer, F.A. *Radar Data Processing, Vol. 2, Advanced Topics and Applications*; Research Studies Press: New York, NY, USA, 1986.
7. Kameda, H.; Tsujimichi, S.; Kosuge, Y. Target Tracking under Dense Environments Using Range Rate Measurements. In Proceedings of the 37th SICE Annual Converence, International Session Papers, Chiba, Japan, 29–31 July 1998; pp. 927–932.
8. Yeom, S.W.; Kirubarajan, T.; Bar-Shalom, Y. Track Segment Association, Fine-Step IMM and Initialization with Doppler for Improved Track Performance. *IEEE Trans. Aerosp. Electron. Syst.* **2004**, *40*, 293–309. [CrossRef]
9. Doucet, A.; Gordon, J.; Krishnamurthy, V. Particle Filter for State Estimation of Jump Markov Linear Systems. *IEEE Trans. Signal Process.* **2001**, *49*, 613–624. [CrossRef]
10. Bar-Shalom, Y. Negative Correlation and Optimal Tracking with Doppler Measurements. *IEEE Trans. Aerosp. Electron. Syst.* **2001**, *37*, 1117–1120. [CrossRef]
11. Gurfil, P.; Kasin, N.J. Two-Step Optimal Estimator for Three-Dimensional Target Tracking. *IEEE Trans. Aerosp. Electron. Syst.* **2005**, *41*, 780–793. [CrossRef]
12. Wang, J.G.; Long, T.; He, P.K. Use of the Radial Velocity Measurement in Target Tracking. *IEEE Trans. Aerosp. Electron. Syst.* **2003**, *39*, 401–413. [CrossRef]
13. Duan, Z.S.; Han, C.Z.; Li, X.R. Sequential Nonlinear Tracking Filter with Range-Rate Measurements in Spherical Coordinates. In Proceedings of the 7th International Conference on Information Fusion, Stockholm, Sweden, 28 June–1 July 2004; pp. 599–605.
14. Julier, S.J.; Uhlmann, J.K. Unscented Filtering and Nonlinear Estimation. *Proc. IEEE* **2004**, *92*, 401–422. [CrossRef]
15. Ristic, B.; Farina, A. Joint Detection and Tracking Using Multi-Static Doppler-Shift Measurements. In Proceedings of the 2012 IEEE International. Conference on Acoustic, Speech and Signal Processing (ICASSP), Kyoto, Japan, 25–30 March 2012; pp. 3881–3884.
16. Ristic, B.; Farina, A. Target Tracking Via Multi-Static Doppler Shifts. *IET Radar Sonar Navig.* **2013**, *7*, 508–516. [CrossRef]
17. Battistelli, G.; Chisci, L.; Fantacci, C.; Farina, A.; Graziano, A. Distributed Multitarget Tracking with Range-Doppler Sensors. In Proceedings of the European Signal Processing Conference (EUSIPCO 2013), Marrakech, Morocco, 9–13 September 2013.
18. Battistelli, G.; Chisci, L.; Fantacci, C.; Farina, A.; Graziano, A. A New Approach for Doppler-Only Target Tracking. In Proceedings of the 16th International Conference on Information Fusion, Istanbul, Turkey, 9–12 July 2013; pp. 1616–1623.
19. Battistelli, G.; Chisci, L.; Fantacci, C.; Farina, A.; Graziano, A. Distributed Tracking with Doppler Sensors. In Proceedings of the the 52nd IEEE Conference on Decision and Control, Firenze, Italy, 10–13 December 2013.
20. Mušicki, D.; La Scala, B.; Evans, R. The Integrated Track Splitting Filter-Efficient Multi-Scan Single Target Tracking in Clutter. *IEEE Trans. Aerosp. Electron. Syst.* **2007**, *43*, 1409–1425.
21. Blair, W.D.; Brandt-Pierce, M. NNJPDA for Tracking Closely-Spaced Rayleigh Targets with Possibly Merged Measurements. In Proceedings of the SPIE Conference on Signal and Data Processing of Small Targets, Denver, CO, USA, 1 October 1999; pp. 396–408.

22. Mitchell, A.E.; Smith, G.E.; Bell, K.L.; Rangaswamy, M. Single Target Tracking with Distributed Cognitive Radar. In Proceedings of the 2017 IEEE Radar Conference (RadarConf), Seattle, WA, USA, 8–12 May 2017; pp. 0285–0288.
23. Nguyen, N.H.; Doğançay, K.; Davis, L.M. Adaptive Waveform and Cartesian Estimate Selection for Multistatic Target Tracking. *Signal Proc.* **2015**, *111*, 13–25. [CrossRef]
24. Bell, K.; Baker, C.; Smith, G.; Johnson, J.; Rangaswamy, M. Cognitive Radar Framework for Target Detection and Tracking. *IEEE J. Sel. Top. Signal Proc.* **2015**, *9*, 1427–1439. [CrossRef]
25. Haykin, S. Cognitive Dynamic Systems. *Proc. IEEE* **2014**, *102*, 1369–1372. [CrossRef]
26. Romero, R.A.; Goodman, N.A. Cognitive Radar Network: Cooperative Adaptive Beamsteering for Integrated Search-And-Track Application. *IEEE Trans. Aerosp. Electron. Syst.* **2013**, *49*, 915–931. [CrossRef]
27. Samadi, S.; Mohammad, K.; Jafar, A. Optimum Range of Angle Tracking Radars: A Theoretical Computing. *Int. J. Electr. Comput. Eng.* **2019**, *9*, 1755–1762. [CrossRef]
28. Wang, X.; Mušicki, D. Low Elevation Sea-Surface Target Tracking. *IEEE Trans. Aerosp. Electron. Syst.* **2007**, *44*, 764–774.
29. Mušicki, D.; Evans, R. Joint Integration Probabilistic Data Association-JIPDA. *IEEE Trans. Aerosp. Electron. Syst.* **2004**, *40*, 1093–1099. [CrossRef]
30. Mušicki, D.; Evans, R.J.; Stankovic, S. Integrated Probabilistic Data Association. *IEEE Trans. Autom. Control* **1994**, *39*, 1237–1241. [CrossRef]
31. Mušicki, D.; La Scala, B. Multi-Target Tracking in Clutter Without Measurement Assignment. *IEEE Trans. Aerosp. Electron. Syst.* **2008**, *44*, 31–43. [CrossRef]
32. Mušicki, D.; Suvorova, S. Tracking in Clutter Using IMM-IPDA Based Algorithms. *IEEE Trans. Aerosp. Electron. Syst.* **2008**, *44*, 394–413. [CrossRef]
33. Mušicki, D.; Wang, X.; Fletcher, F.; Ellem, R. Efficient Active Sonar Multi-Target Tracking. In Proceedings of the IEEE OCEANS'06 Asia Pacific, Singapore, 16–19 May 2006.
34. Blackman, S.; Popoli, R. *Design and Analysis of Modern Tracking Systems*; Artech House: Norwood, MA, USA, 1999.
35. Bar-Shalom, Y.; Chang, K.C.; Blom, H.A.P. Automatic Track Formation in Clutter with a Recursive Algorithm. In Proceedings of the 28th Conference on Decision and Control, Tampa, FL, USA, 13–15 December 1989; pp. 1402–1408.
36. Applied Physics Laboratory, University of Washington. *APL-UW High Frequency Ocean Environmental Acoustic Models Handbook*; APL-UW TR 9407, AEAS 9501; PN: Seattle, WA, USA, October 1994.
37. Mušicki, D.; Morelande, M. Gate Volume Estimation for Target Tracking. In Proceedings of the 7th International Conference on Information Fusion (Fusion 2004), Stockholm, Sweden, 28 June 2004.

Article

Direction of Arrival Estimation Using Two Hydrophones: Frequency Diversity Technique for Passive Sonar

Peng Li [1,2,3,*], Xinhua Zhang [1,3,4] and Wenlong Zhang [4]

1 Acoustic Science and Technology Laboratory, Harbin Engineering University, Harbin 150001, China; xinhua_zh@126.com
2 Key Laboratory of Marine Information Acquisition and Security (Harbin Engineering University), Ministry of Industry and Information Technology, Harbin 150001, China
3 College of Underwater Acoustic Engineering, Harbin Engineering University, Harbin 150001, China
4 Department of Underwater Weaponry & Chemical Defense, Dalian Navy Academy, Dalian 116018, China; 15604285615m0@sina.cn
* Correspondence: arthuralecto@hrbeu.edu.cn; Tel.: +86-180-0366-3672

Received: 28 February 2019; Accepted: 26 April 2019; Published: 29 April 2019

Abstract: The traditional passive azimuth estimation algorithm using two hydrophones, such as cross-correlation time-delay estimation and cross-spectral phase estimation, requires a high signal-to-noise ratio (SNR) to ensure the clarity of the estimated target trajectory. This paper proposes an algorithm to apply the frequency diversity technique to passive azimuth estimation. The algorithm also uses two hydrophones but can obtain clear trajectories at a lower SNR. Firstly, the initial phase of the signal at different frequencies is removed by calculating the cross-spectral density matrix. Then, phase information between frequencies is used for beamforming. In this way, the frequency dimension information is used to improve the signal processing gain. This paper theoretically analyzes the resolution and processing gain of the algorithm. The simulation results show that the proposed algorithm can estimate the target azimuth robustly under the conditions of a single target (SNR = −16 dB) and multiple targets (SNR = −10 dB), while the cross-correlation algorithm cannot. Finally, the algorithm is tested by the swell96 data and the South Sea experimental data. When dealing with rich frequency signals, the performance of the algorithm using two hydrophones is even better than that of the conventional broadband beamforming of the 64-element array. This further validates the effectiveness and advantages of the algorithm.

Keywords: direction of arrival estimation; frequency diversity; passive sonar

1. Introduction

Azimuth estimation is an important research area in passive sonar applications. Since two hydrophones are easy to deploy and throwing buoys is also easy in actual combat, azimuth estimation algorithms, based on the cross-correlation time-delay estimation of two hydrophones are often applied to buoys and autonomous underwater vehicles (AUVs) [1]. However, the cross-correlation algorithm requires a high signal-to-noise ratio (SNR) and can only estimate one target. Using only two sensors to estimate more targets and obtain more accurate estimation results has always been the focus of the research on passive sonar applications.

In passive detection research, many array processing algorithms can improve the performance of azimuth estimation, such as the split aperture method, which can obtain an extremely small size and spacing of the array elements, while avoiding the formation of grating lobes [2], and can also modify the beamforming process, according to the linear phase relationship between two subarrays, to obtain high-precision azimuth estimation results [3]. In addition, the co-prime array algorithm can achieve a higher degree of freedom, with a limited number of elements, thus increasing the number of estimable sources [4]. However, such algorithms need a special array structure first. For example, the split aperture method requires two sub-arrays, with element spacing of $(p-1)\lambda/2$ and $p\lambda/2$, and co-prime arrays, needing two sub-arrays, have M and N sensors, where M and N are co-prime with the appropriate inter-element spacing [5]. For two hydrophones, it is difficult to obtain such spatial information. Therefore, in order to improve the azimuth estimation performance of two hydrophones, only additional information from other dimensions or equivalent spatial information from other dimensions can be added.

In the application of a multiple-input multiple-output (MIMO) radar, there is a frequency diversity array (FDA) technique [6], the idea of which is to combine the spatial information and the frequency information. In 2006, Antonik et al. first proposed the concept of FDA at the International Radar Conference [7]. The algorithm introduces a frequency difference between each array element at the transmitting terminal and combines the distance and the scanning angle to improve the anti-interference ability [8]. In recent years, scholars from various countries have done a great deal of research on FDA, such as improving the practicality of FDA [9], reducing the array cost [10], extending FDA to distance dimensions [11], and applying FDA to the bistatic joint estimation of the distance and azimuth [12]. The application of the FDA algorithm in radars has matured. Researchers have made a comprehensive analysis of the algorithm's performance [10,11,13]. Whatever the improvement of the algorithm, it is always the case that the phase difference changes, caused by the sound path and the frequency, are used to relate the distance and the angle change.

In passive sonar applications, the target is often a broadband source. However, the conventional towed array processing only divides the frequency band into many sub-bands. Then, the azimuth estimation results are calculated and added together. Both the wideband processing method [14] and the time domain beamforming algorithm [15] do not take advantage of the relationship between the frequency, target azimuth and signal phase. Inspired by the FDA technique in radars, this paper applies the idea of frequency diversity to the azimuth estimation of two hydrophones in a passive sonar. The information dimension of the dual-element output signal is improved by the frequency information, thereby realizing a high performance of the azimuth estimation. However, the passive algorithm of the two hydrophones has an important difference from the commonly used algorithm in the MIMO radar. That is, the received signal of the passive sonar is unpredictable, and the initial phase of each frequency point is unknown. Therefore, the algorithm first removes the initial phase on each frequency component of the signal by conjugate processing, which calculates the cross-spectral density between the two elements. A frequency domain vector that can be used for beamforming is constructed using a cross-spectrum, the phase of which changes with the target azimuth and the sensor interval between the different frequency. The corresponding weighted vector is designed to obtain the azimuth estimation result.

The remainder of this paper is organized as follows: In Section 2, we first briefly introduce the cross-correlation method, cross-spectral method and FDA. Then the passive azimuth estimation algorithm of two hydrophones, based on the FDA technique, is proposed, and the processing gain and resolution of the algorithm are analyzed. In Section 3, the algorithm and the traditional algorithm are compared by simulation experiments, and the effectiveness and advantages of the algorithm are verified. Section 4, experimental data processing further proves that the proposed two-hydrophone passive azimuth estimation algorithm using the FDA technique is better than the cross-correlation method and can obtain a clear azimuth history diagram. In addition, the influence of the energy spectrum distribution of the signal on the estimation result is analyzed. The final conclusion is given in Section 5.

2. Theoretical Derivation

This paper is based on the idea of FDA technology and proposes a passive azimuth estimation algorithm applied to two hydrophones. First, in Section 2.1, we briefly review the passive azimuth estimation algorithms commonly used in two hydrophones and the frequency diversity techniques used in radar. The algorithm proposed in this paper is introduced in Section 2.2. The resolution and processing gain of the algorithm are analyzed, and the algorithm is extended.

2.1. Conventional Algorithm

Section 2.1.1 briefly introduces two commonly used azimuth estimation algorithms on two hydrophones: The cross-correlation method and cross-spectrum method. In Section 2.1.2, the FDA algorithm in a MIMO radar is briefly introduced.

2.1.1. Cross-Correlation Method and Cross-Spectral Method

First, the cross-correlation method is introduced [16]: the two-hydrophone receiver model is shown in Figure 1, where d is the array element spacing and θ is the signal incoming wave direction.

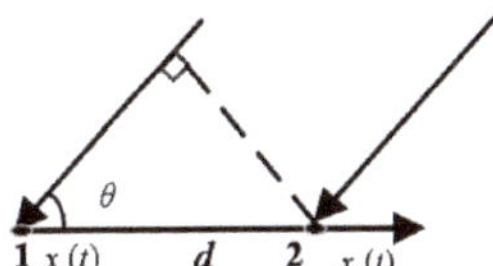

Figure 1. Received signal model using two hydrophones.

$x_1(t)$, $x_2(t)$ are the received signals of hydrophone 1 and hydrophone 2, respectively, and their cross-correlation functions can be expressed as:

$$R_{x_1x_2}(\tau) = E[x_1(t)x_2(t-\tau)] \tag{1}$$

where $E[\bullet]$ is a mathematical expectation. When the noise and the signals are independent of each other, and the SNR is high enough, after calculating the delay τ_0, corresponding to the correlation peak, the direction of arrival (DOA) estimation can be acquired, according to Equation (2):

$$\tau_0 = d\cos\theta/c \tag{2}$$

where c is the speed of sound in water. In addition to the cross-correlation delay estimation algorithm, the commonly used algorithm also has a cross-spectral method [17]. Let the Fourier transform of $x_1(t)$ be $X_1(f)$, and the Fourier transform of $x_2(t)$ can be obtained as $X_1(f)e^{j2\pi f\tau_0}$, according to the delay characteristic of the Fourier transform. Then, the cross-spectrum of hydrophone 1 and the hydrophone 2 can be obtained as follows:

$$Z_X(f) = X_1{}^*(f)X_2(f) = |X_1(f)|^2 e^{j2\pi f\tau} \tag{3}$$

It can be found, from Equation (3) that the time delay τ_0 is included in the phase information of the cross-spectrum, namely:

$$2\pi fd\cos\theta/c = \arctan\left\{\frac{\mathrm{Im}[Z(f)]}{\mathrm{Re}[Z(f)]}\right\} \tag{4}$$

The DOA can be estimated according to Equation (4). However, such algorithms first have requirements on SNR concerning the received array signals. Secondly, for wideband signals, when using cross-correlation time delay estimation, the cross-correlation function graph shows many periodic peaks [18], which further increases the difficulty of peak finding. Therefore, implementing DOA estimation based on two hydrophones at a low SNR is very important. We found that neither of these algorithms effectively utilized the phase relationship between the frequencies. The FDA technique in a MIMO radar will be described below, which effectively utilizes the phase relationship between the frequencies.

2.1.2. FDA Technique

As shown in the Figure 2, the frequency of the waveform radiated from each sensor was incremented by Δf from element to element.

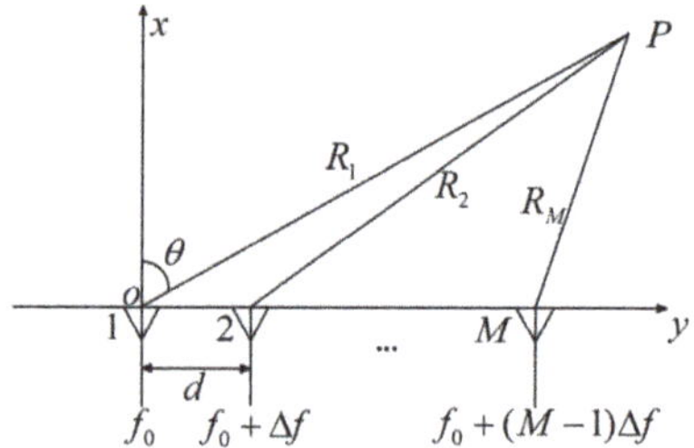

Figure 2. Schematic diagram of an FDA space structure.

By means of quadrature modulation or matched filtering, only the corresponding frequency signal is received. It is easy to obtain the phase difference between the adjacent elements (mth and m+1th), which can be written as:

$$\Delta\varphi = 2\pi f_0 d \sin\theta / c - 2\pi R_m \Delta f / c + 2\pi \Delta f d \sin\theta / c. \tag{5}$$

where R_m is the distance from the sound source to the mth sensor. When the far field condition is met, R_m can be recorded as:

$$R_m = R_1 - d\sin\theta. \tag{6}$$

According to the beamforming principle [19], it can be calculated that the phase shift of Equation (5) cause the beam at some apparent angle θ' [20]:

$$\theta' = \arcsin\left\{\sin\theta - \frac{R_1 \Delta f}{d f_0} + \frac{\Delta f \sin\theta}{f_0}\right\}. \tag{7}$$

Equation (7) associates the scan angle θ', the target azimuth θ and the target distance. Therefore, FDA can estimate the DOA and distance and can also suppress clutter interference.

It should be noted that the DOA estimation of the MIMO radar and two passive hydrophones have two important differences: (1) When FDA is applied in the MIMO radar, it is used for multiple array elements, and the frequency of the transmitted signal varies with the number of elements. In the algorithm of this paper, only two array elements are used, and the received signal is sampled in the frequency domain. (2) In the MIMO radar, the waveform of each transmitted signal is known. Therefore, the initial phase of the received signal for each element at each frequency is controllable. In the algorithm of this paper, since it is applied to a passive sonar, the initial phase of each frequency is unknown. Therefore, the application of the idea of FDA to the DOA estimation of two passive hydrophones has to be greatly changed.

2.2. FDA Technique of Two Hydrophones

2.2.1. Theory

According to the model in Figure 1, the frequency domain expressions of the received signals of the two hydrophones are:

$$\begin{aligned} S_1(f) &= \left|X_1(f)\right| e^{j[2\pi fr/c+\phi(f)]} + N_1(f) \\ S_2(f) &= \left|X_1(f)\right| e^{j[2\pi fd\cos\theta/c+2\pi fr/c+\phi(f)]} + N_2(f) \end{aligned} \tag{8}$$

where $N_1(f)$ and $N_2(f)$ are the ambient noise received by hydrophone 1 and hydrophone 2, respectively. $\phi(f)$ is the random, frequency-dependent phase of the source. It can be observed, from the above equation, that in the phase information of $S_2(f)$, the first item contains the azimuth information of the target, the second term relates to the propagation distance, and the third term is the initial phase of the frequency. Since, in the application condition of the passive sonar, the target distance and the initial phase of the sound source signal are unknown, we first calculate the cross-spectrum of the two sensor signals to remove the phase in the second and third terms:

$$Z(f) = S_1{}^*(f)S_2(f) = Z_X(f) + Z_N(f) \tag{9}$$

where $Z_X(f)$ is the cross-spectrum of the signal, and $Z_N(f)$ is the component related to environmental noise. $Z_X(f)$ and $Z_N(f)$ are denoted as:

$$\begin{aligned} Z_X(f) &= \left|X_1(f)\right|^2 e^{j2\pi fd\cos\theta/c} \\ Z_N(f) &= X_1{}^*(f)N_2(f) + N_1{}^*(f)X_2(f) + N_1{}^*(f)N_2(f) \end{aligned} \tag{10}$$

After obtaining the cross-spectrum $Z(f)$, $Z(f_m)$ is obtained by sampling $Z(f)$ in the frequency domain. According to the idea of the frequency diversity technique, the frequency of the sampling point is f_m, and the frequency increment is Δf. Vector $[Z(f_1), Z(f_2), \ldots, Z(f_M)]$ can be generated as:

$$Z(f_m) = \left|X_1(f_m)\right|^2 e^{j2\pi f_m d\cos\theta/c} + Z_N(f), f_m = f_1 + (m-1)\Delta f. \tag{11}$$

It can be found, from Equation (11), that the phase difference of $Z_X(f_m)$ between the adjacent sampling points is $j2\pi\Delta fd\cos\theta/c$. There is no phase relationship between the various frequencies of ambient noise. Array manifolds are generated by the phase relationship in Equation (12):

$$A_{fm} = e^{-j2\pi m\Delta fd\cos\theta/c}. \tag{12}$$

The beamforming output can be obtained according to the principle of in-phase superposition:

$$\text{Beam}(\theta) = \sum_{m=1}^{M} Z(f_m)A_{fm} \tag{13}$$

2.2.2. Performance Analysis

According to the above analysis, it is easy to obtain the directivity function of the passive two-hydrophone algorithm based on FDA technology:

$$R(\theta) = \left| \frac{\sin\left(\pi M\Delta f\frac{d\sin\theta}{c}\right)}{M\sin\left(\pi\Delta f\frac{d\sin\theta}{c}\right)} \right|. \tag{14}$$

The azimuth resolution, based on half the width of the main lobe, is defined as in Equation (14):

$$\theta_r = \arcsin\left(\frac{c}{N\Delta f d}\right). \tag{15}$$

In order not to obtain a grating lobe, the scanning angle θ needs to satisfy $\sin(\theta) \le \frac{c}{2\Delta f d}$. In general, the scanning angle is −90° to 90°, so when frequency domain sampling is performed on the signal, the frequency interval $\Delta f \le \frac{c}{2d}$ should be satisfied.

According to the beamforming of Equation (13), the output SNR is:

$$SNR_{out} = 10\log\frac{\sum_{m=1}^{M}\left|X_1(f_m)\right|^2}{\sum_{m=1}^{M} A_{f_m} Z_N(f_m)}. \tag{16}$$

It can be seen from Equation (16) that the less related the ambient noise between the frequencies, the higher the output SNR.

2.2.3. Algorithm Extension: Three Hydrophones

From the above theoretical derivation, we can find that the passive two-hydrophone algorithm based on FDA technology is similar to the single-frequency signal processing algorithm of a conventional towed-line array. Similarly, the algorithm can be extended to higher dimensions, for instance, using a wideband signal to obtain a performance similar to the single-frequency processing of a circular array. The specific process is as follows:

Assume that the three hydrophones, a, b, and c, have a radius of R. The angle with the reference abscissa are θ_a, θ_b and θ_c, respectively. The incident angle of the far-field sound source is θ_0. According to the spatial structure of the three hydrophones, as shown in Figure 3, the frequency domain model of the received signals can be obtained as follows:

$$\begin{aligned} S_a(f) &= e^{j2\pi f(-R\cos(\theta_a-\theta_0)+r)/c+\varphi(f)} \\ S_b(f) &= e^{j2\pi f(-R\cos(\theta_b-\theta_0)+r)/c+\varphi(f)} \\ S_c(f) &= e^{j2\pi f(-R\cos(\theta_c-\theta_0)+r)/c+\varphi(f)} \end{aligned} \tag{17}$$

where r is the propagation distance. Similarly, we also calculate the cross-spectrum to remove the initial phase:

$$\begin{aligned} Z_{ab}(f_1) &= S_a(f_1)\cdot S_b{}^*(f_1) \\ Z_{ac}(f_m) &= S_a(f_m)\cdot S_c{}^*(f_m),\ m = 1,\ 2,\ 3,\ \ldots,\ M \end{aligned} \tag{18}$$

where f_1 is the starting frequency, and f_m is the frequency of the sampling point. Bring Equation (17) into Equation (18) and expand the cosine term to obtain:

$$\begin{aligned} Z_{ab}(f_1) &= e^{j2\pi f_1 R((\cos(\theta_b)-\cos(\theta_a))\cos(\theta_0)+(\sin(\theta_b)-\cos(\theta_a))\sin(\theta_0))/c} \\ Z_{ac}(f_m) &= e^{j2\pi f_m R((\cos(\theta_c)-\cos(\theta_a))\cos(\theta_0)+(\sin(\theta_c)-\cos(\theta_a))\sin(\theta_0))/c} \end{aligned} \tag{19}$$

Let $A_1 = \cos(\theta_b) - \cos(\theta_a)$ and $B_1 = \sin(\theta_b) - \sin(\theta_a)$, and then take the conjugate of two cross-spectra and multiply:

$$Y_m = Z_{ab}(f_1)\cdot Z_{ac}{}^*(f_m) \tag{20}$$

Let $\gamma = \arccos\frac{A_3}{\sqrt{A_3{}^2+B_3{}^2}}$, $A_3 = f_1A_1 - f_mA_2$, $B_3 = f_1B_1 - f_mB_2$, and Equation (20) be transformed into:

$$Y_r = e^{j2\pi R\sqrt{A_3{}^2+B_3{}^2}\cos(\gamma-\theta_0))/c} \tag{21}$$

In order to construct a circular array manifold, i.e., $e^{j2\pi R\cos(m\theta_s-\theta_0))/c}$, where $\theta_s = 2\pi/M$. Let $\gamma = m\theta_s$, Equation (21) can be obtained:

$$\cos m\theta_s = \frac{A_3}{\sqrt{A_3{}^2 + B_3{}^2}} \tag{22}$$

According to Equation (23), f_m can be solved:

$$f_m = \frac{B_1\cos(m\theta_s) - A_1\sqrt{1-\cos^2(m\theta_s)}}{B_2\cos(m\theta_s) - A_2\sqrt{1-\cos^2(m\theta_s)}} f_1 \tag{23}$$

Therefore, it is only necessary to know the coordinates of the three array elements, and it is easy to obtain R, θ_a, θ_b, θ_c according to the geometric relationship. The target position can be estimated. Beamforming is shown in Equation (24):

$$\text{Beam}(\theta) = \sum_{m=1}^{M} Z_{ab}(f_1)Z_{ac}{}^*(f_m)e^{j2\pi R\sqrt{A_3{}^2+B_3{}^2}\cos(m\theta_s-\theta_0))/c} \tag{24}$$

In this way, we have implemented a wideband signal based on the azimuth estimates of the three hydrophones. Because the algorithm is designed with reference to the circular array, there is no problem with port and starboard ambiguity.

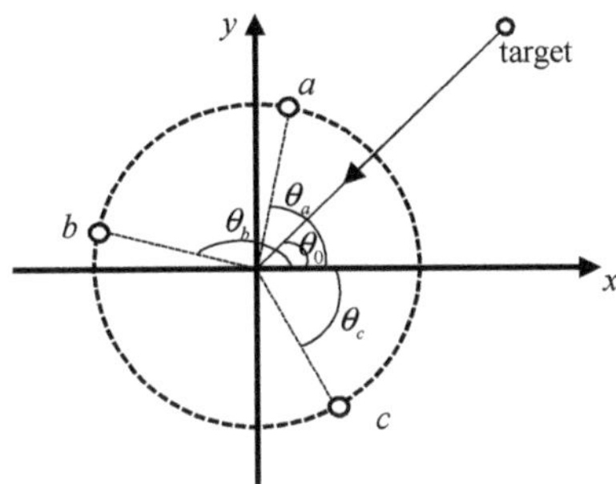

Figure 3. Schematic diagram of the spatial structure of the three hydrophone.

3. Simulation

3.1. Comparison of the Cross-Correlation Algorithm and Frequency Diversity Algorithm

The cross-correlation method and cross-spectrum method have a similar performance under the same signal-to-noise ratio. Moreover, the conventional cross-spectral method is based on the discrete spectrum of the received signal to directly estimate the azimuth. In practical applications, the relative frequency deviation of signals and the leakage of spectrum will lead to an azimuth estimation error. Thus, here, we only compare the proposed algorithm with the cross-correlation method. Assuming that the two hydrophones are 128m apart, the signal is Gaussian white noise, with 100–200 Hz bandpass filtering. First, consider a single target with an incident angle of 40°. The sampling frequency is 4 kHz, and the number of samples is 4096. According to the relevant algorithm processing time length, set $\Delta f = 1$ Hz in the frequency diversity algorithm. The sampling bandwidth $M\Delta f$ is set to 100 Hz according to the signal bandwidth. When the in-band SNR of the received hydrophone signal is set to 0 dB, where the noise is Gaussian white noise, the simulation result is shown in Figure 4.

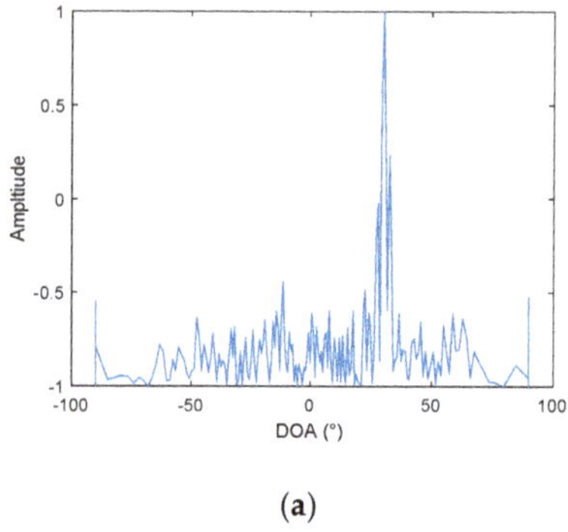

(a)

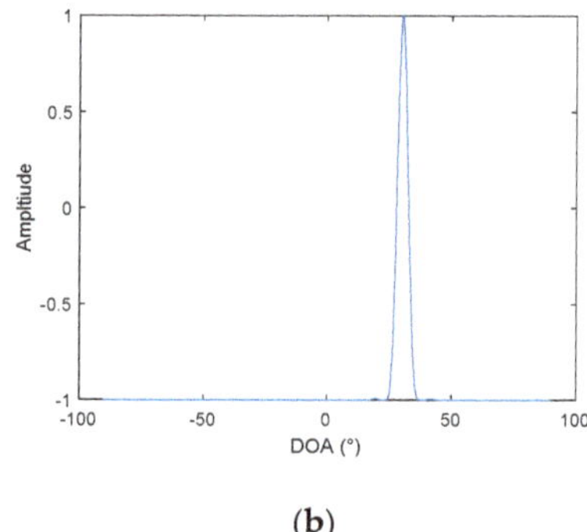

(b)

Figure 4. DOA of two hydrophones, SNR = 0 dB: (**a**) the cross-correlation method and (**b**) frequency diversity algorithm.

When the in-band SNR is set to −16 dB, the simulation results are shown in Figure 5:

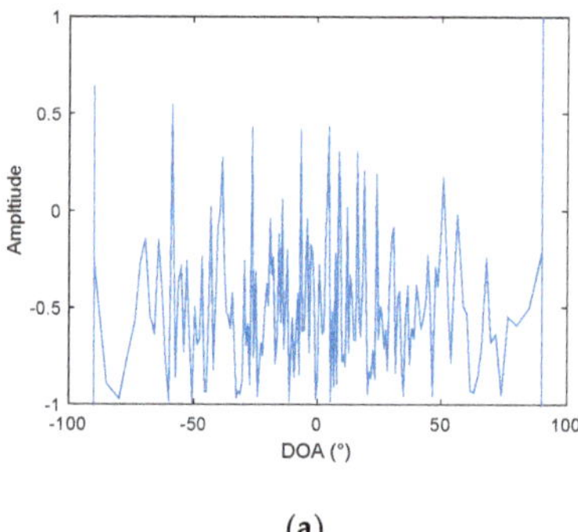

(a)

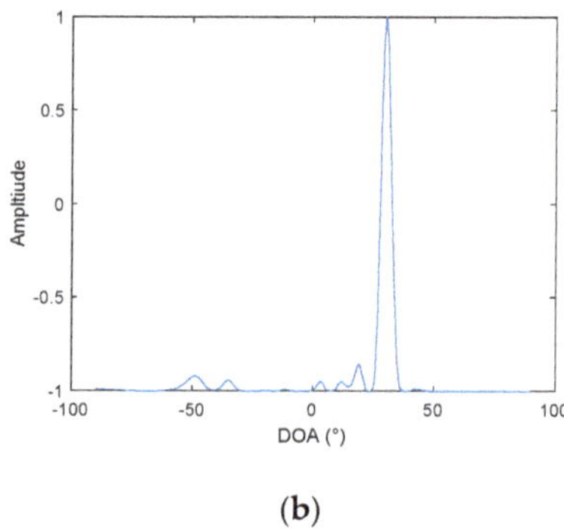

(b)

Figure 5. DOA of two hydrophones, SNR = −16 dB: (**a**) The cross-correlation method and (**b**) frequency diversity algorithm.

It can be seen, from Figures 4 and 5, that the DOA estimation performance of the frequency diversity algorithm is superior to the cross-correlation method under different SNR. When the SNR is reduced to −16 dB, the cross-correlation method can no longer estimate the azimuth of the target, while using the frequency diversity algorithm, and a robust estimation of the target azimuth can still be achieved. Change the number of sound sources to two, and the bearing angles are 30° and 50°. The in-band SNR is set to −10 dB. The simulation results are shown in Figure 6.

When the number of sound sources is changed to two, it can be seen from Figure 6, that when the in-band SNR is reduced to −10 dB, the cross-correlation method can no longer estimate the azimuth of the target. This is because the source signal has a wideband, and the two target signals are not completely independent. Therefore, there are many periodic pseudo peaks in the cross-correlation. Therefore, when the number of targets changes from 1 to 2, the SNR used to compare the performance of the two algorithms is increased from −16 dB to −10 dB. When frequency diversity techniques are used, the target azimuth can still be estimated robustly. The reason is that the frequency diversity technique uses the phase relationship in the signal frequency dimension, so the processing gain is improved, compared to the cross-correlation method, and the resolution is not affected by the correlation between signals. Since the simulated two target signals are band-limited white Gauss noise, whose spectrum is random, in the latter analysis, it can be found that the beamformed output amplitude obtained by the proposed algorithm, is related to the energy distribution in the frequency domain, so the amplitudes of the two sources are different.

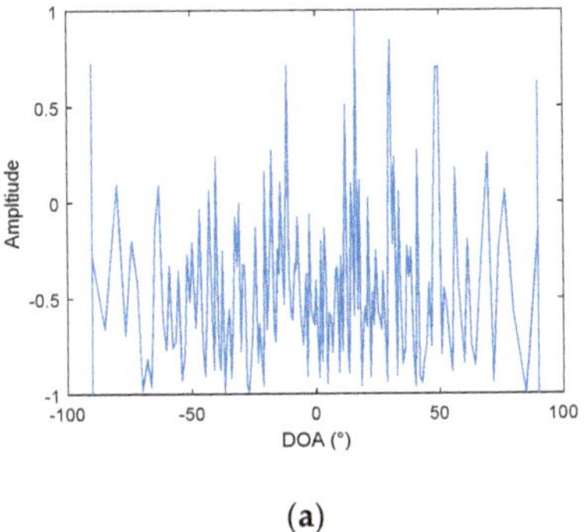

(a)

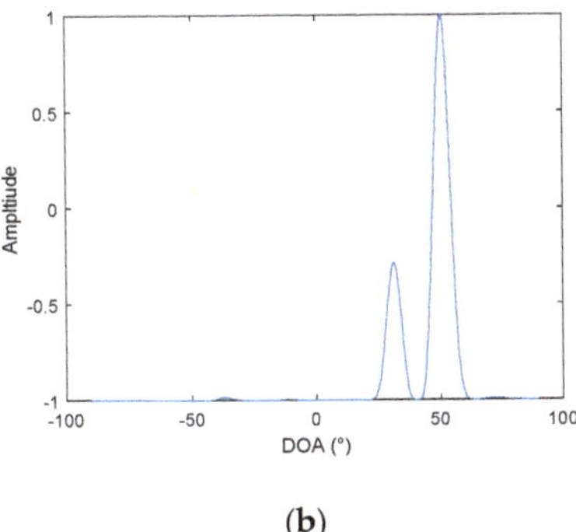

(b)

Figure 6. DOA of two hydrophones, SNR = −10 dB, two targets: (**a**) the cross-correlation method and (**b**) frequency diversity algorithm.

3.2. Resolution of the Frequency Diversity Algorithm

According to Equation (15), increasing the signal processing bandwidth $M\Delta f$ and the sensor interval d can improve the resolution of the algorithm. Figure 7 is a simulation verification of this property. The sampling frequency is 4 kHz, the number of samples is 4096, $\Delta f = 1$ Hz, and the starting frequency $f_0 = 100$ Hz.

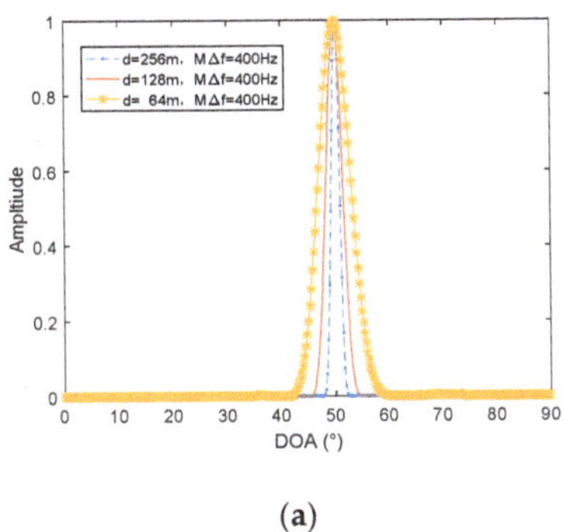

(a)

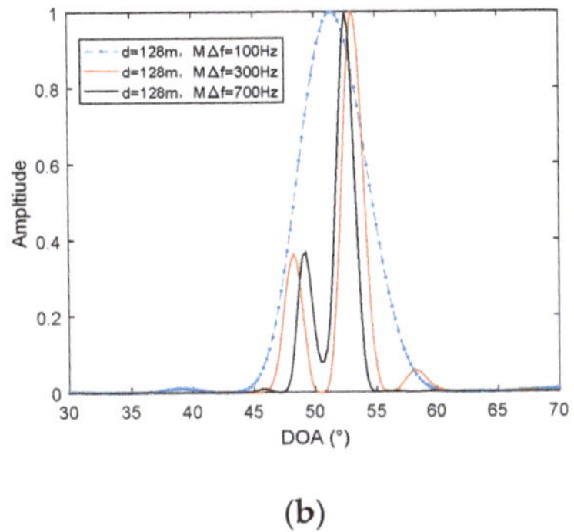

(b)

Figure 7. Azimuth resolution of the proposed algorithm, with different d and $M\Delta f$. (**a**) Sensor interval d and (**b**) processing bandwidth $M\Delta f$.

In Figure 7a, there is only one target at 50°. When the processing bandwidth $M\Delta f$ is set to 400 Hz, as the sensor interval decreases, the width of the main lobe becomes wider, so the resolution decreases. The two targets in Figure 7b have an incoming wave direction of 50° and 52°. When d is set to 128 m, as the processing bandwidth $M\Delta f$ increases, the main lobe width becomes narrower, and the resolution increases. In addition, when $M\Delta f = 100$ Hz, the algorithm cannot separate two targets. When $M\Delta f = 300$ Hz, the algorithm can separate the two targets, but the peaks do not appear exactly at 50° and 52°, but at 48° and 53°. When $M\Delta f = 700$ Hz, the two peaks appear at exactly 50° and 52°, so the azimuth estimation is accurate. In summary, the larger the sensor interval, the wider the processing bandwidth $M\Delta f$, the higher the resolution of the algorithm, and the more accurate the azimuth estimation.

3.3. Simulation of Frequency Diversity Based on Three Hydrophones

Assuming three hydrophones, the positions of which are not in a straight line, the coordinates are: (0,0), (−325 m, 141 m), (−253 m, −213 m). The radius of the circle is determined to be 202 m, according to the position of the three points. The target signal is 50–1000 Hz band-limited white noise, and the incident angle is 40°. Using Equation (23) to calculate f_{m} according to f_1, wherein the sampling point M is set to 512, the azimuth estimate can be obtained according to Equation (24). The result is shown in Figure 8.

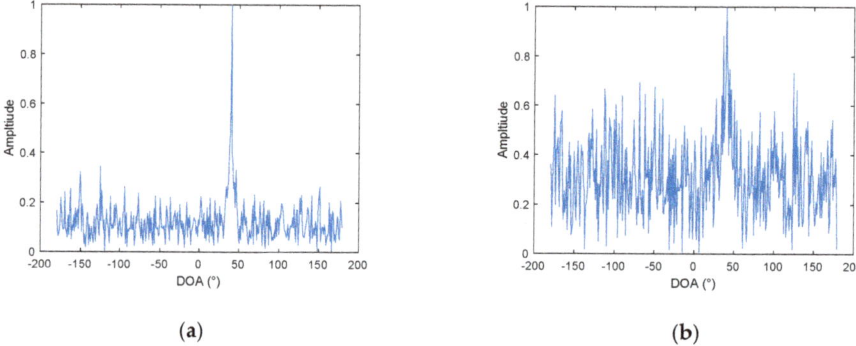

Figure 8. DOA estimation results under different SNR: (**a**) SNR = 0 dB (**b**) SNR = −10 dB.

From Figure 8, it can be seen that, when three hydrophones are used, an azimuth estimation result of −180° to 180° can be obtained, and there is no problem with port and starboard ambiguity. When the position of the three hydrophones changes, the radius of the virtual circle changes. The simulations compared the azimuth estimation results under the three apertures, with radii of 20 m, 202 m, and 2019 m, as can be seen in Figure 9:

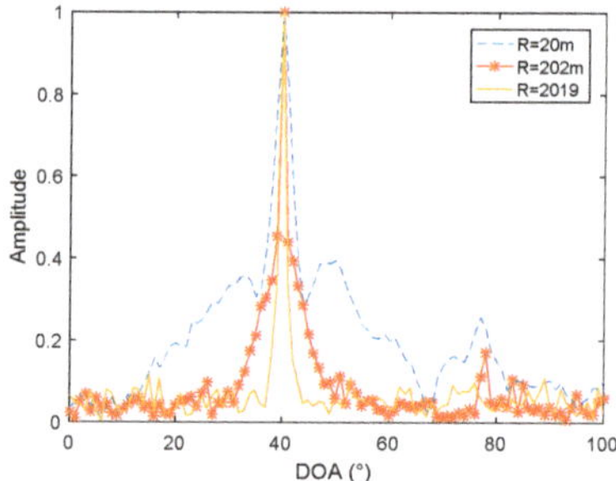

Figure 9. The relationship between the azimuth resolution and the virtual radii of three hydrophones.

As can be seen, in Figure 9, the increase in the radius of the virtual circle is beneficial to the resolution. The farther the distance is placed in the three hydrophones, the better the azimuth estimation performance.

4. Experimental Data Verification

The algorithm is first verified with Swell 96 horizontal south array data [21], using the 14th to 28th array elements. The SWellEx-96 Experiment was conducted between May 10 and 18, 1996, approximately 12 km from the tip of Point Loma near San Diego, California. Acoustic sources, towed from the R/V Sproul, transmitted various broadband and multi-tone signals at frequencies between 50 and 400 Hz.

In order to further compare the performance of the two hydrophone algorithms, conventional array processing is used to obtain the azimuth estimation result as a reference, because the array gain of the processing of multiple array elements leads to a clear trajectory. The processing frequency bandwidth is 20–1000 Hz. The azimuth history diagram is shown in Figure 10a. It can be seen, from the figure, that within this time period (1–3500 s), there are mainly two targets, one with a large span in the azimuth, and one mainly at around 40°.

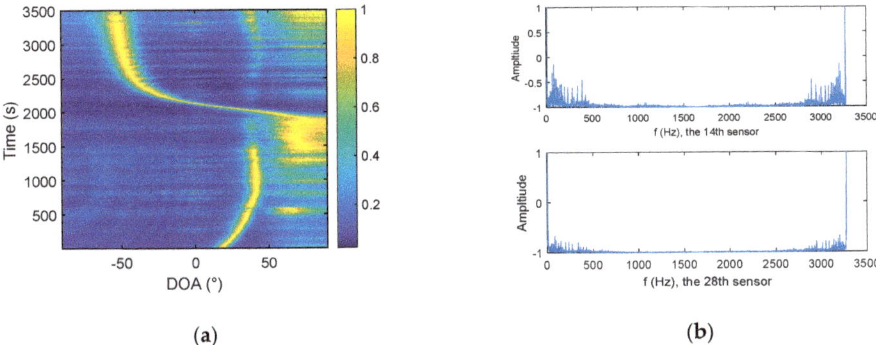

Figure 10. (**a**) Azimuth history diagram of Swell 96 data. (**b**) Signal spectrum of the 14th and 28th sensors.

The 14th array element and the 28th array element are selected as the two hydrophones, and the distance between them is 106 m. Figure 10b is the signal spectrum of the 14th and 28th elements, and the Fourier transform time is from 1000 s to 1001 s. The results of the cross-correlation method and the frequency diversity algorithm are shown in Figure 11. The sampling frequency f_s is 3277 Hz, the number of samples is 3277, and $\Delta f = 1$ Hz. The processing bandwidth $M\Delta f$ is set to 980 Hz, according to the processing bandwidth of 20–1000 Hz.

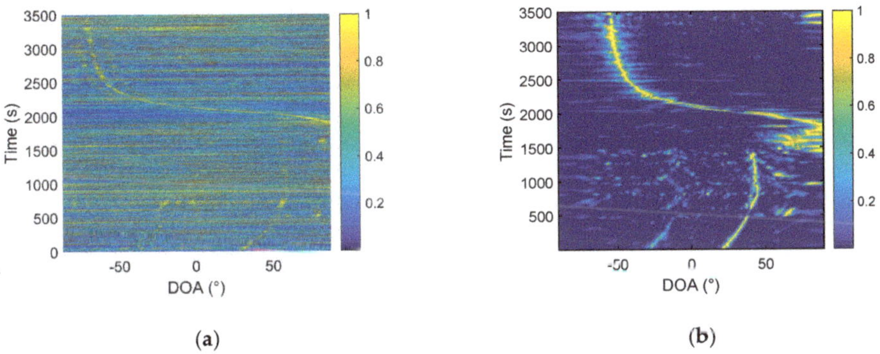

Figure 11. Azimuth history diagram of the two hydrophones: (**a**) The cross-correlation method and (**b**) frequency diversity algorithm.

From the comparison in Figure 11, it can be found that, using the same processing time, the same hydrophone, the target trajectory, estimated by the frequency diversity algorithm, is obviously clearer than that obtained by the cross-correlation algorithm. The algorithm is further verified by the South Sea data. Similarly, the conventional array processing is performed first, and a relatively accurate orientation estimation result is obtained. Then, we compare the cross-correlation method based on the passive two-hydrophone and the frequency diversity algorithm. In the array processing, 64 array elements are selected, with an interval of 4 m, and the processing method uses CBF. The processing frequency band is 20–400 Hz, and the sampling frequency is 2048 Hz. The azimuth estimation results are in Figure 12.

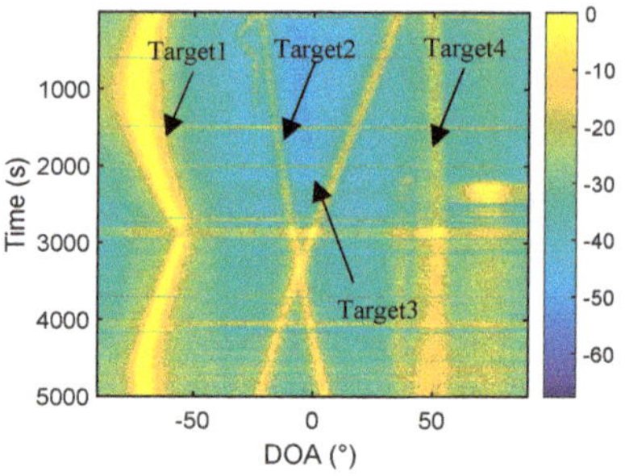

Figure 12. Azimuth history diagram of the 64-element array.

The data are processed using two hydrophones, as shown in Figure 13, and the number of samples is 2048, $\Delta f = 1$ Hz, and $M\Delta f$ is set to 380 Hz.

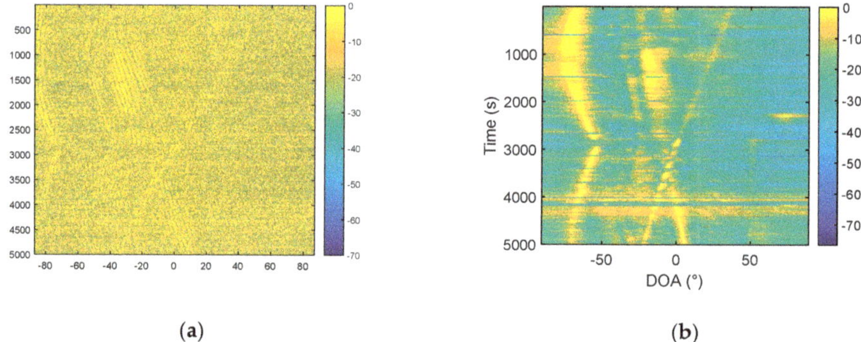

Figure 13. Azimuth history diagram of the two hydrophones: (**a**) the cross-correlation method and (**b**) frequency diversity algorithm.

In Figure 13a, only a little blurred outline can be seen, and the trajectory of the target can hardly be observed. In Figure 13b, the trajectories of target 1, target 2, and target 3 can be clearly observed. The trajectory of target 4 is not clear. It can be explained that the performance of the two-hydrophone algorithm based on frequency diversity technology is significantly higher than that of the cross-correlation algorithm. Moreover, we found, in the experiment, that the frequency diversity algorithm has a higher processing gain, which is easily seen before taking the beam energy by 10 lg. Before taking 10 lg, the beam energy is shown in Figure 14.

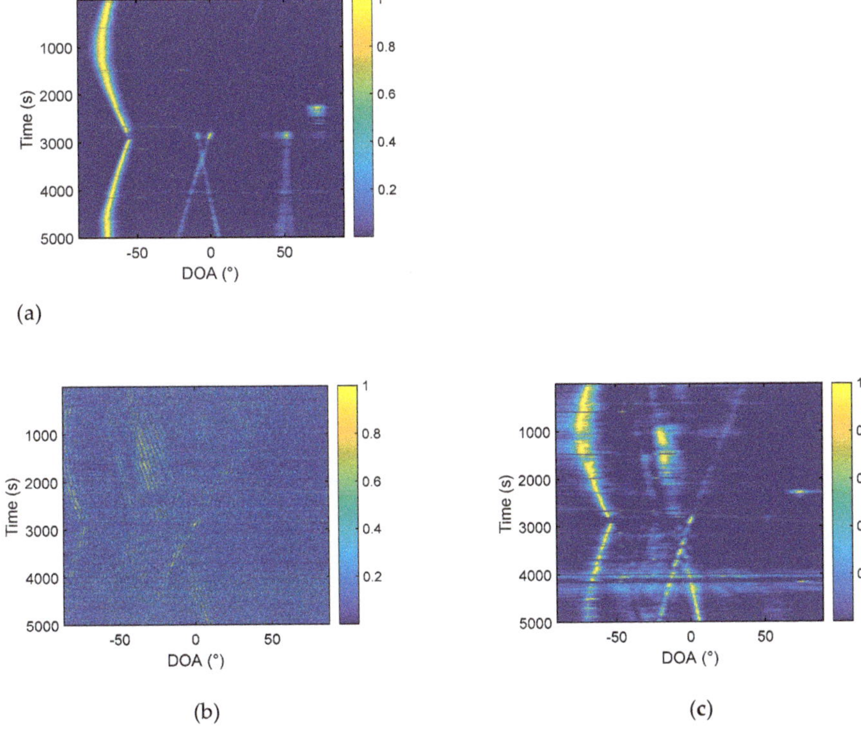

Figure 14. Azimuth history diagram of the two hydrophones: (**a**) The conventional 64-element processing; (**b**) two hydrophones, cross-correlation algorithm; and (**c**) two hydrophones, frequency diversity algorithm.

Comparing Figure 14b with Figure 14c, it is found that the frequency diversity algorithm is better than the cross-correlation algorithm, regardless of whether the log is taken or not. Furthermore, comparing Figure 14c with Figure 14a, it can be found the energy of target 2 and target 3 is significantly improved when the frequency diversity algorithm is used. To further reflect this feature, we take the azimuth estimation result at 4500 s as an example. At this time, target 2 and target 3 are located at 3° and −17°, respectively, and it is apparent, from the comparison of Figure 15, that the energy of target 2 and target 3 is enhanced.

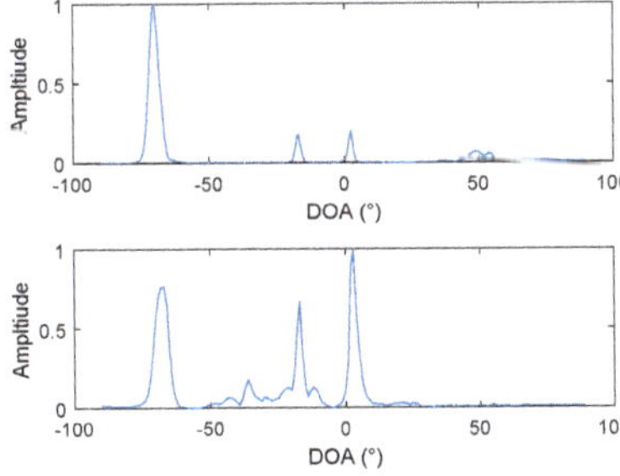

Figure 15. DOA results at 4500s: (**top**): 64-element array processing, and (**bottom**): the two-hydrophone frequency diversity algorithm.

The reason for this phenomenon is that the frequency diversity algorithm of the two hydrophones mainly uses the frequency domain information of the signal. Therefore, rich frequency domain information and uniform frequency domain energy distribution are beneficial for the energy of the beamforming output. The spectrums of Target 1 and Target 4 are shown in Figure 16a, and the spectra of target 2 and target 3 are shown in Figure 16b. From the comparison of Figure 16a,b, the spectrums of target 2 and target 3 are significantly richer than that of target 1 and target 4, and the energy distributions are more uniform. Therefore, in the estimation results of the two hydrophones, the energy of targets 2 and 3 is strengthened. Among them, the spectrum energy distribution of target 4 is the most concentrated so, in the two-hydrophone azimuth estimation, target 4 can hardly be observed.

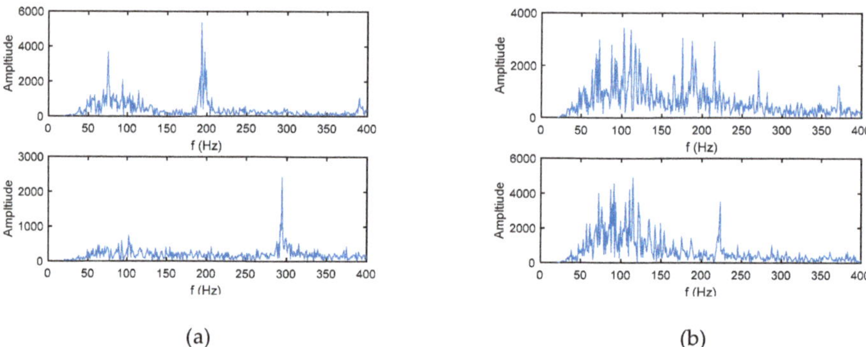

Figure 16. Frequency spectrum of the four targets. (**a**) The upper picture is target 1, and the lower picture is target 4; and (**b**) the upper picture is target 2, and the lower picture is target 3.

5. Conclusions

This paper proposes a frequency diversity algorithm to achieve passive azimuth estimation using two hydrophones. Compared with the traditional cross-correlation method, the algorithm has a high processing gain and can obtain a clear target trajectory. When the energy of the target signal is evenly distributed in the frequency domain, the processing gain of the algorithm using two hydrophones may even exceed the CBF processing gain of multiple array elements. In addition, in the theoretical derivation, the relationship between the resolution, the sensor interval and the processing bandwidth is analyzed. In the simulation and experiment, the feasibility of the algorithm and the advantages, compared with the cross-correlation method, are verified.

The algorithm proposed in this paper can obtain a clear target trajectory using the wideband signals received by only two hydrophones. This is significant for the application of target estimation in the field of buoy and AUV collaborative operations. The idea of using frequency domain information to virtualize array element domain information has important academic value for passive sonar azimuth estimation.

Author Contributions: Conceptualization: P.L. and X.Z.; data curation: X.Z.; formal analysis: P.L.; investigation: P.L. and W.Z.; methodology: P.L.; project administration: X.Z.; resources: X.Z.; supervision: P.L.; writing—original draft: P.L.; writing—review and editing: P.L. and W.Z.

Funding: This research was funded by the National Natural Science Foundation of China, grant number [61271443].

Conflicts of Interest: The authors declare no conflict of interest.

References

1. Piskur, P.; Szymak, P. Algorithms for Passive Detection of Moving Vessels in Marine Environment. *J. Mar. Eng. Technol.* **2017**, *16*, 377–385. [CrossRef]
2. Talman, J.R.; Lockwood, G.R. Evaluation of the Radiation Pattern of a Split Aperture Linear Phased Array for High Frequency Imaging. *IEEE Trans. Ultrason. Ferroelectr. Freq. Control* **2000**, *47*, 117–124. [CrossRef] [PubMed]
3. Qian, T. Application of split-beam processing of line array in underwater acoustic detection. *Tech. Acoust.* **2015**, *34*, 551–555. [CrossRef]
4. Vaidyanathan, P.P.; Pal, P. Sparse Sensing With Co-Prime Samplers and Arrays. *IEEE Transactions on Signal. Process.* **2011**, *59*, 573–586. [CrossRef]
5. Si, W.; Zeng, F.; Hou, C.; Peng, Z. A Sparse-Based Off-Grid DOA Estimation Method for Coprime Arrays. *Sensors* **2018**, *18*, 3025. [CrossRef] [PubMed]
6. Secmen, M.; Demir, S.; Hizal, A.; Eker, T. Frequency Diverse Array Antenna with Periodic Time Modulated Pattern in Range and Angle. In Proceedings of the IEEE Radar Conference, Boston, MA, USA, 17–20 April 2007; pp. 427–430. [CrossRef]
7. Antonik, P.; Wicks, M.C.; Griffiths, H.D.; Baker, C.J. Multi-mission multi-mode waveform diversity. In Proceedings of the IEEE Conference on Radar, Verona, NY, USA, 24–27 April 2006; pp. 580–582.
8. Hu, B.L.; Liao, G.S.; Xu, J.W.; Zhu, S.Q.; Gong, J.L. Study on Space-time Character of Clutter for Forward-looking Frequency Diverse Array Radar. *J. Electron. Inf. Technol.* **2013**, *35*, 2694–2699. [CrossRef]
9. Eker, T.; Demir, S.; Hizal, A. Exploitation of Linear Frequency Modulated Continuous Waveform (LFMCW) for Frequecny Diverse Arrays. *IEEE Trans. Antennas Propag.* **2013**, *61*, 3546–3554. [CrossRef]
10. Huang, J.J.; Tong, K.F.; Baker, C.J. Frequency Diverse Array: Simulation and Design. In Proceedings of the 2009 IEEE Radar Conference, Pasadena, CA, USA, 4–8 May 2009; pp. 1–4.
11. Xu, J.; Zhu, S.; Liao, G. Space-time-range Adaptive Processing for Airborne Radar Systems. *IEEE Sens. J.* **2014**, *15*, 1602–1610. [CrossRef]
12. Sammartino, P.F.; Backer, C.J.; Griffiths, H.D. Frequency Diverse MIMO Techniques for Radar. *IEEE Trans. Aerosp. Electron. Syst.* **2013**, *49*, 201–222. [CrossRef]
13. Wang, W.Q.; Shao, H.Z.; Cai, J.Y. Range-angle-dependent Beamforming by Frequency Diverse Array Antenna. *Int. J. Antennas Propag.* **2012**, *2012*, 1018–1020. [CrossRef]
14. Li, J.; Lin, Q.H.; Kang, C.Y.; Wang, K.; Yang, X.T. DOA Estimation for Underwater Wideband Weak Targets Based on Coherent Signal Subspace and Compressed Sensing. *Sensors* **2018**, *18*, 902. [CrossRef] [PubMed]
15. Li, L.R.; Zhang, X.X.; Liu, H.H. A Target Detection Technology Based on Time Domain Beam Output Signal Characteristics. *Tech. Acoust.* **2018**, *6*, 540–544. [CrossRef]
16. Lai, X.M.; Torp, H. Interpolation methods for time delay estimation using cross-correlation method for blood velocity measurement. *IEEE Trans. Ultrason. Ferroelectr. Freq. Control* **1999**, *46*, 277–290. [CrossRef] [PubMed]
17. Cheng, P.; Chen, J.F.; Ma, C.; Zhang, Z. Optimized cross array passive localization based on crosspower-spectrum phase. In Proceedings of the 2011 IEEE International Conference on Signal Processing, Xian, China, 14–16 September 2011; pp. 1–5. [CrossRef]
18. Yan, X.Y. Research of Sound Source Location Techniques Based on Double Microphone Array. Master's Thesis, YanShan University, Qinhuangdao, China, May 2012.
19. Tian, T.A.; Xue, L. Sonar technology. In *Sonar Beamforming Technology*, 2nd ed.; Tian, T., Ed.; Harbin Engineering University Press: Harbin, China, 2009; pp. 76–79. ISBN 978-7-81133-598-9.
20. Zheng, Y. A Study on the Beamforming of Frequency Diverse Array. *Mod. Radar* **2015**, *37*, 29–33. [CrossRef]
21. The SWellEx-96 Experiment. Available online: http://swellex96.ucsd.edu/index.htm (accessed on 8 April 2015).

Review

Adaptive Echolocation and Flight Behaviors in Bats Can Inspire Technology Innovations for Sonar Tracking and Interception

Clarice Anna Diebold, Angeles Salles and Cynthia F. Moss *

Department of Psychological and Brain Sciences, Johns Hopkins University, Baltimore, MD 21287, USA; clarice.diebold@jhu.edu (C.A.D.); ANGIESALLES@jhu.edu (A.S.)
* Correspondence: cynthia.moss@jhu.edu

Received: 14 March 2020; Accepted: 18 May 2020; Published: 23 May 2020

Abstract: Target tracking and interception in a dynamic world proves to be a fundamental challenge faced by both animals and artificial systems. To track moving objects under natural conditions, agents must employ strategies to mitigate interference and conditions of uncertainty. Animal studies of prey tracking and capture reveal biological solutions, which can inspire new technologies, particularly for operations in complex and noisy environments. By reviewing research on target tracking and interception by echolocating bats, we aim to highlight biological solutions that could inform new approaches to artificial sonar tracking and navigation systems. Most bat species use wideband echolocation signals to navigate dense forests and hunt for evasive insects in the dark. Importantly, bats exhibit rapid adaptations in flight trajectory, sonar beam aim, and echolocation signal design, which appear to be key to the success of these animals in a variety of tasks. The rich suite of adaptive behaviors of echolocating bats could be leveraged in new sonar tracking technologies by implementing dynamic sensorimotor feedback control of wideband sonar signal design, head, and ear movements.

Keywords: biosonar; predictive tracking; tracking algorithms

1. Introduction

Tracking moving targets in noisy and complex environments is a challenge that must be solved by biological organisms and artificial systems alike. Autonomous machines, such as self-driving cars or motorized wheelchairs, make use of iterative algorithms to navigate and map new environments [1]. Sonar offers valuable advantages for environmental mapping and target tracking, particularly in dark environments, and biological solutions can inspire innovation in this technology arena [2].

Diverse animal groups have evolved strategies for tracking moving targets by generating estimates of target motion. Much of the biological research that informs current understanding of target tracking in animals focuses on visually dominant species. Some organisms use a constant target-bearing strategy, such as linear optical trajectory (LOT) strategy, to maintain a fixed relationship between heading angle and a selected target, to eventually intercept a prey item [3,4], while other organisms use predictive internal models to anticipate the motion of erratically moving prey [5,6]. Biological models have served to inspire optimization and tracking algorithms, including cuckoo birds [7], ants [8], and fireflies [9]. Auditory-specialists, like the echolocating bat, provide a powerful biological model for target tracking by sonar. Bats are the only mammals capable of powered flight [10] and can dynamically modify both path planning and echolocation signal design as they track and approach target [11,12]. They also display differences in flight and echolocation behaviors in open and cluttered environments [13]. The rich suite of adaptive behaviors exhibited by echolocating bats operating in different environments can serve to inspire technological advances in sonar tracking and localization algorithms.

Here, we review the bat's dynamic sonar and flight behaviors as they perform natural tasks, with a focus on tracking and pursuit strategies across ecological niches. Our goal is to highlight the dazzling display of bat adaptive behaviors, which engineers could implement in new technological applications and innovations.

2. Echolocation in Bats

Over 1000 species of bats echolocate [14]. The majority of echolocating bats produce signals with the larynx, emitting ultrasonic calls through the mouth or nose. There are some exceptions, such as *Rousettus aegyptiacus*, which emits ultrasonic clicks with the tongue. The discrete sonar signals emitted by echolocating bats reflect from objects in the path of the sound beam and return to the bat in the form of echoes. Laryngeal echolocating bats can emit pulses as short as 0.5 milliseconds, with frequencies that typically range from 25 to 150 kHz, though some bats produce sonar calls at frequencies outside that range [15–17]. Bats use the features of returning echoes to generate 3D representations of their surroundings [18–20].

The anatomical structure of the bat's outer ears functions as two receivers with a specialized skin flap, known as the tragus (see Figure 1). The tragus introduces elevation-dependent spectral changes in echoes, which bats can use for vertical localization [21,22]. Inter-aural differences are used by the bat to estimate the horizontal location of objects with accuracy of ~1.5 deg [23]. Bats can enhance cues for sound localization by moving their head and pinna independently, to amplify interaural differences used to localize sonar targets [24,25]. Finally, bats rely on the time delay between each sonar call and echo return to gauge the distance to a target, showing distance-difference discrimination thresholds of approximately 1 to 3 cm [18,26], depending on the species. Importantly, bats dynamically modify the spectro-temporal features of sonar calls with respect to task (e.g., search, approach, and interception phases of foraging) and the environment (e.g., dense vegetation or open space) [27]. These adaptations rely on a robust audio-motor feedback system that supports advanced navigation and tracking behaviors.

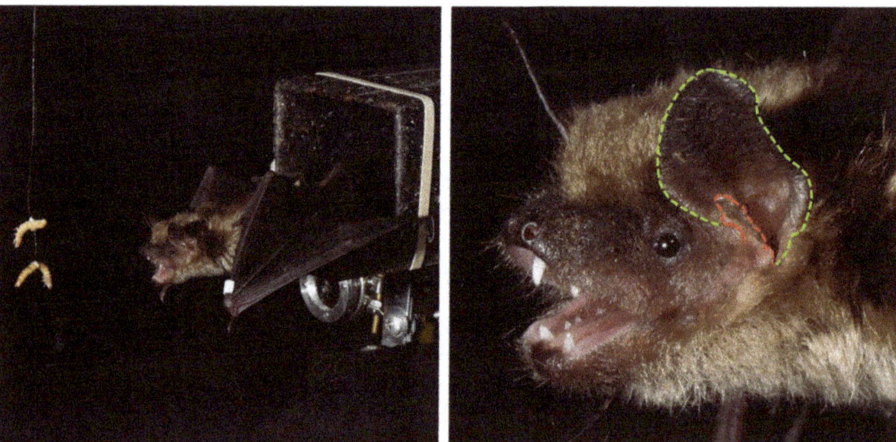

Figure 1. *Eptesicus fuscus* bat. Left panel: Bats are trained to perch on a platform and produce echolocation calls to track and intercept approaching targets (mealworms). This experimental setup allows us to study bat sonar tracking behavior while maintaining careful control of the target motion. Right panel: Closeup of the head of the bat, showing details of the external ear anatomy. A green dashed line delineates the left pinna, which acts as a receiver and can be independently moved to control inter-aural differences, necessary for azimuthal localization of targets [24]. The red dashed line delineates the enlarged tragus, which contributes to target elevation estimation. Photos courtesy of Dr. Brock Fenton.

Different species of bats have evolved specialized sonar signal designs. Call types can broadly be broken into two different categories: frequency modulated (FM) signals and constant frequency (CF) signals. FM signals sweep across a broad range of frequencies and are well suited for target localization, whereas CF signals are narrowband tones that are typically longer in duration than FM signals, and they tend to be used by bats that hunt for fluttering targets in dense vegetation [17,28,29]. CF sonar signals are often combined with FM components (CF-FM), whose bandwidth increases when animals must estimate target distance [18]. Sonar call structures depend on the environment and preferred prey of a bat. FM sweeps alone are employed by most echolocating bats and can vary in bandwidth, according to the task at hand. FM bats that forage in open fields tend to emit narrowband FM search calls with low duty-cycle, to detect prey, and shift to broadband FM signals to intercept and capture evasive insects. Bats that forage in or near clutter emit short, very broadband FM calls, to reduce masking effects by the echoes returning from nearby foliage [30]. CF–FM bats rely on Doppler Shift Compensation (DSC), compensating for the Doppler shift introduced by their own movement by lowering the frequency of emitted calls to stabilize the frequency of returning echoes to a band that they hear best (i.e., detection and frequency discrimination thresholds are lowest) [31–33].

While the call structures described above can aid in tracking targets in cluttered conditions, bats still must contend with masking effects when target echoes are obscured by other sounds. Forward masking occurs when the interfering signals precede the target signal, backward masking occurs when the interfering signals follow the target signal, and simultaneous masking occurs when interfering signals return at the same time as the target signal [17]. To reduce interference from signals in their environment, bats may adjust the duration of their sonar emissions, to reduce overlap of target echoes with their own echolocation broadcasts and clutter echoes [34]. Some species of bats avoid dense clutter conditions altogether [35]. In laryngeal FM echolocators, echoes that return from objects off-axis from the sonar beam axis are weaker and low-pass filtered, allowing the bat to separate clutter echoes from on-axis target echoes [36]. In conditions with multiple objects that return a cascade of echoes for each sonar emission, bats may change flight velocity and path planning, to reduce clutter interference [13,37].

Echolocating bats show additional adaptive sonar behaviors to track objects and avoid obstacles. For example, bats adjust the directional aim of sonar signals to detect and localize objects in the environment [38–40]. Some bats alternate between emitting sounds in groups at short inter-pulse intervals (20–40 ms) and longer inter-pulse intervals (>50 ms) in cluttered environments [37,41–43]. They may also make frequency adjustments in successive echolocation calls, possibly to facilitate pulse-echo assignment when multiple echoes return at different delays from clutter objects extended along the range axis [44].

Along with acoustic interference in reverberant, cluttered habitats, bats must also operate in a cocktail-party-like environment, where they must parse echoes from their own calls and the sonar signals from other bats, to select and track sonar targets, while also listening in on social calls produced by nearby conspecifics [11,45,46]. In acoustically complex environments, bats employ a vast array of behavioral strategies to maximize target information and minimize interference [47]. In the presence of conspecifics, bats may adjust frequencies of signals or cease calling entirely, to reduce sonar jamming [46], or some species, such as *Tadarida brasiliensis*, produce sinusoidal FM calls to jam the echolocation of competing bats for food [48]. Bats have also been shown to eavesdrop on the sounds produced by bats foraging nearby [49]. The ability to quickly modify behavior to counter masking and potential jamming signals is a key adaptation bats exhibit, to minimize signal interference.

Bats are auditory specialists that have evolved a high-resolution active sensing system to represent objects in their surroundings, for the purpose of target tracking and obstacle avoidance. The adaptations of bats from their engagement in natural tasks have inspired sonar technology, but the *full suite of strategies used by bats remains to be exploited in the advance of artificial sonar systems*.

3. Target Tracking by Echolocating Bats

Many predatory bats track moving insect prey while navigating through cluttered environments. This creates an added cognitive challenge: Not only must the bats use intermittent echo returns from stationary objects, such as foliage and buildings, to steer around obstacles, but they must also process echoes from moving prey items to track target trajectories and plan successful interception. As described above, bats use the time delay between each call and echo to estimate target range [18]. However, when tracking prey, the bat's estimate of a moving target's position is obsolete by the time the bat receives information carried by the most recent echo. Delays accumulate from the time it takes for (1) a sonar broadcast to travel to the object, (2) the echo to return to the bat's ears, (3) the brain to process information from the returning echo, and (4) the generation of an appropriate motor response. These delays collectively can add up to as much as 100 ms following each sonar transmission [27]. To accommodate these delays, bats have evolved sophisticated tracking strategies, adapted both to movement patterns of prey and features of the environment.

3.1. Sonar Tracking Behaviors

Sonar tracking strategies in FM-calling aerial hawking insectivores like *Eptesicus fuscus* reveal the fast-dynamic modifications in sonar behavior as the bat approaches a target. In open environments, bats emit long (8–25 ms), low frequency (<30 kHz), narrowband search signals. The shallow FM search signals are produced at a low repetition rate, as infrequently as every-other wingbeat (interpulse intervals ~200 ms). Approach calls are usually broadband signals (duration 2–6 ms), sweeping over 30–120 kHz. As FM bats approach a target, they lock their sonar beam on the prey item and reduce their signal duration and pulse intervals, until they prepare to intercept their target by emitting 0.5–1 ms signals at a high repetition rate (150–200 Hz) [17,50,51]. A similar trend seen in CF–FM bats, with the duration of the CF component and the bandwidth of the FM component of calls modified as the animal approaches a target [52,53]. Environmental conditions, clutter, and prey identity all contribute to further specializations of this insect-pursuit sequence (Figure 2).

3.2. Tracking Evasive Prey

When targets move in linear trajectories, many different organisms, including falcons [54], dogs [3], and fish [55], track moving targets by approaching along a straight trajectory, while keeping constant the angle between the animal's heading and the selected target, as the distance between the two decreases. This strategy is known as a constant bearing (CB) strategy, which is effective for intercepting a target moving along a straight and predictable path. However, the pursuer of an erratically moving target would never converge to the optimum bearing by using the CB strategy.

Many insectivorous bats must contend with prey that can actively maneuver to avoid capture and even jam echolocation. The predator–prey dynamics between bats and insects have revealed an evolutionary arms race that produces extremely specialized behaviors through selective pressures. Many different insects have evolved hearing sensitivities in the ultrasound frequency ranges of the echolocation signals used by predatory bats [56–59]. Some insects have also evolved various evasive flight maneuvers in response to bat signals, from highly stereotyped linear movement away from the bat, demonstrated by Coleopterans (beetles) [56,60,61], to erratic flight trajectories in Lepidopterans (butterflies and moths). Praying mantids have a cyclopean ear to detect bat ultrasound and drop to the ground in response to sonar signals [56]; lacewing moths also cease flying and suddenly plummet downward when they detect the hunting echolocation calls of their main predator, *Pipistrellus pipistrellus*, [57]. This plummeting strategy significantly decreases capture success by the bats [62–64]. Additionally, some insects, such as the tiger moth *Bertholdia trigona*, have developed ultrasonic clicks, which disrupt the bat's ability to successfully track prey by using echolocation [65]. When bats hear tiger moth ultrasonic clicks, they reverse their insect-capture-sequence pattern by elongating call durations and pulse intervals, likely to contend with multiple sound streams [66]. Bats must not only employ their own tracking strategies for capturing moving targets in midair, but also contend with counter strategies that insects have developed specifically to evade capture.

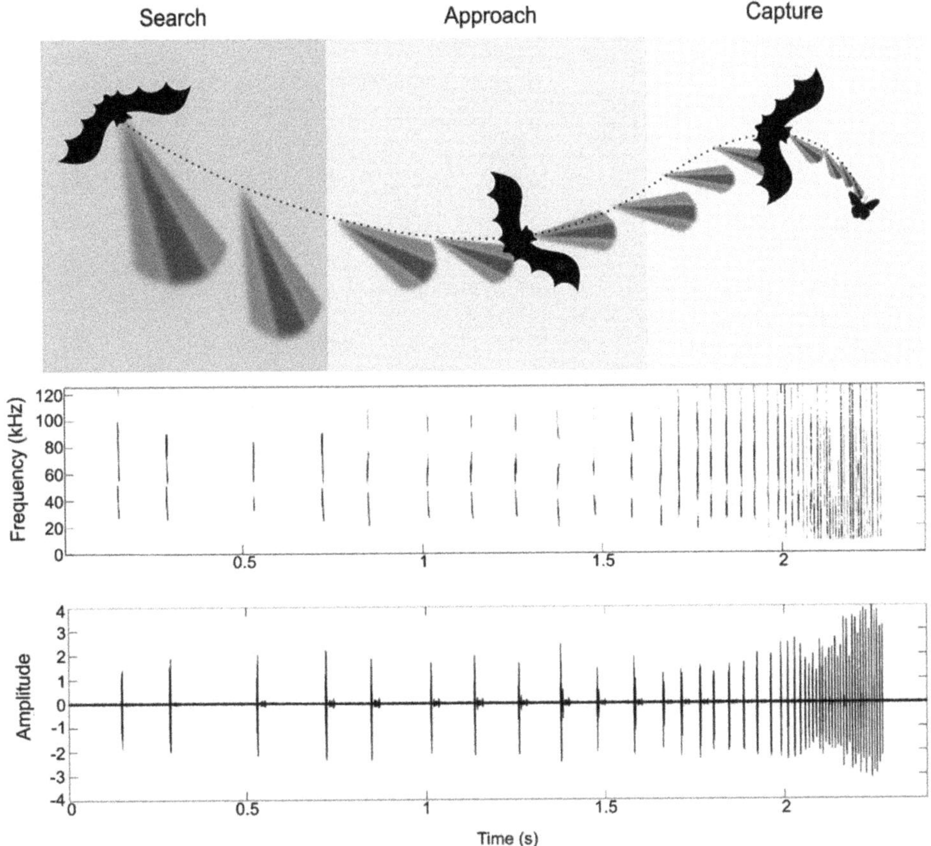

Figure 2. Classic insect-pursuit sequence of an FM bat. The top panel shows a depiction of a bat pursuing an insect. The grayscale fans illustrate the directional aim of the bat's sonar beam, with the darkest regions illustrating the beam axis containing greatest sound energy. In the search phase, bats orient their beam aim to scan the environment in different directions and emit narrowband, long-duration sonar calls. The approach phase commences when echo information about a target returns to the bat; it is characterized by an increase in FM bandwidth, the bat locking its sonar beam aim onto the selected target, and the bat increasing its rate of sonar calls. Finally, when the bat moves to capture the insect, it emits a quick succession of calls, further decreasing the inter-pulse interval, until it intercepts the target. The center panel depicts spectrograms (time frequency representations) of natural echolocation calls from a target tracking sequence of a big brown bat, *Eptesicus fuscus*, in the lab, and shows the approach and capture phases of insect pursuit (low signal-to-noise ratio may have affected the quality of the spectrograms of some signals). The lower panel shows the waveform of the above sequence. Increases in signal amplitude with decreasing target distance are an artifact of the recording conditions. These panels illustrate the change in sonar-call repetition rate in a bat approaching and intercepting a target.

The challenges echolocating bats face in capturing erratically flying insect prey means that a CB strategy would not incorporate the flexibility needed for successful capture. By extension, it has been proposed that bats maintain an optimal bearing by minimizing changes in the absolute direction relative to the target, termed a constant absolute target direction (CATD) strategy [67]. The CATD model posits that an animal generates and updates internal estimations of the distance and direction of the target relative to itself (in the bat through echolocation), to compute a time-optimal strategy for intercepting erratically moving targets (Figure 3). The CATD strategy, analogous to parallel navigation [68], has been demonstrated in predatory robber flies [69] and interpreted as a strategy for motion camouflage in dragonflies [70]. It has also been implemented in models with a sensorimotor feedback system that relies on delays, which may have application for unmanned vehicle control [71].

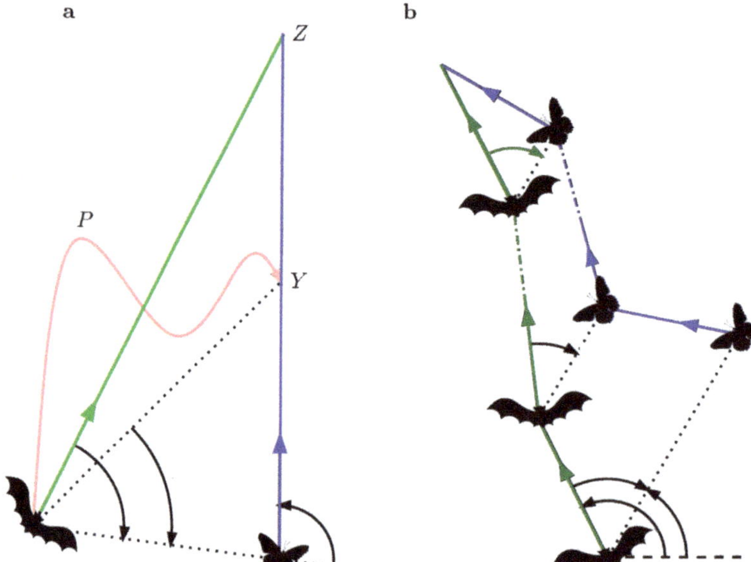

Figure 3. Modified from Ghose et al. 2006 [67]. (**a**) Constant bearing strategy (CB). Target (insect) moves in a straight line, at a constant velocity (blue line), and is pursued by bat that holds a fixed target bearing (green line), aiming where it will intercept the target (interception point Z). An alternative nonlinear path (pink) can be adopted by the bat when pursuing a target with constant linear velocity, resulting in a shorter intercept time at point Y (see Ghose et al. 2006 for further description). (**b**) Constant Absolute Target Direction strategy (CATD). Target (insect) moves erratically, by changing both direction and speed along path (blue). True erratic target motion cannot have a global time-minimum intercept, but can be approximated by infinite constant velocity segments. The bat's path (green) can follow a locally time-optimal path by adjusting its flight trajectory to minimize changes in the absolute direction of the target. This strategy relies on the target position update acquired from returning echoes.

Although the CATD strategy suggests that bats build an internal model of target motion, the echolocating bat's implementation of predictive strategies for target tracking has been a controversial topic. Masters and colleagues previously reported that the big brown bat *Eptesicus fuscus* uses a non-predictive strategy when tracking a moving target, orienting head aim to the location of the last returning echo, rather than the target's actual location [72]. Further studies in bats, however, indicate that a non-predictive model cannot account for the success of bats tracking occluded or evasive targets. Behavioral studies of the fishing bat *Noctilio leporinus* demonstrated that this species could use the movement of an artificial fish before it disappeared under water from the acoustic view of the bat, to accurately plan target interception [73]. Computational modeling of the trajectories of foraging bats have shown that anticipatory motor planning reflects realistic capture performance [74], and that bats attend to the future location of prey items in a sequence, to guide flight-path selection and improve capture rates [75]. Recently, we conclusively demonstrated that *E. fuscus* relies on predictive models of target trajectories when tracking moving targets [76]. Specifically, we have empirical evidence to refute the Masters et al. [72] claim of non-predictive tracking, and show that bats rely on a predictive model to track moving objects and even continue to track targets when echoes are blocked by an occluder during a portion of the target's trajectory. This suggests a strategy bats may employ to contend with insect prey that disappear momentarily behind clutter in the environment. Furthermore, we found that when internal models of target motion were violated by unpredictable changes in velocity, bats quickly modified echolocation behavior, to probe the environment for additional information, in order to update internal models and resume tracking the target. Our behavioral data align with a model in which bats estimate target velocity based on the echo arrival time differences between past sampled locations and further advance head aim by a fixed angle. Bats are able to track evasive or occluded targets by building predictive models of target trajectory and use this information to successfully intercept erratic prey. Future studies will investigate constraints on sonar tracking models that bats use to navigate in complex environments.

4. Adaptive Bat Echolocation Behaviors Inspire Artificial Sonar Tracking Systems

The echolocation and flight behavior of bats have been a source of inspiration for many technological advances, however, many key features of bat sonar tracking have yet to be realized in artificial systems. As described above, bats rapidly modify the spectro-temporal features of echolocation calls, and these adaptive sonar behaviors are fundamental to their performance in navigating complex environments while tracking and intercepting targets. Full implementation of these adaptive sonar behaviors, coupled with the use of wideband sonar signals, offers great potential for future technology applications. In this section, we present some examples in which bats have inspired technology thus far.

In 2010, a standard bat algorithm (BA) was proposed as a metaheuristic algorithm that uses similar processes of echolocating bats for global optimization [77]. The standard BA uses idealized behaviors of echolocating bats, which draws from a limited subset of parameters. These idealized behaviors or rules are as follows:

1. Bats use echolocation to sense distance and can identify and categorize targets relative to background barriers.
2. Bats fly randomly, varying the frequency and intensity of narrowband echolocation signals to detect prey, and can adjust the parameters of their sonar sounds relative to their distance to the target.
3. Call intensity varies, from a large positive value to a minimum constant value.

This algorithm iteratively updates the position and velocity of a virtual bat, using these three idealized rules. This updating allows for a more dynamic and efficient method for optimizing the processing of sensory information, allowing the BA to solve constrained and unconstrained optimization problems better than similar biologically inspired algorithms [78,79]. However, these idealized rules greatly simplify the components of adaptive echolocation, e.g., assuming that bats use a single sonar frequency, which changes with distance. The algorithm does not consider the bat's use of wideband FM signals or task-dependent behavioral measures at a given distance, such as preparing to intercept a target vs. flying by that target. The standard BA has been further developed to incorporate a directional bat algorithm (dBA), which improves performance in different types of environments and conditions, including premature convergence due to low exploration [80]. More recent advances in a binary bat algorithm (bBA) address traffic network determination problems [81] and parameter initialization to improve convergence velocity and accuracy [82]. Further integration of a more complete repertoire of adaptive behaviors of bats would continue to improve this optimization algorithm.

Robotic navigation has employed the Simultaneous Localization and Mapping (SLAM), which constructs and updates a spatial map of a novel, fixed environment, from both allocentric and egocentric perspectives, thus allowing an agent to navigate without *a priori* knowledge of the surroundings. In the last decade, there have been significant strides in creating reliable SLAM algorithms, however, there are still limitations to these approaches. Sensor uncertainty, the processing demands of complex computations, and challenges of dynamic environments contribute to the current limitations of SLAM algorithms [83]. One approach to this problem is RatSLAM, which uses the computational models of the hippocampus in rodents to inform navigation in novel environments with ambiguous landmark information [84]. This biologically inspired approach to SLAM has yielded promising results, with increased place-recognition performance and recovery from path integration errors. Expanding on this biological model, Steckel and Peremans have proposed the use of the echolocating bat for a sonar-based model of SLAM [2], which can operate under conditions where optical information is reduced or unavailable. Previous SLAM systems with sonar capabilities required impractically large numbers of sonar measurement to converge on a functional map [2], but BatSLAM offers a new way to use sonar information more efficiently, to allow autonomous sonar-guided robots to navigate an environment. Like Yang's Bat Algorithm, BatSLAM draws inspiration from the bat's biosonar, to localize the positions of obstacles to generate an experience map and modify motor commands for path integration. Additionally, they use the physical structure of the bat's external ears to allow binaural sound localization, though they do not incorporate adaptive sonar signal design, head movements, or the ability of the bat to dynamically move each pinna independently, to amplify interaural differences (Figure 1). Combining directionality of sonar emissions and binaural echo reception of bats, Steckel and Peremans developed the Echolocation Related Transfer Function (ERTF) for spectro-spatial filtering, realizing a biomimetic sonar system that localizes multiple acoustic objects with wideband sonar [2]. It creates consistent maps of large environments that can converge over time, to relatively accurate metric maps, to support navigation. More recently, there have been advances to BatSLAM, which include odometric information and an acoustic flow model, which allows for a novel 3D sonar sensor [85], as well new optimization of 2D-experience mapping through the addition of an audio-aware perceptual hash with a closed-loop detection algorithm, using fixed CF and FM sonar signals [86]. These new enhancements to BatSLAM enable richer representations of complex environments, however, dynamic bat sonar adjustments offer many additional features that could be incorporated in future versions, for operation in more complex and dynamic environments.

Tracking algorithms have a myriad of uses, from surveillance [87] to biomedical applications [88]. While the accuracy of tracking algorithms has improved significantly in the last decade, they often fail to contend with background noise and clutter, which interferes with localization of a selected target. Kalman filters operate with iterative processes that aid in estimating the position and motion of a target and have been a standard for addressing the challenge of noisy target data. The addition of Kalman filters to tracking algorithms dramatically improves tracking fidelity and reduces interference by local maxima [89], particularly in linear systems. In nonlinear systems, extended Kalman filters also perform an iterative process with increased success [90], but concerns have been raised about inconsistent mapping and a penchant for underestimating covariance [91], particularly in handling sonar and vision data for the bearing of a target [92,93]. Improvements to complex tracking-condition algorithms have proven promising, such as application of multiple Kalman filters, which allows precise tracking of dynamically moving targets [94]. However, target tracking by artificial systems remains a challenge, and target interception success is still low.

Finally, biologically inspired approaches to sonar tracking algorithms have hailed some success, with iterative predictive algorithms approaching performance levels comparable to biological systems [95]. Both for biological and artificial systems, real-world tracking requires sensory input, which is then processed to output accurate pursuit of moving targets (Figure 4). Some models have implemented rudimentary behavioral features of bat flight trajectories and putative predictive tracking [74]. Behavioral research on bat sonar target tracking has provided valuable insights into the strategies these animals employ to perform complex tasks, including differential adaptation of their echolocation behavior with respect to moving targets and stationary obstacles [96], as well as Doppler shift compensation and discrimination [31], all while contending with different environmental constraints and conditions [27,30]. Empirical studies of adaptive and predictive sonar tracking behaviors of bats [76], in conjunction with neurophysiological experiments, will provide insights to the computations employed by echolocating animals to carry out tasks under real-world conditions, and in turn provide further inspiration for new algorithms and neural networks that will improve artificial tracking systems.

Figure 4. Bats as a biological model to inspire tracking algorithms. Tracking moving targets requires sensory information that can be in the form of echoes (left/center panels) or visual stimuli (right panel). This information must then be processed by computations that allow for the prediction of future states (shown in gray). Man-made systems like Autonomous Underwater Vehicles (AUV) use sonar to track different moving targets, such as marine life and wildlife (left panel). Bats acquire discrete sensory information in the form of echo returns from adaptive sonar emissions; echo snapshots are integrated, to enable the prediction of the future position of a moving insect (center panel). These predictive tracking algorithms can be implemented in technologies that use sonar or other sensory modalities, such as drone videography of a bicycle race (right panel).

5. Conclusions

Our review aims to highlight the richness of adaptive sonar behavior and performance exhibited by diverse bat species, which collectively can inspire exciting advances in sonar tracking technology. Bats rely on a highly developed audio-vocal feedback system that supports computation of the distance and direction of objects in their surroundings. Most bats make use of wideband sonar signals and dynamically modify the spectro-temporal features of echolocation sounds in response to sensory information about the location of targets and obstacles. Bats navigate highly complex environments, identify targets relative to surrounding clutter, and are able to anticipate target motion, in order to intercept and capture moving targets in flight. Bat echolocation has inspired sonar technology advances for decades, however, artificial systems have yet to incorporate the full richness of adaptive sonar behaviors for target tracking and interception.

Author Contributions: Conceptualization: C.A.D., A.S., and C.F.M.; Funding acquisition: C.A.D., A.S., C.F.M.; Supervision: C.F.M.; Visualization: C.A.D. and A.S.; Writing—original draft: C.A.D. and A.S.; Writing—review and editing: C.A.D., A.S., and C.F.M. All authors have read and agreed to the published version of the manuscript.

Funding: This work was funded by a National Science Foundation Graduate Research Fellowship awarded to CD (GRFP 2018261398), a Human Frontiers Science Program Fellowship awarded to AS (LT000220/2018), and NSF Brain Initiative (NCS-FO 1734744), AFOSR (FA9550-14-1-0398NIFTI), and ONR (N00014-17-1-2736) grants awarded to CFM.

Acknowledgments: We thank Katie Rabasca for comments on an earlier version of the manuscript.

Conflicts of Interest: The authors declare no conflict of interest.

References

1. Thrun, S. Robotic Mapping: A Survey. In *Exploring Artificial Intelligence in the New Millennium*; Morgan Kaufmann Publishers Inc.: San Francisco, CA, USA, 2003; pp. 1–35.
2. Steckel, J.; Peremans, H. BatSLAM: Simultaneous Localization and Mapping Using Biomimetic Sonar. *PLoS ONE* **2013**, *8*, e54076. [CrossRef] [PubMed]
3. Shaffer, D.M.; Krauchunas, S.M.; Eddy, M.; McBeath, M.K. How Dogs Navigate to Catch Frisbees. *Am. Psychol. Soc.* **2003**, *15*, 437–441.
4. McBeath, M.K.; Shaffer, D.M.; Kaiser, M.K. How Baseball Outfielders Determine Where to Run to Catch Fly Balls. *Science* **1995**, *268*, 569–573. [CrossRef] [PubMed]
5. Olberg, R.M.; Worthington, A.H.; Venator, K.R. Prey Pursuit and Interception in Dragonflies. *J. Comp. Physiol. A* **2000**, *186*, 155–162. [CrossRef]
6. Mischiati, M.; Lin, H.-T.; Herold, P.; Imler, E.; Olberg, R.; Leonardo, A. Internal Models Direct Dragonfly Interception Steering. *Nature* **2014**, *517*, 333–338. [CrossRef]
7. Yang, X.-S.; Deb, S. Engineering Optimisation by Cuckoo Search. *Int. J. Math. Model. Numer. Optim.* **2010**, *1*, 330–343. [CrossRef]
8. Dorigo, M.; Maniezzo, V.; Colorni, A. Ant System: Optimization by a Colony of Cooperating Agents. *IEEE Trans. Syst. Man Cybern. Part B Cybern.* **1996**, *26*, 29–41. [CrossRef]
9. Yang, X.-S. Firefly Algorithm, Stochastic Test Functions and Design Optimisation. *Int. J. Bio-Inspired Comput.* **2010**, *2*, 78–84. [CrossRef]
10. Norberg, U.M. Evolution of Vertebrate Flight: An Aerodynamic Model for the Transition from Gliding to Active Flight. *Am. Nat.* **1985**, *126*, 303–327. [CrossRef]
11. Chiu, C.; Xian, W.; Moss, C.F. Adaptive echolocation behavior in bats for the analysis of auditory scenes. *J. Exp. Boil.* **2009**, *212*, 1392–1404. [CrossRef]
12. Chiu, C.; Reddy, P.V.; Xian, W.; Krishnaprasad, P.S.; Moss, C.F. Effects of competitive prey capture on flight behavior and sonar beam pattern in paired big brown bats, *Eptesicus fuscus*. *J. Exp. Boil.* **2010**, *213*, 3348–3356. [CrossRef] [PubMed]
13. Falk, B.; Jakobsen, L.; Surlykke, A.; Moss, C.F. Bats coordinate sonar and flight behavior as they forage in open and cluttered environments. *J. Exp. Boil.* **2014**, *217*, 4356–4364. [CrossRef] [PubMed]
14. Fenton, M.B.; Simmons, N.B. *Bats*; University of Chicago Press: Chicago, IL, USA, 2015; Available online: https://www.press.uchicago.edu/ucp/books/book/chicago/B/bo17089187.html (accessed on 13 March 2020).
15. Fenton, M.B.; Grinnell, A.; Popper, A.N.; Fay, R.R.; Acoustical Society of America (Eds.) Bat Bioacoustics. In *Springer Handbook of Auditory Research*; ASA Press/Springer: New York, NY, USA, 2016; Volume 54.
16. Webster, F.A.; Brazier, O.G. Experimental Studies on Target Detection, Evaluation and Interception by Echolocating Bats. In *Biological Information Handling Systems and Their Functional Analogs*; Aerospace Medical Res. Lab.: Wright-Patterson Airforce Base, OH, USA, 1965.
17. Schnitzler, H.-U.; Kalko, E.K.V. Echolocation by Insect-Eating Bats: We Define Four Distinct Functional Groups of Bats and Find Differences in Signal Structure That Correlate with the Typical Echolocation Tasks Faced by Each Group. *BioScience* **2001**, *51*, 557–569. [CrossRef]
18. Simmons, J.A. The Resolution of Target Range by Echolocating bats. *J. Acoust. Soc. Am.* **1973**, *54*, 157–173. [CrossRef] [PubMed]
19. Simmons, J.; Moss, C.F.; Ferragamo, M. Convergence of Temporal and Spectral Information into Acoustic Images of Complex Sonar Targets Perceived by the Echolocating Bat, Eptesicus Fuscus. *J. Comp. Physiol. A* **1990**, *166*. [CrossRef] [PubMed]
20. Moss, C.F.; Surlykke, A. Auditory Scene Analysis by Echolocation in Bats. *J. Acoust. Soc. Am.* **2001**, *110*, 2207–2226. [CrossRef]
21. Müller, R. A Numerical Study of the Role of the Tragus in the Big Brown Bat. *J. Acoust. Soc. Am.* **2004**, *116*, 3701–3712. [CrossRef]
22. Chiu, C.; Moss, C.F. The Role of the External Ear in Vertical Sound Localization in the Free Flying Bat, Eptesicus Fuscus. *J. Acoust. Soc. Am.* **2007**, *121*, 2227–2235. [CrossRef]
23. Simmons, J.A.; Kick, S.A.; Lawrence, B.D.; Hale, C.; Bard, C. Acuity of Horizontal Angle Discrimination by the Echolocating Bat, *Eptesicus fuscus*. *J. Comp. Physiol. A* **1983**, *153*, 321–330. [CrossRef]

24. Wohlgemuth, M.; Luo, J.; Moss, C.F. Three-Dimensional Auditory Localization in the Echolocating Bat. *Curr. Opin. Neurobiol.* **2016**, *41*, 78–86. [CrossRef]
25. Wohlgemuth, M.; Kothari, N.B.; Moss, C.F. Action Enhances Acoustic Cues for 3-D Target Localization by Echolocating Bats. *PLoS Boil.* **2016**, *14*, e1002544. [CrossRef] [PubMed]
26. Moss, C.F.; Schnitzler, H.-U. Behavioral Studies of Auditory Information Processing. In *Hearing by Bats; Springer Handbook of Auditory Research*; Popper, A.N., Fay, R.R., Eds.; Springer: New York, NY, USA, 1995; pp. 87–145. [CrossRef]
27. Moss, C.F.; Chiu, C.; Surlykke, A. Adaptive Vocal Behavior Drives Perception by Echolocation in Bats. *Curr. Opin. Neurobiol.* **2011**, *21*, 645–652. [CrossRef] [PubMed]
28. Von Der Emde, G.; Schnitzler, H.-U. Fluttering Target Detection in Hipposiderid Bats. *J. Comp. Physiol. A* **1986**, *159*, 765–772. [CrossRef]
29. Schnitzler, H.-U.; Flieger, E. Detection of Oscillating Target Movements by Echolocation in the Greater Horseshoe Bat. *J. Comp. Physiol. A* **1983**, *153*, 385–391. [CrossRef]
30. Schnitzler, H.-U.; Moss, C.F.; Denzinger, A. From Spatial Orientation to Food Acquisition in Echolocating Bats. *Trends Ecol. Evol.* **2003**, *18*, 386–394. [CrossRef]
31. Schnitzler, H.-U. The Ultrasonic Locating Sounds of the Horseshoe Bats (Chiroptera Rhinolophidae) in Various Orientation Situations. *Z. Vergl. Physiol.* **1968**, *57*, 376–408. [CrossRef]
32. Neuweiler, G.; Bruns, V.; Schüller, G. Ears Adapted for the Detection of motion, or how echolocating bats have exploited the capacities of the mammalian auditory system. *J. Acoust. Soc. Am.* **1980**, *68*, 741–753. [CrossRef]
33. Suga, N. Biosonar and Neural Computation in Bats. *Sci. Am.* **1990**, *262*, 60–68. [CrossRef]
34. Kalko, E.K.V.; Schnitzler, H.-U. The Echolocation and Hunting Behavior of Daubenton's Bat, *Myotis Daubentoni*. *Behav. Ecol. Sociobiol.* **1989**, *24*, 225–238. [CrossRef]
35. Sleep, D.J.H.; Brigham, R.M. An Experimental Test of Clutter Tolerance in Bats. *J. Mammal.* **2003**, *84*, 216–224. [CrossRef]
36. Bates, M.; Simmons, J.A.; Zorikov, T.V. Bats Use Echo Harmonic Structure to Distinguish Their Targets from Background Clutter. *Science* **2011**, *333*, 627–630. [CrossRef] [PubMed]
37. Moss, C.F.; Bohn, K.; Gilkenson, H.; Surlykke, A. Active Listening for Spatial Orientation in a Complex Auditory Scene. *PLoS Boil.* **2006**, *4*, e79. [CrossRef] [PubMed]
38. Yovel, Y.; Falk, B.; Moss, C.F.; Ulanovsky, N. Optimal Localization by Pointing Off Axis. *Science* **2010**, *327*, 701–704. [CrossRef] [PubMed]
39. Ghose, K.; Moss, C.F. The Sonar Beam Pattern of a Flying Bat as it Tracks Tethered Insects. *J. Acoust. Soc. Am.* **2003**, *114*, 1120–1131. [CrossRef]
40. Ghose, K.; Moss, C.F. Steering by Hearing: A Bat's Acoustic Gaze is Linked to Its Flight Motor Output by a Delayed, Adaptive Linear Law. *J. Neurosci.* **2006**, *26*, 1704–1710. [CrossRef]
41. Petrites, A.E.; Eng, O.S.; Mowlds, D.S.; Simmons, J.A.; Delong, C.M. Interpulse Interval Modulation by Echolocating Big Brown Bats (*Eptesicus fuscus*) in Different Densities of Obstacle Clutter. *J. Comp. Physiol. A* **2009**, *195*, 603–617. [CrossRef]
42. Sändig, S.; Schnitzler, H.-U.; Denzinger, A. Echolocation Behaviour of the Big Brown Bat (Eptesicus fuscus) in an Obstacle Avoidance Task of Increasing Difficulty. *J. Exp. Boil.* **2014**, *217*, 2876–2884. [CrossRef]
43. Kothari, N.; Wohlgemuth, M.; Hulgard, K.; Surlykke, A.; Moss, C.F. Timing Matters: Sonar Call Groups Facilitate Target Localization in Bats. *Front. Physiol.* **2014**, *5*, 168. [CrossRef]
44. Hiryu, S.; Bates, M.E.; Simmons, J.A.; Riquimaroux, H. FM Echolocating Bats Shift Frequencies to Avoid Broadcast–Echo Ambiguity in Clutter. *Proc. Natl. Acad. Sci. USA* **2010**, *107*, 7048–7053. [CrossRef]
45. Wright, G.S.; Chiu, C.; Xian, W.; Wilkinson, G.S.; Moss, C.F. Social Calls Predict Foraging Success in Big Brown Bats. *Curr. Boil.* **2014**, *24*, 885–889. [CrossRef]
46. Warnecke, M.; Chiu, C.; Engelberg, J.; Moss, C.F. Active Listening in a Bat Cocktail Party: Adaptive Echolocation and Flight Behaviors of Big Brown Bats, Eptesicus fuscus, Foraging in a Cluttered Acoustic Environment. *Brain Behav. Evol.* **2015**, *86*, 6–16. [CrossRef] [PubMed]
47. Corcoran, A.J.; Moss, C.F. Sensing in a Noisy World: Lessons from Auditory Specialists, Echolocating Bats. *J. Exp. Boil.* **2017**, *220*, 4554–4566. [CrossRef] [PubMed]
48. Corcoran, A.J.; Conner, W.E. Bats Jamming Bats: Food competition through Sonar Interference. *Science* **2014**, *346*, 745–747. [CrossRef] [PubMed]

49. Barclay, R.M.R. Interindividual Use of Echolocation Calls: Eavesdropping by Bats. *Behav. Ecol. Sociobiol.* **1982**, *10*, 271–275. [CrossRef]
50. Surlykke, A.; Moss, C.F. Echolocation Behavior of Big Brown Bats, *Eptesicus fuscus*, in the Field and the Laboratory. *J. Acoust. Soc. Am.* **2000**, *108*, 2419–2429. [CrossRef]
51. Simmons, J.; Fenton, M.; O'Farrell, M. Echolocation and Pursuit of Prey by Bats. *Science* **1979**, *203*, 16–21. [CrossRef]
52. Matsuta, N.; Hiryu, S.; Fujioka, E.; Yamada, Y.; Riquimaroux, H.; Watanabe, Y. Adaptive Beam-Width Control of Echolocation Sounds by CF-FM Bats, *Rhinolophus ferrumequinum* Nippon, during Prey-Capture Flight. *J. Exp. Boil.* **2013**, *216*, 1210–1218. [CrossRef]
53. Fawcett, K.; Jacobs, D.S.; Surlykke, A.; Ratcliffe, J.M. Echolocation in the Bat, *Rhinolophus capensis*: The Influence of Clutter, Conspecifics and Prey on Call Design and Intensity. *Boil. Open* **2015**, *4*, 693–701. [CrossRef]
54. Kane, S.A.; Zamani, M. Falcons Pursue Prey Using Visual Motion Cues: New Perspectives from Animal-Borne Cameras. *J. Exp. Boil.* **2014**, *217*, 225–234. [CrossRef]
55. Lanchester, B.S.; Mark, R.F. Pursuit and Prediction in the Tracking of Moving Food by a Teleost Fish (*Acanthaluteres spilomelanurus*). *J. Exp. Boil.* **1975**, *63*, 627–645.
56. Yager, D.D. Structure, Development, and Evolution of Insect Auditory Systems. *Microsc. Res. Tech.* **1999**, *47*, 380–400. [CrossRef]
57. Miller, L.A.; Surlykke, A. How Some Insects Detect and Avoid Being Eaten by Bats: Tactics and Countertactics of Prey and Predator. *BioScience* **2001**, *51*, 570. [CrossRef]
58. Hoy, R.; Nolen, T.; Brodfuehrer, P. The Neuroethology of Acoustic Startle and Escape in Flying Insects. *J. Exp. Boil.* **1989**, *146*, 287–306.
59. Ter Hofstede, H.M.; Ratcliffe, J.M. Evolutionary Escalation: The Bat–Moth Arms Race. *J. Exp. Boil.* **2016**, *219*, 1589–1602. [CrossRef]
60. Forrest, T.G.; E Farris, H.; Hoy, R.R. Ultrasound Acoustic Startle Response in Scarab Beetles. *J. Exp. Boil.* **1995**, *198*, 2593–2598.
61. Yager, D.D.; Spangler, H.G. Behavioral Response to Ultrasound by the Tiger Beetle Cicindela Marutha Dow Combines Aerodynamic Changes and Sound Production. *J. Exp. Boil.* **1997**, *200*, 11.
62. Miller, L.A.; Olesen, J. Avoidance Behavior in Green Lacewings. *J. Comp. Physiol. A* **1979**, *131*, 113–120. [CrossRef]
63. Miller, L.A. The Orientation and Evasive Behavior of Insects to Bat Cries. In *Exogenous and Endogenous Influences on Metabolic and Neural Control, vol 1.*; Addink, D.F., Spronk, N., Eds.; Pergamon Press: Oxford, UK, 1982; pp. 393–405.
64. Miller, L.A. How Insects Detect and Avoid Bats. In *Neuroethology and Behavioral Physiology*; Huber, F., Markl, H., Eds.; Springer: Berlin/Heidelberg, Germany, 1983; pp. 251–266. [CrossRef]
65. Corcoran, A.J.; Barber, J.; Conner, W.E. Tiger Moth Jams Bat Sonar. *Science* **2009**, *325*, 325–327. [CrossRef]
66. Corcoran, A.J.; Barber, J.; Hristov, N.I.; Conner, W.E.; Lin, H.-T.; Trimmer, B.A. How Do Tiger Moths Jam Bat Sonar? *J. Exp. Boil.* **2011**, *214*, 2416–2425. [CrossRef]
67. Ghose, K.; Horiuchi, T.K.; Krishnaprasad, P.S.; Moss, C.F. Echolocating Bats Use a Nearly Time-Optimal Strategy to Intercept Prey. *PLoS Boil.* **2006**, *4*, e108. [CrossRef]
68. Rafie-Rad, M. Time-Optimal Solutions of Parallel Navigation and Finsler Geodesics. *Nonlinear Anal. Real World Appl.* **2010**, *11*, 3809–3814. [CrossRef]
69. Fabian, S.T.; Sumner, M.E.; Wardill, T.J.; Rossoni, S.; Gonzalez-Bellido, P.T. Interception by Two Predatory Fly Species is Explained by a Proportional Navigation Feedback Controller. *J. R. Soc. Interface* **2018**, *15*, 20180466. [CrossRef] [PubMed]
70. Mizutani, A.; Chahl, J.; Srinivasan, M.V. Motion Camouflage in Dragonflies. *Nature* **2003**, *423*, 604. [CrossRef] [PubMed]
71. Reddy, P.V.; Justh, E.W.; Krishnaprasad, P.S. Motion Camouflage with Sensorimotor Delay. In Proceedings of the 2007 46th IEEE Conference on Decision and Control, New Orleans, LA, USA, 12–14 December 2007; pp. 1660–1665.
72. Masters, W.; Moffat, A.; Simmons, J. Sonar tracking of horizontally moving targets by the big brown bat Eptesicus fuscus. *Science* **1985**, *228*, 1331–1333. [CrossRef]

73. Campbell, K.A.; Suthers, R.A. Predictive Tracking of Horizontally Moving Targets by the Fishing Bat, *Noctilio leporinus*. In *Animal Sonar*; Springer: Boston, MA, USA, 1988; pp. 501–506.
74. Erwin, H.R. Algorithms for Sonar Tracking in Biomimetic Robotics. In *School of Computing and Technology*; University of Sunderland: Sunderland, UK, 2001.
75. Fujioka, E.; Aihara, I.; Sumiya, M.; Aihara, K.; Hiryu, S. Echolocating bats use future-target information for optimal foraging. *Proc. Natl. Acad. Sci. USA* **2016**, *113*, 4848–4852. [CrossRef]
76. Salles, A.; Diebold, C.; Moss, C.F. Prediction strategies for target tracking in the echolocating bat, Eptesicus fuscus. In Proceedings of the Society for Neuroscience Meeting, Chicago, IL, USA, 19–23 October 2019.
77. Yang, X.-S. A New Metaheuristic Bat-Inspired Algorithm. In *Nature Inspired Cooperative Strategies for Optimization (NICSO 2010)*; Studies in Computational Intelligence; González, J.R., Pelta, D.A., Cruz, C., Terrazas, G., Krasnogor, N., Eds.; Springer: Berlin/Heidelberg, Germany, 2010; pp. 65–74. [CrossRef]
78. Yang, X.-S.; Gandomi, A.H. Bat Algorithm: A Novel Approach for Global Engineering Optimization. *Eng. Comput.* **2012**, *29*, 464–483. [CrossRef]
79. Gandomi, A.H.; Yang, X.-S.; Alavi, A.H.; Talatahari, S. Bat Algorithm for Constrained Optimization Tasks. *Neural Comput. Appl.* **2012**, *22*, 1239–1255. [CrossRef]
80. Chakri, A.; Khelif, R.; Benouaret, M.; Yang, X.-S. New directional bat algorithm for continuous optimization problems. *Expert Syst. Appl.* **2017**, *69*, 159–175. [CrossRef]
81. Srivastava, S.; Sahana, S. Application of Bat Algorithm for Transport Network Design Problem. *Appl. Comput. Intell. Soft Comput.* **2019**, *2019*, 1–12. [CrossRef]
82. Ma, X.-X.; Wang, J.-S. Optimized Parameter Settings of Binary Bat Algorithm for Solving Function Optimization Problems. *J. Electr. Comput. Eng.* **2018**, *2018*. [CrossRef]
83. Mohsen, M.; Islam, M.N.; Karimi, R. Simultaneous Localization and Mapping: Issues and Approaches. *Int. J. Comput. Sci. Telecommun.* **2013**, *4*, 1–7.
84. Milford, M.; Wyeth, G.; Prasser, D. RatSLAM: A Hippocampal Model for Simultaneous Localization and Mapping. In Proceedings of the IEEE International Conference on Robotics and Automation, 2004. Proceedings, ICRA '04. 2004, New Orleans, LA, USA, 26 April–1 May 2004; Volume 1, pp. 403–408. [CrossRef]
85. Steckel, J.; Peremans, H. Spatial Sampling Strategy for a 3D Sonar Sensor Supporting BatSLAM. In Proceedings of the 2015 IEEE/RSJ International Conference on Intelligent Robots and Systems (IROS), Hamburg, Germany, 28 September–2 October 2015; pp. 723–728.
86. Chen, M.; Hu, W. Research on BatSLAM Algorithm for UAV Based on Audio Perceptual Hash Closed-Loop Detection. *Int. J. Pattern Recognit. Artif. Intell.* **2018**, *33*, 1959002. [CrossRef]
87. Siebel, N.T. Design and Implementation of People Tracking Algorithms for Visual Surveillance Applications. Ph.D. Thesis, University of Reading, Reading, UK, March 2003.
88. Goobic, A.; Welser, M.; Acton, S.; Ley, K. Biomedical application of target tracking in clutter. In Proceedings of the Conference Record of Thirty-Fifth Asilomar Conference on Signals, Systems and Computers (Cat.No.01CH37256), Pacific Grove, CA, USA, 4–7 November 2001.
89. Pan, J.; Hu, B.; Zhang, J.Q. An Efficient Object Tracking Algorithm with Adaptive Prediction of Initial Searching Point. In *Advances in Image and Video Technology*; Chang, L.-W., Lie, W.-N., Eds.; Springer: Berlin/Heidelberg, Germany, 2006; Volume 4319, pp. 1113–1122. Available online: http://link.springer.com/10.1007/11949534_112 (accessed on 22 January 2020).
90. Julier, S.J.; Uhlmann, J. A counter example to the theory of simultaneous localization and map building. In Proceedings of the 2001 ICRA. IEEE International Conference on Robotics and Automation (Cat. No.01CH37164), Seoul, Korea, 21–26 May 2001; Volume 4, pp. 4238–4243.
91. Huang, G.P.; Mourikis, A.I.; Roumeliotis, S.I. Analysis and improvement of the consistency of extended Kalman filter based SLAM. In Proceedings of the 2008 IEEE International Conference on Robotics and Automation, Pasadena, CA, USA, 19–23 May 2008; Institute of Electrical and Electronics Engineers (IEEE): Piscataway, NJ, USA, 2008; pp. 473–479.
92. Julier, S.J.; Uhlmann, J.K. *A General Method for Approximating Nonlinear Transformations of Probability Distributions*; Technical report for Robotics Research Group: Oxford, UK, 1996.
93. Julier, S.J.; Uhlmann, J. New Extension of the Kalman Filter to Nonlinear Systems. In *Signal Processing, Sensor Fusion, and Target Recognition VI*; International Society for Optics and Photonics: Bellingham, WA, USA, 1997; Volume 3068, pp. 182–193. [CrossRef]

94. Blackman, S. Multiple Hypothesis Tracking for Multiple Target Tracking. *IEEE Aerosp. Electron. Syst. Mag.* **2004**, *19*, 5–18. [CrossRef]
95. Erwin, H.R.; Wilson, W.W.; Moss, C.F. A Computational Sensorimotor Model of Bat Echolocation. *J. Acoust. Soc. Am.* **2001**, *110*, 1176–1187. [CrossRef]
96. Mao, B.; Aytekin, M.; Wilkinson, G.S.; Moss, C.F. Big Brown Bats (*Eptesicus fuscus*) Reveal Diverse Strategies for Sonar Target tracking in Cluttera. *J. Acoust. Soc. Am.* **2016**, *140*, 1839–1849. [CrossRef]

MDPI
St. Alban-Anlage 66
4052 Basel
Switzerland
Tel. +41 61 683 77 34
Fax +41 61 302 89 18
www.mdpi.com

Sensors Editorial Office
E-mail: sensors@mdpi.com
www.mdpi.com/journal/sensors

www.ingramcontent.com/pod-product-compliance
Lightning Source LLC
LaVergne TN
LVHW071549170726
843515LV00008B/2246
9783036535371